Calculus in Vector Spaces

MONOGRAPHS AND TEXTBOOKS IN PURE AND APPLIED MATHEMATICS

Other Volumes in Preparation

Calculus
in
Vector Spaces

LAWRENCE J. CORWIN

Department of Mathematics
Rutgers University
New Brunswick, New Jersey

ROBERT H. SZCZARBA

Department of Mathematics
Yale University
New Haven, Connecticut

MARCEL DEKKER, INC.   NEW YORK AND BASEL

Library of Congress Cataloging in Publication Data

Corwin, Lawrence J
    Calculus in vector spaces.

    (Pure and applied mathematics ; no. 52)
    Includes index.
    1. Calculus. 2. Vector spaces. I. Szczarba,
Robert H.   joint author. II. Title.
QA303.C823   515   79-17037
ISBN 0-8247-6832-9

MARCEL DEKKER, INC.

270 Madison Avenue, New York, New York   10016

Current printing (last digit):

10   9   8   7   6   5   4

PRINTED IN THE UNITED STATES OF AMERICA

TO OUR PARENTS

PREFACE

Two of the most important subjects in the undergraduate mathematics curriculum are linear algebra and multivariable calculus. In addition to being essential to many of the applications of mathematics to the physical, biological, and social sciences, they play a critical role in the development of pure mathematics. The purpose of this book, as the title indicates, is to give a unified, integrated treatment of these subjects.

There are a number of reasons for combining these two fields in one course. The first is that they complement one another extremely well. For example, the derivative of a function of several variables is best thought of as a linear transformation; thereafter otherwise arcane results like the chain rule become almost transparent. Conversely, the spectral theorem for self-adjoint linear transformations can be proved with the help of results on constrained maxima and minima obtained using the calculus. Moreover a topic like quadratic forms is better motivated when it can be presented with an application like the classification of critical points.

A second reason for combining these two courses is that linear algebra, by itself, is sometimes a pedagogical problem. It tends to double as a service course for students in other disciplines and as an introductory course in algebra for mathematics majors. As a result, students often find it dry and abstract. Combining it with multivariable calculus, a more "concrete" subject, provides the necessary motivation. At the same time, this merger clarifies some of the notions of calculus for these students.

This clarification has valuable effects. Students have an unfortunate tendency to compartmentalize their knowledge; they often fail to make important connections between different courses in mathematics. We hope here to make students aware of the value of searching for these links, and to give them some notion of mathematics as a unified subject rather than an assortment of unrelated topics.

This book is based upon the material in a two semester course taught at Yale for the past five years. This course is taught at the sophomore level but generally includes a significant number of well qualified freshmen. The core of the course consists of the material in Chapters 2 through 6 and Chapters 11 through 16 although Chapter 6 is not always done in detail. (Chapter 1 contains background material and can be referred to as needed.) The amount of additional material that can be included depends upon the background of the students, the degree to which rigor is emphasized, and the inclination of the instructor. The Yale course referred to above typically included the first two sections of Chapter 7, Sections 1, 4, and 5 in Chapter 9, and the first six sections of Chapter 10.

As a casual examination will show, we have included proofs of virtually all the results stated in the text. However, even in the course here at Yale, the degree to which these proofs were covered has varied greatly from year to year, indeed, even within a single year. Some of the proofs most likely to be skipped are given in separate sections or in Appendices. Our reason for including proofs is that it is easier to omit an unwanted proof than it is to find an omitted one.

As usual, in a work of this kind, many people contribute to the final result. We would like especially to thank Bob Anderson, Mike Anderson, Paul Blanchard, Bill Massey, Ed Odell, and George Seligman, for their numerous valuable suggestions. In particular, the treatment of Jordan Canonical forms in Appendix 3 is largely due to George Seligman. We would also like to thank Eugene Krc for a careful job of proofreading. Finally, we would like to thank Donna Belli for transforming an almost unintelligible manuscript into typescript.

The Authors

CONTENTS

Calculus in Vector Spaces

CHAPTER 1

SOME PRELIMINARIES

INTRODUCTION

We collect here some basic information about topics which, while not really part of the topics under study, are necessary to it. In addition to providing a convenient reference for this material, this chapter serves to establish some of our notation. Since most of the ideas discussed here will be familiar to the reader, we suggest that this chapter not be worked through in detail to begin with but rather be referred to when necessary.

1. THE RUDIMENTS OF SET THEORY

In this section, we review some of the ideas and notation from set theory. We also briefly discuss the notion of a function.

A *set* is a collection of objects: the objects are the *elements* of the set. One specifies a set by one of two methods:

(a) naming the elements (usually listing them inside curly brackets), or

(b) giving a criterion for membership in the set (again in curly brackets).

Thus the set whose elements are 2,4,6,8, and 10 can be given as

$\{2,4,6,8,10\}$

or as

$\{x:x$ is an even integer and $2 \le x \le 10\}$

or as

$\{x:x$ satisfies $x^5 - 30x^4 + 340x^3 - 1800x^2 + 4384x - 3840 = 0\}$

All of these describe the *same* set; what matters is not the description, but the actual elements of the set.

Sets are usually denoted by letters. The symbol $\in$ is used to denote membership in a set: $x \in S$ means "x is an element of the set S." Similarly, $\notin$ denies membership: $x \notin S$ means "x is not an element of the set S."

The simplest set of all is the *empty set*, usually denoted by ϕ. This set has no elements: it is characterized by the property that $x \notin \phi$, whatever x is.

A set S is called finite if either it is empty or the elements of S can be indexed by the whole numbers less than or equal to some number n - that is, if $S = \{x_1,\ldots,x_n\}$. For instance, $\{2,4,6,8,10\}$ is finite. Sets which are not finite are infinite. One example is R, the set of real numbers.

A set S is called a *subset* of T if every element of S is also an element of T; that is, if $x \in T$ whenever $x \in S$. Thus the subsets of $\{2,4,6\}$ are

$\phi, \{2\}, \{4\}, \{6\}, \{2,4\}, \{2,6\}, \{4,6\}, \{2,4,6\}$

Notice that ϕ is a subset of every set: regardless of the set S, $x \in S$ whenever $x \in \phi$, since x is never in ϕ. (If this bit of logic seems too fancy, read the next section for a further explanation.)

We write "A is a subset of B" as $A \subset B$, and "A is not a subset of B" as "A $\not\subset$ B."

Two sets are *equal* if they contain the same elements. It should be clear that two sets A,B are equal (A = B) exactly when $A \subset B$ and $B \subset A$.

There are three basic ways of making new sets from old; by union, by intersection, and by complementation. The *union* of two sets A and B, written A ∪ B, is defined by

$$A \cup B = \{x : x \in A \text{ or } x \in B\}^*$$

For instance, if $A = \{1,2,3,4,5\}$ and $B = \{2,4,6,8,10\}$, then

$$A \cup B = \{1,2,3,4,5,6,8,10\}$$

The *intersection* of A and B, A ∩ B, is defined by

$$A \cap B = \{x : x \in A \text{ and } x \in B\}$$

If, for instance, $A = \{1,2,3,4,5\}$ and $B = \{2,4,6,8,10\}$, as before, then

$$A \cap B = \{2,4\}$$

The *complement* of B in A, A - B, is defined by

$$A - B = \{x : x \in A \text{ and } x \notin B\}$$

Thus if A and B are as above, $A - B = \{1,3,5\}$ and $B - A = \{6,8,10\}$.

The notions of union and intersection can both be extended. Suppose, for instance, that $A_1, A_2, \ldots, A_n$ are sets. We define

$$\bigcup_{j=1}^{n} A_j = \{x : x \in A_j \text{ for some } j, 1 \leq j \leq n\}$$

$$= \{x : x \in A_1 \text{ or } x \in A_2 \text{ or } \cdots \text{ or } x \in A_n\}$$

and

$$\bigcap_{j=1}^{n} A_j = \{x : x \in A_j \text{ for every } j, 1 \leq j \leq n\}$$

$$= \{x : x \in A_1 \text{ and } x \in A_2 \text{ and } \cdots \text{ and } x \in A_n\}$$

That is, $\bigcup_{j=1}^{n} A_j$ contains every element in at least one of the A's, while $\bigcap_{j=1}^{n} A_j$ consists of the elements common to all the A's.

---

* Or both. In mathematics, "or" usually means "one or the other or both."

We can carry this idea one step further. Suppose that we have
sets $A_1, A_2, A_3, \ldots$. We define

$$\bigcup_{n=1}^{\infty} A_n = \{x : x \in A_n \text{ for some } n = 1, 2, \ldots\}$$

$$= \{x : x \in A_1 \text{ or } x \in A_2 \text{ or } \cdots\}$$

and

$$\bigcap_{n=1}^{\infty} A_n = \{x : x \in A_n \text{ for all } n = 1, 2, \ldots\}$$

$$= \{x : x \in A_1 \text{ and } x \in A_2 \text{ and } \cdots\}$$

In fact, we can extend these notions still further. For instance,
suppose J is any set and we are given, for each $j \in J$, a set A_j. We
can then define

$$\bigcup_{j \in J} A_j = \{x : x \in A_j \text{ for some } j \in J\}$$

$$\bigcap_{j \in J} A_j = \{x : x \in A_j \text{ for all } j \in J\}$$

We sometimes write

$$\bigcup_j A_j \quad , \quad \bigcap_j A_j$$

if the "index" set J is either clear from the context or not important.

There is one special case of complementation that often arises.
In most situations, we are concerned with various subsets of some fixed
set S. (For instance, we might be working with various sets of integers; all of the sets are subsets of the set Z of all integers.) The
complement of a set A in S is then written A': $A' = S - A$. Since we do
not mention S in this notation, it should be clear in advance what the
set S is. If, for instance, $S = \{2,4,6,8,10\}$ and $A = \{2,4,6\}$, then
$A' = \{8,10\}$. But if $S = Z$, the set of all integers, and $A = \{2,4,6\}$,
then $A' = \{x : x \text{ is an integer, and } x \neq 2, 4, \text{ or } 6\}$.

The notions of union, intersection, and complementation are connected by two formulas, known as *De Morgan's laws*:

$$(\cup_j A_j)' = \cap_j A_j' \tag{1.1}$$

$$(\cap_j A_j)' = \cup_j A_j' \tag{1.2}$$

The proofs are similar. For instance, if $x \in (\cup_j A_j)'$, then $x \notin \cup_j A_j$; hence $x \notin A_j$ for every j, or $x \in A_j'$ for every j. Therefore $x \in \cap_j A_j'$, and this proves that $(\cup_j A_j)' \subset \cap_j A_j'$. The reverse inclusion and the proof of (1.2) are left as Exercise 9.

We mention one other useful term about sets. The sets A and B are said to be *disjoint* if $A \cap B = \phi$. Thus {1,2,3} and {4,5,6} are disjoint; {1,2,3} and {3,4,5} are not. Note that A and A' are always disjoint.

If A and B are sets, their *Cartesian product*, $A \times B$, is the set consisting of all ordered pairs (a,b), with $a \in A$ and $b \in B$. If, for instance, A = {1,2} and B = {1,2,3}, then

$$A \times B = \{(1,1),(1,2),(1,3),(2,1),(2,2),(2,3)\}$$

and

$$B \times A = \{(1,1),(2,1),(3,1),(1,2),(2,2),(3,2)\}$$

Notice that two ordered pairs (a,b) and (c,d) are equal exactly when a = c and b = d. Thus the order of listing elements in an ordered pair matters (though the order of listing elements of a set does not).

Similarly, the set $A \times B \times C$ (where C is a third set) consists of ordered triples (a,b,c), with $a \in A$, $b \in B$, and $c \in C$. It should be clear how to generalize this notion.

The reason for the name "Cartesian product" is that Cartesian coordinates in the plane let one regard the plane as $R \times R$. We generally write $R \times R$ as R^2. Similarly, $R^n = R \times R \times \cdots \times R$ (n factors) is the set of ordered n-tuples of real numbers.

Let A and B be sets. A *function* $f: A \to B$ is a rule which associates to each element x in A a unique element f(x) in B. The key points here are that the function must be defined for every x in A and that

$f(x)$ must be uniquely specified. (For instance, if $B = R$, the set of real numbers, and $A = \{x \in R : 0 \leq x \leq 1\}$, then $f(x) = \pm \sqrt{x}$ is not a function because one has a choice of values for $f(x)$. But $f(x) = \sqrt{x}$, the positive square root of x, is a function.)

There are two ways to specify the function $f : A \to B$: by listing the value of $f(x)$ for each $x \in A$, or by describing a rule used to compute $f(x)$. The first method is usually unwieldy, but it makes one important fact clear: a function is defined not by the exact phrasing of the rule, but by the actual correspondence between x and $f(x)$. Thus $f(x) = (x+1)^2$ and $g(x) = x^2 + 2x + 1$ define the same function, since $f(x) = g(x)$ for every real number x.

This may be a good time to bring up a point which usually causes confusion: the way one describes some common functions. Lots of common functions have names (like sin, tan, and log), and the value of sin at x, for instance, is sin x. Lots of others do not. For instance, one common function assigns to x the value x^2. This function is often called "the function x^2", which is fine much of the time but which can cause confusion when one wants to distinguish between a function, (which, as we have seen, is a law of correspondence) and its value at a point. We shall usually say something like "the function f, where $f(x) = x^2$" to describe this function, which is accurate but rather pedantic-sounding. Another way out is to write "the function $x \mapsto x^2$", which is shorter but which does not give the function a name. We could give the function a name like "the squaring function", but that would not solve the problem for the function $x \mapsto x^2 + 3x - 7$. All in all, pedantry seems the best solution.

Finally, a function $f : X \to Y$ is said to be *injective* (or *1-1*) if $f(x) = f(x')$ implies $x = x'$ and *surjective* (or *onto*) if, given any $y \in Y$, there is an $x \in X$ with $f(x) = y$. The function f is *bijective* if it is both injective and surjective.

EXERCISES

1.   Let $A = \{1,2,3,5,7,10\}$, $B = \{1,3,4,6,20,21\}$, and $C = \{2n : n \text{ a positive integer}\}$. Determine the following.

   (a)   A ∪ B   (d)   A ∪ B ∪ C
   (b)   A ∩ C   (e)   A - C
   (c)   (A ∪ B) ∩ C   (f)   C - B

2.   Determine all subsets of X = {1,2,a,b}.

3.   Suppose A and B are sets with A ⊆ B. Prove that

   (a)   A ∪ B = B
   (b)   A ∩ B = A
   (c)   A - B = ϕ

4.   Let A and B be sets. Prove that

   (a)   if A ∪ B = B, then A ⊂ B
   (b)   if A ∩ B = A, then A ⊂ B
   (c)   if A - B = ϕ, then A ⊆ B

5.   Show that if A and B are sets, then

   (a)   A ∩ B ⊆ B, A ∩ B ⊆ A
   (b)   A ⊆ A ∪ B, B ⊆ A ∪ B

6.   Show that if A, B, and C are sets, then

   (a)   A ∩ (B ∪ C) = (A ∩ B) ∪ (A ∩ C)
   (b)   A ∪ (B ∩ C) = (A ∪ B) ∩ (A ∪ C)

7.   Show that if A and B are sets, then

   (a)   A = (A ∩ B) ∪ (A - B)
   (b)   ϕ = (A ∩ B) ∩ (A - B)

8.   Which of the following are functions?

   (a)   F:R → R: F(x) = $\sqrt[3]{x}$
   (b)   G:R → R: F(x) = y, where $y^3 - y + x = 0$
   (c)   H:R → R, where H(x) = the biggest integer ≤ x

9.   Prove the De Morgan laws (Equations (1.1) and (1.2)).

2. SOME LOGIC

Mathematics is distinguished from all other sciences by its insistence
on rigor: every statement must be derived by logical rules from clearly stated assumptions. We cannot go into all the subtleties of logic
here, but we shall give a brief outline of some basic material.

A rigorous treatment of our subject would start with a list of
the axioms which we assume. We shall treat this as unnecessary, taking
it for granted that the reader knows some algebra, geometry, trigonometry, and calculus. We shall often use facts from those fields without special comment.

For the most part, however, we shall be developing new results,
generally by proving lemmas, propositions, theorems, and corollaries.
There is no strong distinction among the terms. We generally reserve
the term "theorem" for the most important results; propositions are
somewhat less important or interesting. A lemma often has no particular interest in itself, but is useful in proving some other result.
A corollary is an easy consequence of a previous result.

Every mathematical statement is assumed to be either true or false.
The statements we deal with will be either simple statements (whose
truth or falsity is presumably known) or statements built of simple
statements in certain easily describable ways. The most important ways
are the following:

(1) Conjunction: if P and Q are statements, then "P and Q" is
a statement. Thus the conjunction of "$\pi > 3$" and "$e > 4$" is "$\pi > 3$
and $e > 4$." The conjunction of two statements is true only when both
statements are true. (Thus "$\pi > 3$ and $e > 4$" is false, since $e \leq 4$.)

(2) Disjunction: if P and Q are statements, then "P or Q" is a
statement, and it is true if either or both of the two statements is
true. (The disjunction of the two statements in the previous example
is "$\pi > 3$ or $e > 4$", and it is true because $\pi > 3$.) In common speech,
"P or Q" often implies that exactly one of P and Q is true; in mathematics, there is no such implication, and both may be true.

(3) Negation: if P is a statement, so is "not P"; "not P" is
false if P is true and true if P is false. "Not P", of course, is the

statement which says the opposite of P. (The negation of "$\pi > 3$" is
"$\pi \not> 3$", or "$\pi \leq 3$.") One sometimes writes "$\sim$P" for "not P".

(4) Implication: if P and Q are statements, then "P implies
Q", or "if P, then Q", is a statement. (It is the sort of statement
one expects to see in theorems.) The statement "if P, then Q" is true
unless P is true and Q is false. (Thus "if P, then Q" means only that
either P is false or Q is true.) In ordinary speech, implications are
expected to involve a rational connection between the two statements;
thus "if it rains hard tomorrow, the reservoir will overflow" is a reasonable sort of implication (though its truth depends on the state of
the reservoir), while "if it rains hard tomorrow, then Napoleon lost
at Waterloo" is not. In mathematics, no such connection is required:
"if e > 4, then $\pi > 3$" is a true implication, though there does not
seem to be any causal connection between the sizes of e and π.

To prove the implication "P implies Q," we need to rule out the
possibility that P is true and Q is false. We can do this directly,
by showing that Q is true whenever P is, or indirectly, by showing
that if Q is false, then P is false, or (also indirectly) by showing
that assuming P is true and Q is false leads to a contradiction.

One sometimes writes "P $\Rightarrow$ Q" for "P implies Q."

(5) Equivalence. "P is equivalent to Q" means "P implies Q and
Q implies P." This holds if P and Q are both true or both false; it
is false if one is true and the other is false. Another common way of
expressing equivalence is "P if and only if Q" and one sometimes writes
P $\Leftrightarrow$ Q as well.

To prove that P $\Leftrightarrow$ Q, one ordinarily needs to prove two statements:
P $\Rightarrow$ Q and Q $\Rightarrow$ P. The statements "P $\Rightarrow$ Q" and Q $\Rightarrow$ P" are called *converses* of one another. The converse of a true statement need *not* be true.
The *contrapositive* of "P $\Rightarrow$ Q" is "not Q $\Rightarrow$ not P", and the contrapositive of a true statement is true. (The reason is that "not Q $\Rightarrow$ not P"
is false only if "not Q" is true and "not P" is false, or if Q is false
and P is true; that is exactly the condition which makes "P $\Rightarrow$ Q" false.)

It is easy to check (see Exercise 3) that if P $\Rightarrow$ Q and Q $\Rightarrow$ R, then
P $\Rightarrow$ R. This fact leads to another way of proving the equivalence of
a number of statements: we can reason in a cycle. For example, one

possible way of proving statements P, Q, and R equivalent is to show
$P \Rightarrow Q \Rightarrow R \Rightarrow P$. (It might well turn out not to be the simplest way.)

We now consider another method of complicating simple statements:
quantification. It arises in calculus, in the definition of continuity at a point: the function f is continuous at a real number x if
for all real numbers $\varepsilon > 0$, there is a number $\delta > 0$ such that if y is
any number with $|x-y| < \delta$, then $|f(x) - f(y)| < \varepsilon$.

There are two quantifiers: the universal ("for all", written "$\forall$")
and the existential ("there exists", or "$\exists$"). By using these symbols
and the implication sign, the definition of continuity given above can
be abbreviated as

$$(\forall \varepsilon > 0)\,(\exists \delta > 0)\,(\forall y)\,|x-y| < \delta \Rightarrow |f(x) - f(y)| < \varepsilon.$$

If we begin with a statement P(x) involving some unspecified term
x, then "For all x, P(x)" is true only if the statement P(x) is true
when we specify x in any way. (Ordinarily x will be restricted, explicitly or otherwise, to some fixed set; in the definition of continuity, for instance, x, y, ε, and δ are tacitly assumed to be real numbers.) One counterexample makes the statement false. Thus "$(\forall x) x^2 > 0$"
is false, since $0^2 \leq 0$. By contrast, "There exists an x such that
P(x)" is true if we can find one number x for which P(x) is true. Thus
"$(\exists x) x^2 > 0$" is true; let x = 1, for instance.

The order of quantifiers is very important. Consider, for instance, the statements

$$(\forall x)\,(\exists y \neq x) y > x$$

and

$$(\exists x)\,(\forall y \neq x) y > x$$

The first statement says that given any number x, there is a bigger
number y; that is, there is no biggest number. The second says that
there is one number x smaller than every other number y; that is, there
is a smallest number. The first statement is true; the second, false.

We have seen that one example can make a statement involving $\forall x$ false or one involving $\exists x$ true. This state of affairs is reflected in the way we work out the negation of a statement involving quantifiers. Suppose, for instance, that the statement $(\forall x)P(x)$ is false. Then we can find some x such that P(x) is false; that is, the statement $(\exists x)$ $\sim P(x)$ is true. Similarly, if the statement $\exists(x)P(x)$ is false, then for every x we choose, P(x) must be false. That means that the statement $(\forall x) \sim P(x)$ is true. In short, to form the negation of a statement involving a quantifier, one changes the quantifier ($\forall$ to $\exists$, or vice versa) and puts the negation inside. For instance, to find the negation of

$$(\forall \varepsilon > 0)(\exists \delta > 0)(\forall y)(|x\text{-}y| < \delta \Rightarrow |f(x) - f(y)| < \varepsilon)$$

proceed from

$$\sim (\forall \varepsilon > 0)(\exists \delta > 0)(\forall y)(|x\text{-}y| < \delta \Rightarrow |f(x) - f(y)| < \varepsilon)$$

through

$$(\exists \varepsilon > 0) \sim (\exists \delta > 0)(\forall y)(|x\text{-}y| < \delta \Rightarrow |f(x) - f(y)| < \varepsilon)$$

and

$$(\exists \varepsilon > 0)(\forall \delta > 0) \sim (\forall y)(|x\text{-}y| < \delta \Rightarrow |f(x) - f(y)| < \varepsilon)$$

to

$$(\exists \varepsilon > 0)(\forall \delta > 0)(\exists y) \sim (|x\text{-}y| < \delta \Rightarrow |f(x) - f(y)| < \varepsilon)$$

or

$$(\exists \varepsilon > 0)(\forall \delta > 0)(\exists y)(|x\text{-}y| < \delta \text{ and } |f(x) - f(y)| \geq \varepsilon)$$

We mentioned that one cannot interchange an existential quantifier and a universal one without possibly changing the meaning of the statement. However, one can interchange two existential quantifiers or two universal quantifiers without trouble.

In principle, every mathematical statement which is not an axiom
or a definition should be proved. In practice, any sufficiently obvious statement is merely asserted, on the grounds that any proof
would be tedious, possibly confusing, and unnecessary (since anybody
could supply one if required). The problem which arises, of course,
is that not everyone agrees about what is obvious. It is a good idea
to provide proofs for statements we dismiss as "obvious"; at the very
least, they test one's understanding of the subject.

EXERCISES

1.   Show that if P is any statement, then "P or $\sim$P" is true and "P
and $\sim$P" is false.

2.   Show that the statements "$\sim$(P and Q)" and "($\sim$P) or ($\sim$Q)" are
equivalent.

3.   Show that if P $\Rightarrow$ Q and Q $\Rightarrow$ R, then P $\Rightarrow$ R.

4.   Let $a_1, a_2, a_n, \ldots$ be a sequence of real numbers. This sequence
converges to the number a if, for any $\varepsilon > 0$, there is an integer N
such that $|a_n - a| < \varepsilon$ whenever n > N. Write out what it means for a
sequence *not* to converge to the number a.

5.   A function f:R $\rightarrow$ R is *uniformly continuous* if, for any $\varepsilon > 0$,
there is a $\delta > 0$ such that for all x, y, if $|x - y| < \delta$ then $|f(x) -
f(y)| < \varepsilon$. Write out what it means for a function not to be uniform-
ly continuous.

3.   MATHEMATICAL INDUCTION

Mathematical induction is a somewhat specialized, but very useful,
method of proof. Suppose that we want to prove a statement true for
every positive integer n. It suffices to show that
      (a)   the statement is true for n = 1;
      (b)   whenever the statement is true for n = k, it is true for
n = k + 1.

An example will probably help. We shall prove the following result:

PROPOSITION 3.1. *For all positive integers* n,

$$1 + 2 + \cdots + n = \frac{n(n+1)}{2} \tag{3.1}$$

Proof. For n = 1, the formula reads

$$1 = \frac{1(1+1)}{2}$$

This result is certainly true, since both sides are 1.
Now suppose that (3.1) holds for n = k:

$$1 + 2 + \cdots + k = \frac{k(k+1)}{2} \tag{3.2}$$

We need to prove the corresponding result for n = k + 1:

$$1 + 2 + \cdots + (k+1) = \frac{(k+1)[(k+1)+1]}{2} \tag{3.3}$$

This is not hard, since we can use (3.2):

$$1 + 2 + \cdots + (k+1) = (1 + 2 + \cdots + k) + (k+1)$$

$$= \frac{k(k+1)}{2} + (k+1)$$

$$= (k+1)(\frac{k}{2} + 1) = (k+1)\frac{(k+2)}{2}$$

which is the same as (3.3). Thus the proposition is proved.
To see intuitively why the proposition holds, notice that we
proved it for n = 1. Letting k = 1, we see that the proposition holds
for n = 2. Now let k = 2; the proposition must hold for n = 3, and so
on. By proceeding in this way, it seems clear that we can prove formula (3.1) for any given positive integer. Mathematical induction
states that we can indeed do this; thus it expresses a basic property
of the set of natural numbers, N.

It will often be convenient in what follows to use the so-called "sigma" notation for sums. If $a_1, \ldots, a_k$ are numbers (or elements in any set in which addition makes sense), we define

$$\sum_{i=1}^{k} a_i = a_1 + a_2 + \cdots + a_k \tag{3.4}$$

The left side of (3.4) is read "the sum of the a_i as i goes from one to k". For example, equation (3.1) can be written

$$\sum_{i=1}^{n} i = \frac{n(n+1)}{2}$$

One can use mathematical induction to prove many other propositions besides formulas (though the usual examples of the method involve formulas). Here is a different example.

PROPOSITION 3.2. *If n is a positive integer, then $x^n - y^n$ is divisible by x - y.*

Proof. When n = 1, $x^1 - y^1 = x - y$ is certainly divisible by x - y. Next, suppose that the proposition holds for n = k; that is, $x^k - y^k$ is divisible by x - y. Then $x^{k+1} - y^{k+1} = x(x^k - y^k) + y^k(x - y)$, and both terms are divisible by x - y. Therefore $x^{k+1} - y^{k+1}$ is divisible by x - y. The proposition follows, by mathematical induction.

EXERCISES

1. Prove that $\sum_{j=1}^{n} j^2 = \frac{n(n+1)(2n+1)}{6}$.

2. Prove that $\sum_{j=0}^{n} r^j = \frac{1-r^{n+1}}{1-r}$ (if $r \neq 1$).

3. Prove that $\sum_{j=1}^{n} j^3 = \frac{n^2(n+1)^2}{4}$.

4. Prove that $n^5 - n$ is divisible by 5 for any positive integer n.

5. What is wrong with the following inductive proof?

Theorem. In any set S with n elements, all the elements are equal.

Proof. If n = 1, there is only one element, and the result is obvious. Now assume the result for n = k; we prove it for n = k + 1. Let the elements of S be $x_1, \ldots, x_{k+1}$.

By the inductive hypothesis, we know that

$$x_1 = x_2 = \cdots = x_k.$$

We also know (again by the inductive hypothesis) that

$$x_2 = x_3 = \cdots = x_{k+1}.$$

Therefore $x_1 = x_2 = \cdots = x_{k+1}$, and the inductive step is proved.

*6. Let n and k be integers and define quantities $\binom{n}{k}$ by

$$\binom{n}{k} = \begin{cases} \dfrac{n!}{k!\,(n-k)!} & \text{if } 0 \le k \le n \\[2mm] 0 & \text{if } k < 0 \text{ or } k > n \end{cases}$$

(Here $n! = 1 \cdot 2 \cdot 3 \cdots (n-1) \cdot n$ if $n > 0$ and $0! = 1$.)

(a) Show that for each non-negative integer n, and for all integers k,

$$\binom{n}{k-1} + \binom{n}{k} = \binom{n+1}{k}$$

(b) Prove by induction that $\binom{n}{k}$ is an integer whenever n is a non-negative integer and k is any integer.

(c) If n is an *arbitrary* integer (i.e., not necessarily non-negative), define $\binom{n}{k} = \dfrac{n(n-1)(n-2)\cdots(n-k+1)}{k!}$ if $k > 0$, $\binom{n}{0} = 1$, and $\binom{n}{k} = 0$ if $k < 0$. Show that this definition agrees with one above for non-negative integers n. Prove that *this* $\binom{n}{k}$ is an integer for all integers n and k.

(d) Prove the *binomial theorem*: If a and b are any numbers, and if n is any non-negative integer, then

$$(a+b)^n = \sum_{k=0}^{n} \binom{n}{k} a^k b^{n-k} (= b^n + \binom{n}{1} ab^{n-1} + \binom{n}{2} a^2 b^{n-2}$$

$$+ \cdots + \binom{n}{n} a^n)$$

(e) Prove *Leibniz' formula*: If n is a non-negative integer, and if f(x) and g(x) are two functions, each differentiable at least n times, set $D^0(f) = f$, $D^k(f) = f^{(k)}(x) = \dfrac{d^k f}{dx^k}$, and likewise for g. Then

$$D^n(f \cdot g) = \sum_{k=0}^{n} \binom{n}{k} D^k(f) D^{n-k}(g)$$

*7. Discover and prove results like (d) and (e) in Exercise 6 above for $(a_1 + a_2 + \cdots + a_r)^n$ and for $D^n(f_1 \cdot f_2 \cdot f_3 \cdots \cdot f_r)$.

*8. (a) Show that if S = number of elements in S = n, the number of subsets T of S with $|T| = k$ is $\binom{n}{k}$.

(b) Show that if $|S| = n$ and if T is another set with $|T| = k$, then the number of mappings of S into T is k^n, and the number of one-to-one ("injective") mappings is

$$\binom{k}{n}!$$

(It might be easier to do (b) before (a).)

9. Let i be the imaginary number with $i^2 = -1$ and prove *De Moivre's theorem* $(\cos \theta + i \sin \theta)^n = \cos n\theta + i \sin n\theta$.

4.   INEQUALITIES AND ABSOLUTE VALUE

In this book we shall often need to manipulate inequalities involving absolute values. We include a brief review here.

To begin with, we give the standard notation for expressing inequality between two numbers:

a < b means "a less than b"

a $\leq$ b means "a less than or equal to b"

a $>$ b means "a greater than b"

a $\geq$ b means "a greater than or equal to b"

These relations have the following properties:

If $a < b$ and $b < c$, then $a < c$ $\qquad$ (4.1)

If $a < b$ and c is any number, then $a + c < b + c$ $\qquad$ (4.2)

If $a < b$ and $c > 0$, then $ac < bc$ $\qquad$ (4.3)

If $a < b$ and $c < 0$, then $ac > bc$ $\qquad$ (4.4)

These relations also hold for $\leq$, $>$, and $\geq$ (but the conditions on c in (4.3) and (4.4) remain as they are).

The *absolute value of a real number* x, $|x|$, is defined by

$|x| = x$ if $x \geq 0$

$|x| = - x$ if $x < 0$

Thus $|3| = 3$, $|-\pi| = \pi$ $|\sqrt{2} - 4| = 4 - \sqrt{2}$, $|0| = 0$. Notice that $-|x| \leq x \leq |x|$.

Since by convention $\sqrt{x}$ is the positive square root of x, we could also define $|x|$ to be $\sqrt{x^2}$. Notice in particular that $|x|^2 = x^2$.

For any real numbers x and y, we have

$$|xy| = |x||y| \qquad (4.5)$$

and

$$|x+y| \leq |x| + |y| \quad \text{(the } triangle\ inequality) \qquad (4.6)$$

Inequality (4.5) is easy to see. The triangle inequality is easily proved by checking cases ($x > 0$, $x < 0$, etc.). A faster method is the following: (4.6) is equivalent to

$$|x+y|^2 \leq (|x| + |y|)^2$$

or

$$x^2 + 2xy + y^2 \le |x|^2 + 2|x||y| + |y^2| = x^2 + 2|x||y| + y^2$$

But since $2xy \le 2|x||y|$, this last inequality is certainly true, and (4.2) follows.

We shall often meet inequalities like $|2x - 7| < 4$ or $|2x - 7| > 4$. As we see now, these inequalities can be reduced to inequalities not involving the absolute value.

PROPOSITION 4.1. *The inequality* $|a| < b$ *is equivalent to the two inequalities* a < b *and* -b < a.

For example, the inequality $|2x - 7| < 4$ is equivalent to the two inequalities

$$-4 < 2x - 7 < 4$$

or

$$3 < 2x < 11 \quad \text{(using (4.2))}$$

or

$$3/2 < x < 11/2 \quad \text{(using (4.3))}$$

Proof of Proposition 4.1. We shall assume b > 0; this is the case we shall need (and the other case is easy enough).

We first show that $|a| < b$ implies -b < a < b. If $a \ge 0$, then $|a| = a$, so a < b. In addition, -b < a since -b < 0. If a < 0, then $|a| = -a$ and the inequality $|a| < b$ becomes -a < b. Using equation (4.4) with c = - 1 we have a > -b or -b < a. Again a < 0, so clearly a < b.

Suppose now that -b < a < b. We need to prove that $|a| < b$. If $a \ge 0$, then $|a| = a$, so a < b implies $|a| < b$. If a < 0, then $|a| = -a$ and -b < a implies b > -a (using equation (4.4)), or $-a = |a| < b$.

In an exactly analogous manner, we can prove the following:

PROPOSITION 4.2. *The inequality* $|a| > b$ *holds if and only if either* a > b *or* a < -b.

We leave the proof to the reader.

Remark. Proposition 4.1 remains true if < is replaced by ≤. Proposition 4.2 remains true if > is replaced by ≥.

We note here for future reference some sets in R which will arise in the future, the *intervals*. We have

$$(a,b) = \{x : a < x < b\}$$

$$(a,b] = \{x : a < x \leq b\}$$

$$[a,b) = \{x : a \leq x < b\}$$

$$[a,b] = \{x : a \leq x \leq b\}$$

$$(a,\infty) = \{x : x > a\}$$

$$[a,\infty) = \{x : x \geq a\}$$

$$(-\infty,a) = \{x : x < a\}$$

$$(-\infty,a] = \{x : x \leq a\}$$

$$(-\infty,\infty) = R$$

The principle is that square brackets include the endpoints, while parentheses do not.

Sets of the form (a,b), (a,∞), or (-∞,a) are called *open* intervals; those of the form [a,b], [a,∞), or (-∞,a] are called *closed* intervals. (The reason for the names will be explained in Section 3.2.) The sets [a,b) and (a,b] are sometimes called *half-open* intervals.

EXERCISES

1.   Prove equation (4.5) directly from the definition of absolute value.

2.   Prove equation (4.6) directly from the definition of absolute value.

3.   Prove Proposition 4.2.

4.   What intervals do the following inequalities describe?

   (a) $|3x - 5| < 8$

   (b) $|6x + 1| \leq 11$

   (c) $|2 - 3x| < 4$

5.   What pairs of intervals do the following inequalities describe?

   (a) $|2x - 7| > 1$

   (b) $|7x + 3| > 14$

   (c) $|4 - 3x| \geq 7$

6.   (a)   Show that if $x_1, \ldots, x_n$ are real numbers, then there exists a largest number: $\exists j (1 \leq j \leq n)$: for all i, $1 \leq i \leq n$, $x_i \leq x_j$. (We write $x_j = \max\{x_1, \ldots, x_n\}$.)

   (b)   Show also that $\exists k (1 \leq k \leq n)$: for all i, $1 \leq i \leq n$, $x_i \geq x_k$. (We write $x_k = \min\{x_1, \ldots, x_n\}$). (Hint: use induction.)

5.   EQUIVALENCE RELATIONS

There is often a need in mathematics to say that two things are equivalent to each other in some sense. For instance, congruent triangles can be regarded as the same for most purposes in geometry; for some purposes, when size is unimportant, similar triangles can be regarded as equivalent. Mathematicians have therefore found it useful to systematize the notion of equivalence.

   An *equivalence relation* on a set S is a relation, $\sim$ on S, which satisfies the following rules:

   (1)   For any a $\in$ S, a $\sim$ a (*reflexive law*).

   (2)   If a $\sim$ b, then b $\sim$ a (*symmetric law*).

   (3)   If a $\sim$ b and b $\sim$ c, then a $\sim$ c (*transitive law*).

Here are some examples:

   1.   Ordinary equality is an equivalence relation; it is easy to see that properties (1) - (3) hold.

2. Suppose that we define a relation on the integers by decreeing that a ≡ b if a - b is even. Then   is an equivalence relation.

3. More generally, let n > 1 be an integer, and write a ≡ b mod n ("a is congruent to b mod n") if a - b is divisible by n. Then congruence mod n is an equivalence relation.

4. The relation a ≤ b is *not* an equivalence relation; property (2) fails.

5. The relation a‖b (a is parallel to b), among lines in the plane, becomes an equivalence relation if one agrees that any line is parallel to itself. Otherwise, property (1) fails.

6. The relation "The angle between a and b is < 30°", again between lines in the plane, is not an equivalence relation; property (3) fails.

Given an equivalence relation ~, and an element x, we define

$$R_x = \{y : x \sim y\}$$

Thus R_x is the set of elements equivalent to x; it is often called the *equivalence class* of x.

PROPOSITION 5.3. *If R_x and R_y are two equivalence classes for the relation ~, then either $R_x \cap R_y = \phi$ or $R_x = R_y$. (Equivalence classes are either disjoint or identical.)*

Proof. Suppose $R_x \cap R_y \neq \phi$; we need to show that $R_x = R_y$. Pick z in $R_x \cap R_y$, and let w be any element of R_x. Then x ~ z and x ~ w; hence (properties (2) and (3)) w ~ z. Since z ∈ R_y, z ~ y; now property (3) says that w ~ y. Therefore w ∈ R_y. It follows that $R_x \subset R_y$. The reverse inclusion is proved in the same way.

COROLLARY. *If ~ is an equivalence relation on a set S, then S is the disjoint union of the subsets of R_x, x ∈ S.*

For many purposes, all elements of an equivalence class behave identically. Indeed, it is a standard mathematical trick to regard the equivalence classes as basic objects and to work with them. We shall see a number of examples of this device in the rest of this book.

EXERCISES

1.   Which of the following are equivalence relations?

   (a)   a ~ b if a is divisible by b (a,b nonzero integers).

   (b)   (a,b) ~ (c,d) if ad = bc ((a,b) and (c,d) are pairs of integers with b and d nonzero).

   (c)   a ~ b if a + b is divisible by 3 (a,b nonzero integers).

   (d)   a ~ b if ab > 0 (a,b real numbers).

   (e)   a ~ b if a and b have the same parents.

2.   Let p be a fixed positive integer. We say that two integers a
and b are *congruent* modulo p (written a ≡ b mod p) if a - b is divisible by p.

   (a)   Prove that the relation on the set Z of integers of being
equivalent modulo p is an equivalence relation.

   (b)   Determine the number of equivalence classes of this equivalence relation.

3.   Can property (1) of equivalence relations be derived from the
other two? (If it can, give a proof. If a derivation is impossible,
give a relation satisfying (2) and (3), but not (1).)

CHAPTER 2

VECTOR SPACES

INTRODUCTION

This book deals with two main topics: linear algebra, or the study of linear functions on vector spaces, and multivariable calculus, or the study of nonlinear functions on vector spaces. The first step for each is the study of vector spaces, and this is the topic of this chapter. We define vector spaces, subspaces, and linear transformations, and prove some simple results about them. We also investigate linear transformations between Euclidean spaces in some detail.

1. THE CARTESIAN PLANE

Most of our work in this course will involve vector spaces in one way or another, and our first order of business is to describe what a vector space is. In this section, we shall give an example of a typical vector space and discuss its properties; in the next, we shall give the definition of a vector space.

The space we shall deal with here is the *Cartesian plane* R^2; it consists of ordered pairs of real numbers. Geometrically, the point (a,b) represents the point on the plane whose x-coordinate is a and whose y-coordinate is b. (See Figure 1.1.) The same notation (a,b) is used for a point in R^2 and an open interval. Usually it is clear from the context which is meant.

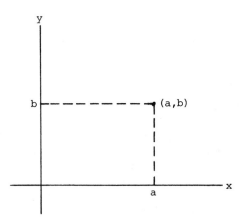

FIGURE 1.1.   Coordinates in the Cartesian plane

It is sometimes useful to think of the point (a,b) in other ways. In many applications, (a,b) is regarded as a line segment stretching from the origin to the point (a,b). (See Figure 1.2.) Different points in R^2 give rise to different line segments; more specifically, these segments differ either in length or in direction. For this reason, one sometimes represents the point (a,b) by any line segment with the same length and direction as the segment from (0,0) to (a,b). For example, if (x,y) is any point in the plane, then the segment from (x,y)

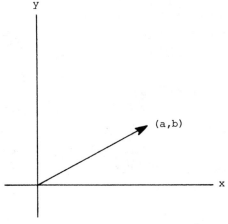

FIGURE 1.2.   A vector in R^2

to (x+a,y+b) has the same length and direction as that from (0,0) to (a,b). (See Figure 1.3.) Such directed line segments are usually called *vectors*; in physics, they represent certain kinds of quantities such as force or velocity.

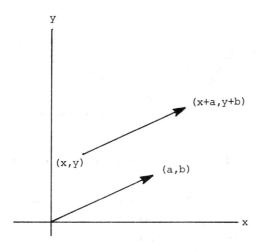

FIGURE 1.3. Vectors in R^2

In what follows, we shall call the elements of R^2 vectors. Instead of always writing a vector as an ordered pair, we shall often represent it by a single letter with an arrow above: $\vec{v}$ = (a,b). We call the numbers a and b the *components* of $\vec{v}$.

    There are a number of algebraic operations which can be performed on vectors in R^2; we shall be concerned in this section with two of them. We define the *sum* of two vectors, (a,b) and (c,d), by

$$(a,b) + (c,d) = (a+c,b+d) \tag{1.1}$$

That is, one adds vectors in R^2 by adding their components. For example, (2,1) + (4,-2) = (6,-1).

    We can interpret the sum geometrically if we think of the vectors as directed line segments: to add (a,b) and (c,d), put the tail of the arrow representing (c,d) at (a,b) and see where the head is. (See Figure 1.4.)

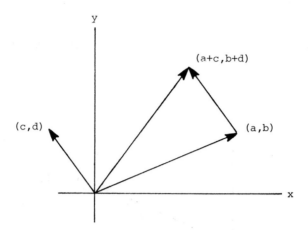

FIGURE 1.4. The sum of two vectors

The other operation is called *scalar multiplication*; in this operation, one multiplies a vector by a real number. The product of (a,b) by the number r is defined by

$$r(a,b) = (ra,rb) \qquad\qquad (1.2)$$

Thus one simply multiplies each component by r. For instance, $7(1,-3) = (7,-21)$.

Interpreting the product geometrically is not hard. The vector $2(a,b)$ for instance, points in the same direction as (a,b), but is twice as long; $\frac{1}{3}(a,b)$ points in the same direction as (a,b), but is a third as long. There is one slightly tricky point: multiplication by a negative number reverses the direction of the vector. Thus $-2(a,b)$ points in the opposite direction from (a,b) and is twice as long. (See Figure 1.5.)

These operations satisfy a number of relations, some of which are the following:

If $\vec{v}_1, \vec{v}_2, \vec{v}_3 \in R^2$, then $(\vec{v}_1 + \vec{v}_2) + \vec{v}_3 = \vec{v}_1 + (\vec{v}_2 + \vec{v}_3)$. $\qquad$ (1.3)

If $\vec{v}, \vec{w} \in R^2$, then $\vec{v} + \vec{w} = \vec{w} + \vec{v}$. $\qquad\qquad$ (1.4)

If $\vec{0} = (0,0)$, then $\vec{v} + \vec{0} = \vec{0} + \vec{v} = \vec{v}$ for any $\vec{v} \in R^2$. $\qquad$ (1.5)

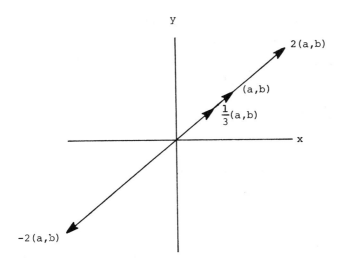

FIGURE 1.5.   Scalar multiples of a vector

If $\vec{v} = (x,y) \in R^2$ and $-\vec{v} = (-x,-y)$, then $\vec{v} + (-\vec{v}) = \vec{0}$.   (1.6)

$1 \cdot \vec{v} = \vec{v}$ for all $\vec{v} \in R^2$.   (1.7)

If $\vec{v}_1, \vec{v}_2 \in R^2$ and $r \in R$, then $r(\vec{v}_1 + \vec{v}_2) = r\vec{v}_1 + r\vec{v}_2$.   (1.8)

If $\vec{v} \in R^2$ and $r_1, r_2 \in R$, then $(r_1 + r_2)\vec{v} = r_1\vec{v} + r_2\vec{v}$.   (1.9)

If $\vec{v} \in R$ and $r_1, r_2 \in R$, then $(r_1 r_2)\vec{v} = r_1(r_2\vec{v})$.   (1.10)

Relation (1.3) is called the *associative law*, (1.4) the *commutative law*, and (1.8) and (1.9) two forms of the *distributive law*.

The proofs of these formulas are all easy; one just evaluates both sides. For instance, if $\vec{v} = (x,y)$, then the left-hand side of (1.10) is

$$r_1 r_2(x,y) = (r_1 r_2 x, r_1 r_2 y)$$

while the right side is

$$r_1(r_2(x,y)) = r_1(r_2 x, r_2 y)$$
$$= (r_1 r_2 x, r_1 r_2 y)$$

Thus the two sides are equal. We leave the verification of the other
formulas as exercises.

   Finally, if $\vec{v}$ and $\vec{w}$ are vectors, we shall write $\vec{v}$ - $\vec{w}$ for $\vec{v}$ + $(-\vec{w})$.
If $\vec{v}$ = (a,b) and $\vec{w}$ = (c,d), then $-\vec{w}$ = (-c,-d), so that $\vec{v}$ - $\vec{w}$ = (a-c,
b-d).

EXERCISES

1.   Perform the following computations:

   (a) (1,4) + (2,8)   (i) 0(3,3)
   (b) (2,3) - (-1,7)   (j) 3(1,2) + (2,7)
   (c) (4,-2) + (-4,8)   (k) -1(-2,3) + 2(4,7)
   (d) (3,-2) - (-2,3)   (ℓ) 2(3,8) + (-2)(3,8)
   (e) (5,-1) + (-5,1)   (m) 3(1,0) + 2(0,1)
   (f) 6(1,-3)   (n) 4(1,3) - 2(2,6)
   (g) 3(-2,7)   (o) 5(1,2) + 3(-1,4)
   (h) -2(4,-5)

2.   Prove formulas (1.3) - (1.6).

3.   Prove formulas (1.7), (1.9), and (1.10).

4.   Prove that

$$a \sum_{i=1}^{k} \vec{v}_i = \sum_{i=1}^{k} a\vec{v}_i$$

where $\vec{v}_1,\ldots,\vec{v}_k \in R^2$ and $a \in R$.

5.   Prove that

$$\left(\sum_{i=1}^{k} a_i \right)\vec{v} = \sum_{i=1}^{k} a_i\vec{v}$$

where $a_1,\ldots,a_k \in R$ and $\vec{v} \in R^2$.

2.   THE DEFINITION OF A VECTOR SPACE

We now give the definition of a vector space and list a number of examples.

A *vector space* is a set V on which two operations are defined.
The first, addition, is a function which assigns to a pair of vectors
$\vec{v},\vec{w} \in V$ their sum, $\vec{v} + \vec{w}$ (also in V); the second, scalar multiplication, assigns to any vector $\vec{v} \in V$ and any real number r the vector
$r \cdot \vec{v} \in V$.   These operations satisfy the following axioms:

$$\text{If } \vec{v}_1, \vec{v}_2, \vec{v}_3 \in V, \text{ then } (\vec{v}_1 + \vec{v}_2) + \vec{v}_3 = \vec{v}_1 + (\vec{v}_2 + \vec{v}_3). \qquad (2.1)$$

$$\text{If } \vec{v}, \vec{w} \in V, \text{ then } \vec{v} + \vec{w} = \vec{w} + \vec{v}. \qquad (2.2)$$

$$\text{There is a vector } \vec{0} \text{ in V such that } \vec{0} + \vec{v} = \vec{v} + \vec{0} = \vec{v},$$
$$\text{for any } \vec{v} \in V. \qquad (2.3)$$

$$\text{If } \vec{v} \in V, \text{ then there is a vector } (-\vec{v}) \in V \text{ such that}$$
$$\vec{v} + (-\vec{v}) = (-\vec{v}) + \vec{v} = \vec{0}. \qquad (2.4)$$

$$1 \cdot \vec{v} = \vec{v} \text{ for all vectors } \vec{v} \in V. \qquad (2.5)$$

$$\text{If } \vec{v}_1, \vec{v}_2 \in V \text{ and } r \in R, \text{ then } r \cdot (\vec{v}_1 + \vec{v}_2) = r \cdot \vec{v}_1 + r \cdot \vec{v}_2. \qquad (2.6)$$

$$\text{If } \vec{v} \in V \text{ and } r_1, r_2 \in R, \text{ then } (r_1 + r_2) \cdot \vec{v} = r_1 \cdot \vec{v} + r_2 \cdot \vec{v}. \qquad (2.7)$$

$$\text{If } \vec{v} \in V \text{ and } r_1, r_2 \in R, \text{ then } (r_1 r_2) \cdot \vec{v} = r_1 \cdot (r_2 \cdot \vec{v}). \qquad (2.8)$$

Note. These vector spaces are sometimes called *real vector
spaces* because the scalar multiplication is defined for real numbers.
Later on, we shall consider vector spaces in which the scalar multiplication is defined for complex numbers.

Here are some examples:

(1) We have already seen that the Cartesian plane, R^2, satisfies
these axioms (if we define addition and scalar multiplication as in
Section 1). Another example is Euclidean 3-space, R^3, consisting of
all ordered triples of real numbers (a typical element is $\vec{v} = (a,b,c)$;
see Figure 2.1), with the operations defined as follows:

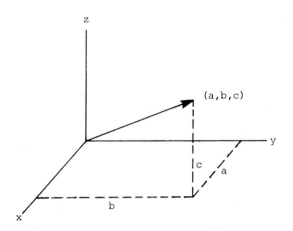

FIGURE 2.1. A vector in R^3

$$(x_1, x_2, x_3) + (y_1, y_2, y_3) = (x_1 + y_1, x_2 + y_2, x_3 + y_3)$$
$$r \cdot (x_1, x_2, x_3) = (rx_1, rx_2, rx_3)$$

Checking that all the axioms hold is easy; in fact, it is almost the same as for R^2.

(2) There is nothing special about 2 or 3 in the first example. In fact, we can define *Euclidean n-space*, R^n (where n is a positive integer) to be the set of all ordered n-tuples $(x_1, \ldots, x_n)$ of real numbers, with componentwise addition,

$$(x_1, \ldots, x_n) + (y_1, \ldots, y_n) = (x_1 + y_1, \ldots, x_n + y_n)$$

and componentwise scalar multiplication,

$$r \cdot (x_1, \ldots, x_n) = (rx_1, \ldots, rx_n)$$

(When n = 2, Euclidean 2-space is what we have been calling the Cartesian plane.) We shall be using this example throughout the course, and it will often be convenient to write (x_i) for $(x_1, \ldots, x_n)$. In particular, addition and scalar multiplication are then defined by

$$(x_i) + (y_i) = (x_i + y_i)$$
$$r(x_i) = (rx_i)$$

We can go even further; define R^∞ to be the vector space consisting of all infinite sequences $(x_1, x_2, \ldots, x_n, \ldots)$ with addition
and scalar multiplication defined by

$$(x_1, \ldots, x_n, \ldots) + (y_1, \ldots, y_n, \ldots) = (x_1 + y_1, \ldots, x_n + y_n, \ldots)$$

$$r(x_1, \ldots, x_n, \ldots) = (rx_1, \ldots, rx_n, \ldots)$$

(3) $C(R)$ (sometimes also called $C^0(R)$, for reasons to be made
clear later) is the vector space of all continuous functions from the
reals to the reals. The operations are defined as follows: if f_1, f_2
are continuous functions from R to R, then $f_1 + f_2$ is the function
given by

$$(f_1 + f_2)(x) = f_1(x) + f_2(x) \tag{2.9}$$

and $r \cdot f_1$ is defined by

$$(r \cdot f_1)(x) = r(f_1(x)) \tag{2.10}$$

This may require some explanation. Recall that a function from
R to R is specified by giving its value for each real number. To
specify $f_1 + f_2$, therefore, we have to know what $(f_1 + f_2)(x)$ is for
each $x \in R$. Formula (2.9) tells us just that; similarly, (2.10) defines $(r \cdot f_1)(x)$ for each $x \in R$. These formulas actually say that the
operations are what one would expect them to be. For instance, if
$f_1(x) = x^2$ and $f_2 = \sin$ (that is $f_2(x) = \sin x$), then $f_1 + f_2$ is the
function whose value at x is $x^2 + \sin x$, and $(2 \cdot f_2)(x) = 2 \sin x$.

There is one other point about this example. To see that $C(R)$
is a vector space with these operations, we need first of all to know
that they are well-defined: that is, we need to know that if f_1 and
f_2 are continuous functions, so are $f_1 + f_2$ and $r \cdot f_1$. These facts
should be familiar from elementary calculus; we shall see a proof of
a more general fact later in the text. Of course, we also need to
check axioms (2.1)-(2.8). (They do hold; see Exercise 2.)

Similarly, we can define $C(X)$ where X is any nonempty real interval.

(4)   The smallest vector space of them all, $\{\vec{0}\}$, has exactly one element, $\vec{0}$.   (Axiom (2.3) guarantees us that every vector space has a zero element.)   The vector space operations are easy to define, since we have no choice:

$$\vec{0} + \vec{0} = \vec{0}$$
$$r \cdot \vec{0} = \vec{0}, \text{ all } r \in R$$

(5)   A $p \times n$ (or p by n) *matrix* is a rectangular array of numbers

$$\begin{pmatrix} a_{11} & a_{12} & \cdots & a_{1n} \\ a_{21} & a_{22} & \cdots & a_{2n} \\ & & \cdots & \\ a_{p1} & a_{p2} & \cdots & a_{pn} \end{pmatrix}$$

We often abbreviate such a matrix by (a_{ij}).   Note that a_{ij} is the entry in the ith row and j^{th} column.

We define the addition and scalar multiplication of $p \times n$ matrices as follows:

$$\begin{pmatrix} a_{11} & \cdots & a_{1n} \\ & \cdots & \\ a_{p1} & \cdots & a_{pn} \end{pmatrix} + \begin{pmatrix} b_{11} & \cdots & b_{1n} \\ & \cdots & \\ b_{p1} & \cdots & b_{pn} \end{pmatrix} = \begin{pmatrix} a_{11}+b_{11} & \cdots & a_{1n}+b_{1n} \\ & \cdots & \\ a_{p1}+b_{p1} & \cdots & a_{pn}+b_{pn} \end{pmatrix}$$

$$r \begin{pmatrix} a_{11} & \cdots & a_{1n} \\ & \cdots & \\ a_{p1} & \cdots & a_{pn} \end{pmatrix} = \begin{pmatrix} ra_{11} & \cdots & ra_{1n} \\ & \cdots & \\ ra_{p1} & \cdots & ra_{pn} \end{pmatrix}$$

or, more briefly,

$$(a_{ij}) + (b_{ij}) = (a_{ij} + b_{ij})$$
$$r(a_{ij}) = (ra_{ij})$$

For example,

$$
\begin{pmatrix} 2 & 0 & -1 \\ 3 & 3 & 7 \end{pmatrix} + \begin{pmatrix} 1 & 5 & 2 \\ 0 & -6 & 6 \end{pmatrix} = \begin{pmatrix} 3 & 5 & 1 \\ 3 & -3 & 13 \end{pmatrix}
$$

$$
3 \begin{pmatrix} 2 & 0 & -1 \\ 3 & 3 & 7 \end{pmatrix} = \begin{pmatrix} 6 & 0 & -3 \\ 9 & 9 & 21 \end{pmatrix}
$$

We denote the vector space of $p \times n$ matrices by $\mathbb{m}(p,n)$.

(6) The vector space $\mathcal{P}_n$ consists of all polynomials (in x) of degree $\leq n$. A typical element is $a_n x^n + a_{n-1} x^{n-1} + \cdots + a_1 x + a_0$. Addition and scalar multiplication are defined as one would expect:

$$
(a_n x^n + \cdots + a_0) + (b_n x^n + \cdots + b_0) = (a_n + b_n)x^n + \cdots + (a_0 + b_0)
$$

$$
r(a_n x^n + \cdots + a_1 x + a_0) = ra_n x^n + \cdots + ra_1 x + ra_0
$$

Besides these examples, we should show some non-examples. Here are some sets with operations which are *not* vector spaces.

(7) The integers do not form a vector space. It is clear how to add integers, but defining scalar multiplication by real numbers is a bit harder. In fact, there is no way to do it and satisfy all the axioms. (See Exercise 8.)

(8) Let W be the set of all continuous real-valued functions $f: R \to [0,1]$, with addition and scalar multiplication as usual:

$$
(f_1 + f_2)(x) = f_1(x) + f_2(x)
$$
$$
(r \cdot f_1)(x) = rf_1(x)
$$

for $x \in R$.

This is *not* a vector space. The trouble is that addition and scalar multiplication are not well-defined. For instance, if $f(x) = (1+x^2)^{-1}$, then 2f is not in W, because $(2f)(0) = 2$ (and functions in W have values between 0 and 1). Thus we do not even have the vector space operations on which to check the axioms.

EXERCISES

1.   Verify that the vector space axioms hold for R^n (with operations as in Example 2).

2.   Verify that $C(R)$ of Example 3 is a vector space.

3.   The following are not vector spaces. In each case, find an axiom that fails to hold.

   (a)   The set R^2 with the following operations:

$$(x_1,x_2) + (y_1,y_2) = (x_1 + y_1, x_2 + y_2)$$
$$r(x_1,x_2) = (rx_1, x_2)$$

   (b)   The set R^3, with the following operations:

$$(x_1,x_2,x_3) + (y_1,y_2,y_3) = (x_1 + y_1, x_2 + y_2, x_3 + y_3)$$
$$r(x_1,x_2,x_3) = (rx_1, 0, rx_3)$$

   (c)   The set R^2, with

$$(x_1,x_2) + (y_1,y_2) = (x_1 + y_1, x_2)$$
$$r(x_1,x_2) = (rx_1, rx_2)$$

   (d) Z^2, the set of all pairs of *integers* (m_1,m_2), with

$$(m_1,m_2) + (n_1,n_2) = (m_1 + n_1, m_2 + n_2)$$
$$r(m_1,m_2) = ([rm_1],[rm_2])$$

where $[x]$ = greatest integer $\leq x$. (Thus $[1] = 1$, $[\frac{3}{2}] = 1$, $[\pi] = 3$, $[-1.2] = -2$.)

   (e)   The set R^2 with

$$(x_1,x_2) + (y_1,y_2) = (x_1 - y_1, x_2 - y_2)$$
$$r(x_1,x_2) = (rx_1, rx_2)$$

(f)   The set R^2 with

$$(x_1,x_2) + (y_1,y_2) = (x_1 + y_1, x_2 + y_2)$$
$$r(x_1,x_2) = (rx_2, rx_1)$$

4.   Which of the following are vector spaces?

(a)   The set R^2 with

$$(x_1,x_2) + (y_1,y_2) = (2x_1 + 2y_1, x_2 + y_2)$$
$$r(x_1,x_2) = (rx_1, rx_2)$$

(b)   The set R^2 with

$$(x_1,x_2) + (y_1,y_2) = (x_1 + y_1, x_2 + y_2)$$
$$r(x_1,x_2) = (-rx_1, -rx_2)$$

(c)   The set of points $V = \{(x_1,x_2,x_3):(x_1,x_2,x_3) \in R^3$, and $x_1 - x_2 + x_3 = 0\}$, with

$$(x_1,x_2,x_3) + (y_1,y_2,y_3) = (x_1 + y_1, x_2 + y_2, x_3 + y_3)$$
$$r(x_1,x_2,x_3) = (rx_1, rx_2, rx_3)$$

(d)   The set P_2 with

$$(p_1 + p_2)(x) = p_1(x) + p_2(x)$$
$$r \cdot p(x) = p(rx)$$

(e)   The set P_2 with

$$(p_1 + p_2)(x) = p_1(x) + p_2(x)$$
$$r \cdot p(x) = p(r) \cdot p(x)$$

(f)   The set R^3 with

$$(x_1,x_2,x_3) + (y_1,y_2,y_3) = (x_2 + y_2, x_3 + y_3, x_1 + y_1)$$
$$r(x_1,x_2,x_3) = (rx_1, rx_2, rx_3)$$

(g) $V = \{\vec{v} \in R^3 : \vec{v}$ can be written as $a(1,0,-1) + b(1,2,1)$, where a,b are real numbers$\}$; addition and scalar multiplication are as in R^3.

(h) $V = \{\vec{v} \in R^3 : \vec{v}$ can be written as $a^2(1,2,3) + b(4,5,6)$, where a,b are real numbers ; addition and scalar multiplication are as in R^3.

5.   Verify that the vector space axioms hold for $\mathbb{m}(n,m)$, defined in Example 5.

6.   Verify that the vector space axioms hold for P_n, defined in Example 6.

7.   Let V_1 and V_2 be vector spaces. We define the *direct product* $V_1 \times V_2$ of V_1 and V_2 to be the vector space with

$$V_1 \times V_2 = \{(\vec{v}_1,\vec{v}_2) : \vec{v}_1 \in V_1 \text{ and } \vec{v}_2 \in V_2\}$$
$$(\vec{v}_1,\vec{v}_2) + (\vec{w}_1,\vec{w}_2) = (\vec{v}_1 + \vec{w}_1, \vec{v}_2 + \vec{w}_2)$$
$$r(\vec{v}_1,\vec{v}_2) = (r\vec{v}_1, r\vec{v}_2)$$

Prove that $V_1 \times V_2$ is a vector space.

*8.   Prove that one cannot define scalar multiplication on the integers so that, with their usual addition, they become a vector space. (Hint: consider $(1/2) \cdot 1$.)

9.   Let V be a vector space, $\vec{v}_1, \ldots, \vec{v}_k \in V$ and $r \in R$. Prove by induction that

$$r\left(\sum_{j=1}^{k} \vec{v}_j\right) = \sum_{j=1}^{k} r\vec{v}_j$$

10.   Let V be a vector space, $\vec{v} \in V$, and $r_1, \ldots, r_k \in R$. Prove by induction that

$$\left(\sum_{j=1}^{k} r_j\right)\vec{v} = \sum_{j=1}^{k} r_j\vec{v}$$

*11. Let V be a vector space, and let $\vec{v}_1, \ldots, \vec{v}_n$ be elements of V, not necessarily distinct. Show that their sum is unambiguously defined. That is, if n = 5 (say), we could form

$$((\vec{v}_4 + \vec{v}_1) + \vec{v}_5) + (\vec{v}_2 + \vec{v}_3)$$

or

$$(((\vec{v}_1 + \vec{v}_2) + \vec{v}_3) + \vec{v}_4) + \vec{v}_5$$

Design and carry out a general inductive proof that the results will always be the same.

*12. Let V be the set of all functions $f: R \to R$ having at least two derivatives everywhere, and such that $f'' - 3f' + 2f = 0$.

(a) Show that V is a vector space.

(b) Show that V is the same as the set of all funcitons of the form $f(x) = ae^x + be^{2x}$, $a, b \in R$. (Hint: $f'' - 3f' + 2f = (f' - f)'$ $- 2(f' - f)$ and, if $g' - \gamma g = h$, then $(e^{-\gamma x} g)' = e^{-\gamma x} h$.)

*13. Let V be the set of all polynomial functions $f(x)$ such that $f(x+3) - 3f(x+2) + 3f(x+1) - f(x) = 0$ for all x.

(a) Show that V is a vector space.

(b) Show that V is the same as $\mathcal{P}_2$, the vector space of polynomials of degree ≤ 2.

(c) Which of (a) and (b) remains valid if we do not restrict f to be a polynomial?

3.   SOME ELEMENTARY PROPERTIES OF VECTOR SPACES

Eventually, we shall make a fairly careful analysis of vector spaces. In this section, we give some easy consequences of the axioms. We begin by showing that a vector space cannot have two zero vectors.

PROPOSITION 3.1.   *The element* $\vec{0}$ *of Axiom 3 is unique; that is, if* $\vec{0}'$ *is an element such that* $\vec{0}' + \vec{v} = \vec{v}$ *for every* $\vec{v} \in V$, *then* $\vec{0}' = \vec{0}$.

Proof. Let $\vec{v} = \vec{0}$. Then

$$\vec{0}' = \vec{0}' + \vec{0} \quad \text{(by axiom 3)}$$

$$= \vec{0} \quad \text{(by hypothesis)}$$

Hence $\vec{0}' = \vec{0}$, as claimed.

We now prove that a vector can have only one inverse.

PROPOSITION 3.2. *Each element of* V *has a unique inverse: if* $\vec{v} + \vec{w} = \vec{0}$, *then* $\vec{w} = -\vec{v}$.

Proof. Suppose that $\vec{v} + \vec{w} = \vec{0}$. Then

$$(-\vec{v}) = (-\vec{v}) + \vec{0} = (-\vec{v}) + (\vec{v} + \vec{w}) \quad \text{(hypothesis)}$$

$$= ((-\vec{v}) + \vec{v}) + \vec{w} \quad \text{(axiom 2)}$$

$$= \vec{0} + \vec{w} \quad \text{(axiom 4)}$$

$$= \vec{w}, \quad \text{as claimed}$$

The next result generalizes Proposition 3.2.

PROPOSITION 3.3. *If* $\vec{v}, \vec{w} \in V$, *the equation* $\vec{x} + \vec{v} = \vec{w}$ *has a unique solution.*

Proof. If $\vec{x} + \vec{v} = \vec{w}$, then

$$\vec{w} + (-\vec{v}) = (\vec{x} + \vec{v}) + (-\vec{v})$$

$$= \vec{x} + (\vec{v} + (-\vec{v}))$$

$$= \vec{x} + \vec{0} = \vec{x}$$

so that $\vec{x}$ must equal $\vec{w} + (-\vec{v})$. In fact, $\vec{x} = \vec{w} + (-\vec{v})$ does solve the equation.

The next two results tell us that multiplication of a vector by 0 and -1 gives the expected results.

PROPOSITION 3.4. *For any* $\vec{v} \in V$, $0 \cdot \vec{v} = \vec{0}$.

Proof. We know from Axiom 3 that $\vec{0}$ is a solution of $\vec{x} + \vec{v} = \vec{v}$.

However,

$$0 \cdot \vec{v} + \vec{v} = 0 \cdot \vec{v} + 1 \cdot \vec{v} \quad \text{(by Axiom (2.5))}$$
$$= (0+1) \cdot \vec{v} \quad \text{(by Axiom (2.7))}$$
$$= 1 \cdot \vec{v} = \vec{v} \quad \text{(by Axiom (2.5))}$$

Thus $0 \cdot \vec{v}$ is also a solution, and the result follows from Proposition 3.3 above.

PROPOSITION 3.5. $(-1) \cdot \vec{v} = -\vec{v}$ *for all* $\vec{v} \in V$.

Proof.   For all $\vec{v}$ in V,

$$(-1) \cdot \vec{v} + \vec{v} = (-1) \cdot \vec{v} + 1 \cdot \vec{v} \quad \text{(by Axiom (2.5))}$$
$$= (-1+1) \cdot \vec{v} \quad \text{(by Axiom (2.7))}$$
$$= 0 \cdot \vec{v} = \vec{0} \quad \text{(by Proposition 3.4)}$$

Thus $(-1) \cdot \vec{v} = -\vec{v}$ by Proposition 3.2.

Proposition 3.4 and Proposition 3.5 show that our notation is reasonable; that is, the vector $\vec{0}$ behaves like the scalar 0, and multiplying $\vec{v}$ by -1 gives $-\vec{v}$. We shall use this to simplify notation in three ways. First of all, we shall omit the parentheses around $-\vec{v}$. Secondly, we shall eliminate the · in multiplication by scalars; from now on, we shall write $r\vec{v}$ instead of $r \cdot \vec{v}$. Thirdly, we shall use - signs more freely with vectors; for instance, we shall write $\vec{v} - \vec{w}$ to mean $\vec{v} + (-\vec{w})$. (The solution to $\vec{x} + \vec{v} = \vec{w}$ will be written as $\vec{x} = \vec{w} - \vec{v}$.)

One more word about notation. We have been using $\vec{0}$ to represent the zero vector in any vector space. When we work with more than one vector space at the same time, this can cause confusion. In these cases, we shall sometimes use a subscript to indicate which zero we are using. Thus $\vec{0}_V$ is the zero vector in V, and so on. A similar remark applies to the symbols for addition and scalar multiplication; "+" can mean different things in different vector spaces, and sometimes we shall write $+_V$ or $+_W$ to emphasize this fact.

EXERCISES

1.   Prove that $(-r)\vec{v} = r(-\vec{v}) = -(r\vec{v})$ for all vectors $\vec{v} \in V$ and all
real numbers r.

2.   Prove that $r\vec{0} = \vec{0}$ for any $r \in R$, where $\vec{0}$ is the zero vector.

3.   Suppose that $\vec{v} \neq \vec{0}$ and that $r\vec{v} = s\vec{v}$. Show that $r = s$.

4.   Suppose that $r \neq 0$ and that $r\vec{v} = r\vec{w}$. Show that $\vec{v} = \vec{w}$.

5.   Suppose that V is a vector space with at least 2 distinct elements. Show that V has infinitely many elements.

6.   Show that if $\vec{w}_1$, $\vec{w}_2$ are given vectors and $r \neq 0$, then the equation $r\vec{v} + \vec{w}_1 = \vec{w}_2$ has exactly one solution $\vec{v}$.

7.   Show that if $\vec{v}_1$, $\vec{v}_2$, and $\vec{w}$ are given vectors with $\vec{v}_1 + \vec{w} = \vec{v}_2 + \vec{w}$,
then $\vec{v}_1 = \vec{v}_2$.

8.   Solve the following equations for vectors in R^3.

   (a) $3\vec{v} + (0,1,2) = (6,4,2)$

   (b) $4\vec{v} - 2(4,8,-5) = (0,-4,2)$

   (c) $3\vec{v} - (2,1,7) = (1,4,2) - \vec{v}$

*9.   Show that Axiom (2.2) is a consequence of the other axioms.
(Hint: expand $(1+1)(\vec{v}+\vec{w})$ in two ways.)

4.   SUBSPACES

We now discuss certain special subsets of a vector space V which inherit the structure of a vector space from V.

   Let V be a vector space. A *subspace* W of V is a nonempty subset of V with the following properties:

$$\text{If } \vec{v}_1, \vec{v}_2 \in W, \text{ then } \vec{v}_1 + \vec{v}_2 \in W. \qquad (4.1)$$

If $\vec{v} \in W$ and $r \in R$, then $r\vec{v} \in W$. (4.2)

More briefly, W is closed under the operations of addition and scalar multiplication.

We shall give some examples in a moment. First, however, we give a result which may help explain the importance of the notion.

PROPOSITION 4.1. *A nonempty subset W of V is a subspace of V if and only if W is itself a vector space (with the same definitions of addition and scalar multiplication as in V).*

Proof. Suppose that W is a subspace of V. We need to check that W satisfies the axioms for a vector space. To begin with, we need well-defined operations of addition and scalar multiplication in W. But the definition of a subspace guarantees that if $\vec{v}_1$ and $\vec{v}_2$ are in W, and r is in R, then $\vec{v}_1 + \vec{v}_2$ and $r\vec{v}_1$ are well-defined elements of W. Thus we can go on to check axioms (2.1)-(2.8) of Section 2.

Axioms (2.1), (2.2), (2.5), (2.6), (2.7), and (2.8) are easy to check: they are true in W because they are true in V. For instance, $\vec{v}_1 + \vec{v}_2 = \vec{v}_2 + \vec{v}_1$ for $\vec{v}_1, \vec{v}_2 \in W$ because $\vec{v}_1, \vec{v}_2 \in W$ implies that $\vec{v}_1, \vec{v}_2 \in V$, which implies that $\vec{v}_1 + \vec{v}_2 = \vec{v}_2 + \vec{v}_1$ (since V is a vector space). That leaves (2.3) and (2.4). We first check (2.3).

We know that W is nonempty. Let $\vec{w}$ be a vector in W. Then $0\vec{w} \in W$, since W is a subspace. But $0\vec{w} = \vec{0}$, by Proposition 3.4. So $\vec{0} \in W$. For any vector $\vec{v} \in W$, $\vec{0} + \vec{v} = \vec{v} + \vec{0} = \vec{v}$, and so axiom (2.3) is satisfied.

Given $\vec{v} \in W$, $-\vec{v} = (-1)\vec{v} \in W$, since W is a subspace. But Proposition 3.5 says that $-\vec{v} + \vec{v} = \vec{v} + -\vec{v} = \vec{0}$, and that takes care of axiom (2.4). Thus we have proved half of the theorem.

Conversely, suppose that W is a vector space (with the same addition and scalar multiplication as for V). Then W is nonempty since axiom (2.3) guarantees that W contains at least one element (namely $\vec{0}_W$; in fact, $\vec{0}_W = \vec{0}_V$ (Exercise 6), but we do not need to know that here.) Also, if $\vec{v}_1$ and $\vec{v}_2$ are in W then $\vec{v}_1 + \vec{v}_2 \in W$, since otherwise addition would not be well-defined in W. Similarly, $r \in R$ and $\vec{v}_1 \in W$

implies $r\vec{v}_1 \in W$, since we know that multiplication by scalars is defined in W. This completes the proof of Proposition 4.1.

Now for some examples:

1. V itself is always a subspace of V. This is easy to check: V is nonempty (since $\vec{0} \in V$), and if $\vec{v}_1, \vec{v}_2 \in V$ and $r \in R$, then $\vec{v}_1 + \vec{v}_2 \in V$ and $r\vec{v}_1 \in V$ because V is a vector space.

2. $\{\vec{0}\}$ is also always a subspace of V. Here we can use Proposition 4.1; $\{\vec{0}\}$ is nonempty, and forms a vector space under the same rules of addition and scalar multiplication as in V. (We know that $\vec{0} + \vec{0} = \vec{0}$, by axiom (2.3), and that $r\vec{0} = \vec{0}$ for any $r \in R$, by Proposition 3.4).

Our first two examples are so universal as to be unexciting, and this is reflected in some terminology: a subspace of V other than V or $\{\vec{0}\}$ is usually called a *proper* subspace. (Presumably this makes V and $\{\vec{0}\}$ improper subspaces, but no one ever uses the term.) The next examples will be of proper subspaces.

3. The set W of vectors of the form $(x_1, x_2, 0)$, where x_1 and x_2 are arbitrary real numbers, is a proper subspace of R^3. The reader should find it easy to check this. Of course, W is just the x_1, x_2 plane in R^3. (See Figure 4.1.)

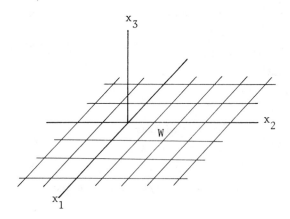

FIGURE 4.1. The $x_1 x_2$ plane in R^3

4. Recall the vector spaces $\mathcal{P}_n$ of Example 6 of Section 2. The spaces $\mathcal{P}_0, \mathcal{P}_1, \ldots, \mathcal{P}_{n-1}$ are all proper subspaces of $\mathcal{P}_n$.

5. Let $C^1(R)$ be the set of all real-valued functions defined on all of R which have continuous first derivatives. Then $C^1(R)$ is a subspace of $C(R)$. It is easy to see that $C^1(R)$ is non-empty; we also have to check that if $f_1, f_2 \in C^1(R)$, then $f_1 + f_2$ and $rf_1 \in C^1(R)$. This is not too hard. For instance, it is a standard result in calculus that $(f_1 + f_2)' = f_1' + f_2'$ and $f_1' + f_2'$ is continuous because it is the sum of two continuous functions. The reasoning for $rf_1 (r \in R)$ is similar.

We can continue this procedure, producing $C^2(R)$ (the subspace of functions defined on all of R which have continuous second derivatives), $C^3(R)$, and so on. (In this notation, it is reasonable to write $C^0(R)$ for $C(R)$.)

6. We already know two subspaces of R^2: $\{0\}$ and R^2 itself. We can obtain others as follows: let $\vec{v} = (x_1, x_2)$ be any nonzero vector in R^2. Then $W = \{\vec{w}: \vec{w} = r\vec{v}$ for some $r \in R\} = \{(rx_1, rx_2): r \in R\}$ is a subspace. (Verify this.) The points of V lying in W form a line in R^2 passing through the origin and $\vec{v}$. (See Figure 4.2.)

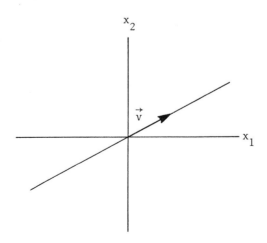

FIGURE 4.2. The line through $\vec{v}$ in R^2

In fact, every proper subspace of R^2 is of this form. Proving this now is an interesting exercise; it will be much easier later. (See Exercise 16.)

7. Let $W = \{\vec{w} = (x_1, x_2, x_3) \in R^3 : x_1 + 2x_2 + 3x_3 = 0, \; 4x_1 + 5x_2 + 6x_3 = 0\}$. Thus W consists of all solutions to the simultaneous linear equations

$$x_1 + 2x_2 + 3x_3 = 0$$
$$4x_1 + 5x_2 + 6x_3 = 0$$

W is a subspace of R^3, as can easily be verified (Exercise 8).

The equations we wrote down were quite arbitrary. More generally, let W consist of the solutions to n homogeneous equations in m unknowns: W is the set of all $(x_1, \ldots, x_m)$ in R^m with

$$a_{11}x_1 + \cdots + a_{1m}x_m = 0$$
$$a_{21}x_1 + \cdots + a_{2m}x_m = 0$$
$$\cdots$$
$$a_{n1}x_1 + \cdots + a_{nm}x_m = 0$$

where (a_{ij}) is an $n \times m$ matrix. Then W is a subspace of R^m. Notice that we do not know of any vector in W other than $\vec{0}$; thus we do not know if W is proper.

8. Let V be any vector space, and let $\vec{v}_1, \ldots, \vec{v}_k$ be vectors in V. Let W be the set of all vectors of the form $r_1\vec{v}_1 + \cdots + r_k\vec{v}_k$, where $r_1, \ldots, r_k$ are real numbers. (Such an expression is called a *linear combination* of the vectors $\vec{v}_1, \ldots, \vec{v}_k$.) Then W is a subspace. (W is nonempty, since $\vec{v}_1 \in W$. W is closed under addition, since if $\vec{w}_1 = r_1\vec{v}_1 + \cdots + r_k\vec{v}_k$ and $\vec{w}_2 = s_1\vec{v}_1 + \cdots + s_k\vec{v}_k$ are typical members of W, then

$$\vec{w}_1 + \vec{w}_2 = (r_1 + s_1)\vec{v}_1 + \cdots + (r_k + s_k)\vec{v}_k$$

is also in W. It is even easier to check scalar multiplication.)

We cannot guarantee that W is proper. If $\vec{v}_1,\ldots,\vec{v}_k$ are all $\vec{0}$, then $W = \{\vec{0}\}$; it is also quite possible for W to equal V, as we shall see.

This method of producing subspaces is quite important, and we shall return to it later.

9. Suppose that W_1 and W_2 are subspaces of V. Then so is $W_1 \cap W_2$.

Proof. First of all, $W_1 \cap W_2 \neq \phi$, since $\vec{0} \in W_1 \cap W_2$. If $\vec{v}_1$ and $\vec{v}_2$ are in $W_1 \cap W_2$, then $\vec{v}_1 \in W_1$, $\vec{v}_2 \in W_1$, and so $\vec{v}_1 + \vec{v}_2 \in W_1$. Similarly, $\vec{v}_1 \in W_2$, $\vec{v}_2 \in W_2$, and so $\vec{v}_1 + \vec{v}_2 \in W_2$. Hence $\vec{v}_1 + \vec{v}_2 \in W_1 \cap W_2$. The proof for scalar multiplication proceeds in the same way. We leave it to the reader.

The fact that we took the intersection of just two subspaces is irrelevant; the same result holds for arbitrary intersections, and the proof is almost identical.

PROPOSITION 4.2. *The intersection of any arbitrary collection of subspaces of V is a subspace: if $\{W_i\}$ is any collection of subspaces of V, $\cap_i W_i$ is a subspace.*

As the proof is essentially the same as the one given above, we leave the details as Exercise 3. Again, we have no guarantee in general that the intersection is a proper subspace.

After all the examples, it is time for some non-examples:

10. The set $W_1 = \{(x_1,x_2) \in R^2 : x_1 \text{ or } x_2 \text{ is } 0\}$ is not a subspace. W is closed under scalar multiplication, but not addition: $(1,0)$ and $(0,1)$ are in W, but $(1,1) = (1,0) + (0,1)$ is not. (See Figure 4.3.)

11. The set $W_2 = \{(x_1,x_2) \in R^2 : x_1 \geq 0\}$ is not a subspace. It is closed under addition, but not scalar multiplication: $(1,0) \in W$, but $(-1,0) = (-1)(1,0)$ is not. (See Figure 4.4.)

12. The set $W_3 = \{(x,x^2,x^3) : x \in R\}$ is not a subspace of R^3; it is not closed under addition or scalar multiplication.

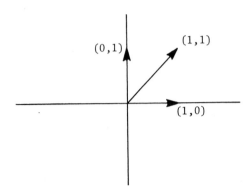

FIGURE 4.3. The set W_1 is not a subspace

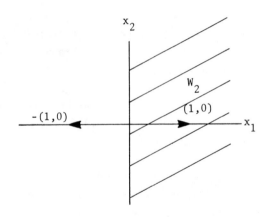

FIGURE 4.4. The set W_2 is not a subspace

13. The set $W_4 = \{(x_1,x_2,1):x_1,x_2 \in R\}$ is not a subspace of R^3, either. (Why?)

14. The union of two subspaces is generally *not* a subspace. The first example above is a case in point. In fact, if W_1 and W_2 are two subspaces of V, neither of which is contained in the other, then $W_1 \cup W_2$ is not a subspace. (See Exercise 4.)

We conclude this section with a slight reformulation of the definition of a subspace. Its main use is that it reduces the number of conditions to be checked (though they become more complicated).

PROPOSITION 4.3. *A nonempty subset* W *of the vector space* V *is a subspace if and only if whenever* $\vec{v}_1, \vec{v}_2 \in W$ *and* $r_1, r_2 \in R$, *then* $r_1\vec{v}_1 + r_2\vec{v}_2 \in W$.

Proof. If W is a subspace, and if $\vec{v}_1, \vec{v}_2 \in W$ and $r_1, r_2 \in R$, then $r_1\vec{v}_1, r_2\vec{v}_2 \in W$ (property (4.2) of the definition), and so $r_1\vec{v}_1 + r_2\vec{v}_2 \in W$ (property (4.1) of the definition).

Conversely, suppose that W is nonempty and has the given property. If $\vec{v}_1, \vec{v}_2 \in W$, then $\vec{v}_1 + \vec{v}_2 = 1\vec{v}_1 + 1\vec{v}_2 \in W$, and if $r \in R$, then $r\vec{v}_1 = r\vec{v}_1 + 0\vec{v}_2 \in W$. Hence W is a subspace.

EXERCISES

1. Which of the following are subspaces of R^3?

  (a) $\{(x_1, x_2, x_3) \in R^3 : x_1 + x_2 + x_3 = 3\}$

  (b) $\{(x_1, x_2, x_3) \in R^3 : x_1 = 0\}$

  (c) $\{(x_1, x_2, x_3) \in R^3 : x_1 = -2x_2\}$

  (d) $\{(x_1, x_2, x_3) \in R^3 : x_1^2 + x_2^2 - x_3 = 0\}$

  (e) $\{(x_1, x_2, x_3) \in R^3 : x_3 = 1\}$

  (f) $\{(x_1, x_2, x_3) \in R^3 : x_1 = 2x_2 = 3x_3\}$

2. Show that $\{(x_1, x_2) \in R^2 : x_2 = ax_1^2 + b\}$ is a subspace of R^2 if and only if $a = b = 0$. What does this mean geometrically?

3. Prove Proposition 4.2.

4. Let W_1 and W_2 be subspaces of the vector space V. Prove (as we claimed in example 5) that if $W_1 \cup W_2$ is a subspace, either $W_1 \subset W_2$ or $W_2 \subset W_1$.

5. Let W be the subset of R^3 consisting of all triples $(x_1, x_2, 1)$, $x_1, x_2 \in R$, and with the vector operations defined as follows:

$$(x_1,x_2,1) + (y_1,y_2,1) = (x_1 + y_1, x_2 + y_2, 1)$$
$$r(x_1,x_2,1) = (rx_1, rx_2, 1)$$

Is W a vector space? Is it a subspace of R^3? Explain.

6.   Let W be a subspace of the vector space V. Show that $\vec{0}_W = \vec{0}_V$.

7.   Show that the set W of Example 6 is indeed a subspace.

8.   Show that the set W of Example 7 is a subspace.

9.   Verify that the set defined in Example 3 is a proper subspace of R^3.

10.   Prove that P_j is a proper subspace of P_n for $j < n$. (See Section. 2, Example 6.)

11.   Let V_1 and V_2 be two subspaces of V. Define the *sum* of V_1 and V_2 to be the subset of V

$$V_1 + V_2 = \{\vec{v}_1 + \vec{v}_2 : \vec{v}_1 \in V_1, \vec{v}_2 \in V_2\}$$

Prove that $V_1 + V_2$ is a subspace of V.

*12.   Let S be a subset (not necessarily a subspace) of a vector space V and define the *span* of S, Span(S) to be the intersection of all sub-spaces of V containing S. (This is well-defined, since V is a sub-space of V containing S).

    (a) Prove that Span(S) is the smallest subspace of V containing S. That is, if V_1 is a subspace of V containing S, then Span(S) $\subseteq V_1$.
    (b) Prove that Span(S) consists of all linear combinations of finite sets of elements of S. (See Example 8.)

13.   (a) If $S \subseteq R^2$ is given by S = {(1,3),(-2,-6)}, is $\vec{v} = (1,1)$ in Span(S)? Is $\vec{v} = (5,15)$ in Span(S)?
    (b) If $S \subseteq R^3$ is given by S = {(1,0,2),(0,1,1),(1,-1,1)}, is $\vec{v} = (1,2,4)$ in Span(S)? Is $\vec{v} = (5,3,3)$ in Span(S)?

14. Show that $S_1 \subset S_2$ implies $\text{Span}(S_1) \subset \text{Span}(S_2)$.

15. In the notation of Exercise 11, prove that

$$V_1 + V_2 = \text{Span}(V_1 \cup V_2)$$

*16. Prove that any proper subspace W of R^2 is of the form

$$W = \{w : w = r\vec{v} \text{ for some } r \in R\}$$

where $\vec{v}$ is a non-zero vector in R^2.

17. Suppose that $V = W_1 \cup W_2$, where W_1 and W_2 are subspaces of V. Show that either $V = W_1$ or $V = W_2$.

*18. Let V be a vector space, and let W be a subspace of V. If $\vec{v}, \vec{w}$ are vectors in V, write $\vec{v} \sim \vec{w}$ if $\vec{v} - \vec{w} \in W$.

    (a) Show that $\sim$ is an equivalence relation. (See Section 1.4)

    (b) Show that if $\vec{v}_1 \sim \vec{w}_1$ and $\vec{v}_2 \sim \vec{w}_2$, then $\vec{v}_1 + \vec{w}_1 \sim \vec{v}_2 + \vec{w}_2$.

    (c) Show that if $\vec{v} \sim \vec{w}$ and r is any real number, then $r\vec{v} \sim r\vec{w}$.

    (d) Let $\bar{v} = \{\vec{w} \in V : \vec{v} \sim \vec{w}\}$, and let $V/W = \{\bar{v} : \vec{v} \in V\}$.

(Notice that if $\vec{v} \sim \vec{w}$, then $\bar{v} = \bar{w}$.) The point of parts (a), (b) and (c) is that we can define addition and scalar multiplication on V/W by

$$\bar{v} + \bar{w} = \overline{(v+w)}$$
$$r\bar{v} = \overline{(rv)}$$

and that these definitions do not depend at all on our choice of $\vec{v}$. (That is, we can replace $\vec{v}$ by $\vec{v}_1$, where $\vec{v} \sim \vec{v}_1$, and we get the same result.) Show that with these definitions, V/W is a vector space. (It is called the *quotient* of V by W.)

5. LINEAR TRANSFORMATIONS

When studying a mathematical structure, one almost invariably also studies certain functions which preserve the structure. In our case,

the functions which preserve the structure of vector spaces are linear transformations, and we take a first look at them in this section.

Let V_1 and V_2 be vector spaces. A *linear transformation* from V_1
to V_2 is a function $T:V_1 \to V_2$ such that

For all vectors $\vec{v}_1, \vec{v}_2 \in V_1$, $T(\vec{v}_1 + \vec{v}_2) = T(\vec{v}_1) + T(\vec{v}_2)$.   (5.1)

For all vectors $\vec{v} \in V_1$ and all $r \in R$, $T(r\vec{v}) = r(T(\vec{v}))$.   (5.2)

(We use the symbol + for the addition in both V_1 and V_2.)

As in the case of subspaces, these two criteria can be reduced
to one.

PROPOSITION 5.1. *A function* $T:V_1 \to V_2$ *is a linear transformation
if and only if for all* $\vec{v}_1, \vec{v}_2 \in V_1$ *and all* $r, s \in R$,

$$T(r\vec{v}_1 + s\vec{v}_2) = r(T(\vec{v}_1)) + s(T(\vec{v}_2))$$ (5.3)

Proof. If T is linear, then

$$T(r\vec{v}_1 + s\vec{v}_2) = T(r\vec{v}_1) + T(s\vec{v}_2)$$
$$= rT(\vec{v}_1) + sT(\vec{v}_2)$$

by the definition of a linear transformation. Conversely, if T satisfies (5.3), then, setting $r = s = 1$, we have

$$T(\vec{v}_1 + \vec{v}_2) = T(\vec{v}_1) + T(\vec{v}_2)$$

Similarly, setting $s = 0$, we see that

$$T(r\vec{v}_1) = T(r\vec{v}_1 + 0\vec{v}_2)$$
$$= rT(\vec{v}_1) + 0T(\vec{v}_2) = rT(\vec{v}_1)$$

Therefore T is linear.

Here are some examples:

1. If V and W are arbitrary vector spaces, we can define a lin-
ear transformation $0:V \to W$ by $0(\vec{v}) = \vec{0}$ for all vectors $\vec{v} \in V$. This
transformation is usually called the *zero transformation*, for obvious
reasons.

2. For any vector space V, the *identity transformation*, $I:V \rightarrow V$, defined by $I(\vec{v}) = \vec{v}$ for all vectors $\vec{v} \in V$, is linear. The defining property of the identity transformation is that it has no effect at all on a vector.

3. We can define a linear transformation $T:R^2 \rightarrow R^3$ by

$$T(x_1,x_2) = (2x_1 - 3x_2, 3x_1 + 4x_2, 5x_1 + 7x_2)$$

(Checking that this is linear is tedious, but straightforward.) There is nothing special about the numbers 2, -3, 3, 4, 5, and 7 in the above definition; any other real numbers would also produce a linear transformation.

4. Let a be any real number and define $T:R \rightarrow R$ by $T(x) = ax$. Then T is a linear transformation. Conversely, suppose $T:R \rightarrow R$ is *any* linear transformation and set $a = T(1)$. Then, for any $x \in R$,

$$\begin{aligned}
T(x) &= T(x \cdot 1) \\
&= xT(1) \qquad \text{(since T is linear)} \\
&= xa = ax
\end{aligned}$$

Thus any linear transformation $T:R \rightarrow R$ is of the form $T(x) = ax$ for some real number a.

5. Let $A = (a_{ij})$ be any $p \times n$ matrix and define a function $T_A:R^n \rightarrow R^p$ by

$$T_A(x_j) = (y_i)$$

where

$$y_i = \sum_{j=1}^{n} a_{ij}x_j$$

For instance, if

$$A = \begin{pmatrix} 2 & -3 \\ 3 & 4 \\ 5 & 7 \end{pmatrix}$$

then $T_A : R^2 \to R^3$ is the linear transformation defined in Example 3.
A straightforward but tedious computation shows that T_A is a linear
transformation for any matrix A.

6. We can define a linear transformation $D = C^1(R) \to C^0(R)$ by

$$Df = f' = \frac{df}{dx}$$

Since every function in $C^1(R)$ has a continuous derivative, Df
is well-defined and is indeed an element of $C^0(R)$. D is linear because of the familiar calculus formula

$$\frac{d}{dx}(af_1 + bf_1) = a\frac{df_1}{dx} + b\frac{df_2}{dx}$$

(where f_1, f_2 are differentiable functions and a,b are real numbers).

7. We can define a linear transformation $T : C^0([0,1]) \to R$ by

$$Tf = \int_0^1 f(x)\,dx$$

($R = R^1$ is, of course, a vector space.) It is not hard to check
that T is linear.

Linear transformations $T : V \to R$ play an important role in the
theory of vector spaces. As a result, they have a special name:
linear functionals.

8. We can define a linear transformation $B : R^3 \to \wp_2$ by letting
$B((a_1, a_2, a_3)) = a_1 x^2 + a_2 x + a_3$. It is not hard to check that B is
linear.

9. Similarly, we can define $C : \wp_2 \to R^3$ by $C(ax^2 + bx + c) = $
(a,b,c). Again, C is linear. B (of the previous example) and C are
each other's inverses, in a sense to be made precise later; it is easy
to see that each undoes whatever the other does.

Now for a few examples of non-linear transformations:

10. $T : R^2 \to R^2$, given by $T((x_1, x_2)) = (x_1^2, x_2^2)$, is not linear:
$T((1,1)) = (1,1)$, but $T((2,2)) = (4,4) \neq 2T((1,1))$.

11. $S:C^0(R) \to R$, defined by $Sf = e^{f(0)}$, is not linear: if $f(x) = x$ and $g(x) = x^2$, then $Sf = 1$, $Sg = 1$, and $S(f+g) = 1 \neq S(f) + S(g)$.

12. $T:R^3 \to R$, given by $T((x_1,x_2,x_3)) = |x_1| + |x_2| + |x_3|$, is not linear: $T((1,0,1)) = 2$, and $T((-1,0,-1)) = 2 \neq -T((1,0,1))$.

Notice that if T is linear, then $T(\vec{0}) = \vec{0}$. (Proof: $T(\vec{0}) = T(0 \cdot \vec{v}) = 0 \cdot T(\vec{v}) = \vec{0}$.) Thus the function S of Example 11 cannot be linear, since $S(\vec{0}) = 1$.

A terminological note: linear transformations are often called *linear operators* or *linear mappings*.

Now for some more definitions. If $T:V_1 \to V_2$ is a linear transformation, the *nullspace* of $T, \eta(T)$, is the set

$$\eta(T) = \{\vec{v} \in V_1 : T(\vec{v}) = \vec{0}\}$$

(This set is sometimes called the *kernel* of T, or *Ker* T.) The *image* of T, $\mathcal{I}(T)$, is the set

$$\mathcal{I}(T) = \{\vec{w} \in V_2 : \vec{w} = T(\vec{v}) \text{ for some } \vec{v} \in V_1\}$$

We shall call V_1 the *domain* of T, and V_2 the *target*. (Some people use the term *range* for what we call the image. Some use it for what we call the target.)

PROPOSITION 5.2. *Let* $T:V_1 \to V_2$ *be a linear transformation. Then the nullspace of T is a subspace of* V_1 *and the image of T is a subspace of* V_2.

Proof. We consider $\eta(T)$ first. $\eta(T)$ is nonempty, since $\vec{0} \in \eta(T)$ (as we noted just above, $T(\vec{0}) = \vec{0}$ if T is linear). If $\vec{v}, \vec{w} \in \eta(T)$ and $r,s \in R$, then

$$
\begin{aligned}
T(r\vec{v} + s\vec{w}) &= rT(\vec{v}) + sT(\vec{w}) \quad \text{(since T is linear)} \\
&= r\vec{0} + s\vec{0} \quad \text{(since } \vec{v},\vec{w} \in \eta(T)) \\
&= \vec{0}
\end{aligned}
$$

Hence $r\vec{v} + s\vec{w} \in \eta(T)$; by Proposition 4.3, $\eta(T)$ is a subspace.

The proof for $\vartheta(T)$ is quite similar. (This should not be surprising; we know so little about vector spaces that there are very few approaches to any question.) Since $T(\vec{0}) = \vec{0}$, $\vartheta(T)$ is not empty. If $\vec{v}_2, \vec{w}_2 \in \vartheta(T)$ and $r,s \in R$, then there are elements $\vec{v}_1, \vec{w}_1 \in V_1$ such that $T(\vec{v}_1) = \vec{v}_2$, $T(\vec{w}_1) = \vec{w}_2$. Then

$$T(r\vec{v}_1 + s\vec{w}_1) = rT(\vec{v}_1) + sT(\vec{w}_1)$$
$$= r\vec{v}_2 + s\vec{w}_2$$

so that $r\vec{v}_2 + s\vec{w}_2 \in \vartheta(T)$. This proves that $\vartheta(T)$ is a subspace.

By using Proposition 5.2, we can see connections between some of our previous examples. Let $A = (a_{ij})$ be an $p \times n$ matrix. Recall (Example 4) that we can define a linear transformation $T_A : R^n \rightarrow R^p$

$$T_A((x_1, \ldots, x_n)) = (y_1, \ldots, y_p)$$

where

$$y_i = \sum_{j=1}^{n} a_{ij} x_j$$

The elements of $h(T_A)$ are therefore the solutions of the simultaneous linear equations

$$a_{11}x_1 + \cdots + a_{1n}x_n = 0$$
$$\cdots$$
$$a_{p1}x_1 + \cdots + a_{pn}x_n = 0$$

If, for instance, $A = \begin{pmatrix} 1 & 2 & 3 \\ 4 & 5 & 6 \end{pmatrix}$, then

$$T_A((x_1, x_2, x_3)) = (x_1 + 2x_2 + 3x_3, 4x_1 + 5x_2 + 6x_3)$$

and

$$h(T_A) = \{(x_1, x_2, x_3) : x_1 + 2x_2 + 3x_3 = 0, 4x_1 + 5x_2 + 6x_3 = 0\}$$

(This is Example 4.7.) Since T_A is linear, $h(T_A)$ is a subspace of R^n. That is, the solutions of homogeneous simultaneous linear equations in p unknowns form a subspace of R^p. This statement is the content of Example 4.7.

Suppose $T:V_1 \to V_2$ is a linear transformation. Recall that if the image of T is all of V_2, then T is said to be *surjective,* or *onto.* (We shall usually use "surjective".) Thus T is surjective if any $\vec{w} \in V_2$ can be written as $T(\vec{v})$ for some $\vec{v}$ in V_1. For example, the transformation $T:R^2 \to R^1$ defined by

$$T(x,y) = x + y$$

is surjective since, for any $x \in R^1$, $T(x,0) = x$. However, the linear transformation $S:R^1 \to R^2$ given by

$$S(x) = (x,x)$$

is *not* surjective since $(1,0)$ is not in the image of S.

Recall also that the linear transformation $T:V_1 \to V_2$ is called *injective,* or *one-one,* if $T(\vec{v}_1) = T(\vec{v}_2)$ implies $\vec{v}_1 = \vec{v}_2$. (We shall use "injective".) For example, the linear transformation $T:R^2 \to R^1$ above is *not* injective (since $T(1,-1) = T(0,0)$), whereas $S:R^1 \to R^2$ is injective.

PROPOSITION 5.3. *Suppose $T:V_1 \to V_2$ is a linear transformation. Then T is injective if and only if $h(T) = \{\vec{0}\}$.*

Proof. If $T(\vec{v}) = T(\vec{v}')$ implies $\vec{v} = \vec{v}'$ for all $\vec{v},\vec{v}'$ in V_1, then $T(\vec{v}) = T(\vec{0}) = \vec{0}$ implies $\vec{v} = \vec{0}$. Thus $h(T) = \{\vec{0}\}$ whenever T is injective.

Conversely, if $h(T) = \{\vec{0}\}$ and $T(\vec{v}) = T(\vec{v}')$, then

$$T(\vec{v}-\vec{v}') = T(\vec{v}) - T(\vec{v}') = \vec{0}$$

It follows that $\vec{v}-\vec{v}' \in h(T)$, so $\vec{v}-\vec{v}' = \vec{0}$. Thus $\vec{v}=\vec{v}'$ and T is injective.

Proposition 5.3 can be regarded as a special case of the following more general result:

PROPOSITION 5.4. *Let* $T:V_1 \to V_2$ *be a linear transformation, and suppose that* $T(\vec{v}_1) = \vec{v}_2$. *Then the solutions of the equation* $T(\vec{v}) = \vec{v}_2$ *are precisely the vectors* $\vec{v} = \vec{v}_1 + \vec{w}_1$, *where* $\vec{w}_1 \in h(T)$.

Proof. If $\vec{v} = \vec{v}_1 + \vec{w}_1$ with $\vec{w}_1 \in h(T)$, then

$$T(\vec{v}) = T(\vec{v}_1) + T(\vec{w}_1)$$
$$= \vec{v}_2 + \vec{0}$$
$$= \vec{v}_2$$

Therefore every vector of the given form satisfies the equation.

Conversely, if $T(\vec{v}) = \vec{v}_2$, let $\vec{w}_1 = \vec{v} - \vec{v}_1$. Then $\vec{v} = \vec{v}_1 + \vec{w}_1$, and

$$T(\vec{w}_1) = T(\vec{v} - \vec{v}_1)$$
$$= T(\vec{v}) - T(\vec{v}_1)$$
$$= \vec{v}_2 - \vec{v}_2$$
$$= \vec{0}$$

Thus $\vec{w}_1 \in h(T)$ and v is of the given form.

Linear transformations which are injective and surjective are also called *isomorphisms* or *bijections* (adjectives: *isomorphic, bijective*). If $T:V \to W$ is an isomorphism, then V and W are the same from the point of view of vector space theory. (They may have different elements, but any true statement about the vector space properties of one is true for the other, too.) More on this later.

EXERCISES

1.   Which of the following are linear transformations?

(a) $T:R^1 \to R^1 : T(x) = x^2$

(b) $B:R^2 \to R^2 : B(x_1, x_2) = (ax_1 + bx_2, cx_1 + dx_2)$

(a,b,c,d are given real numbers).

(c) $F:R^2 \rightarrow R^2:F(x_1,x_2) = (x_1 + x_2,x_2 + 1)$

(d) $A:R \rightarrow R^2: A(x) = (x,x)$

(e) $C:R^2 \rightarrow R^1:C(x_1,x_2) = x_1 x_2$

(f) $S:R^2 \rightarrow R^2:S(x_1,x_2) = (\sin x_1, \cos x_2)$

(g) $Q:R^3 \rightarrow R^3:Q(x_1,x_2,x_3) = (x_2,x_3,x_1)$

(h) $H:\mathcal{P}_1 \rightarrow R^2:H(p) = (p(0),p(1))$, where p is a polynomial of
degree ≤ 1.

2.   Find the nullspace and image of the following linear transformations:

(a) $A:R^2 \rightarrow R^1:A(x_1,x_2) = 2x_1 + 3x_2$

(b) $B:\mathcal{P}_2 \rightarrow R^1:B(p) = p(0)$

(c) $C:R^2 \rightarrow R^2:C(x_1,x_2) = (x_1 - x_2,x_1 + x_2)$

(d) $D:R^2 \rightarrow R^2:D(x_1,x_2) = (x_1 - x_2,x_2 - x_1)$

3.   Verify that the function T_A defined in example 5 is a linear transformation.

4.   What are the nullspace and image of $D:C^1(R) \rightarrow C^0(R)$ (Example 6) defined by $Df = \dfrac{df}{dx}$?

5.   Prove that if T is a linear transformation, then $T(-\vec{v}) = -T(\vec{v})$.

6.   Suppose that $S:V \rightarrow W$ and $T:V \rightarrow W$ are linear transformations. We define the function $(S+T):V \rightarrow W$ by

$$(S+T)(\vec{v}) = S(\vec{v}) + T(\vec{v})$$

Show that S+T is a linear transformation.

7.   If $S:V \rightarrow W$ is a linear transformation and $r \in R$, define the function $rS:V \rightarrow W$ by

$$(rS)(\vec{v}) = r(S\vec{v})$$

Show that rS is a linear transformation.

8.   Let $\mathcal{L}(V,W)$ be the set of all linear transformations from V to W.
Prove that $\mathcal{L}(V,W)$ is a vector space with addition and scalar multiplication defined as in exercises 6 and 7 above.

9.   If $T:V_1 \to V_2$ and $S:V_2 \to V_3$ are linear transformations, we define
the *composite* of S with T, $S{\circ}T:V_1 \to V_3$ by $(S{\circ}T)(\vec{v}) = S(T(\vec{v}))$.

    (a) Show that $S{\circ}T$ is a linear transformation.

    (b) Suppose that $T:R^2 \to R^2$ is given by $T(x_1,x_2) = (2x_1 + x_2,$
$x_1 + 2x_2)$ and $S:R^2 \to R^3$ is given by $S(y_1,y_2) = (y_1 + y_2, y_1 - y_2, y_2)$.
What is $(S{\circ}T)(x_1,x_2)$?

10.   If $T:V_1 \to V_2$ and $S:V_2 \to V_3$ are linear transformations, show that
$\eta(T) \subset \eta(S{\circ}T)$ and $\mathcal{J}(S{\circ}T) \subset \mathcal{J}(S)$.

11.   Let B and C be defined as in examples 8 and 9. Prove that the
composites $B{\circ}C$ and $C{\circ}B$ are the identity transformations on R^3 and P_2
respectively.

12.   Use exercise 10 to prove the following:

    (a) If $S{\circ}T:V_1 \to V_3$ is injective, so is $T:V_1 \to V_2$.

    (b) If $S{\circ}T:V_1 \to V_3$ is surjective, so is $S:V_2 \to V_3$.

13.   Let $T:V \to W$ be a linear transformation. Prove that

$$T\left(\sum_{j=1}^{r} a_j \vec{v}_j \right) = \sum_{j=1}^{r} a_j T(\vec{v}_j)$$

for any $a_1,\ldots,a_r \in R$, $\vec{v}_1,\ldots,\vec{v}_r \in V$.

6. LINEAR TRANSFORMATIONS ON EUCLIDEAN SPACES

We now study, in some detail, linear transformations between Euclidean
spaces. This will provide a preview of results that we shall prove
later for linear transformations between arbitrary vector spaces.

We begin by singling out certain vectors in R^n. Let $\vec{e}_1, \vec{e}_2, \ldots, \vec{e}_n$ $\in R^n$ be defined by

$$\vec{e}_1 = (1,0,0,\ldots,0)$$
$$\vec{e}_2 = (0,1,0,\ldots,0)$$
$$\cdots$$
$$\vec{e}_n = (0,0,\ldots,0,1)$$

We shall sometimes refer to $\vec{e}_1,\ldots,\vec{e}_n$ as the *standard basis* of R^n. Notice that any vector $\vec{v} = (x_1,\ldots,x_n) \in R^n$ is a linear combination of $\vec{e}_1,\ldots,\vec{e}_n$:

$$\vec{v} = \sum_{j=1}^{n} x_j \vec{e}_j$$

In fact, this expression for $\vec{v}$ is unique. For if $\vec{v} = \sum_{j=1}^{n} x_j' \vec{e}_j$ as well, then

$$(0,0,\ldots,0) = \sum_{j=1}^{n} x_j \vec{e}_j - \sum_{j=1}^{n} x_j' \vec{e}_j$$
$$= \sum_{j=1}^{n} (x_j - x_j') \vec{e}_j$$
$$= (x_1 - x_1', x_2 - x_2', \ldots, x_n - x_n')$$

Therefore $x_1 = x_1'$, $x_2 = x_2', \ldots, x_n = x_n'$.

Now let $T: R^n \to R^p$ be a linear transformation. Since $T\vec{e}_1$ is a vector in R^p, it can be written as a linear combination of the standard basis elements in R^p: *

$$T\vec{e}_1 = \sum_{i=1}^{p} a_{i1} \vec{e}_i, \quad a_{11},\ldots,a_{p1} \in R$$

Similarly

---

* For this discussion, we also denote the standard basis in R^p by $\vec{e}_1,\ldots,\vec{e}_p$, since no confusion should result (even though $\vec{e}_1$, say, refers to one vector in R^n and a different one in R^p).

$$Te_j = \sum_{i=1}^{p} a_{ij}\vec{e}_i \; , \; a_{1j},\ldots,a_{pj} \in R$$

for $j = 2,3,\ldots,n$ as well. Then $A = (a_{ij})$ is a $p \times n$ matrix.

PROPOSITION 6.1. *If T and A are as above, then T coincides with the linear transformation T_A defined in Example 3 of Section 5.*

Proof. We need to compute $T\vec{v}$ and $T_A\vec{v}$ for a vector $\vec{v} = (x_1,\ldots,x_n)$ $= \sum_{j=1}^{n} x_j\vec{e}_j \in R^n$. Recall that

$$T_A\vec{v} = (y_1,\ldots,y_p)$$

$$= \sum_{i=1}^{p} y_i\vec{e}_i$$

where

$$y_i = \sum_{j=1}^{n} a_{ij}x_j$$

Thus

$$T_A\vec{v} = \sum_{i=1}^{p} \left(\sum_{j=1}^{n} a_{ij}x_j\vec{e}_i \right)$$

$$= \sum_{i=1}^{p} \sum_{j=1}^{n} a_{ij}x_j\vec{e}_i$$

On the other hand, we have (using the linearity of T)

$$T\vec{v} = T\left(\sum_{j=1}^{n} x_j\vec{e}_j \right)$$

$$= \sum_{j=1}^{n} x_j T(\vec{e}_j)$$

$$= \sum_{j=1}^{n} x_j \left(\sum_{i=1}^{p} a_{ij} \vec{e}_i \right)$$

$$= \sum_{j=1}^{n} \sum_{i=1}^{p} a_{ij} x_j \vec{e}_i$$

Therefore $\vec{Tv} = T_A \vec{v}$ for every $\vec{v} \in R^n$, and $T = T_A$, as claimed.

The discussion above defines a correspondence between linear transformations $T:R^n \to R^p$ and $p \times n$ matrices. We let A_T be the $p \times n$ matrix defined by the linear transformation $T:R^n \to R^p$: if

$$\vec{Te}_j = \sum_{i=1}^{p} a_{ij} \vec{e}_i$$

$j=1,\ldots,n$, then $A_T = (a_{ij})$.

We have seen (in Exercises 6 and 7 of Section 5) that if S and T are two linear transformations from R^n to R^p, then we may define linear transformations $S+T:R^n \to R^p$ and $rT:R^n \to R^p$ (where r is a real number). We have also defined the sum and scalar product of matrices, in Example 5 of Section 2. The next result relates these operations.

PROPOSITION 6.2. *Let* $S,T:R^n \to R^p$ *be linear transformations and* r *a real number. Then*

$$A_{S+T} = A_S + A_T$$
$$A_{rS} = rA_S$$

The proof is straightforward; we leave it as Exercise 4.

A more complicated problem is the following: if $T:R^m \to R^n$ and $S:R^p \to R^m$ are linear transformations, then the composite $T \circ S:R^p \to R^n$, is a linear transformation (Exercise 9 of Section 5). What is $A_{T \circ S}$? To find out, we compute. Let $A_S = (b_{jk})$, where $1 \leq j \leq m$ and $1 \leq k \leq p$; if $\vec{e}_1, \vec{e}_2, \ldots, \vec{e}_p$ is the standard basis for R^p, then

$$S(\vec{e}_k) = \sum_{j=1}^{m} b_{jk} \vec{e}_j \qquad (6.2)$$

Similarly, if $A_T = (a_{ij})$, $1 \leq i \leq n$, $1 \leq j \leq m$, then

$$T(\vec{e}_j) = \sum_{i=1}^{n} a_{ij} \vec{e}_i$$

Then

$$(T \circ S)(\vec{e}_k) = T(S(\vec{e}_k))$$

$$= T\left(\sum_{j=1}^{m} b_{jk} \vec{e}_j\right)$$

$$= \sum_{j=1}^{m} b_{jk} T(\vec{e}_j)$$

$$= \sum_{j=1}^{m} b_{jk}\left(\sum_{i=1}^{n} a_{ij} \vec{e}_i\right)$$

$$= \sum_{i=1}^{n}\left(\sum_{j=1}^{m} a_{ij} b_{jk}\right)\vec{e}_i$$

Now $A_{T \circ S}$ is the matrix (c_{ik}), $1 \leq i \leq n$ and $1 \leq k \leq p$, such that

$$T \circ S(\vec{e}_k) = \sum_{i=1}^{n} c_{ik} \vec{e}_i$$

Therefore

$$c_{ik} = \sum_{j=1}^{m} a_{ij} b_{jk} \qquad\qquad (6.4)$$

Formula (6.4) shows how to compute $A_{T \circ S}$ from A_T and A_S. If, for instance, $S:R^3 \rightarrow R^2$ is given by the matrix

$$A_S = \begin{pmatrix} 1 & 2 & 3 \\ 4 & 5 & 6 \end{pmatrix}$$

and $T:R^2 \rightarrow R^2$ is given by

$$A_T = \begin{pmatrix} 7 & 8 \\ 9 & 0 \end{pmatrix}$$

then

$$A_{T \circ S} = \begin{pmatrix} 7 \cdot 1 + 8 \cdot 4 & 7 \cdot 2 + 8 \cdot 5 & 7 \cdot 3 + 8 \cdot 6 \\ 9 \cdot 1 + 0 \cdot 4 & 9 \cdot 2 + 0 \cdot 5 & 9 \cdot 3 + 0 \cdot 6 \end{pmatrix} = \begin{pmatrix} 39 & 54 & 69 \\ 9 & 18 & 27 \end{pmatrix}$$

Formula (6.4) also gives the definition of matrix multiplication. If $A = (a_{ij})$ is an $n \times m$ matrix and $B = (b_{jk})$ is an $m \times p$ matrix, then the *product* AB is an $n \times p$ matrix, (c_{ik}), defined by

$$c_{ik} = \sum_{j=1}^{m} a_{ij} b_{jk}$$

What this formula tells us is that to compute the entry c_{ik} of the product AB, we multiply the ith row of A by the kth column of B, componentwise and add the result. For example, if

$$A = \begin{pmatrix} 1 & 2 \\ 2 & 3 \end{pmatrix} \qquad\qquad B = \begin{pmatrix} -3 & 1 & 7 \\ 2 & 4 & 0 \end{pmatrix}$$

then the component c_{12} of AB in the first row and second column is obtained by multiplying the first row of A by the second column of B componentwise and adding: $c_{12} = 1 \cdot 1 + 2 \cdot 4$.

Note that in order to form the product AB, the number of columns of A must equal the number of rows of B. When this condition is not met, AB is not defined.

We summarize the results of the discussion above in the following proposition.

PROPOSITION 6.3. *Let* $T:R^m \to R^n$ *be a linear transformation with matrix* A *and* $S:R^p \to R^m$ *a linear transformation with matrix* B. *Then the matrix of the composite* $T \circ S:R^p \to R^n$ *is the product* AB.

Matrix multiplication will recur often and it is worthwhile to become adept at it. Here are a few examples.

(1) $\begin{pmatrix} 1 & 0 \\ 1 & 1 \end{pmatrix} \begin{pmatrix} 2 & 1 \\ 3 & 0 \end{pmatrix} = \begin{pmatrix} 1\cdot 2 + 0\cdot 3 & 1\cdot 1 + 0\cdot 0 \\ 1\cdot 2 + 1\cdot 3 & 1\cdot 1 + 1\cdot 0 \end{pmatrix}$

$$= \begin{pmatrix} 2 & 1 \\ 5 & 1 \end{pmatrix}$$

(2) $\begin{pmatrix} 0 & 1 \\ 1 & 0 \end{pmatrix} \begin{pmatrix} 2 & 3 \\ 4 & 5 \end{pmatrix} = \begin{pmatrix} 0\cdot 2 + 1\cdot 4 & 0\cdot 3 + 1\cdot 5 \\ 1\cdot 2 + 0\cdot 4 & 1\cdot 3 + 0\cdot 5 \end{pmatrix}$

$$= \begin{pmatrix} 4 & 5 \\ 2 & 3 \end{pmatrix}$$

(3) $\begin{pmatrix} 1 & 2 \\ 2 & 3 \end{pmatrix} \begin{pmatrix} 2 & 0 & 7 \\ 1 & -1 & 4 \end{pmatrix} = \begin{pmatrix} 1\cdot 2 + 2\cdot 1 & 1\cdot 0 + 2\cdot(-1) & 1\cdot 7 + 2\cdot 4 \\ 2\cdot 2 + 3\cdot 1 & 2\cdot 0 + 3\cdot(-1) & 2\cdot 7 + 3\cdot 4 \end{pmatrix}$

$$= \begin{pmatrix} 4 & -2 & 15 \\ 7 & -3 & 26 \end{pmatrix}$$

The product

$$\begin{pmatrix} 2 & 0 & 7 \\ 1 & -1 & 4 \end{pmatrix} \begin{pmatrix} 1 & 2 \\ 2 & 3 \end{pmatrix}$$

is not defined, since the first matrix has three columns and the second has only two rows.

(4) $\begin{pmatrix} 3 & -1 & 1 \\ 1 & 0 & 2 \end{pmatrix} \begin{pmatrix} 1 & 7 \\ 6 & -2 \\ 2 & 3 \end{pmatrix} = \begin{pmatrix} 3\cdot 1 + (-1)\cdot 6 + 1\cdot 2 & 3\cdot 7 + (-1)\cdot(-2) + 1\cdot 3 \\ 1\cdot 1 + 0\cdot 6 + 2\cdot 2 & 1\cdot 7 + 0\cdot(-2) + 2\cdot 3 \end{pmatrix}$

$$= \begin{pmatrix} -1 & 26 \\ 5 & 13 \end{pmatrix}$$

We close this section with an interpretation of matrix multiplication in a particular case.

Suppose we consider the elements of R^n as columns or, equivalently, as $n \times 1$ matrices: that is, we write $\vec{v} = (x_1,\ldots,x_m)$ as

$$\begin{pmatrix} x_1 \\ \vdots \\ x_m \end{pmatrix}$$

Then, if $T:R^n \rightarrow R^p$ is a linear transformation with matrix A and if $\vec{v} \in R^n$ is represented by the $n \times 1$ matrix B, then $\vec{Tv}$ is represented by the $p \times 1$ matrix AB. For example, if $T:R^2 \rightarrow R^3$ is the linear transformation with matrix

$$A = \begin{pmatrix} 2 & 1 \\ 0 & 2 \\ 3 & 5 \end{pmatrix}$$

and $\vec{v} = (1,-1)$, then

$$\begin{pmatrix} 2 & 1 \\ 0 & 2 \\ 3 & 5 \end{pmatrix}\begin{pmatrix} 1 \\ -1 \end{pmatrix} = \begin{pmatrix} 1 \\ -2 \\ -2 \end{pmatrix}$$

so that $T(\vec{v}) = (1,-2,-2)$. We leave the verification of this result to the reader. (See Exercise 10.)

EXERCISES

1. Find A_T where

(a) $T:R^2 \rightarrow R$, $T(x_1,x_2) = x_1 - 2x_2$

(b) $T:R^2 \rightarrow R^2$, $T(x_1,x_2) = (3x_1 + x_2, 7x_1 - 3x_2)$

(c) $T:R^2 \rightarrow R^3$, $T(x_1,x_2) = (x_2,x_1 - x_2,x_1 + x_2)$

(d) $T:R^3 \rightarrow R^2$, $T(x_1,x_2,x_3) = (x_1 - x_3,x_1 + x_2 + x_3)$

(e) $T:R^3 \rightarrow R^3$, $T(x_1,x_2,x_3) = (x_3,x_1,x_2)$

(f) $T:R^3 \rightarrow R^3$, $T(x_1,x_2,x_3) = (2x_1 - x_3,x_1 + x_2 + 3x_3,2x_1 - x_2)$

2. Compute the following:

(a) $\begin{pmatrix} 1 & 8 \\ 4 & 3 \end{pmatrix} \begin{pmatrix} 2 & 7 \\ 3 & 4 \end{pmatrix}$

(f) $\begin{pmatrix} 2 & 4 & 8 \\ -6 & 3 & -5 \end{pmatrix} \begin{pmatrix} 3 & -6 \\ -2 & 1 \\ 7 & 5 \end{pmatrix}$

(b) $(1 \quad 3 \quad 7) \begin{pmatrix} 7 \\ 2 \\ 1 \end{pmatrix}$

(g) $\begin{pmatrix} 4 & 5 & 6 & 8 \\ 1 & 3 & 2 & 4 \end{pmatrix} \begin{pmatrix} -9 & -1 & 4 \\ 3 & 2 & 3 \\ 2 & 6 & 0 \\ -7 & 1 & 4 \end{pmatrix}$

(c) $\begin{pmatrix} -2 & 3 & 5 \\ 6 & 1 & 7 \end{pmatrix} \begin{pmatrix} 3 & 2 & 5 \\ 8 & -9 & 4 \\ -6 & 7 & 1 \end{pmatrix}$

(h) $\begin{pmatrix} 3 & 6 \\ -2 & 1 \\ 7 & 5 \end{pmatrix} \begin{pmatrix} 2 & 4 & 8 \\ -6 & 3 & -5 \end{pmatrix}$

(d) $\begin{pmatrix} 7 \\ 2 \\ 1 \end{pmatrix} (1 \quad 3 \quad 7)$

(i) $\begin{pmatrix} 2 & 0 & 0 \\ 0 & 3 & 0 \\ 0 & 0 & 7 \end{pmatrix} \begin{pmatrix} 3 & 2 & 1 & 4 \\ -6 & 4 & 5 & -2 \\ 3 & 1 & 6 & -8 \end{pmatrix}$

(e) $\begin{pmatrix} 3 & 4 \\ -2 & 4 \\ 1 & 4 \end{pmatrix} \begin{pmatrix} 0 & 3 \\ 1 & 2 \end{pmatrix}$

(j) $\begin{pmatrix} 6 & 4 & 3 \\ 1 & 8 & 7 \\ 2 & 2 & 5 \end{pmatrix} \begin{pmatrix} 1 & 6 \\ -4 & 3 \\ 7 & 2 \end{pmatrix}$

3. Let $A = \begin{pmatrix} 1 & 3 \\ 2 & 4 \end{pmatrix}$, $B = \begin{pmatrix} 1 & 4 \\ -2 & 5 \\ 3 & -6 \end{pmatrix}$, and $C = \begin{pmatrix} 3 & -4 & 5 \\ 2 & 1 & 6 \end{pmatrix}$.

Determine which of the following make sense and compute those that do.

(a) $A^2 (= AA)$ (f) AC

(b) B^2 (g) CA

(c) C^2 (h) BC

(d) AB   (i) CB

(e) BA   (j) (CB)A

4. Prove Proposition 6.2.

5.   Find two 2 × 2 matrices A,B with AB ≠ BA.

6.   Find two 2 × 2 matrices, neither of which is the zero matrix but whose product AB is the zero matrix.

7.   Let I be the n × n identity matrix.

    (a)   Show that AI = A for any p × n matrix A.

    (b)   Show that IA = A for any n × p matrix A.

8.   Let A be a q × p matrix, B, C p × n matrices, D an n × m matrix, and r a real number. Prove the following.

    (a) $r(AB) = (rA)B = A(rB)$

    (b) $A(B + C) = AB + AC$

    (c) $(B + C)D = BD + CD$

    (d) $(AB)D = A(BD)$

9.   Let $S:R^p \to R^q$, $T:R^n \to R^p$, and $Q:R^m \to R^n$ be linear transformations and prove that

$$S \circ (T \circ Q) = (S \circ T) \circ Q$$

Use this fact and Proposition 9.1 to prove part (d) of Exercise 8 above.

10.   Let $T:R^n \to R^p$ be a linear transformation with matrix A, and let B be the n × 1 matrix where entries are the components of a vector $\vec{v} \in R^n$. Prove that $T(\vec{v})$ is the vector in R^p whose components are the entries of the n × 1 matrix AB. (This is the assertion made at the end of this section.)

11.   Let $T:R^3 \to R^3$ be the linear transformation with matrix

$$\begin{pmatrix} 2 & 1 & 3 \\ 5 & -2 & 1 \\ 0 & 3 & -3 \end{pmatrix}$$

Use the method described at the end of this section to evaluate the following.

(a) T(1,1,2)   (c) T(2,5,-4)

(b) T(-1,3,0)   (d) T(6,1,2)

12. Let $T:R^4 \to R^5$ be the linear transformation with matrix

$$\begin{pmatrix} 1 & 0 & 2 & 1 \\ -3 & 4 & 1 & 1 \\ 2 & 0 & -1 & 0 \\ 5 & 1 & 1 & 1 \\ 3 & 7 & 0 & -2 \end{pmatrix}$$

Use the method described at the end of this section to evaluate the
following.

(a) T(1,0,2,1)   (c) T(7,4,-4,2)

(b) T(2,1,-2,3)   (d) T(3,6,-3,4)

13. (a) A 3 × 3 matrix of the form

$$\begin{pmatrix} a_{11} & a_{12} & a_{13} \\ 0 & a_{22} & a_{23} \\ 0 & 0 & a_{33} \end{pmatrix}$$

is called *upper triangular*. Show that if A and B are upper triangular
matrices, then so is AB.

*(b) State and prove a similar result for n × n matrices.

14. (a) A 3 × 3 matrix of the form

$$\begin{pmatrix} 0 & a_{12} & a_{13} \\ 0 & 0 & a_{23} \\ 0 & 0 & 0 \end{pmatrix}$$

is called *strictly upper triangular*. Show that if A, B, and C are
strictly upper triangular 3 × 3 matrices, then ABC = 0_3, the 3 × 3
zero matrix.

*(b) State and prove a similar theorem for n × n matrices.

15. A 2 × 2 matrix A is said to be *invertible* if there is a matrix B such that AB = BA = I, the 2 × 2 identity matrix. The matrix B is called an *inverse* for A. Find an inverse for the matrix

$$A = \begin{pmatrix} 2 & 1 \\ 1 & 1 \end{pmatrix}$$

*16. Prove that a 2 × 2 matrix

$$A = \begin{pmatrix} a & b \\ c & d \end{pmatrix}$$

is invertible if and only if ad - bc ≠ 0 and find an inverse in this case. (See Exercise 15.)

*17. Determine which of the following matrices are invertible and find the inverses of those that are.

(a) $\begin{pmatrix} 2 & 3 \\ 3 & 4 \end{pmatrix}$

(c) $\begin{pmatrix} 2 & 3 \\ 2 & 2 \end{pmatrix}$

(b) $\begin{pmatrix} 2 & 1 \\ 4 & 2 \end{pmatrix}$

(d) $\begin{pmatrix} 5 & 2 \\ 6 & 3 \end{pmatrix}$

CHAPTER 3

THE DERIVATIVE

INTRODUCTION

In elementary calculus, the main mathematical tool for studying the
behavior of functions is the derivative. This state of affairs per-
sists when we examine functions from one vector space to another.
When we work in this more general setting, however, we need to change
our notion of the derivative. If $f:R \to R$ is a function, the derivative
of f at a point x_0 is a number $f'(x_0)$ giving the slope of the graph of
f at x_0. However, we can also view $f'(x_0)$ as defining a linear transformation $L:R \to R$, given by

$$L(x) = f'(x_0)x$$

This point of view may seem artificial when dealing with functions of
one variable, but it turns out to be very useful when one does multivariable calculus. For one thing, it simplifies some formulas for computing derivatives. For another, it emphasizes the fundamental idea
underlying the calculus: to study the behavior of nonlinear functions,
we approximate them by linear functions.

In order to define the derivative, we need additional structure,
a norm, on a vector space. We introduce this structure, use it to
define continuity and differentiability, and show how to compute the
derivative of a function between Euclidean spaces.

1.   NORMED VECTOR SPACES

Let V be a vector space. A *norm* on V is a function which assigns to
each vector $\vec{v}$ in V a real number $\|\vec{v}\|$ satisfying the following:

$\|\vec{v}\| \geq 0$ for all $\vec{v}$ in V   (1.1)

$\|\vec{v}\| = 0$ if and only if $\vec{v} = \vec{0}$ (1.2)

$\|a\vec{v}\| = |a| \cdot \|\vec{v}\|$ for any real number a and any $\vec{v}$ in V   (1.3)

$\|\vec{v} + \vec{w}\| \leq \|\vec{v}\| + \|\vec{w}\|$ for all $\vec{v},\vec{w}$ in V   (1.4)

The norm should be thought of as a sort of distance; $\|\vec{v}\|$ is the
distance from $\vec{v}$ to $\vec{0}$. Property (1.4) in the definition of the norm
is called the *triangle inequality* because it expresses the fact that
the length of one side of a triangle is never greater than the sum of
the length of the other two sides. (See Figure 1.1.) Usually (1.4)
is the most difficult property to verify.

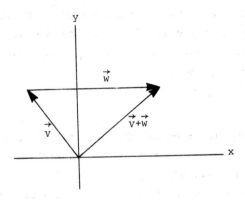

FIGURE 1.1.   A triangle of vectors

A vector space V together with a norm is called a *normed vector space*.
Here are some examples.

(1)   If $V = R^n$, we can define a norm by

$$\|(x_1,\ldots,x_n)\| = (x_1^2 +\cdots+ x_n^2)^{1/2}$$

In particular, if n = 1, $\|x\| = |x|$.

The verification of the first three properties of the definition is trivial. The fourth property can be verified easily for n = 2 but is more difficult for arbitrary n. We shall prove a more general result later on. (See Proposition 4.6.3.) In R^2 and R^3, this norm corresponds to the usual notion of distance because of the Pythagorean theorem.

We shall refer to this norm as the *standard norm on* R^n. Unless we explicitly state otherwise, this will be the norm on R^n in what follows.

(2) We can define another norm on R^n by

$$\|(x_1,\ldots,x_n)\| = |x_1| + |x_2| + \cdots + |x_n|$$

Properties (1.1) through (1.4) in the definition of a norm are immediate.

(3) Still another norm on R^n is given by

$$\|(x_1,\ldots,x_n)\| = \max\{|x_i|:1 \le i \le n\}$$

(See Exercise 6, Section 1.4, for the definition of max.) The verification that this is indeed a norm is straightforward.

(4) Let $V = C^0([0,1])$, the vector space of continuous functions on the interval [0,1], and define a norm on V by

$$\|f\| = \int_0^1 |f(x)|\,dx$$

for f in V. The fact that this is a norm follows from standard properties of the integral of a continuous function. (Once again, only (1.4) causes any difficulty.)

If $\vec{v}$ and $\vec{w}$ are vectors in a normed vector space V, we define the *distance between* $\vec{v}$ *and* $\vec{w}$ to be $\|\vec{v} - \vec{w}\|$. As a consequence of the properties of the norm, we have

$$\|\vec{v} - \vec{w}\| = \|\vec{w} - \vec{v}\| \text{ for all } \vec{v},\vec{w} \text{ in } V \tag{1.5}$$

$$\|\vec{v} - \vec{w}\| \ge 0 \text{ for all } \vec{v},\vec{w} \text{ in } V \tag{1.6}$$

$\|\vec{v} - \vec{w}\| = 0$ if and only if $\vec{v} = \vec{w}$ $\qquad\qquad\qquad$ (1.7)

$\|\vec{v} - \vec{w}\| \leq \|\vec{v} - \vec{v}'\| + \|\vec{v}' - \vec{w}\|$ for all $\vec{v}$, $\vec{v}'$, and $\vec{w}$ in V $\quad$ (1.8)

Another useful inequality is the following.

PROPOSITION 1.1. *For any vectors $\vec{v}, \vec{w}$ in a normed vector space* V, *we have*

$$| (\|\vec{v}\| - \|\vec{w}\|) | \leq \|\vec{v} - \vec{w}\|$$

Proof. There are two cases to consider, depending on whether $\|\vec{v}\| - \|\vec{w}\|$ is ≥ 0 or not.

(a) If $\|\vec{v}\| - \|\vec{w}\| \geq 0$, we need to show that

$$\|\vec{v}\| - \|\vec{w}\| \leq \|\vec{v} - \vec{w}\|$$

This is easy: the triangle inequality says that

$$\|\vec{v}\| \leq \|\vec{v} - \vec{w}\| + \|\vec{w}\|$$

and we simply subtract $\|\vec{w}\|$ from both sides.

(b) If $\|\vec{v}\| - \|\vec{w}\| < 0$, we need to show that

$$\|\vec{w}\| - \|\vec{v}\| \leq \|\vec{v} - \vec{w}\|$$

However, $\|\vec{v} - \vec{w}\| = \|\vec{w} - \vec{v}\|$, and we can prove this inequality exactly as in case (a).

EXERCISES

1. Verify that the norm on R^n defined in Example 2 is in fact a norm.

2. Verify that the norm on R^n defined in Example 3 is in fact a norm.

*3. Verify that the norm on $C^0([0,1])$ defined in Example 4 is in fact a norm.

4. Which of the following define norms on R^3?

(a) $\|(x_1,x_2,x_3)\| = x_1 + x_2 + x_3$

(b) $\|(x_1,x_2,x_3)\| = x_1^2 + x_2^2 + x_3^2$

(c) $\|(x_1,x_2,x_3)\| = |x_1| + |x_2| + 2|x_3|$

(d) $\|(x_1,x_2,x_3)\| = |x_1|$

(e) $\|(x_1,x_2,x_3)\| = \text{Max}(|x_1|,|x_2|) + |x_3|$

(f) $\|(x_1,x_2,x_3)\| = (\sqrt{|x_1|} + \sqrt{|x_2|} + \sqrt{|x_3|})^2$

(g) $\|(x_1,x_2,x_3)\| = 2(x_1^2 + x_2^2 + x_3^2)^{1/2}$

*5.   Two norms $\| \ \|_1$ and $\| \ \|_2$ on a vector space V are said to be *equivalent* if there are constants a,b > 0 such that

$$\|\vec{v}\|_1 \le a\|\vec{v}\|_2 \le b\|\vec{v}\|_1$$

for all $\vec{v} \in V$. Prove that this is an equivalence relation.

6.   Prove that the norms on R^n defined in Examples 1, 2, and 3 are equivalent.

7.   Let $\vec{v}$ be a nonzero vector in the normed vector space V. Prove that there are exactly two multiples of $\vec{v}$ that have norm one.

8.   (a) Show that if $\| \ \|_1$ and $\| \ \|_2$ are two norms on a vector space V, then $\| \ \|$, defined by

$$\|\vec{v}\| = \|\vec{v}\|_1 + \|\vec{v}\|_2$$

is also a norm.

(b) If $\| \ \|$ is a norm, for what values of $a \in R$ is $a \cdot \| \ \|$ a norm?

9.   Let V and W be normed vector spaces and define

$$\|(\vec{v},\vec{w})\| = \|\vec{v}\| + \|\vec{w}\|$$

for $(\vec{v},\vec{w}) \in V \times W$. Prove that this defines a norm on $V \times W$. (See Section 2.2, Exercise 7, for the definition of $V \times W$ as a vector space.)

*10. Let λ be a fixed real number and $\vec{v} = (x_1, \ldots, x_n)$ a vector in R^n. Define

$$\|\vec{v}\|_\lambda = \left(\sum_{j=1}^{n} |x_i|^\lambda \right)^{1/\lambda}$$

(a) For which values of λ is $\|\vec{v}\|_\lambda$ a norm on R^n?

(b) For which values of λ is $\|\vec{v}\|_\lambda$ equivalent to the standard norm on R^n?

11. Let $f:[0,1] \to R$ be continuous and define

$$\|f\|_\infty = \max\{|f(x)| : 0 \le x \le 1\}$$

(Assume that a continuous function on $[0,1]$ has a maximum value; it will be proved in Section 5.5.)

(a) Prove that $\|f\|_\infty$ is a norm on $C([0,1])$.

*(b) Is this norm equivalent to the norm defined in Example 4 above? (See Exercise 5.)

2. OPEN AND CLOSED SETS

Two important kinds of subsets of a normed vector space are the open and closed sets. In this section, we give their definitions and some of their elementary properties.

Let V be a normed vector space, $\vec{v}_0 \in V$, and r any positive real number. The *ball* (or *open ball*) *of radius* r *about* $\vec{v}_0$ is the set

$$B_r(\vec{v}_0) = \{\vec{v} \in V : \|\vec{v} - \vec{v}_0\| < r\}.$$

For example, if $V = R^2$ with the standard norm, then $B_r(\vec{v}_0)$ is a disc with center $\vec{v}_0$ of radius r. (See Figure 2.1.) If $V = R^3$, then $B_r(\vec{v}_0)$ is in fact a ball of radius r with center $\vec{v}_0$. (See Figure 2.2.) Of course, if we change the norm, we change the shape of the open balls. (See Exercise 7.)

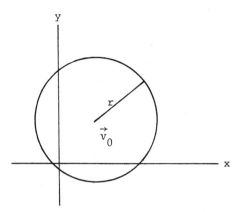

FIGURE 2.1. An open ball in R^2

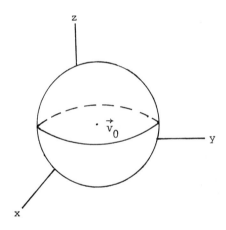

FIGURE 2.2. An open ball in R^3

A subset $U \subset R^n$ is said to be *open* if, for each $\vec{v}_0$ in U, there is an $r > 0$ so that $B_r(\vec{v}_0) \subset U$. That is, all points sufficiently close to $\vec{v}_0$ are also in U. This, in fact, is what makes open sets so useful to us. When we deal with a notion like the differentiability of a function at a point $\vec{v}_0$, we need to examine the behavior of the function at points near $\vec{v}_0$, and we need, therefore, to know that the function is defined everywhere near $\vec{v}_0$. One way of insuring this is to deal with functions defined on open sets.

EXAMPLES

(1) V itself is an open subset of V.

(2) The empty set ϕ is an open subset of V. (Since there are no points $\vec{v}_0$ in ϕ, there is no point at which the criterion could fail to be met.)

(3) The open interval (a,b) (see Section 1.4) is an open subset of $R = R^1$; if $x \in (a,b)$ and $r = \min\{x-a, b-x\}$, then $B_r(x) \subset (a,b)$.

(4) Any open ball is an open set.

Proof of (4). Let $\vec{v}_1$ be a vector in $B_r(\vec{v}_0)$ and set $s = \|\vec{v}_1 - \vec{v}_0\|$. Then $r - s > 0$ and, if $\vec{v} \in B_{r-s}(\vec{v}_1)$,

$$\|\vec{v} - \vec{v}_0\| \leq \|\vec{v} - \vec{v}_1\| + \|\vec{v}_1 - \vec{v}_0\|$$
$$< (r - s) + s = r$$

(See Figure 2.3.) This proves that $B_{r-s}(\vec{v}_1) \subset B_r(\vec{v}_0)$, so $B_r(\vec{v}_0)$ is open.

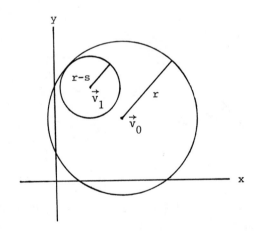

FIGURE 2.3. Open balls are open

(5) The set $X = \{(x_1, x_2) : x_2 = 0\}$ is *not* an open subset of R^2. To see this, we note that any ball about a point in X must contain points outside of X, since it contains points (x_1, x_2) with $x_2 \neq 0$.

We now derive two important properties of open sets.

PROPOSITION 2.1. *The union of an arbitrary collection of open subsets of V is open. The intersection of a finite collection of open subsets of V is open.*

Proof. Let $\{U_j\}$ be any collection of open subsets of V and let $\vec{v}_0$ be a point in $U = \cup U_j$. Then $\vec{v}_0$ is an element of some U_j so, since U_j is open, we can find an $r > 0$ with $B_r(\vec{v}_0) \subset U_j$. However, $U_j \subset U$, so $B_r(\vec{v}_0) \subset U$. This proves that U is open.

Suppose that $U_1, \ldots, U_k$ is a finite collection of open subsets of V and let $\vec{v}_0$ be a point of $U_0 = U_1 \cap U_2 \cap \cdots \cap U_k$. Then $\vec{v}_0 \in U_j$ for all $j = 1, \ldots, k$, so we can find positive numbers $r_1, \ldots, r_k$ with

$$B_{r_j}(\vec{v}_0) \subset U_j$$

$1 \le j \le k$. Set $r = \min_j r_j$, so that $r \le r_j$ for all $j = 1, \ldots, k$. Then

$$B_r(\vec{v}_0) \subset B_{r_j}(\vec{v}_0) \subset U_j$$

for all $j = 1, \ldots, k$, which means that $B_r(\vec{v}_0) \subset U_0$.

A subset C of V is said to be *closed* if its complement, V - C, is open.

EXAMPLES

(6) V itself is a closed subset of V, from Example (2) above.

(7) The empty set is a closed subset of V, from Example (1) above.

(8) Any finite subset of points in V is closed.

Proof. Let $C = \{\vec{v}_1, \ldots, \vec{v}_k\}$ and pick $\vec{v} \in V - C$. Set $r = \min_j \|\vec{v} - \vec{v}_j\|$. Then $B_r(\vec{v}) \subset V - C$, so V - C is open and C is closed.

(9) A *closed ball of radius* r *about* $\vec{v}_0$ in V is defined to be the set

$$\bar{B}_r(\vec{v}_0) = \{\vec{v} \in V : \|\vec{v} - \vec{v}_0\| \le r\}.$$

(Note the $\le$ sign!) It is not hard to show that $\bar{B}_r(\vec{v}_0)$ is a closed set. (See Exercise 4.)

The following result is the analogue of Proposition 2.1 for closed sets. It follows from Proposition 2.1 using the De Morgan laws. (See Section 1.1.) We leave the proof as Exercise 5.

PROPOSITION 2.2. *The intersection of an arbitrary collection of closed subsets of V is closed. The union of a finite collection of closed subsets of V is closed.*

We note that a set can be neither open nor closed. For example, the set

$$(0,1] = \{x \in R^1 : 0 < x \le 1\}$$

is neither open nor closed. (See Exercise 8.)

EXERCISES

1.   Which of the following subsets of R^2 are open?

   (a) $\{(x,y):x > y\}$ (f) $\{(x,y): \|(x,y)\| < 3\}$
   (b) $\{(x,y):x^2 > 7\}$ (g) $\{(x,y): |x| > 2\}$
   (c) $\{(x,y):xy \ne 0\}$ (h) $\{(x,y):y \ge |x|\}$
   (d) $\{(x,y):2x + 3y - 5 < 1\}$ (i) $\{(x,y):x^2 + y^2 = 9\}$
   (e) $\{(x,y):y = 3\}$ (j) $.\{(x,y):y - x \le 6\}$

2.   Which of the following subsets of R^2 are closed?

   (a) $\{(x,y):2x + 3y - 7 \ge 0\}$ (f) $\{(x,y):xy = 1\}$
   (b) $\{(x,y):x \ge y\}$ (g) $\{(x,y):x^2 = y^2\}$
   (c) $\{(x,y):0 < x^2 + y^2 \le 1\}$ (h) $\{(x,y):x \ge 3 \text{ or } y < 2\}$
   (d) $\{(x,y): |x-y| \ne 1\}$ (i) $\{(x,y): \|(x,y)\| < 7\}$
   (e) $\{(x,y):2x - y < 4\}$ (j) $\{(x,y):y \ge |x|\}$

3.   Determine which of the following subsets of R^3 are open, which are closed, and which are neither.

   (a) $\{(x,y,z):x + y + z = 2\}$ (e) $\{(x,y,z):x \ge 0, y > 0, z > 0\}$
   (b) $\{(x,y,z):x^2 + y^2 + z^2 < 1\}$ (f) $\{(x,y,z): |x| + |y| + |z| \ne 0\}$
   (c) $\{(x,y,z):x > 0, y > 0, z > 0\}$ (g) $\{(x,y,z):x \le y \le z\}$
   (d) $\{(x,y,z):x \ge 0, y \ge 0, z \ge 0\}$ (h) $\{(x,y,z):x \le y < z\}$

4.   Prove that a closed ball is a closed set.

5.   Prove Proposition 2.2.

6.   Let U be an open subset of V with $\vec{v}_1,\ldots,\vec{v}_k \in U$.   Prove that $U' = U - \{\vec{v}_1,\ldots,\vec{v}_k\}$ is open.

7.   (a)   Prove that $\|(x,y)\| = |x| + |y|$ defines a norm on R^2.

      (b)   Show that the open ball of radius r about $\vec{v}_0$ has the shape indicated in Figure 2.4.

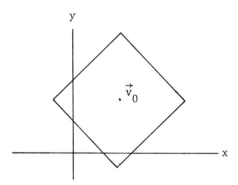

FIGURE 2.4.

8.   Prove that the half open interval $(0,1]$ is neither open nor closed.

9.   (a)   Give an example to show that the intersection of an infinite collection of open sets need not be open.

      (b)   Give an example to show that the union of an infinite collection of closed sets need not be closed.

10.   Give examples to show that the intersection of a closed set and an open set can be open, closed, or neither.

11.   Give examples to show that the union of an open set and a closed set can be open, closed, or neither.

*12. Let S be any subset of a normed vector space V. Prove that there is a largest open set, Int(S), with Int(S) $\subseteq$ S. (That is, if U is any open set with U $\subseteq$ S, then U $\subseteq$ Int(S).) Int(S) is called the *interior* of S.

*13. Determine the interiors of each of the sets of Exercises 1, 2, and 3 above.

*14. Let S be any subset of a normed vector space V. Prove that there is a smallest closed set, Cl(S), with S $\subseteq$ Cl(S). Cl(S) is called the *closure* of S.

*15. Determine the closures of each of. the sets in Exercises 1, 2, and 3 above.

*16. If S is any subset of V, prove that Cl(V-S) = V - Int(S).

*17. Prove that a set S is open if and only if Int(S) = S.

*18. Prove that a set S is closed if and only if $C\ell$(S) = S.

*19. Prove that the open intervals in R^1 (defined in Section 1.4) are open and that the closed intervals are closed.

*20. Let S be any subset of V. Define the *boundary* of S, ∂(S), by ∂(S) = Cl(S) $\cap$ Cl(V-S). Prove that $\vec{v} \in \partial$(S) if and only if, for any r > 0, $B_r(\vec{v}) \cap$ S and $B_r(\vec{v}) \cap$ (V-S) are both nonempty.

*21. Describe the boundaries of the sets in Exercises 1, 2, and 3 above.

*22. Let $\| \ \|_1$ and $\| \ \|_2$ be equivalent norms on the vector space V. (See Exercise 5, Section 1.) Prove that a subset S in V is open relative to $\| \ \|_1$ if and only if it is open relative to $\| \ \|_2$.

*23. Prove the analogue of the assertion in the previous problem for closed sets.

3.   CONTINUOUS FUNCTIONS BETWEEN NORMED VECTOR SPACES

Before considering the notion of continuity for functions between
normed vector spaces, we review this notion for functions $f:R \to R$.

There are many ways of describing the concept of continuity. We
can say, for example, that a function $f:R \to R$ is continuous if its
graph has no "breaks". Thus, the function whose graph is given in
Figure 3.1 is continuous, whereas the function whose graph is given
in Figure 3.2 is not continuous. This description has a great deal
of intuitive value but, unfortunately, is not precise enough to be
usable.

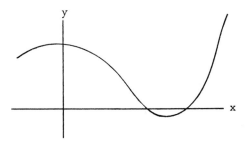

FIGURE 3.1.   The graph of a continuous function

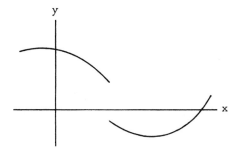

FIGURE 3.2.   The graph of a discontinuous function

Another way of describing continuity is to say that $f:R \to R$ is continuous at a $\in$ R if we can make $f(x)$ as close as we like to $f(a)$ by choosing x close enough to a. This description, too, needs to be made more precise, as follows. We define the function $f:R \to R$ to be *continuous* at a if, for any $\varepsilon > 0$, there is a $\delta > 0$ such that $|f(x) - f(a)| < \varepsilon$ wherever $|x-a| < \delta$.

In terms of the graph of f, this means that the piece of the graph of f between the vertical lines through the points a - δ and a + δ must be entirely between the horizontal lines through the points $f(a)$ - ε and $f(a)$ + ε. (See Figure 3.3).

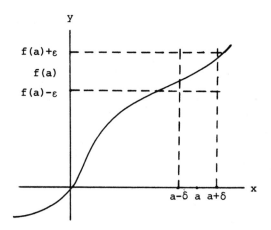

FIGURE 3.3. Continuity of f at a

We now look at some examples.

1.   The function $f(x) = 3x + 2$ is continuous at all a $\in$ R. To prove this, we must show that, for any $\varepsilon > 0$, we can find a $\delta > 0$ such that

$$|f(x) - f(a)| = |3x + 2 - (3a - 2)|$$
$$= |3x - 3a|$$
$$= 3|x - a|$$

is less than ε whenever $|x - a| < \delta$. In other words, if we imagine that someone gives us a small number ε, we must find a number δ so that the above holds. Here the choice of δ is clear; we simply set $\delta = \varepsilon/3$. Then

$$|f(x) - f(a)| = 3|x - a|$$

is certainly less than ε when $|x - a| < \delta = \varepsilon/3$.

2. The function $f(x) = x^2$ is continuous at all $a \in R$. To see this, let ε be any positive number and note that

$$|f(x) - f(a)| = |x^2 - a^2|$$
$$= |x - a| \cdot |x + a|$$

Furthermore, if $|x - a| < 1$, then

$$|x+a| = |x-a + 2a|$$
$$\leq |x-a| + 2|a|$$
$$< 1 + 2|a|$$

Thus

$$|f(x) - f(a)| < (1 + 2|a|) \cdot |x - a|$$

if $|x - a| < 1$. If we choose δ to be the smaller of the two numbers 1, $\varepsilon/(1+2a)$, we see that

$$|f(x) - f(a)| < \varepsilon$$

whenever $|x - a| < \delta$. Notice that here we need to work harder to find δ. One reason is that δ depends on a as well as ε.

3. The function $f:R \to R$ defined by

$$f(x) = \begin{cases} x & \text{if } x \leq 0 \\ x+1 & \text{if } x > 0 \end{cases}$$

is *not* continuous at 0. (See Figure 3.4.)

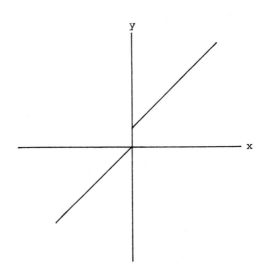

FIGURE 3.4. A function not continuous at 0

    To prove this, we must show the following. (See Section 1.2):
there is a number $\varepsilon > 0$ such that, for all $\delta > 0$, there is an x such
that $|x - 0| < \delta$ and $|f(x) - f(0)| \geq \varepsilon$. If we choose $\varepsilon = 1/2$, we see
that, no matter how small we choose δ, we can always find x > 0 with
$|x - 0| < \delta$. However, for x > 0, $|f(x) - f(0)| = f(x) > 1 > \varepsilon$.

    If we consider $R = R^1$ as a normed vector space with the standard
norm, then $\|x\| = |x|$. This suggests the following extension of the
definition above.

    Let V and W be normed vector spaces, and let U be an open set in
V. A function $f:U \to W$ is *continuous* at $\vec{v}_0 \in U$ if, for any $\varepsilon > 0$, there
is a $\delta > 0$ such that $\|f(\vec{v}) - f(\vec{v}_0)\| < \varepsilon$ whenever $\|\vec{v} - \vec{v}_0\| < \delta$. (Here
we use the notation $\| \ \|$ both for the norm of vectors in V and for the
norm of vectors in W.) Notice that f is defined for all $\vec{v}$ sufficiently
close to $\vec{v}_0$, since U is open.

    If $f:U \to W$ and S is any subset of U, we say that f is *continuous*
on S if f is continuous at each point of S. We say that f is *contin-*
uous if f is continuous at each $\vec{v} \in U$.

    Here are some examples.

4.   Let $f:R^2 \to R$ be defined by $f(x,y) = 2x + y$. Then f is continuous at any point (x_0, y_0) in R^2. To prove this, we let ε be any positive number and note that

$$\|f(x,y) - f(x_0,y_0)\| = |2x + y - (2x_0 + y_0)|$$
$$= |2(x - x_0) + (y - y_0)|$$
$$\leq 2|x - x_0| + |y - y_0|$$

Now,

$$|x - x_0| = ((x - x_0)^2)^{1/2}$$
$$\leq ((x - x_0)^2 + (y - y_0)^2)^{1/2}$$
$$= \|(x,y) - (x_0,y_0)\|$$

and

$$|y - y_0| = ((y - y_0)^2)^{1/2}$$
$$\leq \|(x,y) - (x_0,y_0)\|$$

Thus if we choose $\delta = \varepsilon/3$, we have

$$|x - x_0| < \varepsilon/3$$

and

$$|y - y_0| < \varepsilon/3$$

whenever $\|(x,y) - (x_0,y_0)\| < \delta$. Therefore

$$\|f(x,y) - f(x_0,y_0)\| \leq 2|x - x_0| + |y - y_0|$$
$$< 2\varepsilon/3 + \varepsilon/3 = \varepsilon$$

whenever $\|(x,y) - (x_0,y_0)\| < \delta$, and f is continuous at (x_0,y_0).

The following two examples will be useful later in this chapter.

5.   Define $g:R^n \to R$ by

$$g(x_1,\ldots,x_n) = x_1$$

This function is continuous. To prove this, we need to show that for any $\vec{v}_0 = (a_1,\ldots,a_n) \in R^n$ and $\varepsilon > 0$, there is a $\delta > 0$ such that

$$\|g(\vec{v}) - g(\vec{v}_0)\| = \|g(x_1,\ldots,x_n) - g(a_1,\ldots,a_n)\|$$
$$= |x_1 - a_1|$$

is less than ε whenever $\|\vec{v} - \vec{v}_0\|$ is less than δ. However,

$$|x_1 - a_1| = ((x_1 - a_1)^2)^{1/2}$$
$$\leq ((x_1 - a_1)^2 + \cdots + (x_n - a_n)^2)^{1/2}$$
$$= \|\vec{v} - \vec{v}_0\|$$

Therefore, if we set $\delta = \varepsilon$, we have

$$\|g(\vec{v}) - g(\vec{v}_0)\| = |x_1 - a_1|$$
$$\leq \|\vec{v} - \vec{v}_0\| < \varepsilon$$

wherever $\|\vec{v} - \vec{v}_0\| < \delta$.

Of course, the same argument proves that each of the functions $g_j : R^n \to R$, $1 \leq j \leq n$, defined by

$$g_j(x_1,\ldots,x_n) = x_j$$

is continuous. (See Exercise 11.)

The next example is more difficult.

6. Let $h : R^2 \to R$ be defined by $h(x,y) = xy$. Then h is continuous. To demonstrate this, let ε be any positive number, (a,b) a vector in R^2, and consider the following:

$$|h(x,y) - h(a,b)| = |xy - ab|$$
$$= |xy - ay + ay - ab|$$
$$\leq |xy - ay| + |ay - ab|$$
$$= |x - a| \cdot |y + a| \cdot |y - b|$$

Now suppose that $|y - b| < 1$. Then

$$|y| = |y - b + b|$$
$$\leq |y - b| + |b|$$
$$< 1 + |b|$$

so that

$$|h(x,y) - h(a,b)| < |x-a| \cdot (1 + |b|) + |a| \cdot |y - b| \qquad (3.1)$$

when $|y - b| < 1$. Furthermore,

$$|x - a| = ((x - a)^2)^{1/2}$$
$$\leq ((x - a)^2 + (y - b)^2)^{1/2} \qquad (3.2)$$
$$\leq \|(x,y) - (a,b)\|$$

and, in the same way,

$$|y - b| \leq \|(x,y) - (a,b)\| \qquad (3.3)$$

Thus, if we let δ be the smallest of the three numbers

$$1 , \frac{\varepsilon}{2(1+|b|)} , \frac{\varepsilon}{2|a|}$$

we see from (3.1), (3.2), and (3.3) that

$$|h(x,y) - h(a,b)| < \frac{\varepsilon}{2(1+|b|)}(1 + |b|) + |a|\frac{\varepsilon}{2|a|}$$
$$= \frac{\varepsilon}{2} + \frac{\varepsilon}{2} = \varepsilon$$

whenever $\|(x,y) - (a,b)\| < \delta$. Thus h is continuous.

We conclude this section by introducing another definition; it, too, is an extension of one from elementary calculus. Let V and W be normed vector spaces, U an open subset of the V, and $f:U \to W$ a function. Suppose that $\vec{v}_0 \in U$. We say that $f(\vec{v})$ *approaches the limit* $\vec{w}_0$ *as* $\vec{v}$ *approaches* $\vec{v}_0$, and write

$$\lim_{\vec{v} \to \vec{v}_0} f(\vec{v}) = \vec{w}_0$$

if for every $\varepsilon > 0$ there is a $\delta > 0$ such that if $\|f(\vec{v}) - \vec{w}_0\| < \varepsilon$ whenever $0 < \|\vec{v} - \vec{v}_0\| < \delta$.

Here are two examples.

7.   Define $f:R^2 \rightarrow R^2$ by

$$f(x_1,x_2) = \begin{cases} (x_2,x_1) & \text{if} \quad (x_2,x_1) \neq (0,0) \\ (1,-1) & \text{if} \quad (x_2,x_1) = (0,0) \end{cases}$$

Then $\lim_{\vec{v} \to \vec{0}} f(\vec{v}) = \vec{0}$. For if $\varepsilon > 0$, we set $\delta = \varepsilon$ in the definition:
if $\vec{v} = (x_1,x_2)$ and

$$0 < \|\vec{v}\| = (x_1^2 + x_2^2)^{1/2} < \varepsilon$$

then

$$\|f(\vec{v}) - \vec{0}\| = \|(x_2,x_1) - (0,0)\|$$
$$= (x_2^2 + x_1^2)^{1/2} < \varepsilon$$

as required. (Notice that we do not consider $f(0,0)$ when computing the limit.)

We can use the definition of limit to rephrase the definition of continuity: f is continuous at $\vec{v}_0$ if and only if

$$\lim_{\vec{v} \to \vec{v}_0} f(\vec{v}) = f(\vec{v}_0)$$

(See Exercise 22.) Thus the function f of Example 7 is *not* continuous at $(0,0)$.

8.   A function need not have a limit at a point. A familiar example from elementary calculus is the function $f:R \rightarrow R$ defined by

$$f(x) = \begin{cases} \dfrac{x}{|x|} & \text{if} \quad x \neq 0 \\ 0 & \text{if} \quad x = 0 \end{cases}$$

Then f has no limit at $x = 0$. Intuitively, $f(x)$ approaches 1 as x approaches 0 from the right and $f(x)$ approaches -1 as x approaches 0 from the left. We leave the detailed proof as Exercise 24.

The notion of the limit of a function between normed vector spaces will be useful in Section 5, when we define the derivative.

EXERCISES

1. Prove the following functions continuous at the values of x specified:

  (a) $f(x) = 3x + 3$, all $x \in R$

  (b) $f(x) = 3x^2$, all $x \in R$

  (c) $f(x) = x^2 - x - 3$, all $x \in R$

  (d) $f(x) = \frac{1}{x}$, $x \neq 0$

  (e) $f(x) = x^{1/3}$, all $x \in R$

2. Show that for all values of a and b, $f(x) = ax + b$ and $g(x) = ax^2$ are continuous for all $x \in R$.

3. Let $f:R \to R$ be continuous, let c be a constant, and let $g(x) = f(x) + c$. Show that g is continuous.

*4. Let $f(x) = e^x$, and assume that f is continuous at 0. Prove that f is continuous.

5. Prove the following functions continuous:

  (a) $f:R^2 \to R$, $f(x,y) = 7x + 4y$

  (b) $g:R^2 \to R$, $g(x,y) = 2x - 9y + 3$

  (c) $h:R^2 \to R$, $h(x,y) = 5x - 6y$

  (d) $F:R^2 \to R$, $F(x,y) = 6x$

  (e) $G:R^3 \to R$, $G(x,y,z) = 2x + 3y - 4z$

  (f) $H:R^3 \to R$, $H(x,y,z) = x + \frac{y}{2} + \frac{z}{3} + \frac{1}{4}$

6. Prove the following functions continuous:

  (a) $f:R^2 \to R$, $f(x,y) = 3xy + 2$

  (b) $g:R^2 \to R$, $g(x,y) = x^2 + 2y^2$

(c) $h:R^3 \to R$, $h(x,y,z) = xy + yz$

(d) $F:R^3 \to R$, $F(x,y,z) = xyz$

(e) $G:R^3 \to R^2$, $G(x,y,z) = (x + 2y - z, y + z)$

(f) $H:R^3 \to R^2$, $H(x,y,z) = (x^2 + y, y^2 + yz)$.

*7. Suppose that $f:R \to R$ and $g:R \to R$ are continuous. Prove that $f + g$ is continuous.

*8. Suppose that $f:R \to R$ and $g:R \to R$ are continuous. Prove that fg is continuous. (Here fg is the product of f and g; $(fg)(x) = f(x)g(x)$.)

*9. Define $f(x) = \begin{cases} 0 \text{ if } x \text{ is irrational} \\ 1 \text{ if } x \text{ is rational} \end{cases}$

Prove that f is not continuous at any point.

*10. Set $f(x) = \begin{cases} 0, & \text{if } x \text{ is irrational} \\ \dfrac{1}{q}, & \text{if } x = \dfrac{p}{q} \text{ as a fraction in lowest terms.} \end{cases}$

Show that f is continuous at all irrational points and discontinuous at all rational ones.

11. Define $g_j:R^k \to R$ by $g_j(\vec{v}) = x_j$ (where $\vec{v} = (x_1,\ldots,x_k)$). Prove that g_j is continuous. (See Example 5.)

12. Let $F:R^3 \to R^2$ be given by

$$F(x,y,z) = (f_1(x,y,z), f_2(x,y,z))$$

That is, f_1 and f_2 are the *coordinate functions* of F. Show that F is continuous if and only if f_1 and f_2 are continuous.

*13. Let $\| \ \|_1$ and $\| \ \|_2$ be equivalent norms on the vector space V. (See Exercise 5, Section 1.) Let W be a second normed vector space. Prove that a function $f:V \to W$ is continuous using $\| \ \|_1$ on V if and only if it is continuous using $\| \ \|_2$ on V.

*14.   Prove the analogue of the statement of Exercise 13 for $g:W \to V$.

15.   Let V and W be normed vector spaces and $f:V \to W$ a function.   The *graph of* f, G(f), is the subset of $V \times W$ defined by

$$G(f) = \{(\vec{v},\vec{w}):\vec{w} = f(\vec{v})\}$$

Prove that if f is continuous, then G(f) is a closed subset of $V \times W$ (in the norm $\|(\vec{v},\vec{w})\| = \|\vec{v}\| + \|\vec{w}\|$; see Exercise 9, Section 1.)

16.   Let V and W be normed vector spaces and let $T:V \to W$ be a linear transformation.   Prove that T is continuous on V if and only if T is continuous at $\vec{0}$.

17.   Let V and W be normed vector spaces.   A linear transformation $T:V \to W$ is said to be *bounded* if there is a number $M \geq 0$ such that $\|T\vec{v}\| \leq M \cdot \|\vec{v}\|$ for all $\vec{v} \in V$.   Prove that the following statements are equivalent.

   (a)   T is bounded.
   (b)   There is a number $A \geq 0$ such that $\|T\vec{v}\| \leq A$ whenever $\|\vec{v}\| \leq 1$.
   (c)   There is a number $B \geq 0$ such that $\|T\vec{v}\| \leq B$ whenever $\|\vec{v}\| = 1$.
   (d)   There are numbers $C \geq 0$, $r > 0$ such that $\|T\vec{v}\| \leq C$ whenever $\|\vec{v}\| \leq r$.

18.   Let V and W be normed vector spaces.   Prove that a linear transformation $T:V \to W$ is continuous if and only if it is bounded.

19.   Let V and W be normed vector spaces and B a subset of V.   A function $f:B \to W$ is said to be *bounded* if there is a number $M \geq 0$ such that $\|f(\vec{v})\| \leq M$ for all $\vec{v} \in B$.   (This definition does not agree with the one in Exercise 17.)   Give an example of a continuous function $f:(0,1) \to R$ which is not bounded.

20.   Let V and W be normed vector spaces with U an open subset of V. A function $f:U \to W$ is said to be *locally bounded* if, for each $\vec{v} \in U$, there is an $r > 0$ such that f is bounded on $B_r(\vec{v})$.   Prove that f is locally bounded whenever f is continuous on U.

21. Let U be a subset of the normed vector space V and let $f:U \to R$ be a continuous function. Suppose $f(\vec{v}) \geq 0$ for all $\vec{v} \in U$ and $f(\vec{v}_0)$ = m > 0. Show that there is a number r > 0 such that $f(\vec{v}) > m/2$ for all $\vec{v} \subset B_r(\vec{v}_0)$.

22. Let V, W be normed spaces, and let U be an open set in V. Show that a function $f:U \to W$ is continuous at $\vec{v}_0 \in U$ if and only if

$$\lim_{\vec{v} \to \vec{v}_0} f(\vec{v}) = f(\vec{v}_0)$$

23. Show that the function $f:R \to R$, defined by

$$f(x) = \begin{cases} \dfrac{1}{x} & \text{if } x \neq 0 \\ 0 & \text{if } x = 0 \end{cases}$$

does not have a limit at 0.

24. Show that the function $f:R \to R$ defined by

$$f(x) = \begin{cases} \dfrac{x}{|x|} & \text{if } x \neq 0 \\ 0 & \text{if } x = 0 \end{cases}$$

does not have a limit as x approaches 0.

25. Show that the function $g:R^2 \to R$ defined by

$$g(x,y) = \begin{cases} 1 & \text{if } y = x^2 \\ 0 & \text{otherwise} \end{cases}$$

does not have a limit as (x,y) approaches (0,0).

26. Show that a function can have at most one limit at a point: if

$$\lim_{\vec{v} \to \vec{v}_0} f(\vec{v}) = \vec{w}_0$$

and

$$\lim_{\vec{v} \to \vec{v}_0} f(\vec{v}) = \vec{w}_1$$

then $\vec{w}_0 = \vec{w}_1$.

27. Let U be an open subset of R^n and $f:U \to R^p$ a function with $f(\vec{v})$ = $(f_1(\vec{v}),\ldots,f_p(\vec{v}))$. Prove that the following statements are equivalent:

(a) $\lim_{\vec{v} \to \vec{v}_0} f(\vec{v}) = f(\vec{v}_0)$

(b) $\lim_{\vec{v} \to \vec{v}_0} \|f(\vec{v}) - f(\vec{v}_0)\| = 0$

(c) $\lim_{\vec{v} \to \vec{v}_0} f_j(\vec{v}) = f_j(\vec{v}_0)$ for $j = 1,\ldots,p$

*28. Formulate and prove statements involving the limits of sums, products, and quotients of functions.

4. ELEMENTARY PROPERTIES OF CONTINUOUS FUNCTIONS

In this section, we develop some of the elementary properties of continuous functions and use them to prove that linear transformations between Euclidean spaces are continuous.

Let V and W be normed vector spaces, $f,g:V \to W$ functions and c a real number. We define functions

$$f + g:V \to W$$
$$cf:V \to W$$

by

$$(f+g)(\vec{v}) = f(\vec{v}) + g(\vec{v})$$
$$(cf)(\vec{v}) = c(f(\vec{v}))$$

(See Example 3 of Section 2.2.)

The following two Propositions are analogues of well known results for continuous functions $f:R \to R$.

PROPOSITION 4.1. *Suppose* $f,g:V \to W$ *are continuous at* $\vec{v}_0 \in V$. *Then* $f+g:V \to W$ *is also continuous at* $\vec{v}_0$.

Proof. We need to show that for any $\varepsilon > 0$, there is a $\delta > 0$ such that $\|(f+g)(\vec{v}) - (f+g)(\vec{v}_0)\| < \varepsilon$ whenever $\|\vec{v} - \vec{v}_0\| < \delta$. We have

$$\|(f+g)(\vec{v}) - (f+g)(\vec{v_0})\| = \|f(\vec{v}) + g(\vec{v}) - f(\vec{v_0}) - g(\vec{v_0})\|$$

$$= \|(f(\vec{v}) - f(\vec{v_0})) + (g(\vec{v}) - g(\vec{v_0}))\|$$

$$\leq \|f(\vec{v}) - f(\vec{v_0})\| + \|g(\vec{v}) - g(\vec{v_0})\|$$

Now we use the continuity of f and g at $\vec{v_0}$. Since both $\|f(\vec{v}) - f(\vec{v_0})\|$ and $\|g(\vec{v}) - g(\vec{v_0})\|$ are small when $\|\vec{v} - \vec{v_0}\|$ is small enough, $\|(f+g)(\vec{v}) - (f+g)(\vec{v_0})\|$ will be small when $\|\vec{v} - \vec{v_0}\|$ is small enough. More precisely, since f is continuous at $\vec{v_0}$, we can find $\delta_1 > 0$ so that

$$\|f(\vec{v}) - f(\vec{v_0})\| < \varepsilon/2 \text{ when } \|\vec{v} - \vec{v_0}\| < \delta_1$$

Since g is continuous at $\vec{v_0}$, we can find $\delta_2 > 0$ so that

$$\|g(\vec{v}) - g(\vec{v_0})\| < \varepsilon/2 \text{ when } \|\vec{v} - \vec{v_0}\| < \delta_2$$

Thus if $\delta = \min(\delta_1, \delta_2)$,

$$\|(f+g)(\vec{v}) - (f+g)(\vec{v_0})\| \leq \|f(\vec{v}) - f(\vec{v_0})\| + \|g(\vec{v}) - g(\vec{v_0})\|$$

$$< \frac{\varepsilon}{2} + \frac{\varepsilon}{2} = \varepsilon$$

whenever $\|\vec{v} - \vec{v_0}\| < \delta$, so f + g is continuous at $\vec{v_0}$.

PROPOSITION 4.2. *If* $f:V \to W$ *is continuous at* $\vec{v_0} \in V$ *and c is a real number, then* $cf:V \to W$ *is also continuous at* $\vec{v_0}$.

Proof. We need to show that given any $\varepsilon > 0$, we can find a $\delta > 0$ such that

$$\|(cf)(\vec{v}) - (cf)(\vec{v_0})\| < \varepsilon$$

whenever $\|\vec{v} - \vec{v_0}\| < \delta$. Now

$$\|(cf)(\vec{v}) - (cf)(\vec{v_0})\| = \|c(f(\vec{v})) - c(f(\vec{v_0}))\|$$

$$= |c| \cdot \|f(\vec{v}) - f(\vec{v_0})\|$$

so the assertion is trivial if $c = 0$. Assume $c \neq 0$. Since f is continuous at $\vec{v_0}$, we can find $\delta > 0$ so that

$$\|f(\vec{v}) - f(\vec{v}_0)\| < \varepsilon/|c|$$

when $\|\vec{v} - \vec{v}_0\| < \delta$. It follows immediately that

$$\|(cf)(\vec{v}) - (cf)(\vec{v}_0)\| = |c| \cdot \|f(\vec{v}) - f(\vec{v}_0)\| < \varepsilon$$

when $\|\vec{v} - \vec{v}_0\| < \delta$; thus cf is continuous at $\vec{v}_0$.

An easy induction argument proves the following generalization of Propositions 4.1 and 4.2.

PROPOSITION 4.3. *Let* $g_1,\ldots,g_k:V \to W$ *be continuous at* $\vec{v}_0$, *and let* $a_1,\ldots,a_k$ *be real numbers. Then the function*

$$g = \sum_{j=1}^{k} a_j g_j$$

is also continuous at $\vec{v}_0$.

The proof is left as Exercise 1.

For example, suppose $T:R^n \to R$ is a linear transformation. Then, according to the results of Section 2.6, there are real numbers $a_1,\ldots,a_n$ such that

$$T(x_1,\ldots,x_n) = \sum_{j=1}^{n} a_j x_j$$

Let $g_j:R^n \to R$, $1 \le j \le n$, be defined by

$$g_j(x_1,\ldots,x_n) = x_j$$

These functions are continuous (see Example 5 of the previous section) and T can be written

$$T(x_1,\ldots,x_n) = \sum_{j=1}^{n} a_j g_j(x_1,\ldots,x_n)$$

It follows from Proposition 4.3 that T is continuous. Thus any linear transformation $T:R^n \to R$ is continuous.

The next proposition allows us to generalize this result.

PROPOSITION 4.4. *Let* $f:R^n \to R^p$ *be given by*

$$f(\vec{v}) = (f_1(\vec{v}), f_2(\vec{v}), \ldots, f_p(\vec{v}))$$

where $f_j:R^n \to R$. *Then* f *is continuous at* $\vec{v}_0$ *in* R^n *if and only if each* f_j *is continuous at* $\vec{v}_0$, $1 \le j \le p$.

Proof. We first prove that f is continuous at $\vec{v}_0$ if each f_j is continuous at $\vec{v}_0$.

Suppose we are given $\varepsilon > 0$. We must find a $\delta > 0$ such that $\|f(\vec{v}) - f(\vec{v}_0)\| < \varepsilon$ whenever $\|\vec{v} - \vec{v}_0\| < \delta$. Since each f_j is continuous, we can find $\delta_j > 0$ so that

$$\|\vec{v} - \vec{v}_j\| < \delta_j \Rightarrow |f_j(\vec{v}) - f_j(\vec{v}_0)| < \frac{\varepsilon}{\sqrt{p}}$$

Thus, if $\delta = \min\{\delta_1, \ldots, \delta_p\}$, we have

$$\|\vec{v} - \vec{v}_0\| < \delta \Rightarrow |f_j(\vec{v}) - f_j(\vec{v}_0)| < \frac{\varepsilon}{\sqrt{p}}$$

for all $j = 1, \ldots, m$, so that

$$\left(\sum_{j=1}^{p} (f_j(\vec{v}) - f_j(\vec{v}_0))^2 \right)^{1/2} < \left(\sum_{j=1}^{p} \frac{\varepsilon^2}{p} \right)^{1/2} = \varepsilon$$

It follows that f is continuous at $\vec{v}_0$.

Assume now that f is continuous at $\vec{v}_0$. For any $\varepsilon > 0$, choose $\delta > 0$ so that

$$\|\vec{v} - \vec{v}_0\| < \delta \Rightarrow \|f(\vec{v}) - f(\vec{v}_0)\| < \varepsilon$$

However, for any $j = 1, \ldots, p$,

$$|f_j(\vec{v}) - f_j(\vec{v}_0)| = \left((f_j(\vec{v}) - f_j(\vec{v}_0))^2 \right)^{1/2}$$
$$\le \left(\sum_{i=1}^{p} (f_i(\vec{v}) - f_i(\vec{v}_0))^2 \right)^{1/2}$$
$$= \|f(\vec{v}) - f(\vec{v}_0)\|$$

so

$$\|\vec{v} - \vec{v}_0\| < \delta \Rightarrow |f_j(\vec{v}) - f_j(\vec{v}_0)| < \varepsilon$$

Therefore f_j is continuous at $\vec{v}_0$ for all $j = 1,\ldots,p$.

We can now prove that any linear transformation $T:R^n \to R^p$ is continuous.

PROPOSITION 4.5. *Let* $T:R^n \to R^p$ *be a linear transformation. Then* T *is continuous.*

Proof. According to the results of Section 2.6, we have

$$T(x_1,\ldots,x_n) = (T_1(x_1,\ldots,x_n),\ldots,T_p(x_1,\ldots,x_n))$$

where

$$T_i(x_1,\ldots,x_n) = \sum_{j=1}^{n} a_{ij}x_j$$

$(1 \le i \le p)$ and (a_{ij}) is the matrix of T. We know from Proposition 4.3 that each of the functions T_i is continuous. (See the example following Proposition 4.3.) Thus T is continuous by Proposition 4.4.

We shall later generalize this result to linear transformations between arbitrary normed vector spaces.

EXERCISE

1.   Prove Proposition 4.3.

5.   THE DERIVATIVE

In this section, we give the definition and some examples of the derivative of a function between normed vector spaces. As the definition may seem somewhat mysterious at first, we begin with a review of the one variable case.

Recall that if U is an open subset of R, then the function $f:U \to R$ is said to be *differentiable* at $x_0 \in U$ if

$$\lim_{x \to x_0} \frac{f(x) - f(x_0)}{x - x_0}$$

exists. The value of this limit is called the *derivative of f at* x_0 and is denoted by $f'(x_0)$.

According to the definition of the limit,

$$\lim_{x \to x_0} \frac{f(x) - f(x_0)}{x - x_0}$$

means roughly that the quantity

$$\frac{f(x) - f(x_0)}{x - x_0}$$

is close to $f'(x_0)$ when x is near x_0. Equivalently, we can say that

$$\left| \frac{f(x) - f(x_0)}{x - x_0} - f'(x_0) \right|$$

is small when x is near x_0. Rewriting this last expression, we see that

$$\frac{\left| f(x) - f(x_0) - f'(x_0)(x - x_0) \right|}{\left| x - x_0 \right|} \tag{5.1}$$

is small when x is near x_0. Now, any real number a determines a linear transformation $L : R \to R$ defined by

$$L(x) = ax$$

In particular, $f'(x_0)$ determines a linear transformation $L : R \to R$. Using L, we may rewrite (5.1) as

$$\lim_{x \to x_0} \frac{\left| f(x) - f(x_0) - L(x - x_0) \right|}{\left| x - x_0 \right|} = 0$$

We can now use (5.2) to define differentiability of a function between normed vector spaces. Let V and W be normed vector spaces, U an open subset of V, and $f:U \to W$ a function. We say that f is *differentiable* at $\vec{v}_0 \in U$ if there is a linear transformation $L:V \to W$ such that

$$\lim_{\vec{v} \to \vec{v}_0} \frac{\|f(\vec{v}) - f(\vec{v}_0) - L(\vec{v} - \vec{v}_0)\|}{\|\vec{v} - \vec{v}_0\|} = 0 \tag{5.3}$$

This means that, for any $\varepsilon > 0$, there is a $\delta > 0$ such that

$$\frac{\|f(\vec{v}) - f(\vec{v}_0) - L(\vec{v} - \vec{v}_0)\|}{\|\vec{v} - \vec{v}_0\|} < \varepsilon$$

whenever $0 < \|\vec{v} - \vec{v}_0\| < \delta$. (See Section 3.) Of course, (5.3) also means that for every $\varepsilon > 0$, there is a $\delta > 0$ such that

$$\|f(\vec{v}) - f(\vec{v}_0) - L(\vec{v} - \vec{v}_0)\| < \varepsilon \|\vec{v} - \vec{v}_0\|$$

whenever $0 < \|\vec{v} - \vec{v}_0\| < \delta$.

Notice that we define differentiability only for functions defined on open sets so that, in the limit (5.3), the vector $\vec{v}$ can approach $\vec{v}_0$ from all directions. For example, if n = 1, $\vec{v}$ can approach $\vec{v}_0$ both from the right and from the left.

The basic feature that a function f differentiable at $\vec{v}_0$ has is that it can be approximated near $\vec{v}_0$ by the relatively simple function

$$h(\vec{v}) = L(\vec{v}) + (f(\vec{v}_0) - L(\vec{v}_0)) \tag{5.4}$$

In fact, this approximation is so good that not only does $\|f(\vec{v}) - h(\vec{v})\|$ tend to zero as $\vec{v}$ approaches $\vec{v}_0$, but $\|f(\vec{v}) - h(\vec{v})\|$ multiplied by the large number $1/\|\vec{v} - \vec{v}_0\|$ also tends to zero as $\vec{v}$ approaches $\vec{v}_0$.

The linear transformation L of equation (5.3) is called the *derivative of f at* $\vec{v}_0$. We denote it by $D_{\vec{v}_0} f$ or by $f'(\vec{v}_0)$. As we shall see in the next section, L is uniquely defined by equation (5.3), so that a function differentiable at $\vec{v}_0 \in U$ has exactly one derivative there.

We have seen that if U is an open subset of R^1 and $f:U \to R$ has a derivative $f'(x_0)$ at x_0 in the usual calculus sense, then $f'(x_0)$ is also the derivative in this new sense, once we interpret the number $f'(x_0)$ as the linear transformation

$$L(x) = f'(x_0)x$$

Here are some other examples of derivatives.

(1) If $f:V \to W$ is a constant function and $L:V \to W$ is the zero linear transformation, then

$$f(\vec{v}) - f(\vec{v}_0) - L(\vec{v}-\vec{v}_0) = \vec{0}$$

for all $\vec{v},\vec{v}_0 \in V$. Thus $D_{\vec{v}}f$ is the zero transformation for all $\vec{v} \in V$.

(2) If $f:V \to W$ is a linear transformation and $L = f:V \to W$, then

$$f(\vec{v}) - f(\vec{v}_0) - L(\vec{v}-\vec{v}_0) = \vec{0}$$

for all $\vec{v}_0,\vec{v} \in V$. It follows that $D_{\vec{v}}f = f$ for all $\vec{v} \in V$ whenever f is a linear transformation. That is, a linear transformation is its own derivative. (Warning: this means, in the case of functions from R to R, that the derivative of $f(x) = ax$ is the linear function given by $y = ax$, and hence the linear function corresponding to the real number a. It does *not* mean that the derivative of ax is ax.)

(3) Suppose $h:R^2 \to R^1$ is defined by

$$h(x,y) = xy$$

We shall compute $D_{\vec{v}_0} h$.

We know that the most general linear transformation $L:R^2 \to R^1$ is given by

$$L(x,y) = ax + by$$

(See Section 2.6.) Now let $\vec{v}_0 = (x_0,y_0)$ and $\vec{v} = (x,y)$. To find the derivative of h, we need to produce numbers a and b such that

$$\lim_{\vec{v} \to \vec{v}_0} \frac{|h(\vec{v}) - h(\vec{v}_0) - a(x-x_0) - b(y-y_0)|}{\|\vec{v} - \vec{v}_0\|} = 0$$

or

$$\lim_{(x,y) \to (x_0,y_0)} \frac{|xy - x_0 y_0 - a(x-x_0) - b(y-y_0)|}{((x-x_0)^2 + (y-y_0)^2)^{1/2}} = 0 \qquad (5.5)$$

We can rewrite the numerator of the left side of equation (5.5) as

$$|x(y-y_0) + y_0(x-x_0) - a(x-x_0) - b(y-y_0)|$$

$$= |(x-b)(y-y_0) + (y_0-a)(x-x_0)|$$

$$\leq |x-b| \cdot |y-y_0| + |y_0-a| \cdot |x-x_0|$$

Therefore

$$\frac{|xy - x_0 y_0 - a(x-x_0) - b(y-y_0)|}{((x-x_0)^2 + (y-y_0)^2)^{1/2}} \leq \frac{|x-b| \cdot |y-y_0|}{((x-x_0)^2 + (y-y_0)^2)^{1/2}}$$

$$+ \frac{|y_0-a| \cdot |x-x_0|}{((x-x_0)^2 + (y-y_0)^2)^{1/2}}$$

$$\leq |x-b| + |y_0-a|$$

since

$$|y-y_0| = ((y-y_0)^2)^{1/2}$$

$$\leq ((x-x_0)^2 + (y-y_0)^2)^{1/2}$$

and

$$|x-x_0| = ((x-x_0)^2)^{1/2}$$

$$\leq ((x-x_0)^2 + (y-y_0)^2)^{1/2}$$

It follows that

$$\lim_{(x,y)\to(x_0,y_0)} \frac{|xy-x_0y_0-a(x-x_0)-b(y-y_0)|}{((x-x_0)^2+(y-y_0)^2)^{1/2}}$$

$$\le \lim_{(x,y)\to(x_0,y_0)} (|x-b| + |y_0-a|$$

However,

$$\lim_{(x,y)\to(x_0,y_0)} (|x-b| + |y_0-a|) = 0$$

when $b = x_0$ and $a = y_0$. Thus, $D_{\vec{v}}h$ is the linear transformation taking (x,y) into $y_0x + x_0y$.

As this calculation makes clear, we need a better way of computing derivatives. We already know good methods for functions $f:R \to R$, from elementary calculus. In Section 7, we shall see how to reduce the problem of computing the derivative of a function $f:R^n \to R^p$ to the one variable case.

We say that f is *differentiable on* U if f is differentiable at each $\vec{v} \in$ U. If S is any subset of V (not necessarily open) and $g:S \to W$ is a function, we say that g is *differentiable on* S if there is an open set U containing S and a function $f:U \to W$ such that $f(\vec{v}) = g(\vec{v})$ for $\vec{v} \in$ S and f is differentiable on U. In this case, we define the derivative of g at $\vec{v}_0 \in$ S to be $D_{\vec{v}_0}f$ and denote it by $D_{\vec{v}_0}g$. (The definition may depend on our choice of g, but it will not in the cases where we use it.)

EXERCISES

1.   Determine, directly from the definition, the derivative of the function $f:R^2 \to R^2$ defined by

$$f(x_1,x_2) = (x_1^2 + x_1x_2, x_1 - x_2^2)$$

(Note: Express the derivative of f as T_A for some 2×2 matrix A: see Section 2.6 and Example 3 of this section.)

2. Determine, directly from the definition, the derivative of the function $f:R^3 \to R^2$ defined by

$$f(x_1,x_2,x_3) = (x_1^2 - 2x_1x_2 + x_3, \; x_2^2 + x_3^2)$$

3. Let $f:R \to R^n$ be given by the formula

$$f(x) = (f_1(x), f_2(x), \ldots, f_n(x))$$

where each $f_j:R \to R$ is differentiable, $1 \le j \le n$. Show that

$$(D_{x_0} f)(x) = x\vec{v}_0$$

where $\vec{v}_0 = (f_1'(x_0), \; f_2'(x_0), \ldots, f_n'(x_0))$.

*4. Let $f:R^n \to R$ and $g:R^n \to R^p$ be differentiable at $\vec{v}_0$. Show that $fg:R^n \to R^p$ (defined by $(fg)(\vec{v}) = f(\vec{v})g(\vec{v})$) is differentiable at $\vec{v}_0$, and that

$$D_{\vec{v}_0}(fg)(\vec{v}) = f(\vec{v}_0)(D_{\vec{v}_0} g)(\vec{v}) + ((D_{\vec{v}_0} f)(\vec{v}))g(\vec{v}_0)$$

5. Prove that the function $f:R \to R$ defined by $f(x) = |x|$ is not differentiable at $x = 0$.

6. Let $f:(a,b) \to R$ be differentiable with $f'(c) > 0$, $c \in (a,b)$. Prove that there is a $\delta > 0$ such that the quotient

$$\frac{f(x) - f(c)}{x-c} > 0$$

for $x \in (c-\delta, c+\delta)$, $x \ne c$. (A similar result holds if $f'(c) < 0$.) Note: (a,b) and $(c-\delta, c+\delta)$ denote intervals in R.

7. Let $f:(a,b) \to R$ be a differentiable function and suppose that f has a local maximum (or a local minimum) at $c \in (a,b)$. Prove that $f'(c) = 0$. (Hint: Use Exercise 6.)

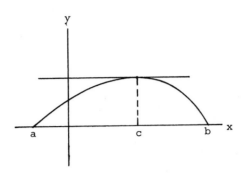

FIGURE 5.1.   Rolle's Theorem

8.   *(Rolle's Theorem)*.   Let f:[a,b] → R be a continuous function
which is differentiable on (a,b), and suppose that f(a) = f(b) = 0.
Prove that there is a number c ∈ (a,b) such that f'(c) = 0. (See
Figure 5.1.) (Assume if necessary that any continuous function on a
closed interval attains its maximum and minimum values. This fact is
proved in Section 5.5.)

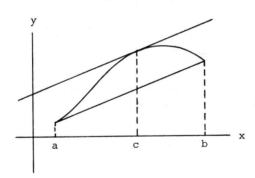

FIGURE 5.2.   The Mean Value Theorem

*9.   *(The Mean Value Theorem)*.   Let f:[a,b] → R be a continuous function which is differentiable on (a,b). Prove that there is a number
c ∈ (a,b) such that

$$f'(c) = \frac{f(b)-f(a)}{b-a}$$

(This equation states that the tangent line to the graph of f at the point c is parallel to the chord joining the points (a,f(a)) and (b,f(b)) in R^2; see Figure 5.2.)

6. ELEMENTARY PROPERTIES OF THE DERIVATIVE[*]

We now derive some of the elementary properties of the derivative. We begin by showing that a differentiable function has a unique derivative.

PROPOSITION 6.1. *Let* U *be an open subset of* V *and* f:U → W *a function differentiable at* $\vec{v}_0$ ∈ U. *Suppose that* L_1, L_2:V → W *are derivatives of* f *at* $\vec{v}_0$. *Then* $L_1 = L_2$.

Proof. Note first of all that

$$\| L_1(\vec{v}-\vec{v}_0) - L_2(\vec{v}-\vec{v}_0) \|$$

$$= \| f(\vec{v}) - f(\vec{v}_0) - L_2(\vec{v}-\vec{v}_0) - (f(\vec{v}) - f(\vec{v}_0) - L_1(\vec{v}-\vec{v}_0)) \|$$

$$\leq \| f(\vec{v}) - f(\vec{v}_0) - L_2(\vec{v}-\vec{v}_0) \| + \| f(\vec{v}) - f(\vec{v}_0) - L_1(\vec{v}-\vec{v}_0)) \|$$

Setting $L = L_1 - L_2$ and $\vec{u} = \vec{v} - \vec{v}_0$, we see that

$$\lim_{\vec{u}\to\vec{0}} \frac{\|L(\vec{u})\|}{\|\vec{u}\|} = 0$$

since both L_1 and L_2 are derivatives for f at $\vec{v}_0$. Thus Proposition 6.1 will be proved if we can show that L is the zero transformation. We proceed by contradiction.

Suppose that $L(\vec{u}_0) \neq \vec{0}$, and set $\varepsilon = \|L(\vec{u}_0)\|/\|\vec{u}_0\|$. (We know that $\vec{u}_0 \neq \vec{0}$, since $L(\vec{0}) = \vec{0}$.) Then, for all $\alpha > 0$, $\|L(\alpha\vec{u}_0)\|/\|\alpha\vec{u}_0\| = \varepsilon$.

[*] Some readers may prefer to skip this section temporarily, going directly to Section 7, where we show how to compute the derivative.

Now we certainly can pick α so that $\|\alpha\vec{u}_0\|$ is less than any $\delta > 0$. This contradicts the fact that $\lim_{\vec{u} \to \vec{0}}(\|L(\vec{u})\|/\|\vec{u}\|) = 0$. Therefore $L(\vec{u}) = \vec{0}$ for all $\vec{u}$ in V, and this proves the proposition.

We next prove an analogue of a well known result for functions of one variable.

Recall that, if $S, T : V \to W$ are linear transformations and c is a real number, then $S + T$ and cT are the linear transformations from V to W defined by

$$(S+T)(\vec{v}) = S(\vec{v}) + T(\vec{v})$$

$$(cT)(\vec{v}) = c(T(\vec{v}))$$

PROPOSITION 6.2. *Let a be a real number and U an open subset of V, and suppose that $f, g : U \to W$ are differentiable at $\vec{v}_0$. Then* af *and f+g are differentiable at $\vec{v}_0$, with*

$$D_{\vec{v}_0}(af) = aD_{\vec{v}_0}f$$

$$D_{\vec{v}_0}(f+g) = D_{\vec{v}_0}f + D_{\vec{v}_0}g$$

Proof. We prove only that af is differentiable at $\vec{v}_0$ with $D_{\vec{v}_0}(af) = aD_{\vec{v}_0}f$, leaving the statement about f+g as Exercise 1.

Set $L = D_{\vec{v}_0}f$. Then

$$\|(af)(\vec{v}) - (af)\vec{v}_0 - (aL)(\vec{v}-\vec{v}_0)\| = \|af(\vec{v}) - af(\vec{v}_0) - aL(\vec{v}-\vec{v}_0)\|$$

$$= |a|\|f(\vec{v}) - f(\vec{v}_0) - L(\vec{v}-\vec{v}_0)\|$$

It follows easily that

$$\lim_{\vec{v}\to\vec{v}_0}\frac{\|(af)(\vec{v}) - (af)(\vec{v}_0) - (aL)(\vec{v}-\vec{v}_0)\|}{\|\vec{v}-\vec{v}_0\|}$$

$$= |a|\lim_{\vec{v}\to\vec{v}_0}\frac{\|f(\vec{v}) - f(\vec{v}_0) - L(\vec{v}-\vec{v}_0)\|}{\|\vec{v}-\vec{v}_0\|} = 0$$

since $L = D_{\vec{v}_0}f$. This completes the proof of the first part of Proposition 6.2.

We know from elementary calculus that a differentiable function $f : R \to R$ is continuous. The next result proves this true for functions between Euclidean spaces.

PROPOSITION 6.3. *Let U be an open subset of* R^n. *If* $f : U \to R^p$ *is differentiable at a point* $\vec{v}_0 \in U$, *then f is continuous at* $\vec{v}_0$.

Proof. Given $\varepsilon > 0$, we know that there is a number $\delta_0 > 0$ such that

$$\frac{\|f(\vec{v}) - f(\vec{v}_0) - D_{\vec{v}_0} f(\vec{v} - \vec{v}_0)\|}{\|\vec{v} - \vec{v}_0\|} < \varepsilon/2$$

if $\|\vec{v} - \vec{v}_0\| < \delta_0$. If $\|\vec{v} - \vec{v}_0\|$ is also less than 1, then

$$\|f(\vec{v}) - f(\vec{v}_0) - D_{\vec{v}_0} f(\vec{v} - \vec{v}_0)\| < \varepsilon/2$$

Since $D_{\vec{v}_0} f$ is linear, Proposition 4.5 implies that $D_{\vec{v}_0} f$ is continuous; thus we can choose δ_1 so that

$$\|D_{\vec{v}_0} f(\vec{v}) - D_{\vec{v}_0} f(\vec{v}_0)\| < \varepsilon/2$$

whenever $\|\vec{v} - \vec{v}_0\| < \delta_1$.

Now let δ be the smallest of 1, δ_0, δ_1. If $\|\vec{v} - \vec{v}_0\| < \delta$, then

$$\|f(\vec{v}) - f(\vec{v}_0)\| \leq \|f(\vec{v}) - f(\vec{v}_0) - D_{\vec{v}_0} f(\vec{v} - \vec{v}_0)\|$$

$$+ \|D_{\vec{v}_0} f(\vec{v} - \vec{v}_0)\|$$

$$< \varepsilon/2 + \|D_{\vec{v}_0} f(\vec{v}) - D_{\vec{v}_0} f(\vec{v}_0)\|$$

$$< \frac{\varepsilon}{2} + \frac{\varepsilon}{2} = \varepsilon$$

Therefore f is continuous at $\vec{v}_0$.

We note that a function continuous at $\vec{v}_0$ need not be differentiable at $\vec{v}_0$. For example, the function $f : R \to R$ defined by $f(x) = |x|$ is everywhere continuous, but not differentiable at $x = 0$. (See Exercise 5, Section 5.)

We extend this result to finite dimensional inner product spaces
in Proposition 4.7.5 and to finite dimensional normed vector spaces
in Chapter 5.   (See Exercise 6 of Section 5.6.)

EXERCISES

1.   Prove the second statement of Proposition 6.2.

2.   (a) Let $f,g:V \to W$ be continuous functions such that $f+g$ is differentiable. Are f and g differentiable? Give a proof or counterexample.

      (b) Suppose that af is differentiable (for $a \neq 0$). Is f? Again,
give a proof or counterexample.

7.   PARTIAL DERIVATIVES AND THE JACOBIAN MATRIX

As we say in Section 5, the computation of the derivative of a function $f:V \to W$ directly from the definition is a complicated matter.
We simplify it here when V and W are Euclidean spaces by reducing the
computation of the derivative of such a function to the case of functions from R to R, a problem studied in elementary calculus. We shall
show later how this applies to arbitrary normed vector spaces.

      Suppose that U is an open subset of R^n and $f:U \to R$ is a function.
The *partial derivative of f with respect to* x_j *at* $\vec{v} = (x_1,\dots,x_n)$
is defined by

$$\frac{\partial f}{\partial x_j} = \lim_{h \to 0} \frac{f(x_1,\dots,x_{j-1},x_j+h,x_{j+1},\dots,x_n) - f(x_1,\dots,x_n)}{h} \qquad (7.1)$$

when the limit exists. If we wish to indicate explicitly the point
at which the partial derivative is taken, we shall write $\frac{\partial f}{\partial x_j}(\vec{v})$ or
$\frac{\partial f}{\partial x_j}\Big|_{\vec{v}}$.

      Note that if $n = 1$, then the partial derivative defined above is
just the ordinary derivative of a function of one variable. In general,

if we fix the point $\vec{v}_0 = (a_1, \ldots, a_n)$ and define a function g of one variable x by

$$g(x) = f(a_1, \ldots, a_{j-1}, x, a_{j+1}, \ldots, a_n) \qquad (7.2)$$

then the partial derivative of f with respect to x_j at the point $\vec{v}$ is the ordinary derivative of g with respect to x evaluated at $x = x_j$. This observation reduces the computation of partial derivatives to that of ordinary differentiation. Thus, to compute $\frac{\partial f}{\partial x_j}$, we treat all variables except x_j as constants and differentiate the resulting function of the single variable x_j in the usual way.

Here are some examples of partial derivatives.

1. If $f:R^2 \rightarrow R$ is given by $f(x_1, x_2) = x_1^2 + x_2$, then $\frac{\partial f}{\partial x_2}$ is obtained by treating the variable x_1 as a constant and differentiating f in the usual way with respect to x_2. Thus

$$\frac{\partial f}{\partial x_2} = 1$$

since $\frac{\partial}{\partial x_2}(x_1^2) = 0$ (x_1 is treated as a constant). Similarly, $\frac{\partial f}{\partial x_1} = 2x_1$.

2. Let $f:R^3 \rightarrow R$ be defined by

$$f(x_1, x_2, x_3) = x_1 x_3^2 + x_1^2 x_2 x_3 + x_2^3$$

Then $\frac{\partial f}{\partial x_1}$ is obtained by treating the variables x_2 and x_3 as constants and differentiating with respect to x_1. Thus

$$\frac{\partial f}{\partial x_1} = x_3^2 + 2x_1 x_2 x_3$$

In the same way, we have

$$\frac{\partial f}{\partial x_2} = x_1^2 x_3 + 3x_2^2 \qquad \frac{\partial f}{\partial x_3} = 2x_1 x_3 + x_1^2 x_2$$

If n = 2, the graph of the function g defined in (7.2) can be obtained by intersecting the graph of f with a plane perpendicular to one of the axes. For instance, if $f(x_1, x_2) = 1 - x_1^2 - x_2^2$ and $g: R \to R$ is defined by

$$g(x) = f(a, x)$$
$$= 1 - a^2 - x^2$$

then the graph of g is the intersection of the graph of f and the plane perpendicular to the x_1-axis through the point $(a, 0, 0)$. (See Figure 7.1.)

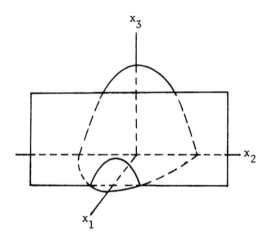

FIGURE 7.1. The graph of $g(x) = f(a, x)$

Suppose now that $f: U \to R^p$ is defined by

$$f(\vec{v}) = (f_1(\vec{v}), \ldots, f_p(\vec{v}))$$

where $f_i: U \to R$, $1 \le i \le p$. For each $\vec{v} \in U$, we define the *Jacobian matrix of f at $\vec{v}$* by

$$J_f(\vec{v}) = (a_{ij})$$

where $a_{ij} = \dfrac{\partial f_i}{\partial x_j}(\vec{v})$, $1 \le i \le p$, $1 \le j \le n$.

Here are some examples of Jacobian matrices.

3. Let $f:R^2 \to R^3$ be given by

$$f(x_1,x_2) = (x_1^2, x_1 x_2, x_2^2)$$

so that $f_1(x_1,x_2) = x_1^2$, $f_2(x_1,x_2) = x_1 x_2$, and $f_3(x_1,x_2) = x_2^2$. Then we have

$$\frac{\partial f_1}{\partial x_1} = 2x_1, \quad \frac{\partial f_2}{\partial x_1} = x_2, \quad \frac{\partial f_3}{\partial x_1} = 0$$

$$\frac{\partial f_1}{\partial x_2} = 0, \quad \frac{\partial f_2}{\partial x_2} = x, \quad \frac{\partial f_3}{\partial x_2} = 2x_2$$

Thus

$$J_f(x_1,x_2) = \begin{pmatrix} 2x_1 & 0 \\ x_2 & x_1 \\ 0 & 2x_2 \end{pmatrix}$$

4. Let $g:R^3 \to R^3$ be given by

$$g(x_1,x_2,x_3) = (x_1 \cos x_2, x_2 e^{x_3}, \sin(x_1 x_3))$$

Then

$$J_g(x_1,x_2,x_3) = \begin{pmatrix} \cos x_2 & -x_1 \sin x_2 & 0 \\ 0 & e^{x_3} & x_2 e^{x_3} \\ x_3 \cos(x_1 x_3) & 0 & x_1 \cos(x_1 x_3) \end{pmatrix}$$

PROPOSITION 7.1. *Let U be an open subset of* R^n *and* $f:U \to R^p$ *a function defined by* $f(\vec{v}) = (f_1(\vec{v}),\ldots,f_p(\vec{v}))$, *where each* $f_i:U \to R$, $1 \le i \le p$. *Then if f is differentiable at a point* $\vec{v}$ *in U, each of the partial derivatives* $\frac{\partial f_i}{\partial x_j}(\vec{v})$ *exists,* $1 \le i \le p$, $1 \le j \le n$. *Furthermore,* $D_{\vec{v}}f:R^n \to R^p$ *is the linear transformation defined by the Jacobian matrix of f at* $\vec{v}$. *(See Section 2.6.)*

Proof. Let $L = D_{\vec{v}}f:R^n \to R^p$, and let $\vec{e}_1,\ldots,\vec{e}_n$ be the standard basis for R^n and $\vec{e}_1,\ldots,\vec{e}_p$ the standard basis for R^p. Suppose that (a_{ij}) is the matrix of L, so that

$$\vec{Le}_j = \sum_{i=1}^{p} a_{ij}\vec{e}_i$$

$1 \le j \le n$. Since f is differentiable at $\vec{v}$, it follows that

$$\lim_{h \to 0} \frac{\|f(\vec{v}+h\vec{e}_j)-f(\vec{v})-L(h\vec{e}_j)\|}{\|h\vec{e}_j\|} = 0$$

for each $j = 1,\ldots,n$. Equivalently, we have

$$\lim_{h \to 0} \frac{\|f(\vec{v}+h\vec{e}_j)-f(\vec{v})-hL(\vec{e}_j)\|}{h} = 0$$

Now

$$\frac{\|f(\vec{v}+h\vec{e}_j)-f(\vec{v})-hL(\vec{e}_j)\|}{h} = \left(\sum_{i=1}^{p}\left(\frac{f_i(\vec{v}+h\vec{e}_j)-f_i(\vec{v})}{h} - a_{ij}\right)^2\right)^{1/2}$$

For the right hand side to tend to 0, each term in the sum must tend
to 0. (See Exercise 11, Section 5.) Therefore

$$\frac{\partial f_i}{\partial x_j}(\vec{v}) = \lim_{h \to 0} \frac{f_i(\vec{v}+h\vec{e}_j)-f_i(\vec{v})}{h} = a_{ij}$$

for all i, j, $1 \le i \le p$, $1 \le j \le n$. This proves that $\dfrac{\partial f_i}{\partial x_j}(\vec{v})$, exists
and equals a_{ij}, completing the proof of Proposition 7.1.

Proposition 7.1 reduces the problem of computing the derivative
of a differentiable function $f:R^n \to R^p$ to that of computing partial
derivatives for the functions $f_1,\ldots,f_p$. As we have seen, a partial
derivative is simply an ordinary derivative with respect to one of the
variables (the others regarded as constant). Computing the derivative
of f thus becomes an exercise in elementary calculus.

We note that the existence of the partial derivatives of a function does not necessarily imply that the function is differentiable
or, for that matter, continuous. For example, let $f:R^2 \to R$ be defined
by

$$f(x_1,x_2) = \begin{cases} \dfrac{x_1 x_2}{x_1^2 + x_2^2} & \text{if } (x_1,x_2) \neq (0,0) \\ \\ 0 & \text{if } (x_1,x_2) = (0,0) \end{cases}$$

This function is not continuous at $(0,0)$ since $f(x,x) = 1/2$ for $x \neq 0$ and $f(0,0) = 0$. Thus, f is not differentiable at $(0,0)$ by Proposition 6.3. However, $f(0,0) = f(h,0) = f(0,h) = 0$ so that both partial derivatives of f exist at $(0,0)$ (and are equal to zero). In Chapter 6 we will show that if the partial derivatives of a function exist *and are continuous*, then the function is differentiable.

EXERCISES

1. Compute the partial derivatives of the following functions.

(a) $f(x_1,x_2) = x_1^2 x_2 + 2x_2^2 - 3$

(b) $f(x_1,x_2) = x_1 \sin(x_1 x_2)$

(c) $f(x_1,x_2) = \dfrac{x_1 + x_2}{x_1^2 + x_2^2}$, $(x_1,x_2) \neq (0,0)$

(d) $f(x_1,x_2,x_3) = x_1 \cos x_2 + x_2 \sin x_3$

(e) $f(x_1,x_2,x_3) = \sqrt{x_1}\, e^{(x_2 + x_1 x_2)}$

2. Compute the Jacobians of the following functions.

(a) $f:R \to R^2$, $f(x_1) = (\cos x_1,\ \sin x_1)$

(b) $f:R^2 \to R^2$, $f(x_1,x_2) = (x_1^2 + x_2,\ 2x_1 x_2 - x_2^2)$

(c) $f:R^2 \to R^3$, $f(x_1,x_2) = (x_1 x_2, x_1^2 + x_2^2 x_1,\ x_2 x_1^2)$

(d) $f:R^3 \to R^2$, $f(x_1,x_2,x_3) = (x_1 e^{x_2}, x_2^2 + x_3 \sin x_1)$

3. Let $f:R^2 \to R$ be defined by

$$f(x_1,x_2) = \begin{cases} \dfrac{x_1 x_2}{x_1^2 + x_2^2} & \text{if } (x_1,x_2) \neq (0,0) \\ \\ 0 & \text{if } (x_1,x_2) = (0,0) \end{cases}$$

(a) Prove that f is not continuous at $(0,0)$.

(b) Prove that both partial derivatives of f exist and equal 0 at $(0,0)$.

(c) Prove that the partial derivatives of f are not continuous at $(0,0)$.

4. (a). Let $f,g: R^n \to R^p$ be differentiable at $\vec{v}_0$ and let f_i and g_i be coordinate functions. Let a be a constant and show that

$$\frac{\partial}{\partial x_j}(f_i + g_i)(\vec{v}_0) = \frac{\partial f_i}{\partial x_j}(\vec{v}_0) + \frac{\partial g_i}{\partial x_j}(\vec{v}_0)$$

and

$$\frac{\partial}{\partial x_j}(af_i)(\vec{v}_0) = a\frac{\partial f_i}{\partial x_j}(\vec{v}_0)$$

(where a is a constant).

(b) Why don't (a) and Proposition 7.1 imply Proposition 6.2?

5. Let $\vec{v}_0$ be a unit vector ($\|\vec{v}_0\| = 1$) in the normed vector space V and let $f:V \to R$ a differentiable function. The *directional derivative of f at $\vec{v}$ in the direction $\vec{v}_0$ is defined to be the number*

$$\lim_{h \to 0} \frac{f(\vec{v}+h\vec{v}_0) - f(\vec{v})}{h}$$

where the limit exists.

(a) Show that if $V = R^n$ and $\vec{v}_0 = \vec{e}_j$, the directional derivative of f at $\vec{v}$ in the direction $\vec{v}_0$ is simply $\frac{\partial f}{\partial x_j}(\vec{v})$.

(b) Show that, if $V = R^2$ and $\vec{v}_0 = (\cos \theta, \sin \theta)$, the directional derivative of f at $\vec{v}$ in the direction $\vec{v}_0$ is

$$\frac{\partial f}{\partial x_1}(\vec{v}) \cos \theta + \frac{\partial f}{\partial x_2}(\vec{v}) \sin \theta$$

*6. Prove that the directional derivative of f at $\vec{v}$ in the direction $\vec{v}_0$ is $(D_{\vec{v}}f)(\vec{v}_0)$.

7.   Let U be an open subset of R^n, and $f:U \to R$ a differentiable function.   The *gradient* of f, grad $f:U \to R^n$, is defined by

$$(\text{grad}(f))(\vec{v}) = (\frac{\partial f}{\partial x_1}(\vec{v}),\ldots,\frac{\partial f}{\partial x_n}(\vec{v}))$$

(We often write ∇f for grad f.)   Prove that the directional derivative
of f at $\vec{v}$ in the direction $\vec{v}_0$ is maximized when

$$\vec{v}_0 = \frac{(\text{grad}(f))(\vec{v})}{\|(\text{grad}(f))(\vec{v})\|}$$

(You will need Theorem 4.6.2 here.)

8.   If V is any normed vector space and $f:V \to R$ is a function differentiable at $\vec{v}_0$, the *tangent plane* to the graph of f at $(\vec{v}_0, f(\vec{v}_0))$
is defined to consist of all pairs $(\vec{v}, y)$ in $V \times R$ satisfying the
equation

$$y - f(\vec{v}_0) = (D_{\vec{v}_0} f)(\vec{v} - \vec{v}_0) \tag{7.3}$$

(Recall that the graph of f is defined to be the subset of $V \times R$
defined by $G(f) = \{(\vec{v}, y) : f(\vec{v}) = y\}$; see Exercise 15, Section 3.)

(a)   Show that (7.3) defines the usual tangent line when $V = R^1$.
(See Figure 7.2.)

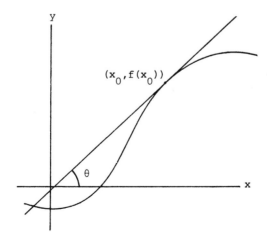

FIGURE 7.2.   The tangent line

(b) Suppose $V = R^n$ and $\vec{v}_0 = (a_1, \ldots, a_n)$. Show that equation
(7.3) can be written

$$y - f(\vec{v}_0) = \sum_{i=1}^{n} \left(\frac{\partial f}{\partial x_i}(\vec{v}_0) \right)(x_i - a_i)$$

(See Figure 7.3.)

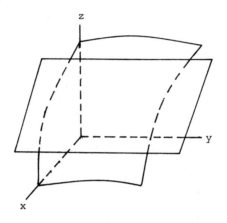

FIGURE 7.3. The tangent plane

CHAPTER 4

THE STRUCTURE OF VECTOR SPACES

INTRODUCTION

Now that we have described some of the important objects of study in this text, we need to investigate them in more detail.

When discussing linear transformations between Euclidean spaces (in Section 6 of Chapter 2), we found it useful to work with the standard basis $\vec{e}_1, \ldots, \vec{e}_n$ or R^n. The key property of this standard basis was that every vector $\vec{v} \in R^n$ was a linear combination of $\vec{e}_1, \ldots, \vec{e}_n$ in exactly one way. The concept of a collection of vectors with this property is extremely useful in arbitrary vector spaces (and not just in R^n); we call such a collection a *basis*. As we see in Section 2, any "finite dimensional" vector space has a basis. The number of elements in a basis for V is called the *dimension* of V. It turns out that this number is independent of the basis chosen.

One of the important uses of the existence of a basis for a vector space is in defining linear transformations. In fact, we show that a linear transformation $T:V \to W$ can be defined by simply assigning values to some basis for V. As a result, we prove that any *finite dimensional" vector space is isomorphic to a Euclidean Space.* This result allows us to assume, when convenient, that the finite dimensional vector spaces we deal with are Euclidean Spaces.

Sections 4, 5, and 6 are concerned with *inner product spaces;* these are vector spaces with the additional structure of an *inner product.* An inner product on V allows us to define a norm on V and also to define the angle between two vectors in V. We also show that any finite dimensional inner product space is isomorphic to a Euclidean Space with the so-called standard inner product.

The final section of the chapter is devoted to cross products on R^3. The results of this section will be needed later in the text.

1. SPANS AND LINEAR INDEPENDENCE

We saw in Section 2.6 that every vector in R^n can be written as a linear combination of the standard basis vectors $\vec{e}_1,\ldots,\vec{e}_n$ in exactly one way. This property was useful for expressing linear transformations as matrices. In this section, we begin the process of finding similar collections of vectors in arbitrary vector spaces. The property that the standard basis enjoys can be divided into two parts: that every vector $\vec{v}$ in R^n is a linear combination of the vectors $\vec{e}_1,\ldots,\vec{e}_n$, and that the expression $\vec{v} = \sum_{j=1}^{n} x_j \vec{e}_j$ is unique. It is more convenient to deal with these two parts separately.

Let V be a vector space, and let S be a nonempty subset (not necessarily a subspace) of V. The *span of* S (written Span(S)) is defined by

$$\text{Span}(S) = \{\vec{v} \in V : \vec{v} = r_1\vec{v}_1 + \cdots + r_k\vec{v}_k, \; r_i \in R, \; \vec{v}_i \in S, 1 \leq i \leq k\}$$

If the set S is the empty set ϕ, we declare Span(S) to be $\{0\}$. Of course, the set S may be infinite. However, infinite linear combinations of vectors are not permitted here or in the rest of this section, for the very good reason that we have no way of defining what they mean. Even when S is infinite, therefore, we are always taking finite linear combinations.

Note that we have already seen the span of a finite set in Example 8 of Section 2.4.

EXAMPLE

If $V = R^3$ and $S = \{(1,2,0),(2,4,0),(0,0,1)\}$, then Span(S) consists of all vectors of the form $r_1(1,2,0) + r_2(2,4,0) + r_3(0,0,1) =$

$(r_1 + 2r_2, 2r_1 + 4r_2, r_3)$. Letting $r_1 + 2r_2 = t_1$ and $r_3 = t_2$, we see easily that $\text{Span}(S) = \{(t_1, 2t_1, t_2) : t_1, t_2 \in R\}$.

PROPOSITION 1.1. $\text{Span}(S)$ *is a subspace of* V. *Furthermore, if* W *is any subspace of* V *containing* S, *then* W *contains* $\text{Span}(S)$.

Thus $\text{Span}(S)$ is the smallest subspace of V containing S.

Proof. We have seen a proof of this proposition in the case where S is finite and nonempty (see Example 2 of Section 2.4). The proof when S is infinite is almost identical, and we leave the details as Exercise 5. If $S = \phi$, then $\text{Span}(S) = \{0\}$ is a subspace of V, and is contained in every other subspace of V. So the proposition holds in this case, too.

In arguments with spans, we sometimes need to argue differently if S is the empty set. (This is reasonable, since $\text{Span}(S)$ was defined differently in that case.) The argument is almost always trivial, and we shall usually omit it in the future.

Remark. One sometimes uses the statement of Proposition 1.1 as the definition of the span: $\text{Span}(S)$ is the smallest subspace of V containing S. This has the advantage that we do not have to treat $S = \phi$ separately. On the other hand, we need to prove that there is some smallest subspace before this definition makes sense.

We next give two properties of the span of a set which will often be useful. We prove only the second of these, leaving the proof of the first as Exercise 6.

LEMMA 1.2.   *If* S_1 *and* S_2 *are subsets of* V *with* $S_1 \subset S_2$, *then* $\text{Span}(S_1) \subset \text{Span}(S_2)$.

LEMMA 1.3.   *Let* $\vec{v}_1, \ldots, \vec{v}_n$ *be any collection of vectors in* V, *and let* $\vec{w}$ *be a linear combination of* $\vec{v}_1, \ldots, \vec{v}_n$. *Then* $\text{Span}(\vec{v}_1, \ldots, \vec{v}_n, \vec{w}) = \text{Span}(\vec{v}_1, \ldots, \vec{v}_n)$. *

---

* When dealing with a finite set $\{\vec{v}_1, \ldots, \vec{v}_n\}$, we often write $\text{Span}(\vec{v}_1, \ldots, \vec{v}_n)$ instead of $\text{Span}(\{\vec{v}_1, \ldots, \vec{v}_n\})$.

Proof of Lemma 1.3. We give two proofs, one fairly concrete and one rather abstract.

Proof 1: $\text{Span}(\vec{v}_1,\ldots,\vec{v}_n) \subset \text{Span}(\vec{v}_1,\ldots,\vec{v}_n,\vec{w})$ by Lemma 1.2; the problem is the reverse inclusion. Since $\vec{w}$ is a linear combination of $\vec{v}_1,\ldots,\vec{v}_n$, we may write $\vec{w} = c_1\vec{v}_1 + \cdots + c_n\vec{v}_n$. If $\vec{v} \in \text{Span}(\vec{v}_1,\ldots,\vec{v}_n,\vec{w})$, then

$$\vec{v} = a_1\vec{v}_1 + \cdots + a_n\vec{v}_n + a\vec{w}$$
$$= a_1\vec{v}_1 + \cdots + a_n\vec{v}_n + a(c_1\vec{v}_1 + \cdots + c_n\vec{v}_n)$$
$$= (a_1 + ac_1)\vec{v}_1 + \cdots + (a_n + ac_n)\vec{v}_n$$

Thus $\vec{v} \in \text{Span}(\vec{v}_1,\ldots,\vec{v}_n)$ and $\text{Span}(\vec{v}_1,\ldots,\vec{v}_n,\vec{w}) \subset \text{Span}(\vec{v}_1,\ldots,\vec{v}_n)$.

Proof 2: Let $S_0 = \{\vec{v}_1,\ldots,\vec{v}_n\}$ and $S_1 = \{\vec{v}_1,\ldots,\vec{v}_n,\vec{w}\}$. As before, it is clear that $\text{Span}(S_0) \subset \text{Span}(S_1)$; we need to prove that $\text{Span}(S_1) \subset \text{Span}(S_0)$. Now $\vec{w} \in \text{Span}(S_0)$ by hypothesis, so that $S_1 \subset \text{Span}(S_0)$. Thus $\text{Span}(S_0)$ is a subspace of V containing S_1. However, $\text{Span}(S_1)$ is the smallest subspace of V containing S_1, by Proposition 1.1. Therefore $\text{Span}(S_1) \subset \text{Span}(S_0)$.

We say, reasonably enough, that a set S *spans* V if $\text{Span}(S) = V$. We can always find some set which spans V - for instance, V itself. We shall be interested, however, in finding a set S which spans V and which is as small as possible in some sense. To make this statement more precise, we define linear dependence.

The set $S = \{\vec{v}_1,\ldots,\vec{v}_k\} \subset V$ is said to be *linearly dependent* if we can find real numbers $r_1,\ldots,r_k$, not all zero, such that $r_1\vec{v}_1 + \cdots + r_k\vec{v}_k = \vec{0}$. (An expression of the form $r_1\vec{v}_1 + \cdots + r_k\vec{v}_k = \vec{0}$ with $r_1,\ldots,r_k$ not all zero is called a *dependence relation* for the vectors $\vec{v}_1,\ldots,\vec{v}_k$.) S is said to be *linearly independent* if it is not linearly dependent. The condition for S to be linearly independent, then, is the following: if $r_1,\ldots,r_k \in R$ are such that $r_1\vec{v}_1 + \cdots + r_k\vec{v}_k = \vec{0}$, then $r_1 = \cdots = r_k = 0$. This condition may seem confusing at first reading; it deserves some study.

An infinite subset S of V is said to be linearly dependent if some finite subset of S is linearly dependent; S is called linearly independent if it is not linearly dependent. Thus S is linearly independent if and only if each finite subset of S is linearly independent. In what follows, we shall be primarily concerned with the finite case. Here are some examples:

(1)   The set $S \subset R^3$ of our previous example,

$$S = \{(1,2,0),(2,4,0),(0,0,1)\}$$

is linearly dependent since $-2(1,2,0) + 1(2,4,0) + 0(0,0,1) = \vec{0}$.

(2)   The set $\{\vec{e}_1,\vec{e}_2\} \subset R^2$ is linearly independent. For if $(0,0) = r_1(1,0) + r_2(0,1) = (r_1,r_2)$, then $r_1 = r_2 = 0$. A similar argument proves that the set $\{\vec{e}_1,\dots,\vec{e}_n\}$ in R^n is linearly independent.

(3)   Any set S containing $\vec{0}$ is linearly dependent; for example, if $S = \{\vec{v}_1,\dots,\vec{v}_k\}$ with $\vec{v}_1 = \vec{0}$, then $1\vec{v}_1 + 0\vec{v}_2 + \dots + 0\vec{v}_k = \vec{0}$.

(4)   Suppose that S has only one element; thus $S = \{\vec{v}\}$. If $\vec{v} = \vec{0}$, S is linearly dependent. If not, S is linearly independent. To prove this, suppose that $r\vec{v} = \vec{0}$. If $r \neq 0$, then $\vec{v} = (\frac{1}{r})\vec{0} = \vec{0}$. So if $\vec{v} \neq \vec{0}$, then $r = 0$. That is, if $\vec{v} \neq \vec{0}$, then $r\vec{v} = \vec{0}$ implies $r = 0$. That means precisely that $\{\vec{v}\}$ is linearly independent.

(5)   The functions f_1,f_2,f_3 where $f_1(x) = 1$, $f_2(x) = x$, and $f_3(x) = x^2$, are linearly independent in $C^0(R)$. Here is one proof: suppose that

$$r_1f_1 + r_2f_2 + r_3f_3 = 0$$

Then for all real numbers x,

$$r_1 + r_2x + r_3x^2 = 0$$

But this is a quadratic equation, which has only two roots if $r_3 \neq 0$. Since every real number is a root, $r_3 = 0$. Thus

$$r_1 + r_2x = 0$$

This equation has only one root if $r_2 \neq 0$; therefore we must have $r_2 = 0$. It follows that

$$r_1 = 0$$

So $r_1 = r_2 = r_3 = 0$, and f_1, f_2, f_3 are linearly independent .

(6) Let $V = R^n$, and let $S = \{\vec{v}_1, \ldots, \vec{v}_k\}$. To say that S spans V is to say that for any $\vec{v} \in V$, one can find numbers $r_1, \ldots, r_k$ such that

$$\vec{v} = r_1\vec{v}_1 + \cdots + r_k\vec{v}_k$$

If this equation is written out in coordinates, one sees that it amounts to solving n equations (one per coordinate) in k unknowns (the r's). Clearly one is more likely to be able to solve them when k is large. We give an illustration below.

To say that S is linearly independent is to say that the only solution to $\vec{0} = r_1\vec{v}_1 + \cdots + r_k\vec{v}_k$ is $r_1 = 0, \ldots, r_k = 0$. The larger k gets, the more likely it is that there are more solutions. For linear independence, then, it is better for k to be small.

For example: if n = 2 and $\vec{v}_1 = (a_1, b_1)$, $\vec{v}_2 = (a_2, b_2)$, and $\vec{v}_3 = (a_3, b_3)$, then $\vec{v}_1, \vec{v}_2, \vec{v}_3$ span R^2 if, for any $\vec{v} = (a, b)$, we can solve the system of equations

$$r_1a_1 + r_2a_2 + r_3a_3 = a$$
$$r_1b_1 + r_2b_2 + r_3b_3 = b \qquad (1.1)$$

for r_1, r_2, r_3. Similarly, if $\vec{v}_1, \vec{v}_2, \vec{v}_3$ are to be linearly dependent, then we can find r_1, r_2, r_3, not all zero, which satisfy the system of homogeneous equations

$$r_1a_1 + r_2a_2 + r_3a_3 = 0$$
$$r_1b_1 + r_2b_2 + r_3b_3 = 0 \qquad (1.2)$$

As we use more vectors, it becomes more likely that we can solve (1.1). Unfortunately, it also becomes likely that (1.2) has nonzero solutions, or that the vectors are linearly dependent. The notions of spanning and linear independence, therefore, tend to work against one another. (Exercises 7 and 8 give more evidence to this effect.)

We have defined the notion of linear dependence and independence for sets of vectors. However, we often talk about vectors being linearly dependent (or independent); "$\vec{v}_1,\ldots,\vec{v}_n$ are linearly dependent" means that the set $\{\vec{v}_1,\ldots,\vec{v}_n\}$ is linearly dependent. (As we noted earlier, we often do the same with spans.)

It is easy to see that a set $S = \{\vec{v}_1,\ldots,\vec{v}_k\}$ is linearly dependent if and only if some $\vec{v}_j$ can be expressed as a linear combination of the others. We conclude this section with a modest but useful extension of this observation.

LEMMA 1.4. *Suppose* $\{\vec{v}_1,\ldots,\vec{v}_n\}$ *is a linearly dependent set of vectors in V, none of which is zero. Then for some* k, $1 < k \leq n$, $\vec{v}_k$ *can be expressed as a linear combination of* $\vec{v}_1,\ldots,\vec{v}_{k-1}$.

Proof. Let k be the smallest integer such that $\vec{v}_1,\ldots,\vec{v}_k$ are linearly dependent. Then k > 1, since no $\vec{v}_j$ is zero, and there is a dependence relation

$$a_1\vec{v}_1 + \cdots + a_k\vec{v}_k = \vec{0}$$

with not all of $a_1,\ldots,a_k$ zero. Now a_k cannot be zero, for then $\vec{v}_1,\ldots,\vec{v}_{k-1}$ would be linearly dependent, which would contradict the definition of k. Thus

$$\vec{v}_k = \frac{-a_1}{a_k}\vec{v}_1 - \frac{a_2}{a_k}\vec{v}_2 - \cdots - \frac{a_{k-1}}{a_k}\vec{v}_{k-1}$$

which proves the lemma.

EXERCISES

1.   Determine which of the following sets of vectors span R^3:

(a) $\{(1,2,1),(-2,0,3),(-1,6,9)\}$

(b) $\{(2,1,1),(2,1,2),(1,0,1)\}$

(c) $\{(1,2,3),(-1,0,0)\}$

(d) $\{(1,0,1),(0,1,0),(1,1,1),(1,-1,1)\}$

(e) $\{(1,2,3),(-1,-2,-3),(2,4,6),(3,5,7)\}$

(f) $\{(0,1,2),(-1,4,5)\}$

(g) $\{(0,1,3),(1,4,6),(-2,6,8),(1,3,9)\}$

(h) $\{(1,0,2),(0,1,5),(0,0,6)\}$

2. Determine which of the following sets of vectors are linearly independent:

(a) $\{(1,0),(1,1),(2,6)\} \subset R^2$

(b) $\{(1,1),(1,-1)\} \subset R^2$

(c) $\{(1,1,0),(0,1,1),(1,0,1)\} \subset R^3$

(d) $\{(1,0,2,1),(3,3,1,-1),(1,1,1,1)\} \subset R^4$

(e) $\{(1,3,5),(2,4,6),(3,5,7)\} \subset R^3$

(f) $\{(1,8,9),(-1,3,3)\} \subset R^3$

(g) $\{(1,0,2),(0,1,5),(0,0,6)\} \subset R^3$

(h) $\{(1,3),(5,4),(2,1)\} \subset R^2$

3. For which values of a and b are the following sets of vectors linearly dependent?

(a) $\{(1,a,3),(2,2,b)\}$

(b) $\{(0,2,-1),(-1,a,b),(2,0,3)\}$

(c) $\{(3,a,1),(1,0,1),(b,0,0)\}$

4. Determine which of the following sets of vectors are linearly independent.

(a) $\{1 - 2x^2,\ x + x^2,\ 2x - 1\} \subset P_3$

(b) $\{2x^3 - 3x^5,\ x^3 - x,\ 3x^5 - x^3 + 4x,\ x^5\} \subset P_5$

(c) $\{\sin x,\ \cos x,\ x^2\} \subset C^0(R)$

(d) $\{e^x, e^{2x}, e^{3x}, x-1\} \subset C^0(R)$

(e) $\left\{ \begin{pmatrix} 1 & 2 \\ 0 & 1 \end{pmatrix},\ \begin{pmatrix} 1 & -6 \\ 1 & -9 \end{pmatrix},\ \begin{pmatrix} -1 & 0 \\ 3 & 3 \end{pmatrix},\ \begin{pmatrix} 2 & 5 \\ -5 & 1 \end{pmatrix} \right\} \subset M(2,2)$

5.   Prove that the span of a subset S of a vector space V is the intersection of all subspaces of V containing S, and is therefore the smallest subspace of V containing S.

6.   Prove Lemma 1.2.

7.   Suppose a subset S spans V and $S \subset T$. Prove that T spans V.

8.   Suppose a subset $S \subset V$ is linearly independent and $T \subset S$. Prove that T is linearly independent.

9.   Suppose S is a linearly independent subset of V and $\vec{v} \notin \text{Span}(S)$. Prove $S \cup \{\vec{v}\}$ is independent.

10.   Prove that a set consisting of two vectors is linearly dependent if and only if one of the vectors is a multiple of the other.

11.   Suppose that $\text{Span}(S) = \text{Span}(T)$ and that $S \subset T$, $S \neq T$. Show that T is linearly dependent.

12.   Suppose that $\vec{u}, \vec{v}, \vec{w}$ are linearly independent vectors of V. Are $\vec{u} + \vec{v}, \vec{u} + \vec{w}, \vec{v} + \vec{w}$ linearly independent?

13.   (a) Show that two vectors $\vec{v}_1 = (x_1, x_2)$ and $\vec{v}_2 = (y_1, y_2) \in R^2$ span R^2 if and only if $x_1 y_2 - x_2 y_1 \neq 0$.

(b) Show that $\vec{v}_1, \vec{v}_2$ are linearly independent if and only if $x_1 y_2 - x_2 y_1 \neq 0$.

(c) Show that any three vectors in R^2 are linearly dependent.

*14.   State and prove the analogue of Lemma 1.3 for infinite sets.

*15.   Show that the functions $1, x, x^2, x^3, \ldots, x^n, \ldots$ are linearly independent in $C^0(R)$.

2. BASES

We continue our effort to abstract the properties of the vectors $\vec{e}_1, \ldots, \vec{e}_n$ in R^n; we begin by combining the notions of the last section. We define a *basis* of V to be a linearly independent set of vectors which spans V.

For example, $\{\vec{e}_1, \vec{e}_2\}$ is a basis for R^2. More generally, $\vec{e}_1, \ldots, \vec{e}_n$ is easily seen to be the basis for R^n. This fact explains the term "standard basis" used for these vectors.

As we have seen, $\{\vec{0}\}$ contains no nonempty linearly independent sets; on the other hand, $\mathrm{Span}(\phi) = \{\vec{0}\}$. Therefore we declare the empty set to be a basis of $\{\vec{0}\}$. This makes $\{\vec{0}\}$ something of a special case in what follows; some theorems do not apply to it without special interpretation, and often the proofs are different. *Henceforth,* $\{\vec{0}\}$ *is tacitly excluded from consideration whenever it causes trouble.*

PROPOSITION 2.1. *A set S is a basis for V if and only if every vector* $\vec{v} \in V$ *can be expressed as a linear combination of elements of S in exactly one way.* [*]

Proof. Suppose S is a basis for V. Then S spans V, and therefore any vector $\vec{v} \in V$ can be written as a linear combination of elements of $S: \vec{v} = \sum_{i=1}^{k} r_i \vec{v}_i$, with $r_i \in R$, $\vec{v}_i \in V$. We have to show that this expression is unique. Suppose that $\vec{v} = \sum_{i=1}^{k} s_i \vec{v}_i$ as well. [†] Then $\vec{0} = \sum_{i=1}^{k} (r_1 - s_i)\vec{v}_i$. Since S is linearly independent, $r_i = s_i$ for each i. Thus the two expressions are actually the same, and this proves uniqueness.

Conversely, suppose that every vector in V can be expressed in exactly one way as a linear combination of elements of S. Then S spans V. Next, suppose that $\vec{v}_1, \ldots, \vec{v}_k \in S$ and $r_1\vec{v}_1 + \cdots + r_k\vec{v}_k = \vec{0}$.

---

[*] Of course, we do not regard $\vec{v}_1$ and $\vec{v}_1 + 0\vec{v}_2$, say, as different expressions for $\vec{v}_1$.

[†] We may as well assume that the new sum involves the same elements of S as the old one, since if we need some new $\vec{v}_j$ for the second sum, we can put $0\vec{v}_j$ in the first sum.

Since $0\vec{v}_1 + \cdots + 0\vec{v}_k = \vec{0}$ and the expression for $\vec{0}$ is unique, $r_1 = r_2$ $= \cdots = r_k = 0$. Therefore S is linearly independent, and therefore S is a basis.

In order for the notion of a basis to be of any value, we must show that the vector spaces we will be dealing with have bases. The next result is the key to this fact.

PROPOSITION 2.2. *Suppose* $S = \{\vec{v}_1, \ldots, \vec{v}_k\}$ *spans* V. *Then we can select a subset of S which is a basis for* V.

Proof. The idea here is to throw out of S, one by one, those elements which can be expressed in terms of the others. Explicitly, we proceed by induction on the number of elements in S.

If $k = 1$ and $\vec{v}_1 \neq 0$, then $\vec{v}_1$ is independent and is therefore a basis for V. If $\vec{v}_1 = \vec{0}$, $V = \{\vec{0}\}$ and the empty subset of S is a basis for V.

Suppose now that $k > 1$ and that the proposition is true whenever the spanning set has $k - 1$ elements. Let $S = \{\vec{v}_1, \ldots, \vec{v}_k\}$ be a spanning set for V. If S contains $\vec{0}$, then $S - \{\vec{0}\}$ spans V and contains $k - 1$ elements; thus, by induction, it contains a basis. If S does not contain $\vec{0}$ and is linearly dependent, we use Lemma 1.4 to find a vector $\vec{v}_j \in S$ which is a linear combination of $\vec{v}_1, \ldots, \vec{v}_{j-1}$. Then, if $S' = S - \{\vec{v}_j\}$, Span $S' = $ Span $S = V$, by Lemma 1.3, and S' contains $k - 1$ elements. Again, by induction, S' contains a basis and the proposition is proved. Finally, if S is linearly independent, S is itself a basis.

We say a vector space V is *finite dimensional* if some finite set spans V. For example, R^n is finite dimensional, since $\{\vec{e}_1, \ldots, \vec{e}_n\}$ spans R^n.

As an immediate consequence of Proposition 2.2, we have the following.

COROLLARY. *Any finite dimensional vector space has a finite basis.*

One way of restating Proposition 2.2 is to say that any finite spanning set for a finite dimensional vector space can be shrunk to

a basis. As we see now, any linearly independent subset of a finite
dimensional vector space can be enlarged to a basis.

PROPOSITION 2.3. *Let* $S = \{\vec{v}_1,\ldots,\vec{v}_k\}$ *be a linearly independent
subset of a finite dimensional vector space* V. *Then there are vectors*
$\vec{v}_{k+1},\ldots,\vec{v}_n$ *such that*

$$\{\vec{v}_1,\ldots,\vec{v}_k,\vec{v}_{k+1},\ldots,\vec{v}_n\}$$

is a basis for V.

Sketch of proof. Since V is finite dimensional, we can find a
finite set $T = \{\vec{w}_1,\ldots,\vec{w}_m\}$ which spans V. Then

$$S \cup T = \{\vec{v}_1,\ldots,\vec{v}_k,\vec{w}_1,\ldots,\vec{w}_m\}$$

also spans V. Using Lemma 1.4 (since S $\cup$ T is clearly linearly dependent), we can find a vector in S $\cup$ T which is a linear combination
of the previous vectors (in the order above). Since S is linearly
independent, this vector must be one of the $\vec{w}$'s, say $\vec{w}_j$. Then S $\cup$ T
- $\{\vec{w}_j\}$ still spans V (by Lemma 1.3) and we can proceed just as in the
proof of Proposition 2.2 to obtain an independent set by throwing away
enough of the $\vec{w}$'s. The details are left to the reader (Exercise 10).

EXERCISES

1.   Select a basis for the appropriate Euclidean space from each of
the following spanning sets:

(a) $\{(1,1,2),(2,2,4),(1,0,2),(0,1,0),(0,0,1),(1,1,1)\}$

(b) $\{(1,0,-2),(-3,0,6),(5,1,1),(8,1,-5),(1,1,9),(3,1,2),(6,4,-1)\}$

(c) $\{(1,0,1,1),(2,0,3,0),(0,0,1,-2),(3,0,4,1),(2,0,4,1),$
      $(3,1,0,1),(6,1,1,5)\}$

(d) $\{(1,2,-2,-2),(2,0,1,5),(4,4,-3,1),(6,8,-7,-3),(5,2,-4,3),$
      $(7,6,-8,-1),(3,1,7,1),(4,-10,7,-1)\}$

2.   Enlarge the independent set S to a basis for the vector space
V, where

(a) $S = \{(1,2)\}$, $V = R^2$

(b) $S = \{(1,0,1)\}$, $V = R^3$

(c) $S = \{(1,0,1),(2,1,1)\}$, $V = R^3$

(d) $S = \{(2,0,5,1),(-1,3,3,1)\}$, $V = R^4$

(e) $S = \{(1,0,1,1),(1,4,3,-1),(2,2,-3,7)\}$, $V = R^4$

(f) $S = \{1 - x^2\}$, $V = P_2$

(g) $S = \{1 - x^2, 1 + x^2 + x^3\}$, $V = P_3$

(h) $S = \left\{ \begin{pmatrix} 2 & 1 \\ 1 & 1 \end{pmatrix} \right\}$, $V = m(2,2)$

(i) $S = \left\{ \begin{pmatrix} 3 & -1 \\ 1 & 2 \end{pmatrix}, \begin{pmatrix} -2 & 2 \\ -5 & 1 \end{pmatrix} \right\}$, $V = m(2,2)$

3.   Prove that the set

$$\{1 + x,\ 1 + x^2, x^2 + x^3, x - x^3\}$$

is a basis for P_3.

4.   (a) Find a basis for the vector space of solutions to the equations

$$3x - 4y + 4z = 0$$
$$5x - 10y + 15z = 0$$
$$x + 2y - 7z = 0$$

(b) Complete the basis of part (a) to a basis for R^3.

5.   (a) Find a basis for the vector space of solutions to the equations

$$2w - 3x + 4y + z = 0$$
$$5w + 9x - 11y + 4z = 0$$
$$5w - 2x + 3y + 3z = 0$$
$$3w + x - y + 2z = 0$$

(b) Complete the basis of part (a) to a basis for R^4.

6.   (a) Find a basis for the kernel and the image of the linear
transformation $T:R^4 \to R^4$ where T is defined by the matrix

$$\begin{pmatrix} 2 & 1 & 1 & -3 \\ -1 & 7 & -5 & 0 \\ 0 & -5 & 3 & 1 \\ 1 & 3 & -1 & -2 \end{pmatrix}$$

(b) Complete each of the bases of part (a) to bases for R^4.

7.   Show that the set $S = \{\vec{v}_1, \vec{v}_2\} \subset R^2$ is a basis if and only if
$\vec{e}_1 = (1,0)$ and $\vec{e}_2 = (0,1)$ are linear combinations of $\vec{v}_1$ and $\vec{v}_2$.

*8.   Show that the set $S = \{\vec{v}_1, \vec{v}_2, \vec{v}_3\} \subset R^3$ is a basis if and only if
$\vec{e}_1, \vec{e}_2$, and $\vec{e}_3$ are linear combinations of $\vec{v}_1, \vec{v}_2$, and $\vec{v}_3$.

9.   The vectors $\vec{v}_1, \ldots, \vec{v}_n \in R^n$ are called *triangular* if

$$\vec{v}_1 = (1, x_{12}, x_{13}, \ldots, x_{1n})$$
$$\vec{v}_2 = (0, 1, x_{23}, \ldots, x_{2n})$$
$$\vec{v}_3 = (0, 0, 1, x_{34}, \ldots, x_{3n})$$
$$\ldots$$
$$\vec{v}_n = (0, 0, \ldots, 1)$$

(That is, if $\vec{v}_j = (x_{j1}, x_{j2}, \ldots, x_{jn})$, then $x_{jk} = 0$ if $k < j$ and $x_{jj} = 1$.)
Show that if $\{\vec{v}_1, \ldots, \vec{v}_n\}$ is a triangular set of vectors in R^n, it is
a basis.

*10.   Write out a formal proof of Proposition 2.3.

11.   Let $\vec{v}$ be any nonzero vector in a finite-dimensional vector space
V. Show that V has a basis containing $\vec{v}$.

12.   An n × n matrix $A = (a_{ij})$ is *symmetric* if $a_{ij} = a_{ji}$ for all
$1 \le i, j \le n$. Prove that the set of symmetric n × n matrices is a
subspace of $\mathbb{m}(n,n)$ and find a basis for it.

13.   An n × n matrix $A = (a_{ij})$ is *skew-symmetric* if $a_{ij} = -a_{ji}$ for
all $1 \le i, j \le n$. Prove that the set of skew-symmetric n × n matrices is a subspace of $\mathbb{m}(n,n)$ and find a basis for it.

3.   BASES AND LINEAR TRANSFORMATIONS

According to the corollary to Proposition 2.2, any finite dimensional vector space has a basis. This is an extremely useful fact. We show here how to use it to define linear transformations between vector spaces. In Chapter 7, we carry these ideas much further.

PROPOSITION 3.1. *Let* $\{\vec{v}_1,\ldots,\vec{v}_n\}$ *be a basis for the vector space* V *and let* $\vec{x}_1,\ldots,\vec{x}_n$ *be arbitrary elements in the vector space* W. *Then there is a unique linear transformation* $T:V \to W$ *with* $T(\vec{v}_i) = \vec{x}_i$, $1 \le i \le n$.

We note that the elements $\vec{x}_1,\ldots,\vec{x}_n$ need not be distinct; in fact, they can all be the same.

Proof. We prove uniqueness first. Suppose then that we have two linear transformations $T,T':V \to W$ with $T(\vec{v}_j) = T'(\vec{v}_j) = \vec{x}_j$, $1 \le j \le n$. Let v be an arbitrary vector in V. Since $\vec{v}_1,\ldots,\vec{v}_n$ is a basis for V, we can write

$$\vec{v} = \sum_{j=1}^{n} r_j\vec{v}_j$$

for some $r_j \in R$, $1 \le j \le n$. But then

$$
\begin{aligned}
T(\vec{v}) &= T(\sum_{j=1}^{n} r_j\vec{v}_j) \\
&= \sum_{j=1}^{n} r_j T(\vec{v}_j) \\
&= \sum_{j=1}^{n} r_j\vec{x}_j \\
&= \sum_{j=1}^{n} r_j T'(\vec{v}_j) \\
&= T'(\sum_{j=1}^{n} r_j\vec{v}_j) = T'(\vec{v})
\end{aligned}
$$

Thus $T(\vec{v}) = T'(\vec{v})$ for all $\vec{v}$ in V, so $T = T'$.

The uniqueness proof shows us how to proceed with existence: to define $T(\vec{v})$ for an arbitrary vector $\vec{v}$, we write

$$\vec{v} = \sum_{j=1}^{n} r_j \vec{v}_j$$

and define

$$T(\vec{v}) = \sum_{j=1}^{n} r_j \vec{x}_j$$

Since there is only one way to write $\vec{v}$ as a linear combination of $\vec{v}_1, \ldots, \vec{v}_n$, $T(\vec{v})$ is well-defined.

We need to check the linearity of T. If

$$\vec{v} = \sum_{j=1}^{n} r_j \vec{v}_j$$

and

$$\vec{v}' = \sum_{j=1}^{n} r'_j \vec{v}_j$$

then, for any $s, t \in R$,

$$
\begin{aligned}
T(s\vec{v} + t\vec{v}') &= T(\sum_{j=1}^{n} (sr_j + tr'_j)\vec{v}_j) \\
&= \sum_{j=1}^{n} (sr_j + tr'_j)\vec{x}_j \\
&= s\sum_{j=1}^{n} r_j \vec{x}_j + t\sum_{j=1}^{n} r'_j \vec{x}_j \\
&= sT(\vec{v}) + tT(\vec{v}')
\end{aligned}
$$

Thus T is a linear transformation. Moreover, $T(\vec{v}_j) = \vec{x}_j$, all j, and therefore T satisfies the requirements of the proposition.

Remarks. (i) The key fact in proving uniqueness is that $\{\vec{v}_1, \ldots, \vec{v}_n\}$ spans V. The key fact in proving existence is that

$\{\vec{v}_1,\ldots,\vec{v}_n\}$ is a linearly independent set. If we could write

$$\vec{v} = \sum_{j=1}^{n} r_j \vec{v}_j = \sum_{j=1}^{n} s_j \vec{v}_j$$

with $r_j \neq s_j$ for some j, then we would have two distinct definitions for $T(\vec{v})$:

$$T(\vec{v}) = \sum_{j=1}^{n} r_j \vec{x}_j$$

$$T(\vec{v}) = \sum_{j=1}^{n} s_j \vec{x}_j$$

and $T(\vec{v})$ would not be well-defined. However, since $\vec{v}_1,\ldots,\vec{v}_n$ form a basis, this unpleasant situation does not exist.

(ii) Note that we did not need to require that W was finite dimensional in Proposition 3.1.

Quite clearly, properties of the elements $\vec{x}_1,\ldots,\vec{x}_n$ of Proposition 3.1 will be reflected in properties of the linear transformation T. An example of this is the following.

PROPOSITION 3.2. *Let* $\{\vec{v}_1,\ldots,\vec{v}_n\}$ *be a basis for V, let* $\vec{x}_1,\ldots,\vec{x}_n$ *be elements of W, and let* T:V → W *be the linear transformation with* $T(\vec{v}_j) = \vec{x}_j$, $1 \leq j \leq n$. *If the* $\vec{x}_1,\ldots,\vec{x}_n$ *are linearly independent vectors, then* T *is injective. If* $\vec{x}_1,\ldots,\vec{x}_n$ *span W, then* T *is surjective.*

Proof. Suppose that $\vec{x}_1,\ldots,\vec{x}_n$ are linearly independent vectors and that $T(\vec{v}) = \vec{0}$, where $\vec{v} = a_1\vec{v}_1 + \cdots + a_n\vec{v}_n$. Then

$$\vec{0} = T(\vec{v}) = T(\sum_{j=1}^{n} a_j \vec{v}_j)$$

$$= \sum_{j=1}^{n} a_j \vec{x}_j$$

Since $\vec{x}_1,\ldots,\vec{x}_n$ are independent, we must have $a_1 = a_2 = \cdots = a_n = 0$; therefore $\vec{v} = \vec{0}$. Thus $h(T) = \{\vec{0}\}$ and T is injective (Proposition 2.5.3).

If $\vec{x}_1,\ldots,\vec{x}_n$ span W, then any $\vec{w} \in W$ can be expressed as a linear combination of $\vec{x}_1,\ldots,\vec{x}_n$. Thus

$$\vec{w} = \sum_{j=1}^{n} a_j \vec{x}_j$$

$$= \sum_{j=1}^{n} a_j T(\vec{v}_j)$$

$$= T(\sum_{j=1}^{n} a_j \vec{v}_j)$$

and $\vec{w} \in \mathcal{J}(T)$. Therefore T is surjective.

Recall that a linear transformation $T:V \to W$ is an *isomorphism* if it is bijective; that is both injective and surjective. We then have the following consequence of Proposition 3.2.

COROLLARY. *Suppose that V and W have bases with the same number of elements; let* $\vec{v}_1,\ldots,\vec{v}_n$ *be a basis for V, and* $\vec{w}_1,\ldots,\vec{w}_n$ *a basis for W, and let* $T:V \to W$ *be the linear transformation with* $T(\vec{v}_j) = \vec{w}_j$, $1 \le j \le n$. *Then T is an isomorphism.*

If $T:V \to W$ is an isomorphism, we can define $T^{-1}:W \to V$ in the usual way; $T^{-1}(\vec{w}) = \vec{v}$ if and only if $T(\vec{v}) = \vec{w}$.

PROPOSITION 3.3. *If* $T:V \to W$ *is an isomorphism, then* $T^{-1}:W \to V$ *is also an isomorphism.*

Proof. The transformation T^{-1} is clearly bijective; we prove T^{-1} linear.

Suppose $\vec{v}_1,\vec{v}_2 \in V$, $\vec{w}_1,\vec{w}_2 \in W$ with $T(\vec{v}_1) = \vec{w}_1$, $T(\vec{v}_2) = \vec{w}_2$. We then have

$$T(a\vec{v}_1 + b\vec{v}_2) = a\vec{w}_1 + b\vec{w}_2$$

for any real numbers a and b. It follows that

$$T^{-1}(a\vec{w}_1 + b\vec{w}_2) = a\vec{v}_1 + b\vec{v}_2$$
$$= aT^{-1}(\vec{w}_1) + bT^{-1}(\vec{w}_2)$$

and the proof is complete.

We say that the vector space V is *isomorphic* to the vector space W if there is an isomorphism $T:V \to W$. Clearly, any vector space V is isomorphic to itself, since the identity mapping is an isomorphism. As a consequence of Proposition 3.1, we see that if V is isomorphic to W, then W is isomorphic to V. We thus can say simply that V and W are isomorphic.

PROPOSITION 3.4. *Any finite dimensional vector space is isomorphic to a Euclidean n-space R^n for some n.*

We shall prove in the next section that this integer n is unique.

Proof. Let V be a finite dimensional vector space with basis $\{\vec{v}_1,\ldots,\vec{v}_n\}$, and let $\{\vec{e}_1,\ldots,\vec{e}_n\}$ be the standard basis for R^n. Define $T:V \to R^n$ to be the unique transformation with $T(\vec{v}_j) = \vec{e}_j$, $1 \le j \le n$. Then T is an isomorphism by the corollary to Proposition 3.2.

The next result is a converse to Proposition 3.2.

PROPOSITION 3.5. *Let $T:V \to W$ be a linear transformation and let $\vec{v}_1,\ldots,\vec{v}_n$ be vectors in V. If T is injective and $\vec{v}_1,\ldots,\vec{v}_n$ are linearly independent, then $T(\vec{v}_1),\ldots,T(\vec{v}_n)$ are linearly independent. If T is surjective and $\vec{v}_1,\ldots,\vec{v}_n$ span V, then $T(\vec{v}_1),\ldots,T(\vec{v}_n)$ span W.*

COROLLARY. *If $T:V \to W$ is an isomorphism and $\vec{v}_1,\ldots,\vec{v}_n$ a basis for V, then $T(\vec{v}_1),\ldots,T(\vec{v}_n)$ is a basis for W.*

Proof. Suppose $a_1 T(\vec{v}_1) + \cdots + a_n T(\vec{v}_n) = \vec{0}$. Then

$$\vec{0} = \sum_{i=1}^{n} a_i T(\vec{v}_i)$$

$$= T(\sum_{i=1}^{n} a_i \vec{v}_i)$$

Since T is injective, $\sum_{i=1}^{n} a_i \vec{v}_i = \vec{0}$, and, since $\{\vec{v}_1,\ldots,\vec{v}_n\}$ is linearly independent, $a_1 = a_2 = \cdots = a_n = 0$. Thus $\{T(\vec{v}_1),\ldots,T(\vec{v}_n)\}$ is linearly independent.

To prove the second part of the proposition, let $\vec{w}$ be any vector in W. Because T is surjective, there is a vector $\vec{v}$ in V with $T(\vec{v}) = \vec{w}$ and, since $\{\vec{v}_1,\ldots,\vec{v}_n\}$ spans V, we can write $\vec{v} = \Sigma_{j=1}^{n} a_j \vec{v}_j$. Thus

$$\vec{w} = T(\vec{v}) = T(\sum_{j=1}^{n} a_j \vec{v}_j)$$

$$= \sum_{j=1}^{n} a_j T(\vec{v}_j)$$

Therefore $\{T(\vec{v}_1),\ldots,T(\vec{v}_n)\}$ spans W.

EXERCISES

1. Define a linear transformation $T:R^2 \to R^3$ taking $\vec{v}$ into $\vec{w}$, where

    (a) $\vec{v} = (1,0)$, $\vec{w} = (1,2,-2)$

    (b) $\vec{v} = (1,1)$, $\vec{w} = (2,0,2)$

    (c) $\vec{v} = (2,-1)$, $\vec{w} = (1,1,0)$

    (d) $\vec{v} = (2,-1)$, $\vec{w} = (1,-1,-2)$

    (e) $\vec{v} = (3,-2)$, $\vec{w} = (7,4,-5)$

2. Define a linear transformation $T:R^3 \to R^3$ taking $\vec{v}_1$ into $\vec{w}_1$ and $\vec{v}_2$ into $\vec{w}_2$, where

    (a) $\vec{v}_1 = (1,0,0)$, $\vec{v}_2 = (0,1,0)$, $\vec{w}_1 = \vec{w}_2 = (2,3,3)$

    (b) $\vec{v}_1 = (1,0,0)$, $\vec{v}_2 = (0,1,0)$, $\vec{w}_1 = (1,0,-1)$, $\vec{w}_2 = (1,2,3)$

    (c) $\vec{v}_1 = (1,1,1)$, $\vec{v}_2 = (2,-1,-1)$, $\vec{w}_1 = (2,-1,3)$, $\vec{w}_2 = (1,1,1)$

    (d) $\vec{v}_1 = (1,-1,2)$, $\vec{v}_2 = (2,1,-1)$, $\vec{w}_1 = (7,4,3)$, $\vec{w}_2 = (-7,3,1)$

    (e) $\vec{v}_1 = (2,3,1)$, $\vec{v}_2 = (1,5,-2)$, $\vec{w}_1 = (4,0,2)$, $\vec{w}_2 = (1,-3,-2)$

3. Define two distinct linear transformations $S,T:R^3 \to R^3$

    (a) taking $(1,0,0)$ into $(2,1,-4)$

    (b) taking $(2,1,3)$ into $(1,1,-2)$

    (c) taking $(1,0,1)$ into $(2,1,3)$ and $(1,1,0)$ into $(7,1,0)$

    (d) taking $(3,1,1)$ into $(1,-1,-1)$ and $(2,0,2)$ into $(2,3,-2)$

4.   Suppose V_1, V_2, V_3 are vector spaces with V_1 isomorphic to V_2 and V_2 isomorphic to V_3. Prove that V_1 is isomorphic to V_3.

*5.   The *dual space* V^* of a vector space V is defined to be the vector space of all linear functionals, that is, linear transformations $V \to R$. Let $\vec{v}_1, \ldots, \vec{v}_n$ be a basis for V and define $\alpha_i \in V^*$, $1 \le i \le n$, to be the unique linear transformation with

$$\alpha_i(\vec{v}_j) = \begin{cases} 1 & \text{if } i = j \\ 0 & \text{if } i \ne j \end{cases}$$

Prove that $\alpha_1, \ldots, \alpha_n$ is a basis for V^*. (This basis is called the *dual basis* to the basis $\vec{v}_1, \ldots, \vec{v}_n$.)

6.   Prove that, if V is finite-dimensional, then V is isomorphic to V^*.

*7.   Let V and W be finite dimensional vector spaces, with $T: V \to W$ a linear transformation. Prove that bases $\{\vec{v}_1, \ldots, \vec{v}_n\}$ for V and $\{\vec{w}_1, \ldots, \vec{w}_m\}$ for W exist such that

$$T(\vec{v}_i) = \begin{cases} \vec{w}_i & \text{for } 1 \le i \le k \\ \vec{0} & \text{for } k < i \le n \end{cases}$$

where k is some integer, $0 \le k \le n$. (If $k = 0$, then $T(\vec{v}_i) = \vec{0}$ for $1 \le i \le n$.)

4.   THE DIMENSION OF A VECTOR SPACE

We think of R^2 as a "two-dimensional" space; similarly, R^3 is three-dimensional, and R^n should have dimension n. One way of making this precise is to notice that the standard basis of R^n has n vectors. In fact, every basis of a finite-dimensional space has the same number of elements, and we can use this fact to define the dimension of a vector space.

The key result we need to prove is the following:

THEOREM 4.1. *Suppose* $\{\vec{v}_1,\ldots,\vec{v}_k\}$ *span* V *and that* $\{\vec{w}_1,\ldots,\vec{w}_r\}$
is a linearly independent subset of V. *Then* $r \leq k$.

COROLLARY. *Any two bases for a finite dimensional vector space
contain the same number of elements.*

Proof of the corollary. If $\alpha = \{\vec{v}_1,\ldots,\vec{v}_n\}$ and $\beta = \{\vec{w}_1,\ldots,\vec{w}_m\}$
are two bases for V then $n \leq m$ since α is linearly independent and
β spans V. However, α also spans V and β is linearly independent, so
$m \leq n$. Thus $m = n$.

We define the *dimension* of any finite dimensional vector space
(written dim V) to be the number of elements in any basis for V. The
corollary above tells us that this notion is well defined. Notice
that R^n does indeed have dimension n under this definition.

Proof of Theorem 4.1. Consider the set of vectors $\{\vec{w}_r,\vec{v}_1,\ldots,\vec{v}_k\}$.
This set spans V and is linearly dependent, since $\vec{w}_r$ can be expressed
as a linear combination of $\vec{v}_1,\ldots,\vec{v}_k$. Applying Lemma 1.4, we see that
some $\vec{v}_j$ can be expressed in terms of $\vec{w}_r,\vec{v}_1,\ldots,\vec{v}_{j-1}$. By re-indexing
the $\vec{v}$'s if necessary, we can assume $j = k$, so that $\{\vec{w}_r,\vec{v}_1,\ldots,\vec{v}_{k-1}\}$
spans V (by Lemma 1.3).

Next we consider the set $\{\vec{w}_{r-1},\vec{w}_r,\vec{v}_1,\ldots,\vec{v}_{k-1}\}$. This set spans
V and is linearly dependent, so we can again apply Lemma 1.4 to ex-
press one of the vectors as a linear combination of the previous vec-
tors. This vector cannot be $\vec{w}_r$, since $\{\vec{w}_1,\ldots,\vec{w}_r\}$ is linearly independent. So, reindexing if necessary, we conclude as above that
$\{\vec{w}_{r-1},\vec{w}_r,\vec{v}_1,\ldots,\vec{v}_{k-2}\}$ span V.

Continuing in this way[*] , adding a $\vec{w}$ to the beginning of the list
and removing a $\vec{v}$, we eventually run out of either the $\vec{v}$'s or the $\vec{w}$'s.
Suppose that the theorem is false - that is, that $r > k$. Then eventually we find that

---

[*] This phrase conceals a use of mathematical induction.

$$\{\vec{w}_{r-k+1}, \ldots, \vec{w}_r, \vec{v}_1\}$$

spans V and is linearly dependent. Thus one of the vectors in this
set can be expressed as a linear combination of the previous vectors.
That vector must be $\vec{v}_1$, since the $\vec{w}$'s are linearly independent. Lemma
1.3 now tells us that $\{\vec{w}_{r-k+1}, \ldots, \vec{w}_r\}$ spans V. In particular, $\vec{w}_{r-k}$
is a linear combination of $\vec{w}_{r-k+1}, \ldots, \vec{w}_r$. This contradicts the assumption that $\{\vec{w}_1, \ldots, \vec{w}_r\}$ is linearly independent.

It follows that $r \leq k$, and the theorem is proved.

We can now prove a strengthened form of Proposition 3.4.

THEOREM 4.2. *Let* V *and* W *be finite dimensional vector spaces.*
Then V *and* W *are isomorphic if and only if* dim V = dim W.

COROLLARY. *Any vector space of dimension* n *is isomorphic to*
Euclidean n-space R^n.

Proof of Theorem 4.2. If V and W are isomorphic, then dim V =
dim W by the Corollary to Proposition 3.5.

If dim V = dim W, we let $\vec{v}_1, \ldots, \vec{v}_n$ be a basis for V and $\vec{w}_1, \ldots, \vec{w}_n$
be a basis for W. The linear transformation T:V → W with $T(\vec{v}_i) = \vec{w}_i$,
$1 \leq i \leq n$, (which exists by Proposition 3.1) is an isomorphism, by
the Corollary to Proposition 3.2.

This last result deserves some comment. It tells us that, once
we choose a basis, any n-dimensional vector space is isomorphic to
R^n. Since isomorphic vector spaces are indistinguishable from the
point of view of linear algebra, we can, when convenient, assume that
the finite dimensional vector spaces we deal with are Euclidean Spaces.
We shall take advantage of this observation frequently in what follows.

To prove that a set of vectors is a basis for a vector space V,
we need to show that it is linearly independent and spans V. In some
special cases, however, we can cut our work in half.

PROPOSITION 4.3. *Let* V *be a vector space of dimension* n, *and let*
$\alpha = \{\vec{v}_1, \ldots, \vec{v}_n\}$ *be a set of* n *vectors.*

(a) *If α is linearly independent, then α spans V (and is a basis).*

(b) *If α spans V, then α is linearly independent (and is a basis).*

In other words, if we take n vectors in an n-dimensional vector space, then either of the two properties of a basis (linear independence or spanning) implies the other.

Proof. We shall just prove (a), leaving (b) as Exercise 1. Suppose that α is linearly independent, but does not span V. Proposition 2.3 says that we can add vectors $\vec{w}_1, \ldots, \vec{w}_k$ so that $\alpha' = \alpha \cup \{\vec{w}_1, \ldots, \vec{w}_k\}$ is a basis for V. But Theorem 3.1 says that α' has n elements. However, α' has $n + k$ elements, and $k > 0$ by hypothesis. This gives a contradiction, and the result follows.

We now prove some results about subspaces of finite dimensional vector spaces.

PROPOSITION 4.4. *Let W be a subspace of the finite dimensional vector space V. Then W is finite dimensional and* $\dim W \leq \dim V$.

Proof. We construct a basis $\vec{w}_1, \ldots, \vec{w}_m$ of W as follows. Let $\vec{w}_1$ be a non-zero vector in W. (The result is trivial if W is the zero subspace.) If $\text{Span}\{\vec{w}_1\} = W$, $\{\vec{w}_1\}$ is a basis for W.

If $\text{Span}\{\vec{w}_1\} \neq W$, let $\vec{w}_2$ be a vector in W, $\vec{w}_2 \notin \text{Span}\{\vec{w}_1\}$. Then, according to Lemma 1.3, $\vec{w}_1, \vec{w}_2$ are independent. (If $\vec{w}_1, \vec{w}_2$ were linearly dependent, then, by Lemma 1.4, $\vec{w}_2$ would be a multiple of $\vec{w}_1$, and $\text{Span}\{\vec{w}_1\}$ would equal $\text{Span}\{\vec{w}_1, \vec{w}_2\}$.) If $\text{Span}\{\vec{w}_1, \vec{w}_2\} = W$, then $\{\vec{w}_1, \vec{w}_2\}$ is a basis for W. If $\text{Span}\{\vec{w}_1, \vec{w}_2\} \neq W$, we choose $\vec{w}_3 \in W$, $\vec{w}_3 \notin \text{Span}\{\vec{w}_1, \vec{w}_2\}$ Again, $\vec{w}_1, \vec{w}_2, \vec{w}_3$ are independent, by the same sort of argument given above.

We continue constructing vectors in this way as long as we can. If $\vec{w}_1, \ldots, \vec{w}_k$ have been constructed, we stop if $\text{Span}\{\vec{w}_1, \ldots, \vec{w}_k\} = W$; if not, we choose $\vec{w}_{k+1} \in W$, $\vec{w}_{k+1} \notin \text{Span}\{\vec{w}_1, \ldots, \vec{w}_k\}$. Since there can be at most $n = \dim V$ independent vectors in V, this process must lead to a basis $\vec{w}_1, \ldots, \vec{w}_m$ for W with $m \leq n$.

The next result is particularly useful.

PROPOSITION 4.5. *Let W be a subspace of the finite dimensional vector space V. Then there is a basis $\vec{v}_1,\ldots,\vec{v}_n$ for V such that $\vec{v}_1,\ldots,\vec{v}_m$ is a basis for W.*

Proof. According to Proposition 4.4, W is a finite dimensional vector space, so it has a basis, $\vec{v}_1,\ldots,\vec{v}_m$. Use the Corollary to Proposition 2.3 to find vectors $\vec{v}_{m+1},\ldots,\vec{v}_n$ such that $\vec{v}_1,\ldots,\vec{v}_m, \vec{v}_{m+1}, \ldots,\vec{v}_n$ form a basis for V.

We can use this last result to prove a theorem about linear transformations, the Rank-Nullity Theorem.

Let $T:V \to W$ be a linear transformation. We define the *rank* of T, rank(T), to be the dimension of the image $\mathcal{J}(T)$ of T and the *nullity* of T, null(T), to be the dimension of the nullspace $\hbar(T)$ of T. Note that T is injective if and only if null(T) = 0 and T is surjective exactly when rank(T) = dim W.

THEOREM 4.6. *Let V and W be finite dimensional vector spaces, with $T:V \to W$ a linear transformation. Then*

rank(T) + null(T) = dim V

Proof. Let $\vec{v}_1,\ldots,\vec{v}_k,\ldots,\vec{v}_n$ be a basis for V such that $\vec{v}_1,\ldots,\vec{v}_k$ give a basis for $\hbar(T)$. (Then null(T) = k.) We claim that $T(\vec{v}_{k+1}),$ $\ldots,T(\vec{v}_n)$ give a basis for $\mathcal{J}(T)$.

First of all, if w is any element in $\mathcal{J}(T)$, we can write $\vec{w} = T(\vec{v})$ for some $\vec{v} \in V$. Now,

$$\vec{v} = \sum_{j=1}^{n} a_j\vec{v}_j$$

so that

$$\vec{w} = T(\vec{v})$$

$$= T(\sum_{j=1}^{n} a_j\vec{v}_j)$$

$$= \sum_{j=1}^{n} a_j T(\vec{v}_j)$$

$$= a_{k+1} T(\vec{v}_{k+1}) + \cdots + a_n T(\vec{v}_n)$$

since $T(\vec{v}_j) = \vec{0}$ for $1 \leq j \leq k$. Thus $T(\vec{v}_{k+1}), \ldots, T(\vec{v}_n)$ span $\mathcal{J}(T)$.
Suppose $b_{k+1} T(\vec{v}_{k+1}) + \cdots + b_n T(\vec{v}_n) = \vec{0}$. Then

$$\vec{0} = \sum_{j=k+1}^{n} b_j T(\vec{v}_j)$$

$$= T\left(\sum_{j=k+1}^{n} b_j \vec{v}_j \right)$$

so that $\sum_{j=k+1}^{n} b_j \vec{v}_j \in h(T)$. Now $\vec{v}_1, \ldots, \vec{v}_k$ is a basis for $h(T)$, and so

$$\sum_{j=k+1}^{n} b_j \vec{v}_j = \sum_{j=1}^{k} a_j \vec{v}_j$$

for some $a_1, \ldots, a_k$. We then have

$$a_1 \vec{v}_1 + \cdots + a_k \vec{v}_k - b_{k+1} \vec{v}_{k+1} - \cdots - b_n \vec{v}_n = \vec{0}$$

However, $\vec{v}_1, \ldots, \vec{v}_n$ are independent, so each of the a_i and b_j is zero.
It follows that $b_{k+1} = \cdots = b_n = 0$; therefore $T(\vec{v}_{k+1}), \ldots, T(\vec{v}_n)$ are
independent.

Thus $T(\vec{v}_1), \ldots, T(\vec{v}_n)$ give a basis for $\mathcal{J}(T)$, which means that
rank$(T) = n - k$. Hence

rank(T) + null$(T) = n = \dim V$

and the theorem is proved.

The Rank-Nullity Theorem is a very useful result. One easy application is the following.

COROLLARY. *Let* $T{:}V \to W$ *be a linear transformation. If* $\dim V >$
$\dim W$, *then* T *cannot be injective. If* $\dim V < \dim W$, *then* T *cannot be*
surjective.

We leave the proof of this result as an exercise.

EXERCISES

1. Prove Part (b) of Proposition 4.3.

2. Prove that the space $\wp$ of all polynomials is not finite dimensional.

3. Prove that R^∞ is not finite dimensional. (See Example 2.2.2.)

4. Let V be the subspace of R^∞ consisting of those sequences $(x_1, x_2, \ldots, x_n, \ldots)$ with the property that only a finite number of the x_i are non-zero. Prove that V is not finite dimensional.

5. Show that $\wp_n$, the vector space of polynomials of degree $\leq n$, has dimension $n + 1$.

6. If V_1 is a subspace of a finite dimensional space V, prove that $\dim V_1 = \dim V$ if and only if $V_1 = V$.

*7. If V_1 and V_2 are finite dimensional subspaces of a vector space V, we define $V_1 + V_2$, the *sum* of V_1 and V_2, to be the span of $V_1 \cup V_2$. (See Exercise 11, Section 2.4.) Prove that

$$\dim V_1 + \dim V_2 - \dim V_1 \cap V_2 = \dim(V_1 + V_2)$$

8. Let V_1 and V_2 be subspaces of the finite dimensional vector space V such that $V = V_1 + V_2$ and $V_1 \cap V_2 = \{\vec{0}\}$. In this case, we say that V is the *direct sum* of V_1 and V_2 and write $V = V_1 \oplus V_2$.

   (a) Prove that any vector $\vec{v} \in V$ can be uniquely written $\vec{v} = \vec{v}_1 + \vec{v}_2$ where $\vec{v}_1 \in V_1$ and $\vec{v}_2 \in V_2$.
   (b) Prove that $\dim V = \dim V_1 + \dim V_2$.
   (c) Suppose V_1 and V_2 are subspaces of V with $V_1 \cap V_2 = \{\vec{0}\}$. Prove that $V = V_1 \oplus V_2$ if and only if $\dim V = \dim V_1 + \dim V_2$.

*9. For $k = 1, 2, \ldots$, let $G_k(x) = \dfrac{x(x+1)\cdots(x+k-1)}{k!}$. (Then $G_k(x)$ is a polynomial of degree k.) Set $G_0(x) = 1$.

   (a) Show that $\Sigma_{i=1}^{n} G_k(i) = G_{k+1}(n)$ for all positive integers k,n.

(b) Show that $\{G_0, G_1, \ldots, G_m\}$ is a basis for $P_m(x)$.

(c) Use the first two parts of this problem to show that if $F_0(x)$ is any polynomial of degree m, then $\sum_{i=1}^{n} F_0(i) = F_1(n)$, where F_1 is a polynomial of degree m + 1.

*10. Let V and W be finite dimensional vector spaces. Prove that $\dim(V \times W) = \dim V + \dim W$. (See Exercise 7, Section 2.2.)

11. Show that, if W is a subspace of V, then

$$\dim V = \dim W + \dim(V/W)$$

(See Exercise 18, Section 2.4.)

12. Determine the rank of each of the following linear transformations:

(a) $T:R^3 \to R^2$, $T(x_1, x_2, x_3) = (2x_1 + x_2, x_2 + x_3)$

(b) $T:R^2 \to R^3$, $T(x_1, x_2) = (x_1, x_2, x_1 - x_2)$

(c) $T:R^3 \to R^3$, $T(x_1, x_2, x_3) = (x_2, x_1, x_2)$

(d) $T:R^3 \to R^3$, $T(x_1, x_2, x_3) = (x_1 + 2x_2 + x_3, x_1 - x_3, 5x_1 + 4x_2 - x_3)$

13. Prove the corollary to Proposition 4.6.

14. Let $T:V \to W$, $S:W \to U$ be linear transformations. Prove that

(a) $\text{rank}(ST) \leq \text{rank } S$

(b) $\text{null}(ST) \geq \text{null } T$

*15. Suppose that V and W are finite dimensional vector spaces and that $T:V \to W$ is injective. Prove that there is a linear transformation $S:W \to V$ such that $ST(\vec{v}) = \vec{v}$ for all $\vec{v}$ in V.

*16. Suppose that V and W are finite dimensional vector spaces and that $T:V \to W$ is surjective. Prove that there is a linear transformation $R:W \to V$ such that $TR(\vec{w}) = \vec{w}$ for all $\vec{w} \in W$.

5.   INNER PRODUCT SPACES

Thus far, we have used R^2 and R^3 as models for determining what vector spaces should be. The definition of a vector space was a generalization of certain algebraic properties of R^2 and R^3, and the definition of a norm was a generalization of length. Now we generalize
the notion of the angle between two vectors.

Let $\vec{v} = (x_1, x_2)$ and $\vec{w} = (z_1, z_2)$ be two points in the cartesian
plane $V = R^2$, as in Figure 5.1. Consider the triangle made by $\vec{v}$, $\vec{w}$,
and the origin. We may regard the sides of this triangle as $\vec{v}$, $\vec{w}$,
and $\vec{v} - \vec{w}$, as shown in the figure. Let θ be the angle between $\vec{v}$ and
$\vec{w}$.

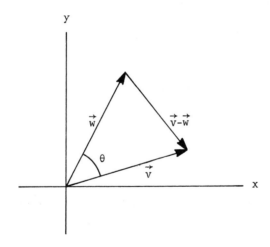

FIGURE 5.1.

According to the law of cosines in trigonometry,

$$\|\vec{v} - \vec{w}\|^2 = \|\vec{v}\|^2 + \|\vec{w}\|^2 - 2\|\vec{v}\|\|\vec{w}\|\cos \theta$$

We can compute most of these terms, since $\vec{v} - \vec{w} = (x_1 - z_1, x_2 - z_2)$:

$$(x_1 - z_1)^2 + (x_2 - z_2)^2 = x_1^2 + x_2^2 + z_1^2 + z_2^2 - 2\|\vec{v}\|\|\vec{w}\|\cos\theta$$

or

$$\cos\theta = \frac{x_1 z_1 + x_2 z_2}{\|\vec{v}\|\|\vec{w}\|}$$

The numerator of the above expression is a rather useful quantity, since it determines $\cos\theta$; it turns out to have some properties which make it more convenient to use than $\cos\theta$. We therefore define the *inner product* of $\vec{v}$ and $\vec{w}$, $< \vec{v},\vec{w} >$, by

$$< \vec{v},\vec{w} > = x_1 z_1 + x_2 z_2$$

This operation is easily seen to have the following properties:

$< \vec{v},\vec{v} > \geq 0$ for all $\vec{v} \in V$ and $\qquad\qquad$ (5.1)
$< \vec{v},\vec{v} > = 0$ if and only if $\vec{v} = \vec{0}$

$< \vec{v},\vec{w} > = < \vec{w},\vec{v} >$ for all $\vec{v},\vec{w} \in V$ $\qquad\qquad$ (5.2)

$< a_1\vec{v}_1 + a_2\vec{v}_2,\vec{w} > = a_1< \vec{v}_1,\vec{w} > + a_2< \vec{v}_2,\vec{w} >$ for all
$\qquad\qquad\qquad\qquad\qquad\qquad\qquad\qquad\qquad\qquad$ (5.3)
$\vec{v}_1,\vec{v}_2,\vec{w} \in V$ and $a_1,a_2 \in R$

We define an *inner product* in an arbitrary real vector space V to be a function which assigns to each pair of vectors $\vec{v},\vec{w}$ in V a real number $< \vec{v},\vec{w} >$, and which satisfies properties (5.1), (5.2), and (5.3).

Remark. Inner products are sometimes called "dot products" or "scalar products". We shall use the term "inner product".

A vector space with a fixed inner product is called an *inner product space*. We now give some examples.

(1) Let $V = R^n$ and define

$$< (x_i),(y_i) > = \sum_{i=1}^{n} x_i y_i$$

We verify (5.1) in this case, leaving (5.2) and (5.3) as an exercise for the reader. For any (x_i) in R^n,

$$< (x_i),(x_i) > = \sum_{i=1}^{n} x_i^2$$

which is always non-negative and vanishes only if each $x_i = 0$.

We call this inner product the *standard inner product* on R^n. Note that for $n = 2$, this is the inner product defined at the beginning of the section.

(2) Let $V = C^0([0,1])$, the vector space of all continuous functions $f:[0,1] \rightarrow R$. Define an inner product on V by

$$< f,g > = \int_0^1 f(x)g(x)dx$$

Both (5.2) and (5.3) follows from elementary properties of the integral and are left to the reader. It is also clear that $< f,f > \geq 0$ for all f in $C^0([0,1])$. That $< f,f > > 0$ whenever f is not the zero function is a consequence of Exercise 21, Section 3.3. (See Exercise 3.)

Any subspace of an inner product space is itself an inner product space. The inner product of two vectors in the subspace is defined to be the inner product of the two vectors as elements of the containing space. Properties (5.1), (5.2), and (5.3) are automatically satisfied.

On the face of it, the definition of an inner product seems asymmetrical in $\vec{v}$ and $\vec{w}$, because property (5.3) is asymmetrical. The following proposition shows this to be an illusion.

PROPOSITION 5.1. *For any vectors* $\vec{v}$, $\vec{w}_1$, $\vec{w}_2$ *in* V *and real numbers* b_1, b_2, *we have*

$$< \vec{v},b_1\vec{w}_1 + b_2\vec{w}_2 > = b_1 < \vec{v},\vec{w}_1 > + b_2 < \vec{v},\vec{w}_2 >$$

Proof.

$$< \vec{v}, b_1\vec{w}_1 + b_2\vec{w}_2 > = < b_1\vec{w}_1 + b_2\vec{w}_2, \vec{v} >$$
$$= b_1 < \vec{w}_1, \vec{v} > + b_2 < \vec{w}_2, \vec{v} >$$
$$= b_1 < \vec{v}, \vec{w}_1 > + b_2 < \vec{v}, \vec{w}_2 >$$

We define the *norm* of a vector $\vec{v}$ in an inner product space V by the equation

$$\|\vec{v}\| = (< \vec{v}, \vec{v} >)^{1/2}$$

Note that this makes sense because of property (5.1). We shall discuss this function in detail in the next section. In particular, we shall prove there that it satisfies the properties of a norm given in Section 3.1.

In the example of R^2, the nonzero vectors $\vec{v}$, $\vec{w}$ are perpendicular (or, equivalently, cos θ = 0) if and only if their inner product is 0. This motivates the following:

Definition. Two vectors $\vec{v}_1$ and $\vec{v}_2$ in an inner product space V are *orthogonal* if $< \vec{v}_1, \vec{v}_2 > = 0$.

Note that $\vec{0}$ is orthogonal to every vector. (For $< \vec{0}, \vec{v} > = < 0 \cdot \vec{0}, \vec{v} > = 0 < \vec{0}, \vec{v} > = 0$.)

If S is any subset of V, we define the *orthogonal complement* $S^\perp$ of S by

$$S^\perp = \{\vec{v} \in V | < \vec{v}, \vec{v}' > = 0 \text{ for all } \vec{v}' \in S\}$$

For instance, if S = $\{(1,1)\} \subset R^2$, then

$$S^\perp = \{(x_1, x_2) \in R^2 : < (x_1, x_2), (1,1) > = 0\}$$
$$= \{(x_1, x_2) \in R^2 : x_1 + x_2 = 0\}$$

(See Figure 5.2.)

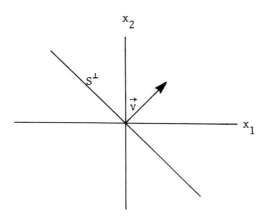

FIGURE 5.2. The orthogonal complement of $S = \{\vec{v}\}$

PROPOSITION 5.2. *If S is a subset of the inner product space V, then $S^{\perp}$ is a subspace of V.*

Proof. $S^{\perp}$ is nonempty, since (as we have remarked) $\vec{0} \in S^{\perp}$. Suppose $\vec{v}_1$ and $\vec{v}_2$ are two vectors in $S^{\perp}$. Then for any $\vec{v}$ in S,

$$< \vec{v}_1 + \vec{v}_2, \vec{v} > = < \vec{v}_1, \vec{v} > + < \vec{v}_2, \vec{v} > \qquad \text{(by 5.3)}$$

$$= 0$$

since both $\vec{v}_1, \vec{v}_2 \in S^{\perp}$. Thus $\vec{v}_1 + \vec{v}_2 \in S^{\perp}$.

Similarly, if $\vec{v} \in S^{\perp}$ and $a \in R$, then for any $\vec{v}' \in S$,

$$< a\vec{v}, \vec{v}' > = a< \vec{v}, \vec{v}' > \qquad \text{(by 5.3)}$$

$$= 0$$

since $\vec{v} \in S^{\perp}$. Thus $a\vec{v} \in S^{\perp}$.

We have shown that $S^{\perp}$ is closed under both addition and multiplication by scalars; by definition, $S^{\perp}$ is a subspace.

EXERCISES

1.   Which of the following define inner products on R^2?

(a) $<(x_1, x_2), (y_1, y_2) > = x_1 y_1 - x_2 y_2$

(b) $<(x_1,x_2),(y_1,y_2)> = 2x_1y_1$

(c) $<(x_1,x_2),(y_1,y_2)> = x_1 + x_2 + y_1 + y_2$

(d) $<(x_1,x_2),(y_1,y_2)> = 2x_1y_1 + x_2y_2 - x_1y_2 - x_2y_1$

(e) $<(x_1,x_2),(y_1,y_2)> = 2x_1y_1 + x_2y_2$

2.   Let a, b be real numbers. Show that

$$< (x_1,x_2),(y_1,y_2)> = ax_1y_1 + bx_2y_2$$

is an inner product on R^2 if and only if $a > 0$ and $b > 0$.

3.   Verify (5.1), (5.2), and (5.3) for Example 2 of this section.
(Exercise 21 of Section 3.3 should be useful in verifying (5.1).)

4.   Let $S \subset R^3$ be the set consisting of the vectors $(-2,1,1)$, $(1,-2,1)$.
Determine $S^{\perp}$.

5.   Find an inner product on R^2 for which $(-1,2)$, and $(2,3)$ are orthogonal.

6.   Let V be an inner product space and let $\vec{v}_0 \in V$ a unit vector
$(\|\vec{v}_0\| = 1)$; define $T:V \to V$ by

$$T(\vec{v}) = \vec{v} - < \vec{v},\vec{v}_0> \vec{v}_0$$

Prove that T is linear and determine its kernel. (T is called the
"orthogonal projection on the orthogonal complement to $\vec{v}_0$". Why?)

7.   Let V be an inner product space, and let $\vec{v},\vec{w}$ be vectors in V.
Prove the *parallelogram law:*

$$\|\vec{v+w}\|^2 + \|\vec{v-w}\|^2 = 2\|\vec{v}\|^2 + 2\|\vec{w}\|^2$$

(Draw a picture to see where the name comes from.)

8.   Prove the *polarization identity:* if $\vec{v},\vec{w}$ are vectors in the inner
product space V, then

$$< \vec{v}, \vec{w} > = \frac{1}{2}(\|\vec{v}+\vec{w}\|^2 - \|\vec{v}\|^2 - \|\vec{w}\|^2)$$

(Thus one can recover the inner product from the norm.)

9. Use the law of cosines for two vectors in R^3 to compute the cosine of the angle between two vectors $\vec{v}, \vec{w}$ in terms of their inner product and their norms.

10. Let V be the vector space of Exercise 4, Section 4, and define $< (x_i), (y_i) > = \Sigma_{i=1}^{\infty} x_i y_i$. Prove that this defines an inner product on V.

*11. Let $\vec{v}_1, \ldots, \vec{v}_n, \vec{w}$ be nonzero vectors in an inner product space such that

   (1) $< \vec{v}_i, \vec{v}_j > \leq 0$ if $i \neq j$
   (2) $< \vec{v}_i, \vec{w} > > 0$ for all i

Show that $\vec{v}_1, \ldots, \vec{v}_n$ are linearly independent. (Hint: suppose that $\Sigma_{j=1}^{n} a_j \vec{v}_j = 0$, with not all a_j zero. Separate the positive and negative terms, getting an expression like $\Sigma_{j \in S} b_j \vec{v}_j = \Sigma_{j \in T} c_j \vec{v}_j$, where $b_j \geq 0$, $c_j \geq 0$, and $S \cap T = \phi$. Show that both sums are 0 and then show that the b's and c's are 0.)

6. THE NORM ON AN INNER PRODUCT SPACE

In the last section, we defined the norm of a vector $\vec{v}$ in an inner product space V by

$$\|\vec{v}\| = (< \vec{v}, \vec{v} >)^{1/2}$$

We need to show that $\vec{v}$ is indeed a norm (as defined in Section 3.1). We begin with the easier properties.

PROPOSITION 6.1. *For any vector* $\vec{v}$ *in* V *and any real number* a, *we have*

(i) $\|\vec{v}\| \geq 0$

(ii) $\|\vec{v}\| = 0$ if and only if $\vec{v} = \vec{0}$

(iii) $\|a\vec{v}\| = |a| \cdot \|v\|$

Proof. The first two properties are immediate consequences of the definition of an inner product. To prove (iii),

$$\|a\vec{v}\| = (< a\vec{v}, a\vec{v} >)^{1/2}$$
$$= (a^2)^{1/2} (< \vec{v}, \vec{v} >)^{1/2} = |a| \cdot |v|$$

as claimed.

We saw in the last section that, for vectors $\vec{u}, \vec{v}$ in R^2,

$$< \vec{u}, \vec{v} > = \|\vec{u}\| \cdot \|\vec{v}\| \cos \theta$$

where θ is the angle between the vectors $\vec{u}$ and $\vec{v}$. Since

$$|\cos \theta| \leq 1$$

it follows that

$$|< \vec{u}, \vec{v} >| \leq \|\vec{u}\| \cdot \|\vec{v}\|$$

for any vector $\vec{u}, \vec{v}$ in R^2. In fact, this inequality holds in any inner product space and is known as the *Schwarz Inequality* (or, sometimes, the Cauchy-Schwarz inequality).

THEOREM 6.2. *Let V be an inner product space. For any vectors $\vec{u}, \vec{v}$ in V,*

$$|< \vec{u}, \vec{v} >| \leq \|\vec{u}\| \cdot \|\vec{v}\|$$

Proof. This proof is straightforward but somewhat artificial. Note first of all that if either $\vec{u}$ or $\vec{v}$ is the zero vector, the inequality above is trivially satisfied. Next observe that, for any real numbers a, b we have

$$0 \leq \|a\vec{u} + b\vec{v}\|^2 = a^2 \|\vec{u}\|^2 + 2ab < \vec{u}, \vec{v} > + b^2 \|\vec{v}\|^2$$

Setting $a = \|\vec{v}\|$, $b = -\|\vec{u}\|$, we obtain

$$0 \leq 2\|\vec{u}\|^2 \cdot \|\vec{v}\|^2 - 2\|\vec{u}\| \cdot \|\vec{v}\|< \vec{u},\vec{v} >$$

Since we can assume $\|\vec{u}\| \neq 0$, $\|\vec{v}\| \neq 0$, we can divide this inequality through by $2\|\vec{u}\| \cdot \|\vec{v}\|$ and obtain

$$< \vec{u},\vec{v} > \, \leq \|\vec{u}\| \cdot \|\vec{v}\|$$

Similarly, if we set $a = \|\vec{v}\|$ and $b = \|\vec{u}\|$, we obtain the inequality

$$-< \vec{u},\vec{v} > \, \leq \|\vec{u}\| \cdot \|\vec{v}\|$$

Since $|< \vec{u},\vec{v} >|$ is either $< \vec{u},\vec{v} >$ or $-< \vec{u},\vec{v} >$, the Schwarz inequality follows from the two inequalities above.

We can now complete the proof that $\|\vec{v}\|$ is a norm by proving the triangle inequality.

PROPOSITION 6.3. *For any two vectors* $\vec{v}_1$ *and* $\vec{v}_2$ *in an inner product space V,*

$$\|\vec{v}_1 + \vec{v}_2\| \leq \|\vec{v}_1\| + \|\vec{v}_2\|$$

Proof. We proceed directly:

$$\begin{aligned}
\|\vec{v}_1 + \vec{v}_2\|^2 &= \|\vec{v}_1\|^2 + \|\vec{v}_2\|^2 + 2< \vec{v}_1,\vec{v}_2 > \quad \text{(by definition)} \\
&\leq \|\vec{v}_1\|^2 + \|\vec{v}_2\|^2 + 2\|\vec{v}_1\| \cdot \|\vec{v}_2\| \quad \text{(by Theorem 6.2)} \\
&= (\|\vec{v}_1\| + \|\vec{v}_2\|)^2
\end{aligned}$$

Since $\|\vec{v}_1 + \vec{v}_2\|$ and $\|\vec{v}_1\| + \|\vec{v}_2\|$ are nonnegative, we can take square roots to obtain the triangle inequality.

This result, together with Proposition 6.1, gives us the following:

COROLLARY. *If* V *is an inner product space, then* $\|\vec{v}\| = (< \vec{v},\vec{v} >)^{1/2}$ *is a norm.*

Now, of course, the results of Section 3.1 apply to this norm.

EXERCISES

1.   Let $\vec{u}$ and $\vec{v}$ be non zero vectors in an inner product space V.
Prove that

$$\|\vec{u} + \vec{v}\|^2 = \|\vec{u}\|^2 + \|\vec{v}\|^2$$

if and only if $\vec{u}$ is orthogonal to $\vec{v}$.

2.   Prove that, for any $f \in C^0([0,1])$,

$$\int_0^1 f^2(x)\,dx \geq (\int_0^1 f(x)\,dx)^2$$

3.   One can determine norms from inner products, but not all norms
arise in this way. Show that the norm of Example 3, Section 3.1, on
R^3,

$$\|(x_1,x_2,x_3)\| = |x_1| + |x_2| + |x_3|$$

does not come from an inner product. (Hint: the polarization identity, Exercise 5.8, shows what the inner product would have to be.)

*4.   Suppose that V has an "inner product" (,) satisfying the follow-
ing properties:

$(\vec{v},\vec{v}) \geq 0$ for all $\vec{v} \in V$;
$(\vec{v},\vec{w}) = (\vec{w},\vec{v})$ for all $\vec{v},\vec{w} \in V$;
$(a_1\vec{v}_1 + a_2\vec{v}_2,\vec{w}) = a_1(\vec{v}_1,\vec{w}) + a_2(\vec{v}_2,\vec{w})$ for all $\vec{v}_1,\vec{v}_2\vec{w} \in V$ and all
$a_1,a_2 \in R$.

(However, $(\vec{v},\vec{v})$ may be 0 even if $\vec{v} \neq \vec{0}$.) Show that the Schwarz inequality still holds for (,). Exercise 8 of Section 5 then shows that
$\{\vec{v} \in V:(\vec{v},\vec{v}) = 0\}$ is a subspace of V. (Hint: one procedure is to
show that if $(\vec{v},\vec{v}) = 0$, then $(\vec{v},\vec{w}) = 0$ for all $\vec{w} \in V$.)

*5.   (a) Show that one gets equality in the Schwarz inequality if
and only if $\vec{v}$, $\vec{w}$ are linearly dependent.

(b) Show that in an inner product space, one gets equality in
the triangle inequality if and only if one of $\vec{v}$, $\vec{w}$ is a nonnegative
multiple of the other.

7.   ORTHONORMAL BASES

In Section 4, we proved the important fact that any finite-dimensional vector space has a basis. We now show that for inner product spaces we can do somewhat more.

Let V be an inner product space. A subset $S = \{\vec{v}_1,\ldots,\vec{v}_k\}$ is said to be *orthonormal* if

$$< \vec{v}_i,\vec{v}_j > = \begin{cases} 1 \text{ if } i = j \\ 0 \text{ if } i \neq j \end{cases}$$

Note. It happens often enough in mathematics that a quantity with two subscripts, a_{ij}, is 1 when $i = j$ and 0 when $i \neq j$, that there is a special notation for it. We define δ_{ij} by

$$\delta_{ij} = \begin{cases} 1 \text{ if } i = j \\ 0 \text{ otherwise} \end{cases}$$

The expression δ_{ij} is called the *Kronecker delta* (after the mathematician Leopold Kronecker, who made more substantial contributions to mathematics than this).

An *orthonormal basis* for V is an orthonormal set which is also a basis.

One great advantage of orthonormal bases is this: we know that if $\vec{u}_1,\ldots,\vec{u}_n$ form a basis for V, then any vector $\vec{v} \in V$ is a linear combination of $\vec{u}_1,\ldots,\vec{u}_n$. But actually finding the expression for v can be a lot of work (involving solving simultaneous linear equations). With orthonormal bases, though, the following theorem solves the problem.

THEOREM 7.1. *Let $\{\vec{u}_1,\ldots,\vec{u}_n\}$ be an orthonormal basis for V, and let $\vec{v} \in V$ be any vector. Then*

$$\vec{v} = \sum_{j=1}^{n} < \vec{v},\vec{u}_j> \vec{u}_j$$

Proof. We know that we can write $\vec{v} = \Sigma_{j=1}^{n} a_j \vec{u}_j$; all we need to do is to compute the a_i. Take the inner product of both sides with $\vec{u}_i$:

$$< \vec{v}, \vec{u}_i > = \sum_{j=1}^{n} a_j < \vec{u}_j, \vec{u}_i >$$

But $< \vec{u}_i, \vec{u}_i > = 1$ and $< \vec{u}_j, \vec{u}_i > = 0$ for $j \neq i$; thus $< \vec{v}, \vec{u}_i > = a_i$, as claimed.

Thoerem 7.1 shows why orthonormal bases are convenient; it does not, of course, show how to obtain them. The next result does that.

PROPOSITION 7.2. *Let* $\{\vec{v}_1, \ldots, \vec{v}_n\}$ *be a basis of* V. *Then there exists an orthonormal basis* $\{\vec{u}_1, \ldots, \vec{u}_n\}$ *of* V *such that for each i with* $1 \leq i \leq n$, $\{\vec{v}_1, \ldots, \vec{v}_i\}$ *and* $\{\vec{u}_1, \ldots, \vec{u}_i\}$ *span the same subspace.*

Proof. The proof is by direct construction of the vectors $\vec{u}_i$, and the procedure is useful enough to have a name: the *Gram-Schmidt* construction. Let $V_i = $ span $\{\vec{v}_1, \ldots, \vec{v}_i\}$. First of all, we need $\vec{u}_1$. As $\vec{u}_1$ must span V_1, it must be a multiple of $\vec{v}_1$; as $\vec{u}_1$ will be part of an orthonormal basis, it must have norm 1. That limits our choice considerably. We let $\vec{u}_1 = \vec{v}_1 / \|\vec{v}_1\|$. (As $\vec{v}_1$ is in a basis, $\vec{v}_1 \neq \vec{0}$, and therefore $\|\vec{v}_1\| \neq 0$.)

To construct $\vec{u}_2$, we need first of all a nonzero vector $\vec{f}_2$ in V_2 orthogonal to $\vec{u}_1$. Figure 7.1 suggests that $\vec{f}_2$ can be constructed by adding a multiple of $-\vec{u}_1$ to $\vec{v}_2$; $\vec{f}_2 = \vec{v}_2 - a_1\vec{u}_1$ for some $a_1 \in$ R. Since $< \vec{f}_2, \vec{u}_1 > = 0$, we have

$$< \vec{v}_2 - a_1\vec{u}_1, \vec{u}_1 > = 0$$

or

$$< \vec{v}_2, \vec{u}_1 > - a_1 < \vec{u}_1, \vec{u}_1 > = 0$$

so that

$$a_1 = < \vec{v}_2, \vec{u}_1 >$$

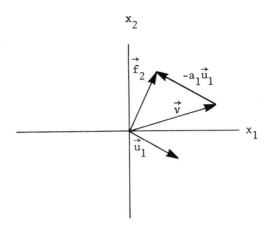

FIGURE 7.1. Constructing a vector orthogonal to $\vec{u}_1$

(since $\|\vec{u}_1\| = 1$). Thus, we let $\vec{f}_2 = \vec{v}_2 - <\vec{v}_2,\vec{u}_1> \vec{u}_1$. Then $\vec{f}_2 \neq \vec{0}$, since if it were $\vec{0}$, then $\vec{v}_2 = <\vec{v}_2,\vec{u}_1> \vec{u}_1$ would be in V_1 and $\vec{v}_1$, $\vec{v}_2$ would not be linearly independent. Let $\vec{u}_2 = \vec{f}_2/\|\vec{f}_2\|$. Then $\vec{u}_1,\vec{u}_2$ are in V_2. Since $\vec{v}_1$, $\vec{v}_2$ are both linear combinations of $\vec{u}_1$, $\vec{u}_2$, $\{\vec{u}_1,\vec{u}_2\}$ spans V_2. Thus $\{\vec{u}_1,\vec{u}_2\}$ is a basis for V_2. In addition, we have constructed it to be an orthonormal basis as well.

The general step is similar. Suppose that we have constructed an orthonormal basis $\{\vec{u}_1,\ldots,\vec{u}_i\}$ for V_i (where $i < n$, since we are done when $i = n$). Let

$$\vec{f}_{i+1} = \vec{v}_{i+1} - \sum_{j=1}^{i} <\vec{v}_{i+1},\vec{u}_j> \vec{u}_j$$

Then $\vec{f}_{i+1} \neq \vec{0}$. For if $\vec{f}_{i+1} = \vec{0}$, we would have $\vec{v}_{i+1} = \sum_{j=1}^{i} <\vec{v}_{i+1},\vec{u}_j> \vec{u}_j$ $\in V_i$, and therefore $\vec{v}_{i+1}$ would be a linear combination of $\vec{v}_1,\ldots,\vec{v}_i$. This would contradict the fact that the $\vec{v}$'s are linearly independent. Moreover, if $1 \leq k \leq i$,

$$< \vec{f}_{i+1},\vec{u}_k > = < \vec{v}_{i+1},\vec{u}_k > - \sum_{j=1}^{i} < \vec{v}_{i+1},\vec{u}_j > < \vec{u}_j,\vec{u}_k >$$

$$= < \vec{v}_{i+1},\vec{u}_k > - < \vec{v}_{i+1},\vec{u}_k > < \vec{u}_k,\vec{u}_k >$$

$$= 0$$

since $< \vec{u}_j, \vec{u}_k > = \delta_{jk}$. Therefore $\vec{f}_{i+1}$ is orthogonal to $\vec{u}_1, \ldots, \vec{u}_i$. Setting $\vec{u}_{i+1} = \vec{f}_{i+1}/\|\vec{f}_{i+1}\|$, we see that the vectors $\vec{u}_1, \ldots, \vec{u}_{i+1}$ are all in V_{i+1}. (Our induction hypothesis says that $\vec{u}_1, \ldots, \vec{u}_i$ are in $V_i \subset V_{i+1}$, and that therefore $\vec{u}_1, \ldots, \vec{u}_i$ are linear combinations of $\vec{v}_1, \ldots, \vec{v}_i$. Our construction of u_{i+1} shows that it is a linear combination of $\vec{v}_1, \ldots, \vec{v}_{i+1}$, and therefore in V_{i+1}.) In fact, they span V_{i+1}. This is not hard to prove directly, but the shortest proof is that $\{\vec{u}_1, \ldots, \vec{u}_{i+1}\}$ form an orthonormal set (by construction), and are therefore linearly independent. (See Exercise 4.) Since there are (i+1) linearly independent vectors in the (i+1)-dimensional space V_{i+1}, they form a basis for V_{i+1}.

The proposition now follows by induction.

COROLLARY. *Let* V_0 *be any subspace of a finite dimensional vector space* V. *Then there is an orthonormal basis* $\{\vec{v}_1, \ldots, \vec{v}_n\}$ *for* V *with the property that* $\{\vec{v}_1, \ldots, \vec{v}_k\}$ *is an orthonormal basis for* V_0.

The proof of this corollary is left as Exercise 5.

For low dimensional vector spaces, the Gram-Schmidt process is a fairly efficient procedure for producing orthonormal bases. As the dimension increases, though, so do the computations. (See Exercise 2.)

Orthonormal bases are a very useful tool, as the next few results may indicate.

PROPOSITION 7.3. *Let* $\{\vec{u}_1, \ldots, \vec{u}_n\}$ *be an orthonormal basis for* V, *and let* $\vec{v} = \Sigma_{i=1}^n a_i \vec{u}_i$, $\vec{w} = \Sigma_{j=1}^n b_j \vec{u}_j$. *Then*

$$< \vec{v}, \vec{w} > = \sum_{i=1}^n a_i b_i$$

Proof. By the basic properties of inner products, we have

$$< \vec{v}, \vec{w} > = < \sum_{i=1}^n a_i \vec{u}_i, \sum_{j=1}^n b_j \vec{u}_j >$$

$$= \sum_{i=1}^n \sum_{j=1}^n a_i b_j < \vec{u}_i, \vec{u}_j >$$

However, $< \vec{u}_i, \vec{u}_j > = \delta_{ij}$, so all terms with $i \neq j$ vanish, and

$$< \vec{v}, \vec{w} > = \sum_{i=1}^{n} a_i b_i$$

COROLLARY 1. *In the above situation,*

$$< \vec{v}, \vec{w} > = \sum_{i=1}^{n} < \vec{v}, \vec{u}_i > < \vec{u}_i, \vec{w} >$$

Proof. Apply Theorem 7.1 to compute the a's and b's.

COROLLARY 2. *In the above situation*

$$\|\vec{v}\| = \left(\sum_{i=1}^{n} a_i^2 \right)^{1/2} = \left(\sum_{i=1}^{n} < \vec{v}, \vec{u}_i >^2 \right)^{1/2}$$

Proof. Set $\vec{w} = \vec{v}$ in the previous corollary.

Proposition 7.3 says that if we use an orthonormal basis, the formula for the inner product is like the standard inner product in R^n. (In R^n with the usual inner product, the vectors $\vec{e}_1, \ldots, \vec{e}_n$ form an orthonormal basis.) The third corollary to Proposition 7.3, in the form

$$\| \sum_{i=1}^{n} a_i \vec{e}_i \|^2 = \sum_{i=1}^{n} a_i^2$$

is a generalization of the Pythagorean Theorem.

Let V and W be inner product spaces. A linear transformation $T:V \rightarrow W$ is an *isometry* if $< T(\vec{v}), T(\vec{v}') > = < \vec{v}, \vec{v}' >$ for all $\vec{v}, \vec{v}'$ in V. We say V and W are *isometrically isomorphic* if there is an isomorphism $T:V \rightarrow W$ which is an isometry.

The next result is an extension of Theorem 4.2.

THEOREM 7.4. *Two finite dimensional inner product spaces V and W are isometrically isomorphic if and only if* dim V = dim W.

Proof. If V and W are isometrically isomorphic, their dimensions are the same (from Theorem 4.2).

To prove the converse, we let $\vec{v}_1,\ldots,\vec{v}_n$ be an orthonormal basis
for V, $\vec{w}_1,\ldots,\vec{w}_n$ be an orthonormal basis for W, and T:V $\rightarrow$ W be the
linear transformation with $T(\vec{v}_i) = \vec{w}_i$. Then T is an isomorphism by
the corollary to Proposition 3.2 and an isometry by Proposition 7.3.

COROLLARY. *Any* n-*dimensional inner product space is isometrically isomorphic to Euclidean* n-*Space* R^n.

Just as in the case of the corollary to Theorem 4.2, this result
allows us to assume that the finite dimensional vector spaces we deal
with in what follows are Euclidean spaces. This will often be extremely convenient.

Suppose now that V and W are finite dimensional inner product
spaces and T:V $\rightarrow$ W is a linear transformation. Let $\alpha = \{\vec{v}_1,\ldots,\vec{v}_n\}$
be an orthonormal basis for V and $\beta = \{\vec{w}_1,\ldots,\vec{w}_m\}$ an orthonormal basis
for W. Then, for each j = 1,...,n, $T(\vec{v}_j)$ is a linear combination
of the basis β. Thus there are numbers $a_{1j},a_{2j},\ldots,a_{mj}$ such that

$$T(\vec{v}) = \sum_{i=1}^{m} a_{ij}\vec{w}_i$$

We call the m × n matrix (a_{ij}) the *matrix of* T *relative to the bases*
α *and* β. Of course, this matrix depends on the bases α and β; we shall
study this dependence in Chapter 7.

We can use this matrix to prove that a linear transformation between finite dimensional inner product spaces is continuous. The
proof is almost identical with the proof of Proposition 3.4.5; we leave
it as Exercise 17.

PROPOSITION 7.5. *Let* V *and* W *be finite dimensional inner product
spaces and* T:V $\rightarrow$ W *a linear transformation. Then* T *is continuous.*

As a result of Proposition 7.5, we are now able to prove Proposition 3.6.3 for functions on finite dimensional inner product spaces.

(The key ingredient in the proof of Proposition 3.6.3 is the continuity of linear transformations.) We leave the details as an exercise.

PROPOSITION 7.6. *Let* V *and* W *be finite dimensional inner product space,* U *an open subset, and* $f:U \to W$ *a function differentiable at* $\vec{v}_0 \in U$. *Then* f *is continuous at* $\vec{v}_0$.

Later on (in the exercises to Section 5.6), we will show that Propositions 7.5 and 7.6 hold for all finite dimensional normed vector spaces.

The assumption that V is finite-dimensional is important. There are linear transformations $T:V \to W$, where V and W are infinite-dimensional normed vector space, which are not continuous. (See Exercises 20 and 21.) For infinite-dimensional vector spaces, therefore, Proposition 3.6.3 does *not* hold unless one also assumes that the derivative is continuous.

EXERCISES

1. Use the Gram-Schmidt method to construct an orthonormal basis for V from the basis α, where

(a) $V = R^2$, $\alpha = \{(1,1),(1,2)\}$

(b) $V = R^3$, $\alpha = \{(1,1,0),(1,0,1),(0,1,1)\}$

(c) $V = R^3$, $\alpha = \{(1,2,1),(-1,0,2),(1,1,1)\}$

(d) $V \subset R^4$ spanned by $\alpha = \{(1,2,0,1),(4,-1,1,1),(-2,-2,1,0)\}$

(e) $V = R^4$, $\alpha = \{(2,0,1,1),(1,0,-1,-1),(1,1,2,2),(-1,2,0,1)\}$

2. Find an orthonormal basis for the vector space P_3 of polynomials of degree ≤ 3 with the inner product

$$< f,g > = \int_0^1 f(x)g(x)dx$$

3. Find an orthonormal basis for R^3 with the inner product defined by

$$< (x_1,x_2,x_3),(y_1,y_2,y_3) > = 2x_1y_1 + x_1y_3 + x_3y_1 + x_2y_2 + x_3y_3$$

4.   Prove that any orthonormal set $\{\vec{v}_1,\dots,\vec{v}_n\}$ is independent. (Hint: take the inner product of any dependence relation with each $\vec{v}_j$.)

5.   Prove the corollary to Proposition 7.2.

6.   Prove that any isometry is injective.

7.   What happens in the Gram-Schmidt procedure if the set of vectors $\{\vec{v}_1,\dots,\vec{v}_n\}$ are linearly dependent?

8.   Let W be a subspace of the inner product space V, and let $\vec{v}_1,\dots,$ $\vec{v}_n$ be an orthonormal basis for V such that $\vec{v}_1,\dots,\vec{v}_k$ is an orthonormal basis for W. Prove that $\vec{v}_{k+1},\dots,\vec{v}_n$ is an orthonormal basis for $W^\perp$.

9.   Let W be a subspace of the finite dimensional inner product space V. Prove that $(W^\perp)^\perp = W$.

*10.   a) Let W be a subspace of the inner product space V. Prove that $V = W \oplus W^\perp$. (See Exercise 8, Section 4.) Thus any vector $\vec{v} \in V$ can be uniquely written $\vec{v} = \vec{w}_1 + \vec{w}_2$ where $\vec{w}_1 \in W$ and $\vec{w}_2 \in W^\perp$.

       b) Define a function $P_W:V \to V$ by $P_W(\vec{v}) = \vec{w}$ where $\vec{v} = \vec{w} + \vec{w}'$, $\vec{w} \in W$, $\vec{w}' \in W^\perp$. Prove that P_W is a linear transformation. (P_W is called the *orthogonal projection* of V onto W.)

11.   Prove that the transformation P_W of the previous exercise satisfies the following.

       (a) The image of P_W is W and the kernel of P_W is $W^\perp$.
       (b) $P_W^2 = P_W$.
       (c) $< P_W(\vec{v}_1),\vec{v}_2 > = < \vec{v}_1,P_W(\vec{v}_2) >$ for all $\vec{v}_1,\vec{v}_2 \in V$.

*12.   Let V be a finite dimensional inner product space and $P:V \to V$ a linear transformation with $P^2 = P$ and $< P\vec{v},\vec{w} > = < \vec{v},P\vec{w} >$ for all $\vec{v},\vec{w} \in V$. Prove that P is the orthogonal projection onto its image.

13.   Use Exercise 8 to obtain a formula for $P_W:V \to V$ (where P_W is as in Exercise 11).

14. Let W be a subspace of the finite-dimensional inner product space V. Show that the linear transformation

$$T:W^{\perp} \to V/W$$

defined by

$$T(\vec{w}) = \bar{w}$$

is an isomorphism. (See Exercise 18, Section 2.4.)

15. Let W be a subspace of the finite-dimensional inner product space V, and let $\vec{v} \in V$. Show that $P_W(\vec{v})$ is the unique closest vector in W to $\vec{v}$.

16. Let V be a finite dimensional inner product space and $\alpha:V \to R$ a linear functional. Prove that there is a unique vector $\vec{v}_0 \in V$ such that $\alpha(\vec{v}) = <\vec{v},\vec{v}_0>$ for all $\vec{v} \in V$. (This is the *Riesz Representation Theorem*.)

17. Prove Proposition 7.5.

18. Suppose $\vec{v}_1,...,\vec{v}_n \in R^n$ are triangular (see Section 2, Exercise 9). What happens when one applies the Gram-Schmidt process to $\vec{v}_n,\vec{v}_{n-1},...,\vec{v}_1$?

*19. Suppose that V is a finite-dimensional inner product space and that $f:V \to V$ satisfies the following two properties:
    (a) $f(\vec{0}) = \vec{0}$;
    (b) $\|f(\vec{v}) - f(\vec{w})\| = \|\vec{v}-\vec{w}\|$ for all $\vec{v},\vec{w} \in V$.

    Prove that f is linear. (Hint: Exercise 8 of Section 5 is useful here.)

*20. (a) Let V be the set of all infinite sequences of real numbers $(x_1,x_2,...)$ such that only finitely many x_j are nonzero. Show that V is a vector space (with addition and scalar multiplication defined componentwise).

(b) For $\vec{x} = (x_1, x_2, \ldots,)$ and $\vec{y} = (y_1, y_2, \ldots)$, define $< \vec{x}, \vec{y} > =$ $\sum_{j=1}^{\infty} x_j y_j$. (Note that the sum has only finitely many nonzero terms.) Show that $<,>$ is an inner product. (See Exercise 4 of Section 4.)

(c) Define $T:V \to V$ by $T((x_1, \ldots, x_n, \ldots) = (x_1, 2x_2, \ldots, nx_n, \ldots)$. Show that T is an unbounded linear transformation (see Exercise 17 of Section 3.3), and that T is therefore not continuous.

*21. Let $V = C^0([0,1])$, with $\|f\|_1 = \int_0^1 |f(x)| dx$, and let $W = C^0([0,1])$, with $\|f\|_\infty$ = maximum value of $|f(x)|$, $0 \leq x \leq 1$. (That $|f|$ has a maximum will be proved in Section 5.5; assume it for now.) Define $T:V \to W$ by

$$Tf = f$$

(so that T is the identity function, but the norm is changing). Show that T is unbounded.

22. (a) Give the details of the proof of Proposition 7.6.

*(b) Show that the above result still holds even if W is not finite-dimensional.

8. THE CROSS PRODUCT IN R^3

In general, it is not possible to introduce an interesting product of vectors in an arbitrary vector space. However, multiplication of vectors is possible in R^1, and also in R^2 (which we can identify with the complex plane; see Section 1, Chapter 10.) We now introduce a multiplication in R^3 which will be useful to us in Chapter 14.

Let $\vec{u} = (a_1, a_2, a_3)$ and $\vec{v} = (b_1, b_2, b_3)$ be vectors in R^3. The cross product of $\vec{u}$ with $\vec{v}$ is defined to be the vector

$$\vec{u} \times \vec{v} = (a_2 b_3 - a_3 b_2, a_3 b_1 - a_1 b_3, a_1 b_2 - a_2 b_1)$$

For example, if $\vec{e}_1, \vec{e}_2$, and $\vec{e}_3$ are the standard vector bases, then

$$\vec{e}_1 \times \vec{e}_2 = \vec{e}_3, \ \vec{e}_2 \times \vec{e}_3 = \vec{e}_1, \ \vec{e}_3 \times \vec{e}_1 = \vec{e}_2$$

The basic properties of the cross product are contained in the following:

PROPOSITION 8.1. *Let a be a real number and* $\vec{u},\vec{v},\vec{w}$ *vectors in* R^3. *Then*

(a) $a(\vec{u} \times \vec{v}) = (a\vec{u}) \times \vec{v} = \vec{u} \times (a\vec{v})$

(b) $(\vec{u} + \vec{w}) \times \vec{v} = (\vec{u} \times \vec{v}) + (\vec{w} \times \vec{v})$

(c) $\vec{u} \times \vec{v} = -\vec{v} \times \vec{u}$

(d) $< \vec{u} \times \vec{v}, \vec{u} > = < \vec{u} \times \vec{v}, \vec{v} > = 0$

Note: (c) shows that the cross product is *not* commutative. It does not satisfy the associative law, either. See Exercises 3, 4, and 5.

Proof. These properties follow by direct computation. For example, to prove $a(\vec{u} \times \vec{v}) = (a\vec{u}) \times \vec{v}$, we use the definition:

$$a(\vec{u} \times \vec{v}) = a(a_2b_3 - a_3b_2, a_3b_1 - a_1b_3, a_1b_2 - a_2b_1)$$
$$= (a(a_2b_3) - a(a_3b_2), a(a_3b_1) - a(a_1b_3), a(a_1b_2) - a(a_2b_1))$$
$$= ((aa_2)b_3 - (aa_3)b_2, (aa_3)b_1 - (aa_1)b_3, (aa_1)b_2 - (aa_2)b_1)$$
$$= (a\vec{u}) \times \vec{v}$$

The remainder of the proof is left to the reader.

COROLLARY. *For any vector* $\vec{v}$ *in* R^3, $\vec{v} \times \vec{v} = 0$.

Proof. By (c) above, $\vec{v} \times \vec{v} = -\vec{v} \times \vec{v}$, and $\vec{0}$ is clearly the only vector with this property.

Another useful property of the cross product is the following "converse" to the corollary above.

PROPOSITION 8.2. *Let* $\vec{u}$ *and* $\vec{v}$ *be vectors in* R^3 *with* $\vec{u} \times \vec{v} = \vec{0}$. *Then either* $\vec{u}$ *is a multiple of* $\vec{v}$ *or* $\vec{v}$ *is a multiple of* $\vec{u}$.

Proof. Let $\vec{u} = (a_1, a_2, a_3)$ and $\vec{v} = (b_1, b_2, b_3)$. By definition, if $\vec{u} \times \vec{v} = \vec{0}$, then

$$a_2 b_3 - a_3 b_2 = a_3 b_1 - a_1 b_3 = a_1 b_2 - a_2 b_1 = 0$$

If both vectors are zero, then either is a multiple of the other.

Suppose $\vec{v} \neq \vec{0}$ and, for simplicity, assume $b_1 \neq 0$. Using the equations above, we see that

$$a_1 = \frac{a_1}{b_1} b_1 \,, \quad a_2 = \frac{a_1}{b_1} b_2 \,, \quad a_3 = \frac{a_1}{b_1} b_3$$

so that $\vec{u} = \frac{a_1}{b_1} \vec{v}$. A similar argument holds if $\vec{u} \neq \vec{0}$.

PROPOSITION 8.3. *Let $\vec{u}$ and $\vec{v}$ be vectors in R^3 and let θ be the angle between $\vec{u}$ and $\vec{v}$, $0 \leq \theta \leq \pi$. Then*

$$\|\vec{u} \times \vec{v}\| = \|\vec{u}\| \cdot \|\vec{v}\| \sin \theta$$

This is exactly the area of the parallelogram spanned by $\vec{u}$ and $\vec{v}$.

Proof. Let $\vec{u} = (a_1, a_2, a_3)$ and $\vec{v} = (b_1, b_2, b_3)$. Then direct computation shows that

$$\|\vec{u} \times \vec{v}\|^2 = (a_1^2 + a_2^2 + a_3^2)(b_1^2 + b_2^2 + b_3^2) - (a_1 b_1 + a_2 b_2 + a_3 b_3)^2$$
$$= \|\vec{u}\|^2 \cdot \|\vec{v}\|^2 - <\vec{u}, \vec{v}>^2$$

Now

$$<\vec{u}, \vec{v}> = \|\vec{u}\| \cdot \|\vec{v}\| \cos \theta$$

so that

$$\|\vec{u} \times \vec{v}\|^2 = \|\vec{u}\|^2 \|\vec{v}\|^2 (1 - \cos^2 \theta)$$
$$= \|\vec{u}\|^2 \cdot \|\vec{v}\|^2 \sin^2 \theta$$

This result leads to another characterization of the cross product $\vec{u} \times \vec{v}$. Let $\vec{n}$ be the unit vector orthogonal to the plane of $\vec{u}$ and $\vec{v}$. We orient $\vec{n}$ so that turning a right hand screw along the axis of $\vec{n}$ from $\vec{u}$ to $\vec{v}$ moves in the direction of $\vec{n}$. Then

$$\vec{u} \times \vec{v} = (\|\vec{u}\| \cdot \|\vec{v}\| \sin \theta) \vec{n}$$

EXERCISES

1. Let $\vec{u} = (0,1,1)$, $\vec{v} = (2,0,1)$, and $\vec{w} = (2,-1,1)$ and compute the following.

(a) $\vec{u} \times \vec{v}$ (e) $< \vec{u}, \vec{v} \times \vec{w} >$

(b) $\vec{u} \times \vec{w}$ (f) $< \vec{w} - \vec{v}, \vec{u} \times (\vec{v} - \vec{w}) >$

(c) $\vec{u} \times (2\vec{u} + \vec{w})$ (g) $\vec{u} \times (\vec{v} \times \vec{w})$

(d) $\vec{v} \times (2\vec{u} - 3\vec{w})$ (h) $\vec{w} \times ((\vec{u} - \vec{v}) \times (\vec{w} - 2\vec{u})$

2. Prove properties (b), (c), and (d) of Proposition 8.1.

3. Show by example that $(\vec{u} \times \vec{v}) \times \vec{w}$ is not necessarily equal to $\vec{u} \times (\vec{v} \times \vec{w})$.

4. Prove $(\vec{u} \times \vec{v}) \times \vec{w} + (\vec{v} \times \vec{w}) \times \vec{u} + (\vec{w} \times \vec{u}) \times \vec{v} = 0$. (This is called the *Jacobi identity*.)

5. Prove $(\vec{u} \times \vec{v}) \times \vec{w} = \vec{u} \times (\vec{v} \times \vec{w})$ if and only if $(\vec{u} \times \vec{w}) \times \vec{v} = 0$.

6. Let V be a 2-dimensional subspace in R^3, and let $\vec{w}$ be a unit vector orthogonal to V. Show that for $\vec{v} \in R^3$,

$$P_V(\vec{v}) = - \vec{w} \times (\vec{w} \times \vec{v})$$

(See Exercise 10, Section 7.)

CHAPTER 5

COMPACT AND CONNECTED SETS

INTRODUCTION

The notion of a compact set, which we introduce in this chapter, is one of the most important in mathematics. It is needed to prove some of the fundamental theorems of analysis, such as the theorem which states that a continuous real-valued function on a closed, bounded interval attains its maximum.

Besides compactness, the chapter is concerned with several other topics. We introduce the notion of a convergent sequence in a normed vector space and discuss least upper bounds and greatest lower bounds of subsets of R. These concepts are needed for the study of compact sets. We also define uniform continuity and connectedness. The notion of connectedness is used to prove another basic theorem, the Intermediate Value Theorem.

The results of this chapter will be needed in a number of the proofs found later, and they lie at the foundations of mathematical analysis. Many of them, however, are easy to accept on faith, and the reader who wishes to can skip this chapter on first reading, referring back when necessary.

1. CONVERGENT SEQUENCES

In this section, we introduce the notion of a convergent sequence in a normed vector space. Our definition is the obvious extension of the corresponding notion for real numbers.

171

Recall that a sequence $a_1, a_2, \ldots, a_n, \ldots$ of real numbers is said to *converge* to the number a if, for any $\varepsilon > 0$, there is an integer N such that $|a_n - a| < \varepsilon$ whenever $n > N$. That is, no matter how small an ε we choose, all of the terms of the sequence from some point on are within ε of a.

For example, the sequence $1, 1/2, 1/3, \ldots, 1/n, \ldots$ converges to 0. For if ε is any positive number, we pick an integer $N > 1/\varepsilon$. Then, whenever $n > N$,

$$\left| \frac{1}{n} - 0 \right| = \frac{1}{n} < \frac{1}{N} < \varepsilon$$

which proves that the sequence converges to 0.

On the other hand, the sequence $1, -1, 1, -1, \ldots$ does not converge at all. Intuitively, this is clear; the terms of the sequence are not getting close to any single number. To prove it, we must show that no matter what a we choose, we can find an $\varepsilon > 0$ such that, for all N, there is a term a_n of this sequence with $n > N$ and $|a_n - a| \geq \varepsilon$. If $a \neq \pm 1$, this is clear; we simply let ε be the smaller of the two numbers $|a - 1|$, $|a + 1|$. Then $|a_n - a| \geq \varepsilon$ for all n.

If $a = 1$, we see that, no matter how large N is, we can find $n > N$ such that $a_n = -1$, so that $|a_n - a| = 2$. Thus we can set $\varepsilon = 2$ in this case. The same argument works if we set $a = -1$.

Since $\|x\| = |x|$ for $x \in R$, we can extend the definition to an arbitrary normed vector space V as follows: let $\vec{v}_1, \vec{v}_2, \ldots, \vec{v}_n, \ldots$ be a sequence of vectors in V. We say that this sequence *converges to a vector* $\vec{v} \in V$ if for any $\varepsilon > 0$, there is an integer N such that $\|\vec{v}_n - \vec{v}\| < \varepsilon$ whenever $n > N$.

We often write a sequence as $\{\vec{v}_n\}_{n=1}^{\infty}$ or simply as $\{\vec{v}_n\}$. If a sequence $\{\vec{v}_n\}$ converges to $\vec{v}$, we call $\vec{v}$ a *limit point* of the sequence and write

$$\lim_{n \to \infty} \vec{v}_n = \vec{v}$$

We now give generalizations of some of the standard results about sequences of real numbers.

PROPOSITION 1.1. *A convergent sequence in a normed vector space V cannot have two distinct limit points.*

Proof. Suppose on the contrary that $\{\vec{v}_n\}$ is a sequence in V with

$$\lim_{n\to\infty} \vec{v}_n = \vec{v}, \lim_{n\to\infty} \vec{v}_n = \vec{v}'$$

and $\vec{v} \neq \vec{v}'$. Then $r = \|\vec{v}-\vec{v}'\| > 0$ and we set $\varepsilon = r/2$. By assumption, we can find integers N and N' such that

$$\|\vec{v}_n - \vec{v}\| < r/2 \text{ when } n > N$$
$$\|\vec{v}_m - \vec{v}'\| < r/2 \text{ when } m > N'$$

Pick some integer p larger than both N and N'. Then

$$r = \|\vec{v}-\vec{v}'\| = \|\vec{v}-\vec{v}_p - (\vec{v}' - \vec{v}_p)\|$$
$$\leq \|\vec{v}-\vec{v}_p\| + \|\vec{v}' - \vec{v}_p\|$$
$$< r/2 + r/2 = r$$

This gives a contradiction, and so the proposition is proved.

If $\{\vec{v}_n\}$ and $\{\vec{w}_n\}$ are sequences in V, we can define the sum of the sequences to be the sequence $\{\vec{v}_n + \vec{w}_n\}$.

PROPOSITION 1.2. *Suppose $\{\vec{v}_n\}$ and $\{\vec{w}_n\}$ are convergent sequences in V. Then the sequence $\{\vec{v}_n + \vec{w}_n\}$ is convergent and*

$$\lim_{n\to\infty}(\vec{v}_n + \vec{w}_n) = \lim_{n\to\infty} \vec{v}_n + \lim_{n\to\infty} \vec{w}_n$$

Similarly, if $\vec{v}_n$ is a sequence in V and b is a real number, we define a sequence $b\vec{v}_n$ whose n^{th} term is $b\vec{v}_n$.

PROPOSITION 1.3. *Let $\{\vec{v}_n\}$ be a convergent sequence in V and b any real number. Then the sequence $\{b\vec{v}_n\}$ is convergent and*

$$\lim(b\vec{v}_n) = b \lim_{n\to\infty} \vec{v}_n$$

The proofs of these two propositions are similar to the proofs of Propositions 3.4.1 and 3.4.2. We give only the proof of Proposition 1.2.

Let ε be any positive real number and let $\lim_{n\to\infty} \vec{v}_n = \vec{v}$, $\lim_{n\to\infty} \vec{w}_n = \vec{w}$. We need to show that there is a N such that

$$\| (\vec{v}_n + \vec{w}_n) - (\vec{v} + \vec{w}) \| < \varepsilon \text{ whenever } n > N$$

Since $\{\vec{v}_n\}$ and $\{\vec{w}_n\}$ are convergent, we can find N_1 and N_2 so that

$$\| \vec{v}_n - \vec{v} \| < \varepsilon/2$$

when $n > N_1$ and

$$\| \vec{w}_n - \vec{w} \| < \varepsilon/2$$

when $n > N_2$. If we set $N = \max(N_1, N_2)$, we have

$$\| (\vec{v}_n + \vec{w}_n) - (\vec{v} + \vec{w}) \| = \| (\vec{v}_n - \vec{v}) + (\vec{w}_n - \vec{w}) \|$$
$$\leq \| \vec{v}_n - \vec{v} \| + \| \vec{w}_n - \vec{w} \|$$
$$< \varepsilon/2 + \varepsilon/2 = \varepsilon$$

when $n > N$. Thus $\lim_{n\to\infty} (\vec{v}_n + \vec{w}_n) = \vec{v} + \vec{w}$.

Let $\{\vec{v}_n\}$ be a sequence in V and $n_1, n_2, \ldots$ positive integers with

$$n_1 < n_2 < \cdots < n_m < \cdots$$

The sequence $\{\vec{w}_m\}$ defined by

$$\vec{w}_m = \vec{v}_{n_m}$$

is called a *subsequence* of the sequence $\{\vec{v}_n\}$. This terminology is reasonable, since $\{\vec{w}_m\}$ is obtained by omitting some of the terms of the sequence $\{\vec{v}_n\}$. For example, the sequence

$$\frac{1}{2}, \frac{1}{4}, \frac{1}{6}, \ldots, \frac{1}{2m}, \ldots$$

is the subsequence of the sequence

$$1, \frac{1}{2}, \frac{1}{3}, \frac{1}{4}, \ldots, \frac{1}{n}, \ldots$$

obtained by omitting the "odd" terms. More explicitly, the sequence $b_m = 1/2m$ is a subsequence of the sequence $a_n = 1/n$ corresponding to the positive integers $n_m = 2m$, or the integers

$$2 < 4 < 6 < \cdots$$

PROPOSITION 1.4. *Any subsequence of a convergent sequence is convergent and has the same limit point.*

Proof. Let $\{\vec{w}_m\}$ be a subsequence of the sequence $\{\vec{v}_n\}$ corresponding to the positive integers

$$n_1 < n_2 < \cdots < n_m < \cdots$$

Suppose $\lim_{n \to \infty} \vec{v}_n = \vec{v}$, let ε be any positive number, and choose N so that

$$\|\vec{v}_n - \vec{v}\| < \varepsilon$$

whenever $n > N$. Since $\vec{w}_m = \vec{v}_{n_m}$ and $n_m \geq n$, we see that

$$\|\vec{w}_m - \vec{v}\| < \varepsilon$$

whenever $m > N$. Thus $\lim_{m \to \infty} \vec{w}_m = \vec{v}$.

We say a subset S of V is *bounded* if there is a real number M such that $\|\vec{v}\| \leq M$ for all $\vec{v} \in S$. For example, the set $\{1, \frac{1}{2}, \frac{1}{3}, \ldots\}$ is bounded, whereas the set $\{1, 2, 3, \ldots\}$ is not bounded.

PROPOSITION 1.5. *Let $\{\vec{v}_n\}$ be a convergent sequence in* V. *Then the set $\{\vec{v}_1, \vec{v}_2, \ldots, \vec{v}_n, \ldots\}$ is bounded.*

Proof. The idea of the proof is to use the fact that the sequence $\{\vec{v}_n\}$ is convergent to take care of all but a finite number of terms of the sequence. Since any finite set is clearly bounded, this will be sufficient.

Suppose $\lim_{n \to \infty} \vec{v}_n = \vec{v}_0$ and pick N so that $\|\vec{v}_n - \vec{v}_0\| < 1$ when n > N. It follows that

$$
\begin{aligned}
\|\vec{v}_n\| &= \|\vec{v}_n - \vec{v}_0 + \vec{v}_0\| \\
&\leq \|\vec{v}_n - \vec{v}_0\| + \|\vec{v}_0\| \\
&\leq 1 + \|\vec{v}_0\|
\end{aligned}
$$

for n > N. Let M be defined by

$$
M = \max\{\|\vec{v}_1\|, \|\vec{v}_2\|, \ldots, \|\vec{v}_N\|, 1 + \|\vec{v}_0\|\}
$$

Then $\|\vec{v}_n\| \leq M$ for all n.

The following proposition is quite useful. Its proof is straight-forward and is left to the reader.

PROPOSITION 1.6. *Let* $\{\vec{v}_n\}$ *be a sequence in* V. *Then* $\{\vec{v}_n\}$ *converges to* $\vec{v}_0$ *if and only if the sequence* $\{\|\vec{v}_n - \vec{v}_0\|\}$ *converges to* 0 *in* R.

We can also use sequences to characterize closed sets.

THEOREM 1.7. *Let* C *be a subset of* V. *Then* C *is closed if and only if whenever* $\vec{v}_1, \vec{v}_2, \ldots$ *is a sequence of vectors in* C *with* $\lim_{j \to \infty} \vec{v}_j = \vec{v}$ *in* V, *then* $\vec{v}$ *must be in* C.

Proof. Suppose C closed; let $\{\vec{v}_1, \vec{v}_2, \ldots\}$ be a sequence in C with $\lim_{n \to \infty} \vec{v}_n = \vec{v}$. We need to show that $\vec{v} \in C$. Suppose not; then $\vec{v} \in V - C$. Since C is closed, V - C is open and there is an r > 0 with $B_r(\vec{v}) \subset V - C$. But this implies that $\|\vec{v}_n - \vec{v}\| \geq r$ for all n, which contradicts the fact that $\lim_{n \to \infty} \vec{v}_n = \vec{v}$. This proves the "only if" part of the theorem.

Now suppose $C \subset V$ is such that whenever $\vec{v}_1, \ldots, \vec{v}_n, \ldots$ is a sequence in C with $\lim_{j \to \infty} \vec{v}_j = \vec{v}$ in V, then $\vec{v} \in C$. If C is not closed, then V - C is not open, and we can therefore find a vector $\vec{v} \in V - C$ with the property that no ball about $\vec{v}$ is contained in V - C. That is, every ball about $\vec{v}$ contains points of C. It follows that for each integer n, we can find a vector $\vec{v}_n$ in $B_{1/n}(\vec{v})$ which is also in C.

Thus $\lim\limits_{n\to\infty} \vec{v}_n = \vec{v}$ but $\vec{v} \notin C$. This contradicts our hypothesis about C, and the result follows.

EXERCISES

1. Which of the following sequences converge ? Give proofs.

   (a) $1, \frac{1}{2}, -1, \frac{1}{3}, 1, \frac{1}{4}, -1, \frac{1}{5}, \ldots$

   (b) $2, \frac{1}{2}, 1\frac{1}{3}, \frac{3}{4}, 1\frac{1}{5}, \frac{5}{6}, 1\frac{1}{7}, \ldots$

   (c) $\frac{1}{3}, \frac{2}{5}, \frac{3}{7}, \frac{4}{9}, \frac{5}{11}, \ldots$

   (d) $1, -2, 3, -4, 5, -6, \ldots$

   (e) $1 + \frac{1}{1}, 1 + \frac{1}{4}, 1 + \frac{1}{9}, 1 + \frac{1}{16}, \ldots$

   (f) $1\frac{1}{2}, 2\frac{1}{4}, 1\frac{1}{8}, 2\frac{1}{16}, 1\frac{1}{32}, \ldots$

   (g) $1, 2^2, 3^3, 4^4, \ldots$

   (h) $\frac{1^{100}}{3^1}, \frac{2^{100}}{3^2}, \frac{3^{100}}{3^3}, \frac{4^{100}}{3^4}, \frac{5^{100}}{3^5}, \ldots$

2. Prove Proposition 1.3.

3. Prove Proposition 1.6.

4. Let a_n be a convergent sequence in R and $\vec{v}$ a fixed vector in V. Prove that the sequence $\{a_n\vec{v}\}$ converges and that

$$\lim_{n\to\infty} (a_n\vec{v}) = (\lim_{n\to\infty} a_n)\vec{v}$$

5. Define a sequence of real numbers by $a_n = r^n$. Prove that this sequence converges to 0 if $|r| < 1$, converges to 1 if $r = 1$, and diverges otherwise.

6. Let $\{\vec{v}_n\}$ and $\{\vec{w}_n\}$ be sequences in V with

$$\lim_{n\to\infty} \vec{v}_n = \vec{v}_0, \quad \lim_{n\to\infty} \vec{w}_n = \vec{w}_0$$

Prove that the sequence

$$\vec{v}_1, \vec{w}_1, \vec{v}_2, \vec{w}_2, \vec{v}_3, \vec{w}_3, \ldots$$

diverges if $\vec{v}_0 \neq \vec{w}_0$ and converges if $\vec{v}_0 = \vec{w}_0$.

*7.   Let V be an inner product space and let $\{\vec{v}_n\}$, $\{\vec{w}_n\}$ be sequences in V with

$$\lim_{n\to\infty} \vec{v}_n = \vec{v}_0, \quad \lim_{n\to\infty} \vec{w}_n = \vec{w}_0$$

Define a sequence $\{a_n\}$ of real numbers by

$$a_n = <\vec{v}_n, \vec{w}_n>$$

and prove that $\lim_{n\to\infty} a_n = <\vec{v}_0, \vec{w}_0>$

8.   Let $\vec{v}_n = (x_n, y_n, z_n)$ be a sequence in R^3. Prove that

$$\lim_{n\to\infty} \vec{v}_n = \vec{v} = (x, y, z)$$

if and only if

$$\lim_{n\to\infty} x_n = x, \quad \lim_{n\to\infty} y_n = y, \quad \lim_{n\to\infty} z_n = z$$

9.   A sequence $\{\vec{v}_n\}$ in V is called a *Cauchy sequence* if for any $\varepsilon > 0$ there is an integer N such that $\|\vec{v}_n - \vec{v}_m\| < \varepsilon$ whenever $n, m > N$. Prove that any convergent sequence is a Cauchy sequence.

10.   If $\{\vec{v}_n\}$ is a Cauchy sequence, prove that the set

$$\{\vec{v}_1, \vec{v}_2, \ldots\}$$

is bounded.

11. Let $\{x_n\}$ be a convergent sequence of real numbers converging to a number x. Show that if $a \geq x_n$ for all n, then $a \geq x$; and that if $b \leq x_n$ for all n, then $b \leq x$. Is it always true that if $a > x_n$ for all n, then $a > x$?

*12. State and prove the analogue of Exercise 8 for sequences in R^n.

*13. State and prove the analogue of Exercise 8 for sequences in a finite dimensional inner product space V.

14. (a) Show that if U is an open subset of the normed vector space V, then for every vector $\vec{v} \in U$ there is a sequence $\{\vec{v}_n\}$ of vectors in U such that $\vec{v}_n \neq \vec{v}$ for all n and $\lim_{n \to \infty} \vec{v}_n = \vec{v}$.

(b) Suppose that $U \subset V$ has the property that, for every point $\vec{v} \in U$ there is a sequence $\{\vec{v}_n\}$ of vectors as in part (a). Is U open?

*15. Suppose that $\{\vec{v}_n\}$ is a sequence of vectors in a normed vector space V with the following property: there is a vector $\vec{v} \in V$ such that if $\{\vec{v}_{n_k}\}$ is any subsequence of $\{\vec{v}_n\}$, some subsequence of $\{\vec{v}_{n_k}\}$ converges to $\vec{v}$. Show that $\lim_{n \to \infty} \vec{v}_n = \vec{v}$.

16. Suppose that $\| \ \|_{(1)}$ and $\| \ \|_{(2)}$ are equivalent norms on the vector space V. (See Section 3.1, Exercise 5, for the definition.) Show that the sequence $\{\vec{v}_n\}$ converges to a vector $\vec{v}$ in the $\| \ \|_1$ norm if and only if it converges to $\vec{v}$ in the $\| \ \|_2$ norm.

*17. Suppose that $\{\vec{v}_n\}$ is a sequence of vectors in a normed vector space such that no subsequence of $\{\vec{v}_n\}$ converges. Show that:

(a) no vector is repeated infinitely many times in the sequence;

(b) $\{\vec{v}_n\}$ (as a subset of V) is closed.

2. FURTHER PROPERTIES OF CONTINUOUS FUNCTIONS

We begin with a useful criterion for continuity which uses convergent sequences.

THEOREM 2.1. *A function* $f:V \to W$ *between normed vector spaces is continuous at* $\vec{v}_0 \subset V$ *if and only if whenever* $\{\vec{v}_n\}$ *is a convergent sequence in V with limit point* $\vec{v}_0$, *then* $\{f(\vec{v}_n)\}$ *is a convergent sequence in W with limit point* $f(\vec{v}_0)$.

Proof. Suppose first of all that f is continuous at $\vec{v}_0$ and let $\{\vec{v}_n\}$ be a convergent sequence with limit point $\vec{v}_0$. We must show that

$$\lim_{n \to \infty} f(\vec{v}_n) = f(\vec{v}_0)$$

Equivalently, we must show that for any $\varepsilon > 0$, there is an integer N such that

$$\|f(\vec{v}_n) - f(\vec{v}_0)\| < \varepsilon$$

whenever $n > N$.

Suppose then that we are given $\varepsilon > 0$. Since f is continuous at $\vec{v}_0$, we can find δ so that

$$\|f(\vec{v}) - f(\vec{v}_0)\| < \varepsilon$$

when $\|\vec{v} - \vec{v}_0\| < \delta$. Furthermore, $\lim_{n \to \infty} \vec{v}_n = \vec{v}_0$, so we can find an integer N such that

$$\|\vec{v}_n - \vec{v}_0\| < \delta$$

whenever $n > N$. Thus

$$n > N \Rightarrow \|\vec{v}_n - \vec{v}_0\| < \delta$$
$$\Rightarrow \|f(\vec{v}_n) - f(\vec{v}_0)\| < \varepsilon$$

and $f(\vec{v}_n)$ converges to $f(\vec{v}_0)$.

We now need to prove the converse, that if $f:V \to W$ takes sequences converging to $\vec{v}_0$ into sequences converging to $f(\vec{v}_0)$, then f is continuous at $\vec{v}_0$. We begin by transforming this statement into a logically equivalent statement. (Use Section 1.2 as a reference for these logical manipulations.) First of all, we may instead prove the contrapositive of the converse; that is, we shall prove that if f is not continuous at $\vec{v}_0$, then there is a sequence $\{\vec{v}_n\}$ in V converging to $\vec{v}_0$ such that $\{f(\vec{v}_n)\}$ does *not* converge to $f(\vec{v}_0)$ in W.

Next, we need to interpret the statement that f is not continuous at $\vec{v}_0$. We know (from the definition) that it means that the statement

$$(\forall \varepsilon > 0)(\exists \delta > 0)(\forall \vec{v} \in V) \text{ if } \|\vec{v} - \vec{v}_0\| < \delta, \text{ then } \|f(\vec{v}) - f(\vec{v}_0)\| < \varepsilon$$

is *false*. We gave a rule in Section 1.2 for negating such statements: change $\forall$ to $\exists$ and $\exists$ to $\forall$, and deny the final statement. Thus "f is not continuous at $\vec{v}_0$" means

$$\text{"}(\exists \varepsilon > 0)(\forall \delta > 0)(\exists \vec{v} \in V): \text{ it is false that if } \|\vec{v} - \vec{v}_0\| < \delta,$$
$$\text{then } \|f(\vec{v}) - f(\vec{v}_0)\| < \varepsilon \text{ "}$$

or

"There is an $\varepsilon_0 > 0$ such that for any $\delta > 0$, we can pick $\vec{v} \in V$ such that $\|\vec{v} - \vec{v}_0\| < \delta$, but $\|f(\vec{v}) - f(\vec{v}_0)\| \geq \varepsilon_0$ "

In particular, for any integer n, we can find a vector $\vec{v}_n$ in V with

$$\|\vec{v}_0 - \vec{v}_n\| < 1/n \text{ and } \|f(\vec{v}_0) - f(\vec{v}_n)\| \geq \varepsilon_0$$

It follows from Proposition 1.6 that the sequence $\{\vec{v}_n\}$ converges to $\vec{v}_0$. However, $\{f(\vec{v}_n)\}$ cannot converge to $f(\vec{v}_0)$, since $\|f(\vec{v}_n) - f(\vec{v}_0)\| \geq \varepsilon_0 > 0$ for all n. This concludes the proof of Theorem 2.1.

Armed with this result, we go back to deriving properties of continuous functions.

Suppose $g:V_1 \rightarrow V_2$, $f:V_2 \rightarrow V_3$ are functions. We define the composite of f with g to be the function $f \circ g : V_1 \rightarrow V_3$ where $(f \circ g)(\vec{v}) = f(g(\vec{v}))$.

PROPOSITION 2.2. *Let* V_1, V_2, *and* V_3 *be normed vector spaces and* $g:V_1 \rightarrow V_2$, $f:V_2 \rightarrow V_3$ *functions. If g is continuous at* $\vec{v}_0 \in V_1$ *and f is continuous at* $g(\vec{v}_0)$ *in* V_2, *then* $f \circ g$ *is continuous at* $\vec{v}_0$.

Proof. We use Theorem 2.1 here. We show that if $\{\vec{v}_n\}$ is a sequence in V_1 converging to $\vec{v}_0$, then the sequence $\{(f \circ g)(\vec{v}_n)\}$ converges to $(f \circ g)(\vec{v})$ in V_3.

Since g is continuous at $\vec{v}_0$, it follows from Proposition 2.1 that the sequence $\{g(\vec{v}_n)\}$ converges to $g(\vec{v}_0)$ in V. Similarly, since f is continuous at $g(\vec{v}_0)$, the sequence $\{f(g(\vec{v}_n))\}$ converges to $f(g(\vec{v}_0))$. But this last assertion means precisely that the sequence $\{(f \circ g)(\vec{v}_n)\}$ converges to $(f \circ g)(\vec{v}_0)$ in V_3. Applying Theorem 2.1 again, we see that $f \circ g$ is continuous at $\vec{v}_0$.

We say a function $f : V \to W$ is *locally bounded* at $\vec{v}_0 \in V$ if there are positive numbers M and δ such that $\|f(\vec{v})\| \leq M$ whenever $\|\vec{v} - \vec{v}_0\| < \delta$. For instance, if $V = R^1$, this says that values of $f(\vec{v})$ are no greater than M in absolute value when $\vec{v}$ is near $\vec{v}_0$, which is what locally bounded ought to mean.

PROPOSITION 2.3. *Let* $f : V \to W$ *be continuous at* $\vec{v}_0 \in V$. *Then* f *is locally bounded at* $\vec{v}_0$.

Proof. Suppose f is not locally bounded at $\vec{v}_0$. Then, for every $\varepsilon > 0$ and M > 0, there is a $\vec{v} \in V$ with $\|\vec{v} - \vec{v}_0\| < \varepsilon$ and $\|f(\vec{v})\| > M$. In particular, for every positive integer n we can find $\vec{v}_n \in V$ with $\|\vec{v}_n - \vec{v}_0\| < 1/n$ and $\|f(\vec{v}_n)\| > n$. It follows from Proposition 1.6 that the sequence $\{\vec{v}_n\}$ converges to $\vec{v}_0$. Furthermore, since the sequence $\{f(\vec{v}_n)\}$ is not bounded, it follows from Proposition 1.5 that it cannot be convergent. Thus f cannot be continuous at $\vec{v}_0$, by Theorem 2.1.

We close this section with two additional characterizations of continuity involving open and closed sets.

PROPOSITION 2.4. *A function* $f : V \to W$ *is continuous if and only if* $f^{-1}(U)$ *is an open subset of* V *whenever* U *is an open subset of* W.

Proof. Recall that $f^{-1}(U)$ is defined by

$$f^{-1}(U) = \{\vec{v} \in V : f(\vec{v}) \in U\}$$

Suppose first of all that f is continuous and U is an open sub-set of W. Let $\vec{v}_0$ be a vector in $f^{-1}(U)$; then $\vec{w}_0 = f(\vec{v}_0) \in U$. Since U is open, we can find an $\varepsilon > 0$ so that $B_\varepsilon(\vec{w}_0) \subset U$. Since f is continuous at $\vec{v}_0$, we can find a $\delta > 0$ so that

$$\|f(\vec{v}) - f(\vec{v}_0)\| < \varepsilon$$

whenever $\|\vec{v} - \vec{v}_0\| < \delta$, or, equivalently,

$$f(\vec{v}) \in B_\varepsilon(\vec{w}_0)$$

whenever $\vec{v} \in B_\delta(\vec{v}_0)$. Thus

$$B_\delta(\vec{v}_0) \subset f^{-1}(B_\varepsilon(\vec{w}_0)) \subset f^{-1}(U)$$

and $f^{-1}(U)$ is open.

Conversely, suppose $f^{-1}(U)$ is an open subset of V whenever U is an open subset of W. Let $\vec{v}_0$ be a vector in V with $\vec{w}_0 = f(\vec{v}_0)$. Then for any $\varepsilon > 0$, $B_\varepsilon(\vec{w}_0)$ is open in W, so $f^{-1}(B_\varepsilon(\vec{w}_0))$ is open in V. Since $\vec{v}_0 \in f^{-1}(B_\varepsilon(\vec{w}_0))$, we can find a $\delta > 0$ so that

$$B_\delta(\vec{v}_0) \subset f^{-1}(B_\varepsilon(\vec{w}_0))$$

This is the same as saying that $\|\vec{v} - \vec{v}_0\| < \delta$ implies that $\|f(\vec{v}) - f(\vec{v}_0)\| < \varepsilon$, so f is continuous at $\vec{v}_0$. Since $\vec{v}_0$ was arbitrarily chosen in V, f is continuous on V and the proposition is proved.

PROPOSITION 2.5. *A function* $f:V \to W$ *is continuous if and only if* $f^{-1}(C)$ *is a closed subset of* V *whenever* C *is a closed subset of* W.

This follows easily from Proposition 2.4 and the fact that $f^{-1}(W - U) = V - f^{-1}(U)$. We leave the details to the reader as Exercise 6.

EXERCISES

1.   Let V be an inner product space and $\vec{v}_0 \in V$ a fixed vector. De-fine $h:V \to V$ by

$$h(\vec{v}) = < \vec{v},\vec{v}_0> \vec{v}$$

Prove that h is continuous.

2.   Let V be a normed vector space, W an inner product space, and f,g:V $\to$ W functions continuous at $\vec{v}_0$ $\in$ V. Prove that F:V $\to$ R, defined by

$$F(\vec{v}) = < f(\vec{v}),g(\vec{v}) >$$

is continuous at $\vec{v}_0$. (Hint: See Exercise 8 of Section 4.5.)

3.   Give an example to show that a function locally bounded at $\vec{v}_0$ need not be continuous at $\vec{v}_0$.

4.   If f:V $\to$ W is continuous and U is open in V, is f(U) necessarily open in W?

5.   If f:V $\to$ W is continuous and C is closed in V, is f(C) necessarily closed in W?

6.   Prove Proposition 2.5.

7.   Prove Proposition 2.2 directly from the definition of continuity (without using Theorem 2.1).

8.   Prove that f:V $\to$ W is continuous at $\vec{v}_0$ $\in$ V if and only if for any open set U_1 of W containing $f(\vec{v}_0)$, there is an open set U_0 in V containing $\vec{v}_0$ such that $f(U_0) \subset U_1$.

9.   Suppose that f:V $\to$ R^3 and g:V $\to$ R^3 are continuous. Define F:V $\to$ R^3 by F(v) = f(v) $\times$ g(v). (See Section 4.8 for the definition of the cross product.) Is F continuous? Why (or why not)?

*10.   Prove the following analogue of Proposition 2.4: Let U be an open subset of V and f:U $\to$ W a function. Prove that f is continuous if and only if $f^{-1}(U_1)$ is an open subset of V whenever U_1 is an open subset of W.

*11. Prove the following analogue of Proposition 2.5: Let X be a
closed subset of V and f:X → W a function. Prove that f is continuous
if and only if f^{-1}(C) is a closed subset of V whenever C is a closed
subset of W.

3. COMPACT SETS

A subset C of the normed vector space V is said to be *compact** if every
sequence of points in C has a subsequence which converges to a limit
point in C.

For example, the open interval

$$(0,1) = \{x \in R^1 : 0 < x < 1\}$$

is *not* compact since the sequence a_n = 1/n in (0,1) has no subsequence
converging to a point *in (0,1)*. This is true because the sequence
converges to 0. Consequently, every subsequence of this sequence converges to 0, and 0 ∉ (0,1). (See Proposition 1.4.)

Another set which is not compact is the set $C = R^1 \subset R^1$. This is
because the sequence a_n = n has no convergent subsequences.

To give examples of sets which are compact is more difficult.
We give one (not very interesting) example here.

PROPOSITION 3.1. *Any finite subset of V is compact.*

Proof. Suppose C = $\{\vec{v}_1, \ldots, \vec{v}_k\}$; let $\{\vec{a}_n\}$ be a sequence in C.
Since C is finite, one of the elements of C must appear an infinite
number of times in this sequence. More explicitly, there is a sequence of integers

$$0 < i_1 < i_2 < \cdots < i_r < \cdots$$

with $\vec{a}_{i_1} = \vec{a}_{i_2} = \cdots = \vec{a}_{i_r} = \cdots = \vec{v}_j$ for some j, 1 ≤ j ≤ k. This gives
us our convergent subsequence with limit point $\vec{v}_j$ in C.

*One sometimes says that C is *sequentially* or *countably* compact.

Proposition 3.1 hardly provides evidence that compact sets are worth studying. We shall describe some more interesting compact sets in a moment. First, however, we prove two simple properties of compact sets.

PROPOSITION 3.2. *Any compact subset* C *of a normed vector space* V *is closed.*

Proof. We use the characterization of closed sets given in Proposition 1.7. In particular, we must show that if $\{\vec{v}_n\}$ is a sequence in C which converges to a vector $\vec{v}$ in V, then $\vec{v}$ is in C. Since C is compact, there is a subsequence of the sequence $\{\vec{v}_n\}$ converging to a vector $\vec{v}'$ in C. Since the sequence $\{\vec{v}_n\}$ is convergent, any subsequence must converge to $\vec{v}$, by Proposition 1.4. It follows that $\vec{v} = \vec{v}'$ is in C, so C is closed.

Recall that a C of a normed vector space V is said to be *bounded* if there is a real number r > 0 such that $\|\vec{v}\| < r$ for all $\vec{v}$ in C. Equivalently, C is bounded if there is an r > 0 with $C \subset B_r(\vec{0})$.

PROPOSITION 3.3. *Any compact set is bounded.*

Proof. Suppose C is a subset of a normed vector space V which is not bounded. Then C is contained in no ball about $\vec{0}$ in V. It follows that, for every integer n, we can find a vector $\vec{w}_n$ in C which is not in $B_n(\vec{0})$. Since no subsequence of the sequence $\{\vec{w}_n\}$ is bounded, it follows from Proposition 1.5 that no subsequence of this sequence can be convergent. Thus C cannot be compact if C is not bounded.

The two results above tell us that a compact set in a normed vector space V is both closed and bounded. In Section 6, we shall show that the converse is true if V is also finite dimensional; namely, that any closed and bounded subset of V is compact. The next result will be useful in proving this.

PROPOSITION 3.4. *Suppose* C *is a compact subset of the normed vector space* V *and* B *is a closed subset of* V *contained in* C. *Then* B *is compact.*

Proof. Let $\{\vec{v}_n\}$ be a sequence in B. We must show that this sequence has a subsequence converging to a vector in B.

Since B is contained in C, $\{\vec{v}_n\}$ is a sequence in C, so we can find a subsequence $\vec{v}_{j_1}, \vec{v}_{j_2}, \ldots$ converging to a vector $\vec{v}$ in C. However, the subsequence $\vec{v}_{j_1}, \vec{v}_{j_2}, \ldots$ is a sequence in the closed set B which converges in V, so its limit point $\vec{v}$ must be in B. This completes the proof.

EXERCISES

Note. The reader may wish to read the next few sections to get some experience with the procedures used to prove statements about compact sets before doing these exercises.

1.   Prove that the union of a finite collection of compact sets is compact. Is the union of an infinite collection of compact sets necessarily compact?

2.   Prove that the intersection of any collection of compact sets is compact.

3.   Prove that if C_1 and C_2 are compact sets in R, then $C_1 \times C_2$ is compact in $R \times R = R^2$. (See Section 1.1 for the definition).

4.   State and prove the analogue of Exercise 3 for R^n.

*5.   Let $C_1, C_2, \ldots, C_n, \ldots$ be a decreasing collection of non-empty compact subsets of V (that is, $C_{n+1} \subset C_n$, all $n \geq 1$). Prove that the intersection $\cap\, C_n$ is non-empty. (Hint: choose $\vec{v}_n \in C_n$, and show that some subsequences of the $\vec{v}_n$ converges to a point $\vec{v}$. Prove that $\vec{v}$ is in the intersection $\cap\, C_n$.)

*6.   Give an example to show that the conclusion of Exercise 5 does not hold if we only assume that the sets C_n are closed sets.

*7.   Prove that the set $C = \{0,1,\frac{1}{2},\frac{1}{3},\frac{1}{4},\ldots\} \subset R$ is compact. (Hint:
let $a_1,a_2,\ldots$, be any sequence of elements of C. If there are elements
of the sequence of arbitrarily large denominators, show that there
is a subsequence converging to 0. If not, show that Proposition 3.1
applies.)

*8.   Let V be a normed vector space with C a compact subset of V.
Prove that for any $\varepsilon > 0$, there is a finite set $F_\varepsilon \subset C$ such that

$$C \subset \bigcup_{\vec{v} \in F_\varepsilon} B_\varepsilon(\vec{v})$$

The remaining problems will involve the notion of an open covering
of a set.

Let V be an inner product space and C a subset of V. An *open
covering* of C is a collection $U = \{U_\lambda : \lambda \in \Lambda\}$ of open subsets of V
such that

$$C \subset \bigcup_{\lambda \in \Lambda} U_\lambda$$

We say that U is *finite* if Λ is finite and *countable* if Λ is countable (i.e., Λ is the natural numbers). An open covering V of C is a
subcovering of the open covering U of C if each open set in V is also
one of the open sets in U.

*9.   Let V be a normed vector space and C a subset of V. Prove that
C is compact if and only if every countable open covering of C has a
finite subcovering. (That is, if $U = \{U_j\}$ is any countable open cov-
ering of C, then there are indices $j_1,\ldots,j_m$ with $C \subset U_{j_1} \cup U_{j_2} \cup \cdots$
$\cup U_{j_m}$.) (Hint: if $\{U_j\}$ has no such subcovering, then $\bigcup_{j=1}^{n} U_j = C$,
all n. Pick $\vec{v}_n \in C - \bigcup_{j=1}^{n} U_j$. Let a subsequence $\{\vec{v}_{n_k}\}$ of $\{\vec{v}_n\}$ converge
to $\vec{v}$; if $\vec{v} \in U_N$, show that $\vec{v}_{n_k} \in U_N$ for arbitrarily large numbers $\{n_k\}$.
Deduce a contradiction.

Conversely, suppose that $\{\vec{v}_n\}$ is a sequence such that no subsequence converges to any point in C. Show that if any subsequence of

$\{\vec{v}_n\}$ converges, then C is not closed. If no subsequence converges in V, then show that $U_j = V - \{\vec{v}_j, \vec{v}_{j+1}, \ldots\}$ is open and that $\{U_j\}$ is a covering with no finite subcovering. (See Exercise 17 of Section 1.)

*10. Let $U = \{U_\lambda : \lambda \in \Lambda\}$ be an open covering of the compact set $C \subset V$. Prove that there is an $r > 0$ such that, for any $\vec{v} \in C$, $B_r(\vec{v}) \subset U_\lambda$ for some $\lambda \in \Lambda$. (This result is called the *Lebesgue Covering Lemma*.) (Hint: if not, then for every n there is a $\vec{v}_n$ such that $B_{1/n}(\vec{v}_n) \not\subset U_\lambda$ for any λ. Pick a subsequence $\{\vec{v}_{n_k}\}$ converging to $\vec{v}$; since $\vec{v}$ is in some U_λ, $\exists N$ with $B_{1/N}(\vec{v}) \subset U_\lambda$. Show that for some n_k, $B_{1/n_k}(\vec{v}_{n_k}) \subset B_{1/N}(\vec{v})$, and get a contradiction.)

*11. Prove that $C \subset V$ is compact if and only if every open covering of C has a finite subcovering. (This result is called the *Heine-Borel Theorem*.) (Hint: use Exercises 9 and 7 in one direction, and 8 in the other.)

*12. We say that a set C has the *finite intersection property* if whenever $C_1, C_2, \ldots, C_m, \ldots$ is a sequence of closed sets such that

$$C \cap \bigcap_{j=1}^{m} C_j \neq \phi$$

for all m, then

$$C \cap \bigcap_{j=1}^{\infty} C_j \neq 0$$

Prove that C is compact if and only if it has the finite intersection property.

4. UPPER AND LOWER BOUNDS

The material in this section will be used later in this chapter, but is somewhat different in content from the other sections. The reader may wish to read the basic definitions and then go ahead to Section 5, referring back as necessary.

Let X be a subset of the real numbers. An *upper bound* for X is a number M with the property that $x \leq M$ for all x in X. A *lower bound* for X is a number m with $m \leq x$ for all x in X. We say that X is *bounded from above* if X has some upper bound and *bounded from below* if X has a lower bound.

EXAMPLES

(1) Let X = $\{x \in R : 0 \leq x < 1\}$. The number 2 is an upper bound for X, since every element of X is ≤ 2. The numbers 1 and 5 are also upper bounds for X. Similarly, -7 is a lower bound for X, since every element of X is ≥ -7. The number is neither an upper bound for X (since $1/2 < 3/4 \in X$) nor a lower bound (since $1/2 > 1/4 \in X$).

(2) Let X = $\{x \in R : x$ is an integer$\}$. Then X has no upper bound, since there is no real number greater than every integer. Similarly, X has no lower bound.

We say that M is a *least upper bound* for X if

(i) M is an upper bound for X

and

(ii) if M' is any other upper bound for X, then $M \leq M'$

Similarly, m is a *greatest lower bound* for X if

(i) m is a lower bound for X

and

(ii) if m' is any other lower bound for X, then $m' \leq m$

EXAMPLES

(3) Let X = $\{x \in R : 0 \leq x < 1\}$. The number 2 is an upper bound for X, since every element of X is ≤ 2. The number 1 is also an upper bound for X, and it is clear that 1 is the smallest upper bound we can choose. Therefore 1 is the least upper bound. Similarly, 0 is the greatest lower bound.

(4) Let $X = \{x \in R : x$ is rational and $x^3 < 2\}$. Then every rational negative number is in X, so no number is smaller than every number in X. Thus X has no lower bound. The number 2 is an upper bound for X, as is $\sqrt[3]{2}$. (The elements of X must be rational, but the upper bounds need not be.) In fact, $\sqrt[3]{2}$ is the least upper bound of X. (See Exercise 4.)

Our first result about upper bounds is fairly easy.

PROPOSITION 4.1. *If a set* $X \subset R$ *has a least upper bound, it is unique. If* X *has a greatest lower bound, it is unique.*

Proof. We prove the statement about least upper bounds, leaving the other half as an exercise. Suppose that x_1 and x_2 are two least upper bounds for X. Then x_1 is an upper bound, by (i) of the definition, so that, from (ii) of the definition,

$$x_1 \leq x_2$$

Similarly, x_2 is an upper bound, and therefore

$$x_2 \leq x_1$$

It follows that $x_1 = x_2$, as claimed.

We denote the least upper bound and greatest lower bound of X by lub X and glb X when they exist. Many books also write sup X (for *supremum* of X) instead of lub X, and inf X (*infimum* of X) for glb X.

It is often important to know that glb's and lub's exist; the purpose of the next theorem is to state that they do.

THEOREM 4.2. *If* $X \subset R$ *is nonempty and bounded from above, then* X *has a least upper bound. If* X *is bounded from below, then* X *has a greatest lower bound.*

To prove Theorem 4.2, we need to use special properties of the real numbers. For example, it is not necessarily true that if X is a nonempty set of *rational* numbers bounded from above, then X has a least upper bound which is *rational*. (See Example 4 above.) We need some sort of statement that the set of real numbers, unlike the rationals, does not have any "holes." Such a statement either must be taken as an axiom about the real numbers or must be proved in the course of providing a construction of the real numbers. Since we are not constructing the real numbers, we give an axiom.

We begin with a definition. A sequence $\{a_n\}$ of numbers in R is called a *Cauchy* sequence if for every $\varepsilon > 0$ there is an integer N such that $|a_m - a_n| < \varepsilon$ wherever $m, n > N$. (See Exercises 9 and 10 of Section 1.) Thus the points of a Cauchy sequence get closer and closer together as n gets large.

Our axiom can now be stated as follows:

Completeness of the real numbers. Let $\{a_n\}$ be a Cauchy sequence of real numbers. Then there is a real number a_0 such that

$$\lim_{n \to \infty} a_n = a_0$$

Armed with this axiom, we can prove Theorem 4.2. We prove only the first half; the proof of the other half is similar.

To begin with, notice we can pick real numbers a_0, b_0 such that a_0 is not an upper bound and b_0 is an upper bound. We can find b_0 because we have assumed that X is bounded above. As for a_0, let x be any element of X (X is nonempty), and let $a_0 = x - 1$. Of course, $b_0 > a_0$.

We proceed to define two sequences, $\{a_n\}$ and $\{b_n\}$. Let $c_0 = (a_0 + b_0)/2$. If c_0 is an upper bound for X, let $b_1 = c_0$ and $a_1 = a_0$. If c_0 is not an upper bound for X, let $b_1 = b_0$ and $a_1 = c_0$. Either way, we have defined a_1 and b_1 so that $a_0 \le a_1$, $b_0 \ge b_1$, a_1 is not an upper bound for X, b_1 is an upper bound for X, $a_1 < b_1$, and $b_1 - a_1 = (1/2)(b_0 - a_0)$. Next, let $c_1 = (a_1 + b_1)/2$. If c_1 is an upper bound for X, let $b_2 = c_1$ and $a_2 = a_1$; if not, let $b_2 = b_1$ and $a_2 = c_1$. And so on.

In this way, we get sequences $\{a_n\}$ and $\{b_n\}$ such that:

$\{a_n\}$ is nondecreasing and $\{b_n\}$ is nonincreasing   (4.1) the elements of $\{b_n\}$ are upper bounds of X, and the elements of $\{a_n\}$ are not $\hfill (4.2)$

for all n, $a_n < b_n$, and $b_n - a_n = \dfrac{1}{2^n} (b_0 - a_0)$ $\hfill (4.3)$

If $m > n$, then $|a_m - a_n| = a_m - a_n \leq b_m - a_n \leq b_n - a_n = 2^{-n}(b_0 - a_0)$. From this, it is easy to see that $\{a_n\}$ is a Cauchy sequence. Thus it converges to a limit, which we shall call a. Similarly, $\{b_n\}$ converges to a limit b. Since $b - a = \lim\limits_{n \to \infty} b_n - a_n = \lim\limits_{n \to \infty} 2^{-n}(b_0 - a_0) = 0$, $a = b$.

We are now ready to show that a is the least upper bound of X. First of all, we must show that it is an upper bound. If not, we can find $x \in X$ with $x > a$. Let $x - a = \varepsilon > 0$. Since $b_n \to a$, we can find an n with $|b_n - a| < \varepsilon$. Then $b_n < x$, which contradicts the fact that b_n is an upper bound.

To finish, we must show that if y is any upper bound, then $a \leq y$. This is not hard. Since a_n is not an upper bound, there is an $x_n \in X$ with $a_n < x_n$. But $y \geq x_n$; thus $y \geq a_n$, for all n. Taking limits (see Exercise 11 of Section 1), we see that $y \geq a$. This proves the theorem.

Note. Our choice of a completeness axiom for the real numbers was somewhat arbitrary; there are various other axioms which work as well. One equivalent statement is that of Theorem 4.2: any nonempty set which is bounded from above has a least upper bound. Another one which is often used says that if $\{x_n\}$ is a nondecreasing sequence of real numbers which is bounded above, then $\{x_n\}$ converges. Each has some advantages. We chose the statement about Cauchy sequences in part because it will be the most convenient form for use in Section 6 of this chapter.

The reader may want to prove each of the forms of the Completeness Axioms from the others. See Exercises 12 and 13 for examples.

EXERCISES

1.   Determine glb X and lub X when they exist, where

   (a) $X = \{x \in R : x^5 < 32\}$

   (b) $X = \{x \in R : |x| \geq 1\}$

   (c) $X = \{x \in R : x > 0 \text{ and } \sin x = 1/2\}$

   (d) $X = \{x \in R : |2x - 4| < 3\}$

   (e) $X = \{x \in R : e^x \leq 3\}$

   (f) $X = \{x \in R : x^2 + x < 2\}$

   (g) $X = \{x \in R : x^3 - 3x^2 + 2x \geq 0\}$

2.   Prove the second half of Proposition 4.1.

3.   Let $\{a_n\}$ be a sequence of real numbers such that $a_j \leq a_{j+1}$ for all j and such that for some number M, $a_j \leq M$ for all j. Let $a = \text{lub}\{a_1, a_2, \ldots, a_n, \ldots\}$. Prove that $a = \lim_{n \to \infty} a_n$.

4.   Prove that $\sqrt[3]{2}$ is the least upper bound for the set

   $\{x \in R : x \text{ is rational and } x^3 < 2\}$

5.   Prove the second half of Theorem 4.2.

6.   (a)   Let X be a bounded subset of R, and let $a \in R$ be positive. Set $aX = \{ax : x \in X\}$. Prove that

   $\text{lub}(aX) = a \text{ lub } X, \quad \text{glb}(aX) = a \text{ glb } X$

   (b)   If a is negative, prove that

   $\text{lub}(aX) = a \text{ glb } X, \quad \text{glb}(aX) = a \text{ lub } X$

7.   Let A and B be non empty subsets of R with $A \subset B$. Prove that

   $\text{lub } A \leq \text{lub } B$
   $\text{glb } A \geq \text{glb } B$

*8.   Prove the Archimedean Principle: given positive real numbers a
and b, then there exists an integer n such that na > b.   (Hint:
otherwise, {a,2a,...} is bounded above.)

*9.   Let f:[a,b] → R be a bounded function. We define the *oscillation*
of f *at* t, ω(f,t), by

$$\omega(f,t) = \lim_{\delta \to 0} (\text{lub}\{f(s):|s-t| < \delta\} - \text{glb}\{f(s):|s-t| < \delta\})$$

Prove this limit exists.

*10.   Prove that the bounded function f:[a,b] → R is continuous at t
if and only if ω(f,t) = 0.

*11.   Let $\{a_n\}$ be an infinite sequence of real numbers and let L be
the set of all points of R which are limit points of some convergent
subsequence of $\{a_n\}$. Suppose L is bounded. Then the *upper limit* of
the sequence a_n is defined by

$$\lim_{n \to \infty} \sup a_n = \text{lub } L$$

and the *lower limit* of the sequence $\{a_n\}$ is defined by

$$\lim_{n \to \infty} \inf a_n = \text{glb } L$$

   (a) Prove that $\lim_{a \to \infty} \sup a_n = a$ if and only if $a \in L$ and whenever
x > a, there is an integer N such that $s_n < x$ for $n \geq N$.

   (b) State and prove the analogous characterization of the lower
limit.

*12.   Prove the Completeness Axiom, assuming the truth of Theorem 4.2.

*13.   Prove that if $\{x_n\}$ is a nondecreasing sequence of real numbers
which is bounded above, then $\{x_n\}$ converges.

5. CONTINUOUS FUNCTIONS ON COMPACT SETS

It may not be clear yet what the point of defining compact sets is.
To give some motivation, we prove an important theorem here: a continuous real-valued function defined on a compact set attains its maximum and minimum. This theorem generalizes one often quoted in calculus courses: a continuous function defined on a closed, bounded interval attains its maximum and minimum. To know that our theorem generalizes the other one, we need to know that closed, bounded intervals
are compact, a fact that we shall prove in the next section.

We begin with a useful result about closed sets.

PROPOSITION 5.1. *If* $X \subset R$ *is closed and bounded, then* lub X
and glb X *are elements of* X.

Proof. We prove that $M = \text{lub } X$ is in X, leaving the proof of the
fact that glb X is in X to the reader.

For any integer $n > 0$, the number $M - 1/n$ cannot be an upper bound
for X, since M is the least upper bound of X. Thus for each such n,
we can find a real number $a_n \in X$ with

$$M - 1/n \leq a_n \leq M$$

It follows easily that $\lim_{n \to \infty} a_n = M$. Since X is closed, M is in X by
Theorem 1.7.

Our next result relates compact sets and continuous functions.
It states that the notion of compactness is preserved by continuous
functions.

PROPOSITION 5.2. *Let* $f : V \to W$ *be a continuous mapping and* C *a
compact subset of* V. *Then the image of* C *under* f,

$$f(C) = \{f(\vec{v}) : \vec{v} \in C\}$$

is a compact subset of W.

Proof. Let $\{\vec{w}_n\}$ be a sequence of vectors in $f(C)$. We must show that there is a subsequence of this sequence converging in $f(C)$.

Since $\vec{w}_n$ is in $f(C)$, there is a vector $\vec{v}_n$ in C with $f(\vec{v}_n) = \vec{w}_n$. The vectors $\vec{v}_n$ form a sequence in the compact set C, so there is a subsequence $\{\vec{v}_{n_j}\}$ converging to a vector $\vec{v}$ in C. Applying Theorem 2.1, we see that the subsequence $\vec{w}_{n_j} = f(\vec{v}_{n_j})$ converges to the vector $f(\vec{v})$, which is clearly in $f(C)$. Therefore $f(C)$ is compact.

We can now state and prove the main theorem of this section.

THEOREM 5.3. *Let C be a compact subset of a normed vector space V and* $f:C \rightarrow R$ *a continuous function. Then f attains its maximum and minimum on* C.

More explicitly, there are vectors $\vec{v}_0$ and $\vec{v}_1$ in C such that

$$f(\vec{v}_0) \leq f(\vec{v}) \leq f(\vec{v}_1)$$

for all $\vec{v}$ in C.

Proof. According to Proposition 5.2, $f(C)$ is a compact subset of R, and so, by Propositions 3.2 and 3.3, is closed and bounded. It follows from Propositions 4.2 and 5.1 that lub $f(C)$ and glb $f(C)$ exist and are in $f(C)$. Pick $\vec{v}_0$, $\vec{v}_1$ such that

$$f(\vec{v}_0) = \text{glb } f(C), \quad f(\vec{v}_1) = \text{lub } f(C)$$

Then

$$f(\vec{v}_0) \leq f(\vec{v}) \leq f(\vec{v}_1)$$

for all $\vec{v}$ in C, and the theorem is proved.

COROLLARY. *Let* [a,b] *be a closed interval and* $f:[a,b] \rightarrow R$ *a continuous function. Then f attains its maximum and minimum on* [a,b].

Strictly speaking, we should not call this result a corollary, since it depends on a fact we have not yet proved: the set [a,b] is compact. This will be proved in the next section.

Theorem 5.3 states the existence of a maximum and a minimum for continuous functions on compact sets; it does not, however, give any information on how to find them. We shall examine this problem (for differentiable functions) in Chapter 8.

Greatest lower bounds and least upper bounds often arise in connection with functions. Suppose that $S \subseteq V$ and $f:S \to R$ is a function; we may consider

$$M = \text{lub } \{f(\vec{v}) : \vec{v} \in S\}$$

and

$$m = \text{glb } \{f(\vec{v}) : \vec{v} \in S\}$$

Neither M nor m need exist, but if f is a bounded function (and S is not the empty set), then M and m both exist (by Theorem 3.2). If f takes on a maximum value, then M is that maximum; similarly, m is the minimum value if f has a minimum. Neither the maximum nor the minimum need exist. For example, if $f:(-1,2) \to R$ is defined by $f(x) = x^2$, then

$$f(S) = \{f(x) : x \in (-1,2)\} = [0,4)$$

Here, f has a minimum but no maximum. However, $f(S)$ has a least upper bound and a greatest lower bound; in fact, lub $f(S) = 4$ and glb $f(S) = 0$.

We conclude this section with a useful result.

PROPOSITION 5.4. *Let* $S \subseteq V$, $f,g:S \to R$ *be bounded functions, and*
$$M = \text{lub}\{f(\vec{v}) : \vec{v} \in S\}, \quad m = \text{glb}\{f(\vec{v}) : \vec{v} \in S\}$$
$$M' = \text{lub}\{g(\vec{v}) : \vec{v} \in S\}, \quad m' = \text{glb}\{g(\vec{v}) : \vec{v} \in S\}$$
$$M'' = \text{lub}\{f(\vec{v}) + g(\vec{v}) : \vec{v} \in S\}, \quad m'' = \text{glb}\{f(\vec{v}) + g(\vec{v}) : \vec{v} \in S\}$$
Then $M'' \leq M' + M$ *and* $m'' \geq m' + m$.

Proof. For every $\vec{v} \in S$, $M \geq f(\vec{v})$ and $M' \geq g(\vec{v})$. Therefore

$$M + M' \geq f(\vec{v}) + g(\vec{v})$$

which shows that M + M' is an upper bound for $\{f(\vec{v}) + g(\vec{v}) : \vec{v} \in S\}$.
It follows that $M'' \leq M + M'$. The other part is similar.
The situation is more complicated for products; see Exercise 5.

EXERCISES

1.   Find a continuous function f on the set $S = \{\vec{v} \in R^3 : \|\vec{v}\| < 1\}$ which
has neither a maximum nor a minimum. (Note that S is not compact.)

2.   Let S be a set in V with the property that every continuous function on S has a maximum and a minimum. Show that S is compact. You
may use the fact (to be proved in the next section) that every closed
and bounded set is compact. (Hint: first prove that S is bounded.
If S is not closed, use Theorem 1.7 to find a sequence $\vec{v}_1, \vec{v}_2, \ldots$ of
points in S with $\lim_{n \to \infty} \vec{v}_n = \vec{v}_0 \notin S$. Let $f(\vec{v}) = \|\vec{v} - \vec{v}_0\|$.)

*3.   Let $C \subseteq V$ be compact, and let $f:C \to R$ be a continuous function
with $f(\vec{v}) > 0$ for all $\vec{v} \in C$. Show that there is a number $\alpha > 0$ such
that $f(\vec{v}) > \alpha$ for all $\vec{v} \in C$. Give an example to show that the conclusion need not hold if C is not compact.

*4.   The following is a sketch of a different proof of Theorem 5.3.
Fill in the details. In what follows, C is compact and $f:C \to R$ is
continuous. Use *no* properties of compactness except the definition.

    (a) Show that f(C) is bounded above. (Hint: if not, then there
are points $\vec{v}_n \in C$ with $f(\vec{v}_n) > n$. Choose a subsequence of the $\vec{v}_n$ which
converges to $\vec{v}_0 \in C$. Find a contradiction from the fact that f is continuous at $\vec{v}_0$.)

    (b) Since f(C) is bounded above, M = lub f(C) exists. Show that
there is a point $\vec{v} \in C$ with $f(\vec{v}) = M$. (Hint: if not, show that $g(\vec{v}) = 1/(M - f(\vec{v}))$ is not bounded above, and use (a) to get a contradiction.)

    (c) Parts (a) and (b) prove that f takes on a maximum value on C.
Prove that f has a minimum.

5.   Let $S \subseteq V$ and $f,g:S \to R$ be bounded *non-negative* functions. Show
that

$$\text{lub}\{f(\vec{v})g(\vec{v}):\vec{v} \in S\} \leq \text{lub}\{f(\vec{v}):\vec{v} \in S\} \cdot \text{lub}\{g(\vec{v}):\vec{v} \in S\}$$

and

$$\text{glb}\{f(\vec{v})g(\vec{v}):\vec{v} \in S\} \geq \text{glb}\{f(\vec{v}):\vec{v} \in S\} \cdot \text{glb}\{g(\vec{v}):\vec{v} \in S\}$$

(See Exercise 6 of the previous section.)

6. A CHARACTERIZATION OF COMPACT SETS

As promised earlier, we now characterize the compact subsets of finite dimensional inner product spaces. The same characterization holds for finite-dimensional normed vector spaces, but we leave the proof of this fact to the exercises.

THEOREM 6.1. *A subset C of a finite dimensional inner product space V is compact if and only if it is closed and bounded.*

We have already proved half of this theorem, namely that a compact set is closed and bounded. What needs to be shown is that any closed and bounded subset of V is compact. We begin with a special case.

THEOREM 6.2. (The Bolzano-Weierstrass Theorem). *Any closed interval*

$$[a,b] = \{x \in R : a \leq x \leq b\}$$

is compact.

Proof. This theorem, like Theorem 4.2, needs the completeness property of the real numbers: every Cauchy sequence converges. (See Section 4.2.) We need to prove that if $a_1, a_2, \ldots,$ is a sequence in [a,b], then some subsequence converges. We do this by finding a subsequence which is a Cauchy sequence.

Let $c = (a+b)/2$, and consider the two intervals $I_1 = [a,c]$ and $I_1' = [c,b]$ formed by dividing [a,b] in half. Each of these intervals has

length $(b-a)/2$. Furthermore, either I_1 or I_1' contains infinitely many of the a's. Choose one with infinitely many of the a's, and choose a_{i_1} to be a point of the sequence lying in this subinterval.
Now divide the interval just chosen into two subintervals of equal length, which we call I_2, I_2'. Each has length $(b-a)/4$. Again, choose one of these subintervals with infinitely many a's, and choose a term a_{i_2} in this interval with $i_2 > i_1$. (Since we have infinitely many terms to choose from, we can surely choose one with a subscript $> i_1$.) Repeat the procedure; that is, split this subinterval into two equal subintervals, choose one containing an infinite number of terms, and choose a_{i_3} in this new subinterval with $i_3 > i_2 > i_1$. Continue in this way.

An example may help. Let the interval be $[0,2]$, and let the sequence be $1/2, 3/2, 1/3, 4/3, 1/4, 5/4, \ldots$. We begin by considering the intervals $[0,1]$ and $[1,2]$. Each of these has infinitely many terms of the sequence (the odd-numbered terms are in $[0,1]$ and the even-numbered terms in $[1,2]$). We may therefore choose either interval; say we choose $[1,2]$. We also need to pick some point of the sequence in $[1,2]$, and we pick $4/3$. (We could equally well pick $3/2$.) Next, we divide $[1,2]$ in half, getting $[1,3/2]$, and $[3/2,2]$. The first of these intervals has infinitely many points of the sequence (the even-numbered terms), while the second has only one point. Thus we must choose $[1,3/2]$ to be our interval, and we choose a point in the sequence which is beyond $4/3$ and is in $[1,3/2]$. For example, $6/5$ will do. (So will $5/4$ or $7/6$, for instance.) Next, we cut $[1,3/2]$ in half, getting $[1,5/4]$ and $[5/4,3/2]$, and so on.

By this procedure, we get a subsequence $a_{i_1}, a_{i_2}, \ldots$ of the original sequence with the property that all the terms a_{i_j} with $j \geq n$ are contained in the nth subinterval chosen. This subinterval is of length $(b-a)/2^n$. Therefore we know that if $j, k > n$, then $| a_{i_j} - a_{i_k} |$ $\leq (b-a)/2^n$. Given any $\varepsilon > 0$, we can choose N so that $(b-a)/2^N < \varepsilon$. Then if $m, n > N$,

$$\left|a_{i_m} - a_{i_n}\right| \le \frac{b-a}{2^N} < \varepsilon$$

and so the sequence $a_{i_1}, a_{i_2}, \ldots$ is a Cauchy sequence. By the completeness axiom, this sequence converges to some real number x. Because [a,b] is closed, $x \in [a,b]$. We have therefore proved that [a,b] is compact, as claimed. (Note, incidentally, that we have also finished the proof of the Corollary to Theorem 5.3.)

Let $\vec{v}_1, \ldots, \vec{v}_n$ be an orthonormal basis for the inner product space V and define for any $r \ge 0$ the set

$$A(r) = \{\vec{v} = \sum_{j=1}^{n} a_j \vec{v}_j : |a_j| \le r \text{ for } j = 1, 2, \ldots, n\}$$

The next result generalizes Proposition 6.2.

PROPOSITION 6.3. *For any* $r \ge 0$*, the set* A(r) *is compact.*

Proof. Let $\{\vec{w}_j\}$ be a sequence in A(r). Then

$$\vec{w}_j = \sum_{i=1}^{n} a_{ji} \vec{v}_i, \quad j = 1, 2, \ldots$$

where $|a_{ji}| \le r$. Consider the sequence $\{a_{j1}\}$ of real numbers. Since $|a_{j1}| \le r$ for all j, each a_{j1} is contained in the closed interval [-r,r] and we can find a subsequence converging to a number a_1 in [-r,r]. This defines a subsequence of the sequence $\{\vec{w}_j\}$ with the property that the sequence of first components converges to a_1. For simplicity, we throw away the $\vec{w}_j$ not appearing in this subsequence and reindex so that

$$\lim_{j \to \infty} a_{j1} = a_1$$

Now consider the sequence of second components $\{a_{j2}\}$. A similar argument shows that we can find a subsequence of this sequence converging to a_2 in [-r,r]. This defines a subsequence of the sequence

$\{\vec{w}_j\}$ whose second components converge to a_2. Note that the sequence of first components of this subsequence still converge to a_1 since it is a subsequence of a convergent sequence with limit a_1. (See Proposition 1.4.)

Continuing in this way, we eventually obtain a subsequence of the original sequence $\{\vec{w}_j\}$ with the property that all of its component sequences converge to numbers in $[-r,r]$. It now follows easily that this subsequence converges to a vector in $A(r)$. (See Exercise 13 of Section 1.)

We can now complete the proof of Theorem 6.1. Let C be a closed and bounded subset of V. Then there is an $r > 0$ such that $C \subseteq B_r(\vec{0})$. It is easily seen that $B_r(\vec{0}) \subseteq A(r)$, so that C is a closed subset of a compact set and is therefore compact (Proposition 3.4). Thus Theorem 6.1 is proved.

EXERCISES

1.   Which of the following are compact sets?

(a) $A = \{x \in R : |x-3| \leq 7\}$

(b) $B = \{x \in R : 0 < x \leq 10\}$

(c) $C = \{(x,y) \in R^2 : 0 < x^2 + y^2 \leq 1\}$

(d) $D = \{(x,y) \in R^2 : |x| + |y| \leq 2\}$

(e) $E = \{(x,y) \in R^2 : x^2 + y^2 = 1\}$

(f) $F = \{(x,y) \in R^2 : xy = 1\}$

(g) $G = \{(x,y) \in R^2 : 1 \leq x^2 + y^2 \leq 2\}$

(h) $H = \{(x,y,z) \in R^3 : x^2 + 2y^2 + 3z^2 \leq 2\}$

(i) $I = \{(x,y,z) \in R^3 : x^2 + y^2 \leq 2 \text{ and } z = xy\}$

(j) $J = \{(x,y,z) \in R^3 : |x-2| + |y-1| \leq 2\}$

(k) $K = \{(x,y,z) \in R^3 : x^2 + y^2 - z = 0 \text{ and } x^2 + y^2 + z - 2 = 0\}$

*2.   Show that if $\{a_n\}$ is an infinite sequence of real numbers in $[a,b]$, then $\lim \sup\{a_n\} \in [a,b]$ and $\lim \sup\{a_n\}$ is the limit of a subsequence of $\{a_n\}$. (This provides an alternate proof of Theorem 6.2; see Exercise 11 of Section 4 for the definition of $\lim \sup a_n$.)

*3. Let V be the set of all sequences (a_n) of real numbers for which only a finite number of terms are non-zero. Define addition, scalar multiplication, and an inner product on V by

$$(a_n) + (b_n) = (a_n + b_n)$$
$$m(a_n) = (ma_n)$$
$$< (a_n),(b_n) > = \sum_{n=1}^{\infty} a_n b_n$$

(See Exercise 18 of Section 4.7.) Then V is easily seen to be an inner product space. Prove that $\bar{B}_1(\vec{0})$ is not compact in spite of the fact that it is closed and bounded. (This shows that the assumption in Theorem 6.1 that V is finite dimensional is really necessary.)

The purpose of the next two exercises is to extend Theorem 6.1 to all finite-dimensional normed vector spaces.

4. Let $\| \ \|_1$ and $\| \ \|_2$ be equivalent norms on the vector space V, and let C be a subset of V. (See Exercise 5 of Section 3.1 for the definition of equivalent norms.)

(a) Show that C is closed and bounded in the $\| \ \|_1$ norm if and only if C is closed and bounded in the $\| \ \|_2$ norm. (See Exercise 22 of Section 3.2.)

(b) Show that C is compact in the $\| \ \|_1$ norm if and only if C is compact in the $\| \ \|_2$ norm. (See Exercise 16 of Section 1.)

*5. Let V be a finite-dimensional normed vector space, and denote the norm of $\vec{v}$ by $\|\vec{v}\|_1$. Let $\{\vec{v}_1, \vec{v}_2, \ldots, \vec{v}_n\}$ be a basis of V, with $\|\vec{v}_j\|_1 = 1$ for each j. (The $\vec{v}_j$ are *not* necessarily an orthonormal basis; in fact, the term makes no sense unless $\| \ \|_1$ comes from an inner product.) We now define a new norm, as follows: if $\vec{v} = \sum_{j=1}^{n} a_j \vec{v}_j$ and $\vec{w} = \sum_{j=1}^{n} b_j \vec{v}_j$, define

$$< \vec{v}, \vec{w} > = \sum_{j=1}^{n} a_j b_j$$

(a) Show that $<\,,\,>$ gives an inner product on V. Denote the norm defined by this inner product by $\|\ \|$.

(b) Show that $\{\vec{v}_1,\vec{v}_2,\ldots,\vec{v}_n\}$ is an orthonormal basis for the inner product $<\,,\,>$.

(c) Show that if we consider V to have the $\|\ \|$ norm, then the function $f:V \to R$,

$$f(\vec{v}) = \|\vec{v}\|_1$$

is continuous.

(d) Show that the $\|\ \|_1$ and $\|\ \|$ norms are equivalent. (Hint: consider f on $S = \{\vec{v} \in V : \|\vec{v}\| = 1\}$.)

(e) Show that a set $C \subset V$ is compact in the $\|\ \|_1$ norm if and only if C is closed and bounded in the $\|\ \|_1$ norm. (See Exercise 4.)

6. Use Exercise 5 to show that:

(a) If V, W are finite-dimensional normed vector spaces and $T:V \to W$ is linear, then T is continuous. (See Proposition 4.7.5 and Exercises 13, 17, and 18 of Section 3.3.)

(b) If V, W are as in (a), U is an open subset of V, and $f:U \to W$ is a function which is differentiable at $\vec{v}_0 \in U$, then f is continuous at $\vec{v}_0$. (See Proposition 3.6.3 and the discussion at the end of Section 4.7.)

*7. Prove Theorem 6.2 by proving that if $\{U_\lambda\}$ is any open covering of $[a,b]$, then there is a finite subcovering. (Then Exercise 11 of Section 3 applies. Hint: let $S = \{x \in [a,b]:$ the interval $[a,x]$ can be covered by finitely many U_λ. Let $c = $ lub S; then $c \in$ some U_{λ_0}. Assume $c < b$ and get a contradiction.)

7. UNIFORM CONTINUITY

We now develop the notion of uniform continuity for functions between normed vector spaces. This material will be needed later for the study of the integral.

Let X be a subset of V, and $f:X \to W$ a function. We say that f is *uniformly continuous on* X if for any $\varepsilon > 0$ there is a $\delta > 0$ such that

$$\|f(\vec{v}) - f(\vec{v}')\| < \varepsilon$$

for all $\vec{v}, \vec{v}'$ in X with $\|\vec{v} - \vec{v}'\| < \varepsilon$.

Clearly any uniformly continuous function is continuous. The difference in this definition is that the number δ depends only on ε, not on the point under consideration. In the definition of continuity, we state only that given ε and $\vec{v}$, there is a δ (which may depend on ε and $\vec{v}$) such that $\|f(\vec{v}) - f(\vec{v}')\| < \varepsilon$ whenever $\|\vec{v} - \vec{v}'\| < \delta$. For example, the function

$$f:(0,1) \to R$$

defined by $f(x) = 1/x$ is continuous but *not* uniformly continuous. To prove this we must show that there is an $\varepsilon > 0$ such that, no matter how small a δ we choose, we can find x,x' in (0,1) with $|x - x'| < \delta$ but $|f(x) - f(x')| \geq \varepsilon$. Set $\varepsilon = 1/2$ and, for any $\delta > 0$, choose an integer n so that $1/n < \delta$. Then if $x = 1/n$ and $x' = 1/(n+1)$,

$$|x - x'| = \frac{1}{n(n+1)} < \frac{1}{n} < \delta$$

but

$$|f(x) - f(x')| = 1 > \frac{1}{2}$$

Thus f is not uniformly continuous.

The next result shows that the notions of continuity and uniform continuity coincide on compact sets.

THEOREM 7.1. *Suppose that* C *is a compact subset of* V. *Then any continuous function* $f:C \to W$ *is uniformly continuous.*

Proof. We proceed by contradiction, and our first problem is how to state that f is not uniformly continuous. Again, we use the procedure given in Section 1.2. Write "f is uniformly continuous"

as

"$(\forall \varepsilon > 0)(\exists \delta > 0)(\forall \vec{v}, \vec{w} \in C): \|\vec{v} - \vec{w}\| < \delta \Rightarrow \|f(\vec{v}) - f(\vec{w})\| < \varepsilon$"

Then one obtains the statement for "f is not uniformly continuous" by interchanging the $\forall$'s and $\exists$'s and denying the conclusion:

"$(\exists \varepsilon > 0)(\forall \delta > 0)(\exists \vec{v}, \vec{w} \in C): \|\vec{v} - \vec{w}\| < \delta$ and $\|f(\vec{v}) - f(\vec{w})\| \geq \varepsilon$"

or

"there exists an $\varepsilon > 0$ such that for every $\delta > 0$ we can find $\vec{v}, \vec{w}$ in C with $\|\vec{v} - \vec{w}\| < \delta$ and $\|f(\vec{v}) - f(\vec{w})\| \geq \varepsilon$."

Having worked out what we are assuming, we can search for the contradiction. Choose the ε guaranteed by our last statement: then (setting $\delta = \frac{1}{n}$) we can find points $\vec{v}_n, \vec{w}_n \in C$ such that $\|\vec{v}_n - \vec{w}_n\| < \frac{1}{n}$ and $\|f(\vec{v}_n) - f(\vec{w}_n)\| \geq \varepsilon$. Because C is compact, some subsequence of $\{\vec{v}_n\}$ converges to a point in C; suppose that $\{\vec{v}_{n_k}\}$ converges to $\vec{v}_0 \in C$. Since f is continuous at $\vec{v}_0$, we can find $\delta > 0$ such that if $\|\vec{v} - \vec{v}_0\| < \delta$, then $\|f(\vec{v}) - f(\vec{v}_0)\| < \varepsilon/2$.

Now choose k so large that

(a) $\|\vec{v}_0 - \vec{v}_{n_k}\| < \frac{\delta}{2}$

(b) $\frac{1}{n_k} < \frac{\delta}{2}$

(We can certainly satisfy these conditions, since $\vec{v}_{n_k}$ converges to $\vec{v}_0$.) Then, since $\|\vec{v}_{n_k} - \vec{w}_{n_k}\| < 1/n_k$

$$\|\vec{v}_0 - \vec{w}_{n_k}\| < \frac{\delta}{2} + \frac{1}{n_k} < \delta$$

Thus

$$\|f(\vec{v}_0) - f(\vec{v}_{n_k})\| < \frac{\varepsilon}{2}$$

and

$$\|f(\vec{v}_0) - f(\vec{w}_{n_k})\| < \frac{\varepsilon}{2}$$

Therefore

$$\|f(\vec{v}_{n_k}) - f(\vec{w}_{n_k})\| < \varepsilon$$

which contradicts our choice of $\vec{v}_{n_k}$, $\vec{w}_{n_k}$.   The theorem follows.

Note. The form of the proof is one which is often useful when dealing with compact sets. To prove a theorem about compact sets, assume that it is false. Pick a sequence of points in the compact set making the theorem false. Extract a convergent subsequence. Then get a contradiction by looking at the limit of the subsequence. The reader may wish to try out this method on Exercises 8, 9, and 10 of Section 3. (See Exercise 4 of Section 5 for another example.) future sections.

EXERCISES

1.   Prove directly from the definition that $f:R \to R$ defined by $f(x) = 7x + 4$ is uniformly continuous.

2.   Prove directly from the definition that $f:R \to R$ defined by $f(x) = ax + b$ is uniformly continuous.

3.   Prove that a bounded linear transformation $T:V \to W$ is uniformly continuous.

4.   Prove directly from the definition that the function $f = [-a,a] \to R$ defined by $f(x) = x^2$ is uniformly continuous.

5.   Prove that the function $f:R \to R$ defined by $f(x) = x^2$ is not uniformly continuous.

6.   Verify that the function $F:(0,1) \to R$ defined by $f(x) = \frac{1}{x}$ is continuous.

*7.   Give another proof of Theorem 7.1 by using the Lebesgue covering lemma (Exercise 10 of Section 3). Hint: given f, ε, and x $\in$ C, we can find $\delta(x)$ such that if y $\in$ C, $\|y - x\| < \delta(x) \Rightarrow f(y) - f(x) < \varepsilon/2$. Let $\{B_{\delta(x)}(x):x \in C\}$ be a covering for C.

8.   CONNECTED SETS

We now introduce two related concepts, connectedness and pathwise connectedness. These will be used to prove the Intermediate Value Theorem.

A subset S in a normed vector space V is said to be *separated* if there are open subsets U_1 and U_2 of V such that $S \subset U_1 \cup U_2$, $U_1 \cap U_2 \cap S = \phi$ and neither $S \cap U_1$ or $S \cap U_2$ is empty. A set $S \subset V$ is *connected* if it cannot be separated.

For example, the set S = $\{x \in R:x \neq 0\}$ is separated; let $U_1 =$ $(-\infty,0)$ and $U_2 = (0,\infty)$. On the other hand, we have the following.

PROPOSITION 8.1. *Any closed interval* [a,b] $\subset$ R *is connected.*

Proof. Suppose not. Then [a,b] $\subset U_1 \cup U_2$ where U_1 and U_2 are open subsets of R such that $U_1 \cap$ [a,b] $\neq \phi$, $U_2 \cap$ [a,b] $\neq \phi$ and $U_1 \cap$ $U_2 \cap$ [a,b] = ϕ. Assume that a $\in U_1$ and define c $\in$ R by

$$c = glb(U_2 \cap [a,b])$$

(If a $\in U_2$, the proof proceeds similarly.) Then c $\in$ [a,b], since [a,b] is a closed set.

Suppose c $\in U_1$. Then, since U_1 is open, there is an $\varepsilon > 0$ such that $B_\varepsilon(c) = (c - \varepsilon,c + \varepsilon) \subset U_1$. Now c is a lower bound for $U_2 \cap$ [a,b], so x $\geq$ c for all x $\in U_2$. However, since $(c - \varepsilon,c + \varepsilon) \subset U_1$ and $U_1 \cap U_2 = \phi$, we must have x $\geq$ c + ε for all x $\in U_2 \cap$ [a,b]. This contradicts the fact that c is the *greatest* lower bound of $U_2 \cap$ [a,b]. Thus c cannot be in U_1.

On the other hand, if c $\in U_2$, then $(c - \varepsilon,c + \varepsilon) \subset U_2$ for some $\varepsilon > 0$. Since a $\in U_1$, c $\neq$ a and we can choose ε small enough so that $(c - \varepsilon,c + \varepsilon) \subset U_2 \cap$ [a,b]. In particular, the points between c - ε

and c are in $U_2 \cap [a,b]$. This contradicts the fact that c is a lower
bound for $U_2 \cap [a,b]$.

It follows that $c \notin U_1 \cup U_2$, so that $[a,b] \not\subseteq U_1 \cup U_2$. This contradicts our original assumption that $[a,b]$ was separated.

We note that any interval in R can be shown to be connected by
a similar argument. This fact, together with the next result, charac-
terizes the connected subsets of R.

PROPOSITION 8.2. *Let S be a connected subset of R and $a,b \in S$*
with $a < b$. If $z \in R$ with $a < z < b$, then $z \in S$.

It follows that the connected subsets of R are exactly the intervals. (See Exercise 9.)

Proof. If $z \notin S$, define $U_1 = (-\infty, z)$ and $U_2 = (z, \infty)$. Then $U_1 \cap$
$U_2 = \phi$, $S \subseteq U_1 \cup U_2$, $a \in U_1 \cap S$, $b \in U_2 \cap S$. Thus S is separated.

We next show that we can characterize separated sets in terms of
closed sets.

PROPOSITION 8.3. *A subset $S \subseteq V$ is separated if and only if there*
are closed sets C_1 and C_2 in V with $S \subseteq C_1 \cup C_2$, $C_1 \cap C_2 \cap S = \phi$,
$C_1 \cap S \neq \phi$, and $C_2 \cap S \neq \phi$.

Proof. Suppose S separated by open sets U_1 and U_2. Define closed sets $C_1 = V - U_1$, $C_2 = V - U_2$. It follows easily that C_1 and C_2
satisfy the conditions of the proposition.

Conversely, if C_1 and C_2 are closed sets satisfying the conditions
of the proposition, then the open sets $U_1 = V - C_1$ and $U_2 = V - C_2$
separate S. We leave the details as an exercise.

A subset $S \subseteq V$ is said to be *pathwise connected* if for any two
points $\vec{v}, \vec{w} \in S$, there is a continuous function $\alpha : [0,1] \to S$ such that
$\alpha(0) = \vec{v}$ and $\alpha(1) = \vec{w}$. (The function α is called a *path* from $\vec{v}$ to $\vec{w}$.)

PROPOSITION 8.4. *Any pathwise connected subset S of V is connected.*

Proof. We shall prove the contrapositive; if S is not connected, then S is not pathwise connected.

Suppose S is separated. Then, according to Proposition 8.3, we can find closed sets C_1 and C_2 in V such that $S \subset C_1 \cup C_2$, $S \cap C_1 \cap C_2 = \phi$, $S \cap C_1 \neq \phi$, and $S \cap C_2 \neq \phi$. Let $\vec{v}$ be a point of $S \cap C_1$ and $\vec{w}$ a point of $S \cap C_2$. If $\alpha:[0,1] \to S$ is a path from $\vec{v}$ to $\vec{w}$, we define subsets $A_1, A_2 \subset [0,1]$ by $A_1 = \alpha^{-1}(C_1)$, $A_2 = \alpha^{-1}(C_2)$. Then A_1 and A_2 are closed subsets of R (see Proposition 2.5.) contained in $[0,1]$, $0 \in A_1$ (since $f(0) = \vec{v} \in C_1$) and $1 \in A_2$ (since $f(1) = \vec{w} \in C_2$). Furthermore, if t is any point in $[0,1]$, $f(t)$ is in C_1 or C_2. If $f(t) \in C_1$, then $t \in A_1$; if $f(t) \in C_2$, then $t \in A_2$. Thus $[0,1] \subset A_1 \cup A_2$. Finally,

$$A_1 \cap A_2 = \alpha^{-1}(C_1) \cap \alpha^{-1}(C_2)$$
$$= \alpha^{-1}(C_1 \cap C_2)$$
$$= \alpha^{-1}(\phi)$$
$$= \phi$$

Thus A_1 and A_2 separate $[0,1]$, which is not possible since $[0,1]$ is connected. It follows that there can be no path from $\vec{v}$ to $\vec{w}$ and S is not pathwise connected.

The next result relates pathwise connectivity and continuous functions.

PROPOSITION 8.5. *Let S be a pathwise connected subset of V and f:V → W a continuous function. Then f(S) is a pathwise connected subset of W.*

Proof. Let $\vec{v}'$, $\vec{w}'$ be any two points of $f(S)$. Then there are points $\vec{v}, \vec{w} \in S$ such that $f(\vec{v}) = \vec{v}'$, $f(\vec{w}) = \vec{w}'$. Since S is pathwise connected, we can find a path $\alpha:[0,1] \to S$ from $\vec{v}$ to $\vec{w}$. Then the composite $f \circ \alpha:[0,1] \to f(S)$ is a path from $\vec{v}'$ to $\vec{w}'$.

We can now prove the *Intermediate Value Theorem*.

THEOREM 8.6. *Let* f:[a,b] → R *be a continuous function and let*
z ∈ R *be any number between* f(a) *and* f(b). *Then there is a* c ∈ [a,b]
with f(c) = z.

Proof.   This follows immediately from Propositions 8.5 and 8.2.
Since [a,b] is pathwise connected, so is f([a,b]). Therefore f([a,b])
is connected, and Proposition 8.2 says that c ∈ f([a,b]).

Theorem 8.6 is another result which is commonly assumed in elementary calculus. It says, for instance, that if f:[0,1] → R is continuous, with f(0) < 0 and f(1) > 0, then there exists a point a in
[0,1] with f(a) = 0. (See Figure 8.1.) Of course, the theorem gives
no information about how to find a.

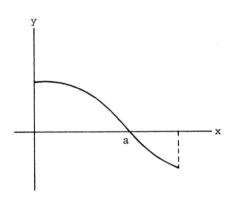

FIGURE 8.1.

EXERCISES

1.   Determine which of the following sets are connected.

(a) $A = \{x \in R: |x - 2| > 3\}$

(b) $B = \{(x,y) \in R^2 : 0 < x^2 + y^2 \leq 1\}$

(c) $C = \{(x,y) \in R^2 : y \geq x^2\}$

(d) $D = \{(x,y) \in R^2 : xy > 1\}$

(e) $E = \{(x,y) \in R^2 : y = x^2\}$

(f) $F = \{(x,y) \in R^2 : x^2 + y^2 = 1\}$

(g) $G = \{(x,y) \in R^2 : x^2 + y^2 = 1 \text{ and } |x| < 1\}$

(h) $H = \{(x,y) \in R^2 : xy = 1\}$

(i) $I = \{(x,y) \in R^2 : |x - 2| > 3\}$

(j) $J = \{(x,y,z) \in R^3 : |x - 1| + |y - 2| > 1\}$

(k) $K = \{(x,y,z) \in R^3 : z - x^2 - y^2 \geq 0\}$

2. (a) Give an example of connected sets S_1, S_2 in R^2 such that $S_1 \cup S_2$ is not connected.

(b) Give an example of connected sets S_1 and S_2 in R^2 such that $S_1 \cap S_2$ is not connected.

3. Let S_1, S_2 be connected sets in V with $S_1 \cap S_2 \neq \phi$. Prove that $S_1 \cup S_2$ is connected.

4. Prove that any interval is connected.

5. A subset $C \subset V$ is *convex* if whenever $\vec{v}, \vec{w} \in C$, then the line segment from $\vec{v}$ to $\vec{w}$ is also in C. (Explicitly, $t\vec{v} + (1-t)\vec{w} \in C$ for all $t \in [0,1]$.) Prove that any convex set is pathwise connected.

6. Give a detailed proof of Proposition 8.3.

7. Let S be a connected subset of V and $f:S \to R$ a continuous function. Let $\vec{v}_0$, $\vec{v}_1$ be two points of S and $z \in R$ with $f(\vec{v}_0) < z < f(\vec{v}_1)$. Prove that there is a $\vec{v} \in S$ such that $f(\vec{v}) = z$.

8. Let $f:[0,1] \to [0,1]$ be continuous. Show that there exists x in $[0,1]$ with $f(x) = x$. (Hint: consider $g(x) = f(x) - x$.) This result is the *Brouwer Fixed Point Theorem* for $[0,1]$.

*9. Let $S \subset R$ be nonempty, and suppose that whenever $x \in S$, $y \in S$, and $x < z < y$, then $z \in S$. Prove that S is an interval. (Hint: if S is bounded above and below, let $a = \text{glb } S$ and $b = \text{lub } S$, and prove

that $(a,b) \subset S \subset [a,b]$. If S is bounded above but not below, let
$b = $ lub S and prove that $S = (-\infty,b)$ or $[-\infty,b]$. Do the other cases
similarly.)

10. Prove that a subset $S \subset V$ is separated if and only if there is
an open set $U \subset V$ and a closed set $C \subset V$ such that

(a) $C \cap S = U \cap S$

(b) $C \cap S \neq \phi$

(c) $C \cap S \neq S$

*11. Prove that a connected open subset of R^n is pathwise connected.

*12. Let $S \subset R^2$ be the set defined by

$$S = \{(x,y):y = \sin 1/x,\ 0 < x \leq 1\} \cup \{(0,y):-1 \leq y \leq 1\}$$

Prove that S is connected but not pathwise connected.

13. In this exercise, we sketch a different proof of Theorem 8.6.
Let $f:[a,b] \to R$ be continuous, and assume that $f(a) < z < f(b)$. Let
$S = \{x \in [a,b]:f(x) \leq z\}$.

(a) Show that S has a least upper bound, c.

(b) Assume that $f(c) \neq z$ and derive a contradiction.

(c) Now give a proof for the case where $f(a) > z > f(b)$.

CHAPTER 6

THE CHAIN RULE, HIGHER DERIVATIVES, AND TAYLOR'S THEOREM

INTRODUCTION

In this chapter, we continue with our study of the derivative of a function between finite dimensional inner product spaces.

We first consider the derivative of the composite of two functions. Our result, the chain rule, expresses the derivative of the composite of two functions in terms of the derivatives of the component functions. Briefly stated, the derivative of the composite is the composite of the derivatives.

We next introduce the higher derivatives of a function. Just as in the case of real valued functions of one variable, higher derivatives play an important role in the study of functions between Euclidean spaces. The higher derivatives can again be considered as linear transformations; however, our emphasis will be on the higher order partial derivatives.

The basic result of this chapter is Taylor's Theorem. This result gives conditions under which a function can be approximated near a point by a polynomial and will be useful to us when we study maxima and minima later in the text.

We assume, throughout this chapter, that all of the spaces that occur are finite dimensional inner product spaces. However, most of what we prove holds in greater generality; we leave the extension to the more general case to the interested reader.

1. THE CHAIN RULE

In this section, we state the chain rule for determining the derivative of the composite of two differentiable functions and give some examples to show how it is used. The proof is given in the next section.

Let V, W', W be finite dimensional inner product spaces, U an open subset of V and U' an open subset of W'.

THE CHAIN RULE. *Let* $f: U \to W'$ *be differentiable at* $\vec{v}_0$, *with* $f(U) \subseteq U'$ *and let* $g: U' \to W$ *be differentiable at* $\vec{w}_0 = f(\vec{v}_0)$. *Then the composite function* $g \circ f: U \to W$ *is differentiable at* $\vec{v}_0$ *and* $D_{\vec{v}_0}(g \circ f)$ *is the composite of* $D_{\vec{v}_0} f$ *with* $D_{\vec{w}_0} g$:

$$D_{\vec{v}_0}(g \circ f) = (D_{\vec{w}_0} g)(D_{\vec{v}_0} f)$$

As stated above, we prove this assertion in the next section.

Suppose now that $V = R^n$, $W' = R^p$, and $W = R^k$. We know from Proposition 3.7.1 that, relative to the standard bases, the matrix of $D_{\vec{w}_0}(g)$ is the Jacobian $J_g(\vec{w}_0)$ and the matrix of $D_{\vec{v}_0}(f)$ is the Jacobian $J_f(\vec{v}_0)$. Since the matrix of the composite of two linear transformations is the product of the matrices (by Proposition 2.6.3), the chain rule can be written

$$J_{g \circ f}(\vec{v}_0) = J_g(f(\vec{v}_0)) J_f(\vec{v}_0)$$

More explicitly, let $(x_1, \ldots, x_n)$ be coordinates in R^n, $(y_1, \ldots, y_p)$ coordinates in R^p, and suppose that

$$f(x_1, \ldots, x_n) = (f_1(x_1, \ldots, x_n), \ldots, f_p(x_1, \ldots, x_n))$$

$$g(y_1, \ldots, y_p) = (g_1(y_1, \ldots, y_p), \ldots, g_k(y_1, \ldots, y_p))$$

Then

$$
J_{g \circ f}(\vec{v}_0) =
\begin{pmatrix}
\dfrac{\partial g_1}{\partial y_1} & \cdots & \dfrac{\partial g_1}{\partial y_p} \\
\vdots & & \vdots \\
\dfrac{\partial g_k}{\partial y_1} & \cdots & \dfrac{\partial g_k}{\partial y_p}
\end{pmatrix}
\begin{pmatrix}
\dfrac{\partial f_1}{\partial x_1} & \cdots & \dfrac{\partial f_1}{\partial x_n} \\
\vdots & & \vdots \\
\dfrac{\partial f_p}{\partial x_1} & \cdots & \dfrac{\partial f_p}{\partial x_n}
\end{pmatrix}
$$

where the entries in the first matrix are evaluated at $f(\vec{v}_0)$ and the entries in the second matrix are evaluated at $\vec{v}_0$.

EXAMPLES

1. If $n = p = k = 1$, then each of the matrices involved is a one by one matrix - that is, a number. In this case, we have

$$
\frac{d(g \circ f)}{dx}(x_0) = \frac{dg}{dy}(f(x_0))\frac{df}{dx}(x_0)
$$

or, with different notation,

$$
(g \circ f)'(x_0) = g'(f(x_0))f'(x_0)
$$

This is the standard one variable chain rule.

2. If $n = p = 2$ and $k = 1$, we have

$$
J_{g \circ f} =
\begin{pmatrix}
\dfrac{\partial g}{\partial y_1} & \dfrac{\partial g}{\partial y_2}
\end{pmatrix}
\begin{pmatrix}
\dfrac{\partial f_1}{\partial x_1} & \dfrac{\partial f_1}{\partial x_2} \\
\dfrac{\partial f_2}{\partial x_1} & \dfrac{\partial f_2}{\partial x_2}
\end{pmatrix}
$$

$$
= \begin{pmatrix}
\dfrac{\partial g}{\partial y_1}\dfrac{\partial f_1}{\partial x_1} + \dfrac{\partial g}{\partial y_2}\dfrac{\partial f_2}{\partial x_1} & \dfrac{\partial g}{\partial y_1}\dfrac{\partial f_1}{\partial x_2} + \dfrac{\partial g}{\partial y_2}\dfrac{\partial f_2}{\partial x_2}
\end{pmatrix}
$$

where all partial derivatives of g are evaluated at $f(\vec{v}_0)$ and all partial derivatives of f_1 and f_2 are evaluated at $\vec{v}_0$. Since

$$J_{g \circ f}(\vec{v}_0) = \left(\dfrac{\partial (g \circ f)}{\partial x_1}(\vec{v}_0) \qquad \dfrac{\partial (g \circ f)}{\partial x_2}(\vec{v}_0) \right)$$

by definition of the Jacobian, we have

$$\frac{\partial (g \circ f)}{\partial x_1}(\vec{v}_0) = (\frac{\partial g}{\partial y_1}(f(\vec{v}_0))(\frac{\partial f_1}{\partial x_1}(\vec{v}_0))) + (\frac{\partial g}{\partial y_2}(f(\vec{v}_0)))(\frac{\partial f_2}{\partial x_1}(\vec{v}_0))$$

and

$$\frac{\partial (g \circ f)}{\partial x_2}(\vec{v}_0) = (\frac{\partial g}{\partial y_1}(f(\vec{v}_0))(\frac{\partial f_1}{\partial x_2}(\vec{v}_0))) + (\frac{\partial g}{\partial y_2}(f(\vec{v}_0)))(\frac{\partial f_2}{\partial x_2}(\vec{v}_0))$$

For example, if $f(x_1, x_2) = (x_1 \sin x_2, x_1 x_2^2)$ and $g(y_1, y_2) = 2y_1 y_2 + 4$, then

$$\frac{\partial f_1}{\partial x_1} = \sin x_2, \quad \frac{\partial f_1}{\partial x_2} = x_1 \cos x_2$$

$$\frac{\partial f_2}{\partial x_1} = x_2^2, \quad \frac{\partial f_2}{\partial x_2} = 2x_1 x_2$$

$$\frac{\partial g}{\partial y_1} = 2y_2, \quad \frac{\partial g}{\partial y_2} = 2y_1$$

It follows that

$$\frac{\partial g}{\partial y_1}(f(x_1, x_2)) = 2x_1 x_2^2, \quad \frac{\partial g}{\partial y_2}(f(x_1, x_2)) = 2x_1 \sin x_2$$

so that

$$\frac{\partial (g \circ f)}{\partial x_1} = 2x_1 x_2^2 \sin x_2 + (2x_1 \sin x_2) x_2^2$$

$$\frac{\partial (g \circ f)}{\partial x_2} = 2x_1 x_2^2 x_1 \cdot \cos x_2 + (2x_1 \sin x_2) 2x_1 x_2$$

This can be checked by noting that

$$(g \circ f)(x_1, x_2) = 2(x_1 \sin x_2)(x_1 x_2^2) + 4$$

and computing the partial derivatives directly.

3.   Let $n = 2$, $m = k = 3$, and define f and g by

$$f(x_1,x_2) = (x_2^2,\ x_1x_2,\ x_1^3)$$

$$g(y_1,y_2,y_3) = (y_1 - y_3^2,\ 2y_1^2 - y_2,\ y_1y_2y_3)$$

Then

$$J_f(x_1,x_2) = \begin{pmatrix} 0 & 2x_2 \\ x_2 & x_1 \\ 3x_1^2 & 0 \end{pmatrix}$$

and

$$J_g(y_1,y_2,y_3) = \begin{pmatrix} 1 & 0 & -2y_3 \\ 4y_1 & -1 & 0 \\ y_2y_3 & y_1y_3 & y_1y_2 \end{pmatrix}$$

Thus

$$J_{g\circ f}(x_1,x_2) = J_g(f(x_1,x_2))J_f(x_1,x_2)$$

$$= \begin{pmatrix} 1 & 0 & -2x_1^3 \\ 4x_2^2 & -1 & 0 \\ x_1^4x_2 & x_1^3x_2^2 & x_1x_2^3 \end{pmatrix} \begin{pmatrix} 0 & 2x_2 \\ x_2 & x_1 \\ 3x_1^2 & 0 \end{pmatrix}$$

$$= \begin{pmatrix} -6x_1^5 & 2x_2 \\ -x_2 & 8x_2^3-x_1 \\ 4x_1^3x_2^3 & 3x_1^4x_2^2 \end{pmatrix}$$

Again, this can be checked by writing out $g\circ f$ explicitly and computing the partial derivatives.

For more general finite dimensional inner product spaces than R^n, we can perform calculations in essentially the same way. Let V, W be finite dimensional inner product spaces, let U be an open subset of V, and let $F:U \to W$ be a differentiable function. Pick orthonormal bases $\{\vec{v}_1,\ldots,\vec{v}_n\}$ for V and $\{\vec{w}_1,\ldots,\vec{w}_p\}$ for W. If $\vec{v} = \Sigma_{j=1}^n x_j \vec{v}_j \in U$, then, for appropriate $f_1,\ldots,f_p$,

$$F(\vec{v}) = F(\sum_{i=1}^n x_i \vec{v}_i)$$

$$= \sum_{j=1}^p f_j (\sum_{i=1}^n x_i \vec{v}_i)\vec{w}_j$$

We may thus think of $F,f_1,\ldots,f_p$ as functions of $x_1,\ldots,x_n$. If we wish, we may now regard V as R^n and W as R^p (using the corollary to Proposition 4.7.4). Alternatively, we can define "partial derivatives" in V by

$$\partial_j f_i (\vec{v}) = \lim_{h \to 0} \frac{f_i (\vec{v} + h\vec{v}_j) - f_i (\vec{v})}{h}$$

(If we think of f_i as a function of $x_1,\ldots,x_n$, then $\partial_j f_i (\vec{v}) = \frac{\partial f_i}{\partial x_i}(\vec{v})$.)
One can now check that the matrix for $D_{\vec{v}}f$ (with respect to the given bases) is

$$\begin{pmatrix} \partial_1 f_1(\vec{v}) & \cdots & \partial_n f_1(\vec{v}) \\ & \cdots & \\ \partial_p f_1(\vec{v}) & \cdots & \partial_p f_n(\vec{v}) \end{pmatrix} \qquad (1.1)$$

just as in Proposition 3.7.1. (See Exercise 10.) In this way, we can reduce calculations involving derivatives on inner product spaces to calculations with partial derivatives and matrices. Now we can calculate with the chain rule on inner product spaces exactly as we have done with Euclidean spaces.

EXERCISES

1. Define functions f,g,h,F,G,H,K by

$$f(x,y) = x^2 + xy - y^2$$

$$g(x,y) = x \sin y$$

$$h(x,y,z) = xyz^2$$

$$F(x,y) = (xy^2, xy)$$

$$G(x,y) = (x+y, x-y, xy)$$

$$H(x,y,z) = (xyz, ze^{xy})$$

$$K(x,y,z) = (xy, yz, zx)$$

Compute the following derivatives, using the chain rule:

(a) $(f \circ F)'$ (f) $(K \circ G)'$

(b) $(g \circ H)'$ (g) $(f \circ H)'$

(c) $(h \circ G)'$ (h) $(G \circ H)'$

(d) $(h \circ K)'$ (i) $(H \circ G)'$

(e) $(F \circ H)'$ (j) $(F \circ F)'$

Check the answers by computing the composite functions and differentiating.

2. Let $f:R \to R$ be differentiable and define $g:R^2 \to R$ by $g(x,y) = f(x + y)$. Prove that $\frac{\partial g}{\partial x} - \frac{\partial g}{\partial y} = 0$.

3. Let V be a finite dimensional inner product space and let $f:V \to R$ be a function differentiable at $\vec{v}_0$. Define $g:V \to R$ by $g(\vec{v}) = (f(\vec{v}))^n$, where n is a positive integer. Prove that g is differentiable at $\vec{v}_0$ and that

$$D_{\vec{v}_0} g = n(f(\vec{v}_0))^{n-1} D_{\vec{v}_0} f$$

4. Let $f:R^2 \to R$ be a function and define

$$F(r,\theta) = f(r \cos \theta, r \sin \theta)$$

This is the function f in *polar coordinates*. (See Figure 1.1.) Compute $\dfrac{\partial F}{\partial r}$ and $\dfrac{\partial F}{\partial \theta}$ in terms of the partial derivative of f.

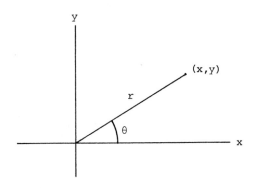

FIGURE 1.1.   Polar coordinates

5.   Let $f: R^3 \to R$ be a function and define

$$F(\rho, \theta, \varphi) = f(\rho \sin \varphi \cos \theta, \ \rho \sin \varphi \sin \theta, \ \rho \cos \varphi)$$

This is the function f in *spherical coordinates*.   (See Figure 1.2.) Compute the partial derivatives of F in terms of those of f.

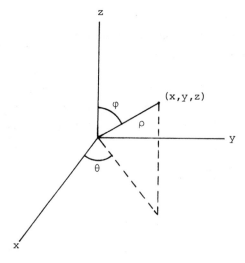

FIGURE 1.2.   Spherical coordinates

6.   Suppose $f:R^1 \to R^4$ and $g:R^4 \to R$ are differentiable.   Derive a formula for $(g \circ f)'$ in terms of the derivatives of f and g.

*7.   A function $f:V \to R$ is said to be *homogeneous of degree* k if $f(t\vec{v})$ $= t^k f(\vec{v})$ for any real number $t > 0$.   Prove that, if $f:R^n \to R$ is homogeneous of degree k and differentiable, then

$$\sum_{i=1}^{n} x_i \frac{\partial f}{\partial x_i}(x_1,\ldots,x_n) = kf(x_1,\ldots,x_n)$$

8.   (a) Let u,v be functions from R^1 to R^1. Define functions $f:R^1 \to R^2$ and $g:R^2 \to R^1$ by

$$f(x) = (u(x),v(x))$$
$$g(x_1,x_2) = x_1 x_2$$

Compute $(g \circ f)'$ by the chain rule and thus derive the product rule for functions of one variable.

(b) Let $U,V:R^n \to R$ be differentiable functions and define $f:R^n \to R^2$ by $f(\vec{v}) = (U(\vec{v}),V(\vec{v}))$. Compute $(g \circ f)'$ by the chain rule; conclude that UV is differentiable, and find its derivative.

9.   Let V be a finite-dimensional inner product space, and let $f:V \to R$ be defined by

$$f(\vec{v}) = \|\vec{v}\|^2 = <\vec{v},\vec{v}>$$

Show that f is differentiable and that $D_{\vec{v}}(f)(\vec{v}_0) = 2<\vec{v}_0,\vec{v}>$. (Either pick an orthonormal basis for V and compute derivatives as described in this section, or proceed directly from the definition of the derivative.)

*10.   (a) Let V and W be finite-dimensional inner product spaces, and let $U \subset V$ be open. Suppose that

$$F:U \to W$$

and

$G:U \rightarrow W$

are differentiable at $\vec{v}_0$. Show that $h:U \rightarrow R$, defined by

$$h(\vec{v}) = < F(\vec{v}), G(\vec{v}) >$$

is differentiable at v_0, and that

$$(D_{\vec{v}}h)(\vec{v}_0) = < D_{\vec{v}}F(\vec{v}_0), G(\vec{v}) > + < F(\vec{v}), D_{\vec{v}}G(\vec{v}_0) >$$

(b) Suppose that $f:U \rightarrow W$ is differentiable at $\vec{v}_0$, and define $g:U \rightarrow R$ by

$$g(\vec{v}) = < f(\vec{v}), f(\vec{v}) >$$

Show that g is differentiable at $\vec{v}_0$ and that

$$(D_{\vec{v}_0} g)(\vec{v}) = 2< f(\vec{v}_0), (D_{\vec{v}_0} f)(\vec{v}) >$$

(c) Suppose that $A:V \rightarrow V$ is a linear transformation. Define

$$g(\vec{v}) = < A\vec{v}, \vec{v} >$$

Show that $(D_{\vec{v}}g)(\vec{v}_0) = < A\vec{v}, \vec{v}_0 > + < \vec{v}, A\vec{v}_0 >$

11. Verify equation (1.1).

2.   PROOF OF THE CHAIN RULE

We begin with some preliminary results which are of interest in their own right.

Recall that a linear transformation $T:V \rightarrow W$ between inner product spaces is said to be *bounded* if there is a number $M > 0$ such that

$$\|T(\vec{v})\| \leq M\|\vec{v}\|$$

for all $\vec{v} \in V$. (See Exercise 17, Section 3.3.) The number M is called a *bound* for T.

PROPOSITION 2.1. *Let* V *and* W *be finite dimensional inner product spaces and* $T:V \rightarrow W$ *a linear transformation. Then* T *is bounded.*

Proof. We know by Proposition 4.7.5 that T is continuous. Thus we can find a $\delta > 0$ so that $\|T(\vec{v})\| < 1$ whenever $\|\vec{v}\| < \delta$. Set $M = 2/\delta$. Then, for any $\vec{v} \in V$,

$$\left\|\frac{\delta\vec{v}}{2\|\vec{v}\|}\right\| = \frac{\delta}{2} < \delta$$

so that

$$\left\|T\left(\frac{\delta\vec{v}}{2\|\vec{v}\|}\right)\right\| < 1$$

However,

$$\left\|T\left(\frac{\delta\vec{v}}{2\|\vec{v}\|}\right)\right\| = \left\|\frac{\delta}{2\|\vec{v}\|}\,T(\vec{v})\right\|$$

$$= \frac{\delta}{2\|\vec{v}\|}\|T(\vec{v})\|$$

Since $\left\|T\left(\frac{\delta\vec{v}}{2\|\vec{v}\|}\right)\right\| < 1$, it follows that

$$\|T(\vec{v})\| \leq \frac{2}{\delta}\|\vec{v}\| = M\|\vec{v}\|$$

for all $\vec{v} \in V$.

Now suppose U is an open subset of V and $f:U \rightarrow W$ is differentiable at $\vec{v}_0$ in U. Then, by the definition of differentiability, f can be approximated near $\vec{v}_0$ (up to a constant)[*] by the linear transformation $D_{\vec{v}_0} f:V \rightarrow W$. As a result, we might expect that an analogue of Proposition 2.1 would hold for f near $\vec{v}_0$.

PROPOSITION 2.2. *Let* V *and* W *be finite dimensional inner product spaces,* U *an open subset of* V, *and* $f:U \rightarrow W$ *a function differentiable at* $\vec{v}_0$. *Then there are positive real numbers* N *and* δ *such that*

---

[*] We say "up to a constant" because we really approximate $f(\vec{v}) - f(\vec{v}_0)$.

$$\|f(\vec{v}) - f(\vec{v}_0)\| \leq N\|\vec{v}-\vec{v}_0\| \tag{2.1}$$

whenever $\|\vec{v}-\vec{v}_0\| < \delta$.

A function satisfying equation (2.1) is said to satisfy the *local Lipschitz condition* at $\vec{v}_0$.

Proof of Proposition 2.2. Let $L = D_{\vec{v}_0} f$. Then

$$\|f(\vec{v}) - f(\vec{v}_0)\| = \|f(\vec{v}) - f(\vec{v}_0) - L(\vec{v}-\vec{v}_0) + L(\vec{v}-\vec{v}_0)\|$$
$$\leq \|f(\vec{v}) - f(\vec{v}_0) - L(\vec{v}-\vec{v}_0)\| + \|L(\vec{v}-\vec{v}_0)\|$$

Since f is differentiable at $\vec{v}_0$, we can choose δ so that

$$\|f(\vec{v}) - f(\vec{v}_0) - L(\vec{v}-\vec{v}_0)\| < \|\vec{v}-\vec{v}_0\|$$

for $\|\vec{v}-\vec{v}_0\| \leq \delta$. (We choose $\varepsilon = 1$ in the definition of differentiability.) Furthermore, since V is finite dimensional, we can use Proposition 2.1 to find an $M > 0$ such that

$$\|L(\vec{v}-\vec{v}_0)\| < M\|\vec{v}-\vec{v}_0\|$$

for all $\vec{v},\vec{v}_0$ in V. Now, defining $N = 1 + M$, we have

$$\|\vec{v}-\vec{v}_0\| < \delta \Rightarrow \|f(\vec{v}) - f(\vec{v}_0)\| \leq N\|\vec{v}-\vec{v}_0\|$$

completing the proof of Proposition 2.2.

We can now prove the chain rule. Let $T = D_{\vec{v}_0} f$ and $S = D_{\vec{w}_0} g$. We must show that, for any $\varepsilon > 0$, we can find $\delta > 0$ with

$$\|g{\circ}f(\vec{v}) - g{\circ}f(\vec{v}_0) - S{\circ}T(\vec{v}-\vec{v}_0)\| < \varepsilon\|\vec{v}-\vec{v}_0\|$$

whenever $\|\vec{v}-\vec{v}_0\| < \delta$. To begin with,

$$\|g{\circ}f(\vec{v}) - g{\circ}f(\vec{v}_0) - S{\circ}T(\vec{v}-\vec{v}_0)\|$$
$$= \|g{\circ}f(\vec{v}) - g{\circ}f(\vec{v}_0) - S(f(\vec{v}) - f(\vec{v}_0)) +$$
$$+ S(f(\vec{v}) - f(\vec{v}_0) - T(\vec{v}-\vec{v}_0))\|$$

$$\leq \|g \circ f(\vec{v}) - g \circ f(\vec{v}_0) - S(f(\vec{v}) - f(\vec{v}_0))\|$$

$$+ \|S(f(\vec{v}) - f(\vec{v}_0) - T(\vec{v} - \vec{v}_0))\| \qquad (2.2)$$

We shall deal with each of these two terms separately.

Since f is differentiable at $\vec{v}_0$, we know by Proposition 2.2 that we can find real numbers k and $\delta_1 > 0$ so that

$$\|f(\vec{v}) - f(\vec{v}_0)\| < k\|\vec{v} - \vec{v}_0\| \qquad (2.3)$$

when $\|\vec{v} - \vec{v}_0\| < \delta_1$. Since g is differentiable at $\vec{w}_0 = f(\vec{v}_0)$ with $S = D_{\vec{w}_0} g$, we can find $\delta_2 > 0$ with

$$\|g(\vec{w}) - g(\vec{w}_0) - S(\vec{w} - \vec{w}_0)\| < \frac{\varepsilon}{2k}\|\vec{w} - \vec{w}_0\| \qquad (2.4)$$

when $\|\vec{w} - \vec{w}_0\| < \delta_2$. Finally, we know that f is continuous at $\vec{v}_0$, since it is differentiable there (by Proposition 4.7.5); thus there is a $\delta_3 > 0$ such that

$$\|\vec{v} - \vec{v}_0\| < \delta_3 \Rightarrow \|f(\vec{v}) - f(\vec{v}_0)\| < \delta_2 \qquad (2.5)$$

Now, if $\|\vec{v} - \vec{v}_0\| < \min\{\delta_1, \delta_3\}$, we have (from equations (2.4) and (2.5))

$$\|g(f(\vec{v})) - g(f(\vec{v}_0)) - S(f(\vec{v}) - f(\vec{v}_0))\| < \frac{\varepsilon}{2k}\|f(\vec{v}) - f(\vec{v}_0)\|$$

$$< \frac{\varepsilon}{2k}(k\|\vec{v} - \vec{v}_0\|) \quad \text{(from (2.3))}$$

$$= \frac{\varepsilon}{2}\|\vec{v} - \vec{v}_0\|$$

This takes care of the first term in the expression (2.2) above. To deal with the second term, we first use Proposition 2.1 to find an $M > 0$ with $\|S\vec{w}\| < M\|\vec{w}\|$. Then

$$\|S(f(\vec{v}) - f(\vec{v}_0) - T(\vec{v} - \vec{v}_0))\| < M\|f(\vec{v}) - f(\vec{v}_0) - T(\vec{v} - \vec{v}_0)\|$$

Furthermore, since f is differentiable at $\vec{v}_0$, we can find $\delta_4 > 0$ so that

$$\|f(\vec{v}) - f(\vec{v}_0) - T(\vec{v} - \vec{v}_0)\| < \frac{\varepsilon}{2M}\|\vec{v} - \vec{v}_0\|$$

whenever $\|\vec{v}-\vec{v}_0\| < \delta_4$. It follows that

$$\|S(f(\vec{v}) - f(\vec{v}_0) - T(\vec{v}-\vec{v}_0))\| < \frac{\varepsilon}{2}\|\vec{v}-\vec{v}_0\|$$

whenever $\|\vec{v}-\vec{v}_0\| < \delta_4$.

If we now set $\delta = \min(\delta_1,\delta_3,\delta_4)$, we see from the inequality (2.2) above that

$$\|g \circ f(\vec{v}) - g \circ f(\vec{v}_0) - S \circ T(\vec{v}-\vec{v}_0)\| < (\varepsilon/2 + \varepsilon/2)\|\vec{v}-\vec{v}_0\|$$

$$= \varepsilon \|\vec{v}-\vec{v}_0\|$$

whenever $\|\vec{v}-\vec{v}_0\| < \delta$. This finishes the proof of the chain rule.

EXERCISES

1.   Use Theorem 5.6.1 and Proposition 5.5.2 to give another proof of Proposition 2.1. (Hint: Consider the image of $B_1(\vec{0})$ under T.) For a generalization, see Exercise 18 of Section 3.3.

2.   Show that any function $f:V \to W$ satisfying the local Lipschitz condition (2.1) is continuous at $\vec{v}_0$.

3.   Let $f:R \to R$ be a function such that

$$|f(x) - f(y)| < M \cdot |x - y|$$

for some number M and all $x,y \in R$. Prove that f is uniformly continuous. (In this case, f is said to satisfy the *Lipschitz condition*.)

4.   Give an example of a continuous function $f:(-1,1) \to R$ satisfying (2.1) at $\vec{v}_0 = 0$ but not differentiable there. (Thus the local Lipschitz condition is weaker than differentiability.)

5.   Let $f:R \to R$ be a function satisfying the condition

$$|f(x) - f(y)| < M|x - y|^n$$

for some $n > 1$. Prove that f is constant.

3.   HIGHER DERIVATIVES

One of the more useful notions in the calculus of one variable is the higher derivative. We now develop the analogous notion for functions between inner product spaces.

Let V and W be finite dimensional inner product spaces, U an open subset of V, and f:U → W a function differentiable at each point of U. We define the *differential* of f to be the function

$$Df:U \to \mathcal{L}(V,W)$$

given by $(Df)(\vec{v}) = D_{\vec{v}}f$. Here $\mathcal{L}(V,W)$ is the vector space of linear transformations·from V to W.

In fact, $\mathcal{L}(V,W)$ can be made into an inner product space as follows. Let T,S:V → W be linear transformations, $\alpha = \{\vec{v}_1,\ldots,\vec{v}_n\}$ an orthonormal basis for V, and $\beta = \{\vec{w}_1,\ldots,\vec{w}_p\}$ an orthonormal basis for W. Then, if $A = (a_{ij})$ and $B = (b_{ij})$ are the matrices of S and T relative to these bases (so that $T(\vec{v}_j) = \Sigma_{i=1}^p a_{ij}\vec{w}_i$ and $S(\vec{v}_j) = \Sigma_{i=1}^p b_{ij}\vec{w}_i$; see Section 4.7), we define the inner product of T and S by

$$< T,S > = \sum_{j=1}^n \sum_{i=1}^p a_{ij}b_{ij}$$

We shall see a way of defining this inner product without choosing bases in Section 5 of Chapter 7.

Suppose now that the function Df is differentiable at $\vec{v}$ in U. We denote its derivative $D_{\vec{v}}(Df)$ by $D_{\vec{v}}^2f$ and call it the *second derivative* of f at $\vec{v}$. Then $D_{\vec{v}}^2f$ is a linear transformation from V into $\mathcal{L}(V,W)$.

Suppose further that $D_{\vec{v}}^2f$ exists for all $\vec{v}$ in U. As above, we can define a *second differential* of f on U,

$$D^2f:U \to \mathcal{L}(V,\mathcal{L}(V,W))$$

by $(D^2f)(\vec{v}) = D_{\vec{v}}^2f:V \to \mathcal{L}(V,W)$. If this function has a derivative at $\vec{v}$ in U, we define the third derivative of f at $\vec{v}$, $D_{\vec{v}}^3f$, to be $D_{\vec{v}}(D^2f)$. (Again, $\mathcal{L}(V,\mathcal{L}(V,W))$ has the inner product defined just as above.)

Quite clearly we can continue this process of defining higher derivatives as long as the lower differentials exist and are differentiable. Just as in the case of the first derivative, we can express the matrices of these higher derivatives in terms of partial derivatives. We shall concentrate on the second derivative of a function $f:U \to R^p$, where U is an open subset of R^n. The extension to arbitrary inner product spaces, is, as before, straightforward; one picks orthonormal bases, as in Section 1. The treatment of derivatives of order higher than the second is also straightforward but increasingly complicated. We leave these matters to the energetic reader.

We begin with the case p = 1. Let U be an open subset of R^n and $f:U \to R$ a function. Suppose each of the partial derivatives of f exists on U and, further, that each of the partial derivatives of the functions

$$\frac{\partial f}{\partial x_i} : U \to R$$

also exists on U. We denote $\frac{\partial}{\partial x_j}(\frac{\partial f}{\partial x_i})$ by

$$\frac{\partial^2 f}{\partial x_j \partial x_i}$$

and call these the *second order partial derivatives* of f.

Of course, one can define higher order partial derivatives in the same way. For instance,

$$\frac{\partial^3 f}{\partial x_1 \partial x_2^2} = \frac{\partial}{\partial x_1}(\frac{\partial}{\partial x_2}(\frac{\partial f}{\partial x_2}))$$

is a partial derivative of f of order three. Here are some examples.

1.   Let $f:R^2 \to R$ be defined by $f(x_1,x_2) = x_1 x_2$. Then

$$\frac{\partial f}{\partial x_1} = x_2, \; \frac{\partial f}{\partial x_2} = x_1$$

and

$$\frac{\partial^2 f}{\partial x_1^2} = \frac{\partial^2 f}{\partial x_2^2} = 0, \qquad \frac{\partial^2 f}{\partial x_1 \partial x_2} = \frac{\partial^2 f}{\partial x_2 \partial x_1} = 1$$

In this case, all partial derivatives of order greater than two vanish.

2.   Let $f:R^3 \to R$ be defined by $f(x_1,x_2,x_3) = x_1 \sin (x_2 x_3)$.   Then

$$\frac{\partial^2 f}{\partial x_3 \partial x_1} = \frac{\partial}{\partial x_3} \sin (x_2 x_3)$$

$$= x_2 \cos (x_2 x_3)$$

and

$$\frac{\partial^3 f}{\partial x_2^2 \partial x_1} = \frac{\partial^2}{\partial x_2^2} \sin (x_2 x_3)$$

$$= \frac{\partial}{\partial x_2} x_3 \cos (x_2 x_3)$$

$$= - x_3^2 \sin (x_2 x_3)$$

Our goal now is to describe the second derivative

$$D_{\vec{V}}^2 f : R^n \to \mathcal{L}(R^n, \mathcal{L}(R^n, R))$$

of a function $f:U \to R$ in terms of the second order partial derivatives of f.

To begin with, let $T_1, \ldots, T_n$ be the elements of $\mathcal{L}(R^n, R)$ defined by

$$T_i (x_1, \ldots, x_n) = x_i$$

$1 \le i \le n$.   It is easily seen that these linear transformations form an orthonormal basis for $\mathcal{L}(R^n, R)$.   Furthermore, if $f:U \to R$ is differentiable, then

$$Df:U \to \mathcal{L}(R^n, R)$$

is given by

$$(Df)(\vec{v}) = \sum_{i=1}^{n} (\frac{\partial f}{\partial x_i}(\vec{v}))T_i$$

If we identify $\mathcal{L}(R^n,R)$ with R^n under the isometry taking the orthonormal basis $T_1,\ldots,T_n$ into the standard basis $\vec{e}_1,\ldots,\vec{e}_n$, the differential of f becomes a function from U into R^n:

$$(Df)(\vec{v}) = (\frac{\partial f}{\partial x_1}(\vec{v}),\ldots,\frac{\partial f}{\partial x_n}(\vec{v}))$$

This last function is traditionally called the *gradient of* f (see Exercise 7, Section 3.7) and denoted by grad f (or, sometimes, by ∇f):

$$\nabla f(\vec{v}) = (\text{grad } f)(\vec{v}) = (\frac{\partial f}{\partial x_1}(\vec{v}),\ldots,\frac{\partial f}{\partial x_n}(\vec{v}))$$

According to the definition, the second derivative of f, $D^2_{\vec{v}}f:R^n \to \mathcal{L}(R^n,R)$ is the derivative of the differential of f. Equivalently, the second derivative of f is the derivative of the gradient of f:

$$D^2_{\vec{v}}f = D_{\vec{v}}(\text{grad } f)$$

Proposition 3.5.1 now states that if grad $f:U \to R^n$ is differentiable, then all the partial derivatives of the component functions of grad f exist and the matrix of $D_{\vec{v}}(\text{grad } f)$ is the Jacobian of grad f. In this case, the partial derivatives of the component functions of grad f are the second order partial derivatives of f and the Jacobian of grad f is the $n \times n$ matrix

$$\begin{pmatrix} \dfrac{\partial^2 f}{\partial x_1 \partial x_1} & \cdots & \dfrac{\partial^2 f}{\partial x_n \partial x_1} \\ & \cdots & \\ \dfrac{\partial^2 f}{\partial x_1 \partial x_n} & \cdots & \dfrac{\partial^2 f}{\partial x_n \partial x_n} \end{pmatrix} \tag{3.1}$$

Interpreting this in terms of the orthonormal basis $T_1,\ldots,T_n$ for $\mathcal{L}(R^n,R)$, we have

$$(D_{\vec{v}}^2 f)(a_1,\ldots,a_n) = \sum_{j=1}^{n} \sum_{i=1}^{n} a_i \left(\frac{\partial^2 f}{\partial x_i \partial x_j}(\vec{v})\right) T_j$$

The matrix (3.1) is called the *Hessian* of f and denoted by H(f). Its value at $\vec{v}$ will be written as $H(f)(\vec{v})$ or $H_{\vec{v}}(f)$. Here is an example.

3. Let $g: R^3 \to R$ be defined by $g(x_1,x_2,x_3) = x_1 \sin(x_2 x_3)$. We compute $H(g)(\vec{v})$ for $\vec{v} = (3,2,\pi/6)$. To begin with,

$$\frac{\partial g}{\partial x_1} = \sin(x_2 x_3)$$

$$\frac{\partial g}{\partial x_2} = x_1 x_3 \cos(x_2 x_3)$$

$$\frac{\partial g}{\partial x_3} = x_1 x_2 \cos(x_2 x_3)$$

It follows that

$$\frac{\partial^2 g}{\partial x_1^2} = 0, \quad \frac{\partial^2 g}{\partial x_2^2} = -x_1 x_3^2 \sin(x_2 x_3), \quad \frac{\partial^2 g}{\partial x_3^2} = -x_1 x_2^2 \sin(x_2^3)$$

$$\frac{\partial^2 g}{\partial x_1 \partial x_2} = x_3 \cos(x_2 x_3) = \frac{\partial^2 g}{\partial x_2 \partial x_1}$$

$$\frac{\partial^2 g}{\partial x_1 \partial x_3} = x_2 \cos(x_2 x_3) = \frac{\partial^2 g}{\partial x_3 \partial x_1}$$

$$\frac{\partial^2 g}{\partial x_2 \partial x_3} = x_1 \cos(x_2 x_3) - x_1 x_2 x_3 \sin(x_2 x_3) = \frac{\partial^2 g}{\partial x_3 \partial x_2}$$

Evaluating these partial derivatives at $\vec{v} = (3,2,\frac{\pi}{6})$, we see that

$$H_{\vec{v}}(g) = \begin{pmatrix} 0 & \frac{\pi}{12} & 1 \\ \frac{\pi}{12} & \frac{-\pi^2\sqrt{3}}{24} & \frac{3-\pi\sqrt{3}}{2} \\ 1 & \frac{3+\sqrt{3}}{2} & -6\sqrt{3} \end{pmatrix}$$

We summarize the discussion above in the form of a proposition.

PROPOSITION 3.1. *Let U be an open subset of* R^n *and* $f:U \to R$ *a function whose second derivative*

$$D_{\vec{v}}^2 f: R^n \to \mathcal{L}(R^n, R)$$

exists at the point $\vec{v} \in U$. *Then all second order partial derivatives of* f *exist at* $\vec{v}$. *Furthermore, the matrix of the linear transformation* $D_{\vec{v}}^2 f$ *(relative to the standard basis for* R^n *and the basis* $T_1, \ldots, T_n$ *for* $\mathcal{L}(R^n, R)$*) is the Hessian of* f *evaluated at* $\vec{v}$.

Consider now the more general situation where $f:U \to R^p$ is defined by $f(\vec{v}) = (f_1(\vec{v}), \ldots, f_p(\vec{v}))$. We define linear transformations $T_{ij}:R^n \to R^p$ by

$$T_{ij}(x_1, \ldots, x_n) = x_i \vec{e}_j$$

$1 \le i \le n$, $1 \le j \le p$, where $\{\vec{e}_1, \ldots, \vec{e}_p\}$ is the standard basis for R^p. It is easily seen that this is an orthonormal basis for $\mathcal{L}(R^n, R^p)$.

PROPOSITION 3.2. *Suppose* $f:U \to R^p$ *is differentiable and that the second derivative of* f *exists at* $\vec{v}$ *in U. Then all second partial derivatives of the functions* $f_1, \ldots, f_p$ *exist and* $D_{\vec{v}}^2 f: R^n \to \mathcal{L}(R^n, R^p)$ *is given by*

$$(D_{\vec{v}}^2 f)(a_1, \ldots, a_n) = \sum_{i=1}^{n} \sum_{j=1}^{n} \sum_{k=1}^{m} a_i \frac{\partial^2 f_k}{\partial x_i \partial x_j}(\vec{v}) T_{jk}$$

The proof of this result is almost identical with that of Proposition 3.1 and is left to the reader.

Computing $D_{\vec{v}}^2 f$ thus involves computing all the second partial derivatives of f. Similarly, the higher derivatives of f are computed by evaluating higher order partial derivatives. The amount of work involved soon becomes considerable. One useful result that reduces the computations greatly is the following.

THEOREM 3.3. *Let U be an open subset of* R^n *and* $f:U \to R$ *a function such that for some fixed* i,j, $1 \le i < j \le n$,

$$\frac{\partial^2 f}{\partial x_i \partial x_j} \; , \; \frac{\partial^2 f}{\partial x_j \partial x_i}$$

exist on U *and are continuous at a point* $\vec{v}$ *in* U. *Then*

$$\frac{\partial^2 f}{\partial x_i \partial x_j}(\vec{v}) = \frac{\partial^2 f}{\partial x_j \partial x_i}(\vec{v})$$

We prove this result in Section 8. (See Exercise 11 for an example where the order of differentation *does* matter.)

We conclude this section with an example showing how the chain rule is used to compute higher order derivatives of composite functions. Although this is in essence a straightforward application of the chain rule, it can be quite complicated.

Let $f: R^1 \to R^1$ and $g: R^2 \to R^1$ be functions whose second derivatives exist. We wish to compute the second order partial derivative $\dfrac{\partial^2 (f \circ g)}{\partial x^2}$ of the composite function $f \, g: R^2 \to R^1$.

To begin with, let (x,y) be coordinates in the domain of g and z the coordinate in the domain of f. Then

$$\frac{\partial (f \circ g)}{\partial x}(x,y) = \frac{df}{dz}(g(x,y))\frac{\partial g}{\partial x}(x,y)$$

$$\frac{\partial (f \circ g)}{\partial y}(x,y) = \frac{df}{dz}(g(x,y))\frac{\partial g}{\partial y}(x,y)$$

by the chain rule. Now

$$\frac{\partial^2 (f \circ g)}{\partial x^2}(x,y) = \frac{\partial}{\partial x}\left(\frac{\partial (f \circ g)}{\partial x}\right)(x,y)$$

$$= \left(\frac{\partial}{\partial x}\left(\frac{df}{dz}(g(x,y))\right)\right)\left(\frac{\partial g}{\partial x}(x,y)\right) \qquad (3.2)$$

$$+ \left(\frac{df}{dz}(g(x,y))\right)\left(\frac{\partial}{\partial x}\left(\frac{\partial g}{\partial x}(x,y)\right)\right)$$

by the product rule for differentiation. The second of the two terms on the right side of equation (3.2) is easily identified with

$$\frac{df}{dz}(g(x,y)) \cdot \frac{\partial^2 g}{\partial x^2}(x,y)$$

We compute the first term.

Note that $\frac{df}{dz}(g(x,y))$ is the composite of two functions, namely $\frac{df}{dz}:R \rightarrow R$ and $g:R^2 \rightarrow R$. Therefore, we need to use the chain rule to differentiate it :

$$\frac{\partial}{\partial x}(\frac{df}{dz}(g(x,y))) = (\frac{d}{dz}(\frac{df}{dz}))(g(x,y))\frac{\partial g}{\partial x}(x,y)$$

$$= \frac{d^2 f}{dz^2}(g(x,y))\frac{\partial g}{\partial x}(x,y)$$

It follows that

$$\frac{\partial^2 (f \circ g)}{\partial x^2}(x,y) = \frac{d^2 f}{dz^2}(g(x,y))(\frac{\partial g}{\partial x}(x,y))^2$$

$$+ \frac{df}{dz}(g(x,y))\frac{\partial^2 g}{\partial x^2}(x,y)$$

For example, if $f(z) = z^3$ and $g(x,y) = x^2 + y$, then

$$\frac{\partial^2 (f \circ g)}{\partial x^2}(x,y) = 6(x^2 + y)(2x)^2 + 3(x^2 + y)^2 \cdot 2$$

$$= 24(x^4 + yx^2) + 6(x^4 + 2x^2 y + y^2)$$

$$= 30x^4 + 36x^2 y + 6y^2$$

Of course, this can be verified by first computing the composite $f \circ g$ and then differentiating it.

EXERCISES

1.   Compute $\frac{\partial^2 f}{\partial x^2}$, $\frac{\partial^2 f}{\partial x \partial y}$ and $\frac{\partial^3 f}{\partial x \partial y^2}$ for the following functions.

(a) $f(x,y) = x^2 y - y^3$

(b) $f(x,y) = x \sin y$

(c) $f(x,y) = xy\, e^{x^2}$

(d) $f(x,y) = \arctan(x + y)$

(e) $f(x,y) = (xy + y^2)^{1/3}$

(f) $f(x,y) = 2y \tan(xy)$

(g) $f(x,y) = x \arcsin(x^2 y)$

2.   Compute $H(f)(0,0)$ for the functions in Exercise 1.

3.   Compute $H(f)(\vec{v}_0)$ where

   (a) $f:R^3 \to R$, $f(x_1,x_2,x_3) = x_1 x_2 - x_3^3$, $\vec{v}_0 = (0,0,0)$

   (b)   f as in (a), $\vec{v}_0 = (1,3,-1)$

   (c) $f:R^3 \to R$, $f(x_1,x_2,x_3) = x_1 \sin(x_2 x_3)$, $\vec{v}_0 = (0,0,0)$

   (d)   f as in (c), $\vec{v}_0 = (3,\frac{\pi}{6},-2)$

   (e) $f:R^4 \to R$, $f(x_1,x_2,x_3,x_4) = x_1 x_3 - x_2 x_4$, $\vec{v}_0 = (0,0,0,0)$

   (f)   f as in (e), $\vec{v}_0 = (2,-1,1,3)$

4.   Let $f,g:R \to R$ be differentiable functions and define $h:R^2 \to R$ by $h(x,t) = f(x - at) + g(x + at)$, where a is a constant. Prove that h satisfies the *wave equation*:

$$a^2 \frac{\partial^2 h}{\partial x^2} = \frac{\partial^2 h}{\partial t^2}$$

(This equation is satisfied by vibrating strings and by air in a pipe.)

5.   Let $f:R \to R$ and $g:R^2 \to R$ be functions whose second derivatives exist. Compute all second order partial derivatives of the composite $f \circ g$ in terms of the derivatives of f and g. (See the example at the end of the section.)

6.   Let $f:R^n \to R$, $g:R \to R^n$ be functions whose second derivatives exist. Compute the second derivative of the composite $f \circ g:R \to R$ in terms of the derivatives of f and g.

7.   If $f:R^2 \to R$ is a function, then $F(r,\theta) = f(r \cos \theta, r \sin \theta)$ is the function f in polar coordinates. (See Exercise 4, Section 1.)

Suppose that the second derivative of the function f exists. Compute all second order partial derivatives of $F(r,\theta)$ in terms of the partial derivatives of f.

8.   Let $f:R^n \to R$ be a twice differentiable function. The *Laplacian* of f, Δf, is defined by

$$\Delta f = \sum_{i=1}^{n} \frac{\partial^2 f}{\partial x_i^2}$$

If $n = 2$ and $F(r,\theta) = f(r \cos \theta, r \sin \theta)$, prove that

$$\Delta f = \frac{\partial^2 F}{\partial r^2} + \frac{1}{r^2} \frac{\partial^2 F}{\partial \theta^2} + \frac{1}{r} \frac{\partial F}{\partial r}$$

9.   If $f:R^3 \to R$ is a function, then

$$F(\rho,\theta,\varphi) = f(\rho \sin \varphi \cos \theta, \rho \sin \varphi \sin \theta, \rho \cos \varphi)$$

is the function f in spherical coordinates. (See Exercise 5, Section 1.) Determine the second order partial derivatives of F in terms of the partial derivatives of f.

10.   With notation as in Exercise 9, derive a formula for the Laplacian of f in terms of the derivatives of F. (See Exercise 8.)

11.   Consider the function $f:R^2 \to R$ defined by

$$f(x,y) = \begin{cases} xy\dfrac{(x^2 - y^2)}{x^2 + y^2} & \text{if} \quad (x,y) \neq (0,0) \\ 0 & \text{if} \quad (x,y) = (0,0) \end{cases}$$

Show that $\frac{\partial f}{\partial x}$ and $\frac{\partial f}{\partial y}$ exist and are continuous, that $\frac{\partial^2 f}{\partial x \partial y}$ and $\frac{\partial^2 f}{\partial y \partial x}$ exist at every point, and that

$$\frac{\partial^2 f}{\partial x \partial y}\bigg|(0,0) \neq \frac{\partial^2 f}{\partial y \partial x}\bigg|(0,0)$$

What conclusion can you draw about the functions $\frac{\partial^2 f}{\partial x \partial y}$ and $\frac{\partial^2 f}{\partial y \partial x}$?

12. Use induction to prove the following extension of Theorem 3.3.
Let U be an open subset of R^n and $f:U \to R$ a function. Suppose that
all partial derivatives of f of order $\leq r$ exist and are continuous
at $\vec{v} \in U$. Prove that the order of differentation of these partial
derivatives does not matter. For example, if r = 3, then

$$\frac{\partial^3 f}{\partial x_1 \partial x_2 \partial x_3}(\vec{v}) = \frac{\partial^3 f}{\partial x_2 \partial x_3 \partial x_1}(\vec{v})$$

4.   TAYLOR'S THEOREM FOR FUNCTIONS OF ONE VARIABLE

As we saw in Section 3.5, the derivative $D_{\vec{v}_0} f$ of a function $f:R^n \to R^p$
at $\vec{v}_0$ provides us with a good linear approximation to f near $\vec{v}_0$;

$$h(\vec{v}) = D_{\vec{v}_0}(\vec{v}-\vec{v}_0) + f(\vec{v}_0)$$

If f has higher derivatives, we might expect to get an even better
approximation using polynomials instead of linear functions. We show
now that this is indeed the case. We restrict outselves to p = 1 and
begin with n = 1, since the results for n > 1 can be reduced to this
case.

Let U be an open interval in R, and let $f:U \to R$ be a function
such that the jth derivative, $f^{(j)}$, exists on U for $1 \leq j \leq m + 1$.
(Of course, $f^{(j)}$ is continuous for $j \leq m$, but $f^{(m+1)}$ need not be.)
Define the mth *degree Taylor polynomial of f about a by*

$$P_m(x) = f(a) + f'(a)(x-a) + \frac{f''(a)}{2}(x-a)^2 + \cdots + \frac{f^{(m)}(a)}{m!}(x-a)^m$$

$$= \sum_{j=0}^{m} \frac{f^{(j)}(a)}{j!}(x-a)^j$$

and the *remainder* by

$$R_m(x) = f(x) - P_m(x)$$

The following result is known as *Taylor's Remainder Theorem* for
functions of one variable.

THEOREM 4.1. *Let* U *be an open interval in* R. *Pick* a $\in$ U, *and let* f:U $\rightarrow$ R *be a function having* (m + 1) *derivatives on* U. *Then for any* x *in* U, *there is a* c $\in$ U *between* x *and* a *such that*

$$R_{m+1}(x) = \frac{f^{(m+1)}(c)}{(m+1)!}(x-a)^{m+1}$$

Proof. For definiteness, assume a < x. (If a = x, the theorem is trivial; if a > x, the proof is essentially the same as for a < x.) Define a number K by the equation

$$f(x) - f(a) - \sum_{j=1}^{m} \frac{f^{(j)}(a)}{j!}(x-a)^j - \frac{K}{(m+1)!}(x-a)^{m+1} = 0 \qquad (4.1)$$

We need to prove that $K = f^{(m+1)}(c)$ for some c between x and a.

Let g be a function defined by

$$g(t) = f(t) - f(a) - \sum_{j=1}^{m} \frac{f^{(j)}(c)}{j!}(t-a)^j - \frac{K}{(m+1)!}(t-a)^{m+1}$$

Then

$$g'(t) = f'(t) - f'(a) - \sum_{j=1}^{m-1} \frac{f^{(j)}(c)}{j!}(t-a)^j - \frac{K}{m!}(t-a)^m$$

and, similarly,

$$g^{(p)}(t) = f^{(p)}(t) - f^{(p)}(a) - \sum_{j=1}^{m-p} \frac{f^{(j)}(c)}{j!}(t-a)^j \qquad (4.2)$$

$$- \frac{K}{(m+1-p)!}(t-a)^{m+1-p}$$

for $1 \le p \le m$. This is proved easily by induction; we omit the details. (See Exercise 3.)

In particular,

$$g(a) = g'(a) = \cdots = g^{(m)}(a) = 0$$

Furthermore, $g(x) = 0$, so by Rolle's Theorem (Exercise 8, Section 3.5), there is a number c_1, $a < c_1 < x$, such that $g'(c_1) = 0$. Similarly, since $g'(c_1) = g'(a) = 0$, there is a number c_2, $a < c_2 < c_1$, with $g''(c_2) = 0$. Continuing inductively, we eventually find a number $c = c_{m+1}$ between a and x with $g^{(m+1)}(c) = 0$. However

$$g^{(m+1)}(t) = f^{(m+1)}(t) - K$$

so $f^{(m+1)}(c) = K$ and we are finished.

Here are some examples.

1. Let $f(x) = e^x$ and $a = 0$. Then $f^{(k)}(x) = e^x$ and $f^{(k)}(0) = 1$ for all k. Therefore, the mth degree Taylor polynomial of the function e^x about $x = 0$ is

$$P_m(x) = \sum_{j=0}^{m} \frac{x^j}{j!}$$

$$= 1 + x + \frac{x^2}{2!} + \frac{x^3}{3!} + \cdots + \frac{x^m}{m!}$$

and the remainder is given by

$$R_m(x) = \frac{e^c}{(m+1)!} x^{m+1}$$

for some c between 0 and x.

This expression for the remainder is useful in determining how good an approximation $P_m(x)$ is for e^x. If, for example $0 \le x \le 1$, then $c \le 1$ and $e^c \le 3$. Therefore

$$\left| e^x - P_m(x) \right| = \left| R_m(x) \right|$$

$$\le \frac{3x^{m+1}}{(m+1)!}$$

In particular, if $x = .1$ and $n = 5$, we have

$$\left| e^{.1} - P_5(.1) \right| \le \frac{3 \cdot 10^{-6}}{6!} = \frac{10^{-7}}{24}$$

2.   Let $f(x) = \sin x$ and $a = 0$. Then

$$f^{(k)}(x) = \begin{cases} \sin x & \text{if } k = 4\ell \\ \cos x & \text{if } k = 4\ell + 1 \\ -\sin x & \text{if } k = 4\ell + 2 \\ -\cos x & \text{if } k = 4\ell + 3 \end{cases}$$

where ℓ is a non negative integer. Thus

$$f^{(k)}(0) = \begin{cases} 0 & \text{if } k \text{ is even} \\ 1 & \text{if } k = 4\ell + 1 \\ -1 & \text{if } k = 4\ell + 3 \end{cases}$$

so that, if $m = 2p + 1$, the mth degree Taylor polynomial of $\sin x$ about $a = 0$ is

$$P_m(x) = x - \frac{x^3}{3!} + \frac{x^5}{5!} - \ldots \pm \frac{x^m}{m!}$$

$$= \sum_{k=0}^{p} (-1)^k \frac{x^{2k+1}}{(2k+1)!}$$

A similar expression holds if m is even. The remainder is given by

$$R_m(x) = \begin{cases} \dfrac{\pm\sin(c)}{(m+1)!}x^{m+1} & \text{if } m \text{ is odd} \\ \dfrac{\pm\cos(c)}{(m+1)!}x^{m+1} & \text{if } m \text{ is even} \end{cases}$$

where c is between 0 and x. In particular,

$$\left| \sin x - P_m(x) \right| = \left| R_m(x) \right|$$

$$\leq \frac{\left| x^{m+1} \right|}{(m+1)!}$$

for all x.

3.   Let $f(x) = \cos x$ and $a = \pi/4$. Then

$$f^{(k)}(a) = \begin{cases} \dfrac{-\sqrt{2}}{2} & \text{if } k = 4\ell + 1 \text{ or } 4\ell + 2 \\[2ex] \dfrac{\sqrt{2}}{2} & \text{if } k = 4\ell \quad\quad \text{or } 4\ell + 3 \end{cases}$$

Thus the mth degree Taylor polynomial for cos x about a = $\pi/4$ is

$$P(x) = \frac{\sqrt{2}}{2}\left(1 - (x - \pi/4) - \frac{(x - \pi/4)^2}{2!}\right.$$

$$\left. + \frac{(x - \pi/4)^3}{3!} + \ldots \pm \frac{(x - \pi/4)^m}{m!}\right)$$

and the remainder is

$$R_m(x) = \begin{cases} \dfrac{\pm \cos(c)(x-\pi/4)^{m+1}}{(m+1)!} & \text{if } m = 4\ell - 1 \text{ or } 4\ell + 1 \\[2ex] \dfrac{\pm \sin(c)(x-\pi/4)^{m+1}}{(m+1)!} & \text{if } m = 4\ell \text{ or } m = 4\ell + 2 \end{cases}$$

where c is between x and a. In particular,

$$\left|\cos x - P_m(x)\right| \le \frac{|x-\pi/4|^{m+1}}{(m+1)!}$$

so that $P_m(x)$ provides a good approximation to cos x for values of x near $\pi/4$.

EXERCISES

1.   Determine the mth degree Taylor polynomial of the function f(x) about a and the remainder where

(a) $f(x) = \log(1+x)$, a = 0, m arbitrary

(b) $f(x) = \cos x$, a = 0, m arbitrary

(c) $f(x) = \arctan x$, a = 0, m arbitrary

(d) $f(x) = \sin x$, a = $\pi/6$, m = 5

(e) $f(x) = e^x$, a = 2, m = 5

(f) $f(x) = (1+x)^r$, r > 0, a = 0, m = 1

(g) $f(x) = (1+x)^r$, r > 0, a = 0, m = 3

2.   Determine the following with error less than 10^{-3}.

  (a) $\sin(.2)$ (b) $e^{.3}$ (c) $\sqrt{2}$

  (d) $\log(1.2)$ (e) $\cos(43°)$

3.   Prove equation (4.2).

4.   There are various other forms of Taylor's Theorem with remainder, in which one gives a different expression for the remainder term. We give one in this exercise and one in the next.

  Show that if f is as in Theorem 4.1 and $f^{(n+1)}$ is also continuous, then

$$f(x) = f(a) + f'(a)(x-a) + \cdots + \frac{f^{(m)}(a)}{m!}(x-a)^m$$

$$+ \int_a^x \frac{f^{(n+1)}(t)(t-a)^n}{n!} dt$$

(Hint: write $f(x) - f(a) = \int_a^x f'(t)dt$ and integrate by parts.)

5.   Let f be as in Theorem 4.1 and let α be an integer ≥ 1. Pick $a,b \in U$, and define A by

$$f(b) = P_m(b) + A(b-a)^\alpha$$

Now consider

$$g(x) = f(b) - \sum_{j=0}^m \frac{f^{(j)}(x)}{j!}(b-x)^j - A(x-a)^\alpha$$

Use Rolle's theorem to show that there is a ξ between a and b with

$$R_m(b) = \frac{f^{(m+1)}(\xi)}{\alpha m!}(b-a)^\alpha(b-\xi)^{m+1-\alpha}$$

6.   (a) Prove one case of L'Hôpital's rule: if f,g have continuous derivatives in U and $f(a) = g(a) = 0$ $(a \in U)$, $g'(a) \neq 0$, then

$$\lim_{x \to a} \frac{f(x)}{g(x)} = \frac{f'(a)}{g'(a)}$$

If $g'(a) = 0$ and $f'(a) \neq 0$, then $\lim\limits_{x \to a} \dfrac{f(x)}{g(x)}$ does not exist.

(b)   Now suppose that f and g have n continuous derivatives, and suppose that

$$f(a) = f'(a) = \cdots = f^{(n-1)}(a) = 0$$

$$g(a) = g'(a) = \cdots = g^{(n-1)}(a) = 0$$

with $g^{(n)}(a) \neq 0$.   Show that

$$\lim_{x \to a} \frac{f(x)}{g(x)} = \frac{f^{(n)}(a)}{g^{(n)}(a)}$$

*7.   (a)   Prove the following variant of the Mean Value Theorem: if f,g are differentiable functions on the open interval $U \subset R^1$, and if $a,b \in U$, then there is a ξ between a and b such that

$$(f(b) - f(a))g'(\xi) = (g(b) - g(a))f'(\xi)$$

(b)   Prove the following more precise form of L'Hôpital's Rule: if f,g are as in (a) and if $f(a) = g(a) = 0$, then

$$\lim_{x \to a} \frac{f(x)}{g(x)} = \lim_{x \to a} \frac{f'(x)}{g'(x)}$$

in the sense that if the right hand side has a limit, then so does the left hand side, and they are equal.

5.   TAYLOR'S THEOREM FOR FUNCTIONS OF TWO VARIABLES

The general form of Taylor's Theorem is notationally quite complicated. We preview it here for functions of two variables, deferring the general case to the next section.

Let U be an open subset of R^2 and $f:U \to R$ be a function whose partial derivatives of order $\leq m$ exist and are continuous.   Let $\vec{a} = (a_1, a_2)$ be a vector in U.   The *Taylor Polynomial of f of degree m about* $\vec{a}$ is defined by

$$P_m(x_1,x_2) = \sum_{i+j \leq m} \frac{1}{i!j!} \left(\frac{\partial^{i+j} f}{\partial x_1^i \partial x_2^j}(\vec{a}) \right) (x_1 - a_1)^i (x_2 - a_2)^j$$

For example,

$$P_1(x_1,x_2) = f(\vec{a}) + \frac{\partial f}{\partial x_1}(\vec{a})(x_1 - a_1) + \frac{\partial f}{\partial x_2}(\vec{a})(x_2 - a_2)$$

and

$$
\begin{aligned}
P_2(x_1,x_2) &= f(\vec{a}) + \frac{\partial f}{\partial x_1}(\vec{a})(x_1 - a_1) + \frac{\partial f}{\partial x_2}(\vec{a})(x_2 - a_2) \\
&\quad + \frac{1}{2}\frac{\partial^2 f}{\partial x_1^2}(\vec{a})(x_1 - a_1)^2 + \frac{\partial^2 f}{\partial x_1 \partial x_2}(\vec{a})(x_1 - a_1)(x_2 - a_2) \\
&\quad + \frac{1}{2}\frac{\partial^2 f}{\partial x_2^2}(\vec{a})(x_2 - a_2)^2 \\
&= P_1(x_1,x_2) + \frac{1}{2}[\frac{\partial^2 f}{\partial x_1^2}(\vec{a})(x_1 - a_1)^2 + \\
&\quad + 2\frac{\partial^2 f}{\partial x_1 \partial x_2}(\vec{a})(x_1 - a_1)(x_2 - a_2) \\
&\quad + \frac{\partial^2 f}{\partial x_2^2}(\vec{a})(x_2 - a_2)^2]
\end{aligned}
$$

Similarly $P_3(x_1,x_2)$ is obtained by adding the term

$$
\begin{aligned}
\frac{1}{6}\frac{\partial^3 f}{\partial x_1^3}&(\vec{a})(x_1 - a_1)^3 + \frac{1}{2}\frac{\partial^3 f}{\partial x_1^2 \partial x_2}(\vec{a})(x_1 - a_1)^2(x_2 - a_2) \\
&\quad + \frac{1}{2}\frac{\partial^3 f}{\partial x_1 \partial x_2^2}(\vec{a})(x_1 - a_1)(x_2 - a_2)^2 \\
&\quad + \frac{1}{6}\frac{\partial^3 f}{\partial x_2^3}(\vec{a})(x_2 - a_2)^3
\end{aligned}
$$

to $P_2(x_1,x_2)$. (We use Exercise 12 of Section 3 here.)

Here are some particular cases.

1. Let $f(x_1,x_2) = x_1^2 x_2 + 1$ and $\vec{a} = (0,0)$. Then

$$\frac{\partial f}{\partial x_1} = 2x_1x_2 \qquad \frac{\partial f}{\partial x_2} = x_1^2$$

$$\frac{\partial^2 f}{\partial x_1^2} = 2x_2 \qquad \frac{\partial^2 f}{\partial x_2^2} = 0$$

$$\frac{\partial^2 f}{\partial x_1 \partial x_2} = 2x_1 = \frac{\partial^2 f}{\partial x_2 \partial x_1}$$

and

$$\frac{\partial^3 f}{\partial x_1^2 \partial x_2} = 2$$

whereas all other third order derivatives vanish.   In addition, all partial derivatives of order ≥ 4 vanish.

Evaluating the partial derivatives above at $\vec{a} = (0,0)$, we see that

$$P_1(x_1,x_2) = f(0,0) = 1$$

$$P_2(x_1,x_2) = 1$$

$$P_3(x_1,x_3) = 1 + \frac{1}{2}(\frac{\partial^3 f}{\partial x_1^2 \partial x_2}(\vec{a}))x_1^2 x_2$$

$$= 1 + x_1^2 x_2$$

$$= f(x_1,x_2)$$

and $P_m(x_1,x_2) = f(x_1,x_2)$ for all $m \geq 4$.

We generalize the result of this example in Exercises 2 and 3.

2.   Let $f(x_1,x_2) = \sin(x_1 x_2)$ and $\vec{a} = (0,0)$.   Then

$$\frac{\partial f}{\partial x_1} = x_2 \cos(x_1 x_2)$$

$$\frac{\partial f}{\partial x_2} = x_1 \cos(x_1 x_2)$$

$$\frac{\partial^2 f}{\partial x_1^2} = - x_2^2 \sin(x_1 x_2)$$

$$\frac{\partial^2 f}{\partial x_1 \partial x_2} = \cos(x_1 x_2) - x_1 x_2 \sin(x_1 x_2) = \frac{\partial^2 f}{\partial x_2 \partial x_1}$$

$$\frac{\partial^2 f}{\partial x_2^2} = - x_1^2 \sin(x_1 x_2)$$

All of these derivatives vanish at $\vec{a} = (0,0)$ except for $\dfrac{\partial^2 f}{\partial x_1 \partial x_2}$, which equals one at $\vec{a}$. Thus

$$P_2(x_1,x_2) = x_1 x_2$$

It is also easy to see that all of the third order derivatives of f vanish at $\vec{a}$. Thus $P_3(x_1,x_2)$ is also equal to $x_1 x_2$.

3.   Let $f(x_1,x_2) = \sin(x_1 x_2)$ as in the previous example and a = $(2,\frac{\pi}{4})$. Then

$$f(\vec{a}) = \sin(\frac{\pi}{2}) = 1$$

$$\frac{\partial f}{\partial x_1}(\vec{a}) = \frac{\pi}{4} \cos \frac{\pi}{2} = 0$$

$$\frac{\partial f}{\partial x_2}(\vec{a}) = 0$$

$$\frac{\partial f^2}{\partial x_1^2}(\vec{a}) = - \frac{\pi^2}{16} \sin \frac{\pi}{2} = \frac{\pi^2}{16}$$

$$\frac{\partial^2 f}{\partial x_1 \partial x_2}(\vec{a}) = \frac{\partial^2 f}{\partial x_2 \partial x_1}(\vec{a})$$

$$= \cos \frac{\pi}{2} - \frac{\pi}{2} \sin \frac{\pi}{2} = - \frac{\pi}{2}$$

and

$$\frac{\partial^2 f}{\partial x_2^2}(\vec{a}) = - 4 \sin \frac{\pi}{2} = - 4$$

Therefore

$$P_2(x_1,x_2) = 1 - \frac{\pi^2}{32}(x_1 - 2)^2 - \frac{\pi}{2}(x_1 - 2)(x_2 - \frac{\pi}{4})$$
$$- 4(x_2 - \frac{\pi}{4})^2$$

Just as in the previous section, we define the *remainder* by

$$R_m(\vec{v}) = f(\vec{v}) - P_m(\vec{v})$$

We also have a Taylor's Remainder Theorem in this case.

THEOREM 5.1. *Let U be an open subset of* R^2*, let* $\vec{a} = (a_1,a_2)$
$\in$ *U, and let* $f:U \to R$ *be a function whose partial derivatives of order*
$\le m + 1$ *exist. Let* $\vec{v} = (x_1,x_2) \in U$ *be such that all vectors on the*
line segment from $\vec{v}$ *to* $\vec{a}$ *lie in U. (See Figure 5.1.) Then there is*
a vector $\vec{c}$ *on this line segment such that*

$$R_m(\vec{v}) = \sum_{i+j=m+1} \frac{1}{i!\,j!}\left(\frac{\partial^{i+j}f}{\partial x_1^i \partial x_2^j}(\vec{c})\right)(x_1 - a_1)^i(x_2 - a_2)^j$$

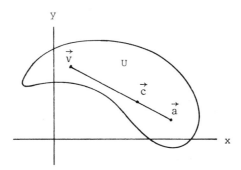

FIGURE 5.1.

Since Theorem 6.1, proved in the next section, is a generalization of this result, we omit the proof of Theorem 5.1.

EXERCISES

1.   Write down the Taylor polynomial of degree m about $\vec{a} \in R^2$ where

(a) $f(x_1,x_2) = 3 + x_1 x_2 + x_2^3$, $\vec{a} = (0,0)$, $m = 3$

(b) $f(x_1,x_2) = 3 + x_1 x_2 + x_2^3$, $\vec{a} = (0,0)$, $m = 57$

(c) $f(x_1,x_2) = 3 + x_1 x_2 + x_2^3$, $\vec{a} = (1,-3)$, $m = 4$

(d) $f(x_1,x_2) = \cos(x_1 x_2)$, $\vec{a} = (0,0)$, $m = 2$

(e) $f(x_1,x_2) = \cos(x_1 x_2)$, $\vec{a} = (0,0)$, $m = 3$

(f) $f(x_1,x_2) = \cos(x_1 x_2)$, $\vec{a} = (2,\frac{\pi}{4})$, $m = 2$

(g) $f(x_1,x_2) = x_1 \tan(x_1 x_2)$, $\vec{a} = (2,\frac{\pi}{8})$, $m = 2$

(h) $f(x_1,x_2) = e^{x_1 \sin x_2}$, $\vec{a} = (0,\frac{\pi}{3})$, $m = 2$

2.   Let $f(x_1,x_2)$ be a polynomial of degree 2,

$$f(x_1,x_2) = \sum_{i+j \leq 2} a_{ij} x_1^i x_2^j$$

$$= a_{00} + a_{10} x_1 + a_{01} x_2 + a_{20} x_1^2 + a_{11} x_1 x_2 + a_{02} x_2^2$$

Prove that the Taylor polynomial of f about $\vec{a} = (0,0)$ of degree m coincides with f whenever $m \geq 2$. What can be said when $m < 2$?

3.   Let $f(x_1,x_2)$ be a polynomial of degree r,

$$f(x_1,x_2) = \sum_{i+j \leq r} a_{ij} x_1^i x_2^j$$

Prove that the Taylor polynomial of f about $\vec{a} = (0,0)$ of degree m coincides with f whenever $m \geq r$. What can be said when $m < r$?

4.   Let $f:R^2 \to R$ be a function having continuous partial derivatives of order ≤ 2 and let $g(x_1,x_2)$ be the polynomial of degree 2 such that

$$\frac{\partial^{i+j} f}{\partial x_1^i \partial x_2^j}(0,0) = \frac{\partial^{i+j} g}{\partial x_1^i \partial x_2^j}(0,0) \tag{5.1}$$

for $i + j \leq 2$. Prove that g is the Taylor polynomial of f of degree 2 about $\vec{a} = (0,0)$.

5. Let $f:R^2 \rightarrow R$ be a function having continuous partial derivatives of order $\leq m$ and let g be a polynomial of degree m such that (5.1) holds for $i + j \leq m$. Prove that g is the Taylor polynomial of f of degree m about $\vec{a} = (0,0)$.

6. State and prove the analogue of Exercise 5 for arbitrary $\vec{a} = (a_1, a_2)$.

6. TAYLOR'S THEOREM FOR FUNCTIONS OF n VARIABLES

We now state and prove the general form of Taylor's Remainder Theorem. We begin with some notation.

Fix an integer $n > 0$ and let $J = (j_1, j_2, \ldots, j_n)$ be a sequence of non-negative integers. We define the *norm* of J by $|J| = j_1 + j_2 + \cdots + j_n$ and J! by $J! = j_1! j_2! \cdots j_n!$. If $|J| = k$, we define the differential operator D_J by

$$D_J = \frac{\partial^k}{\partial x_1^{j_1} \partial x_2^{j_2} \cdots \partial x_n^{j_n}}$$

Thus, if f is a function of n-variables,

$$D_J f = \frac{\partial^k f}{\partial x_1^{j_1} \partial x_2^{j_2} \cdots \partial x_n^{j_n}}$$

If $\vec{v} = (x_1, x_2, \ldots, x_n)$, $\vec{a} = (a_1, a_2, \ldots, a_n) \in R^n$, we define $(\vec{v} - \vec{a})^J$ by

$$(\vec{v} - \vec{a})^J = (x_1 - a_1)^{j_1} (x_2 - a_2)^{j_2} \cdots (x_n - a_n)^{j_n}$$

In particular, if $n = 2$ and $J = (1,3)$, then

$$|J| = 4$$

$$J! = 1! \cdot 3! = 6$$

$$D_J f = \frac{\partial^4 f}{\partial x_1 \partial x_2^3}$$

and

$$(\vec{v} - \vec{a})^J = (x_1 - a_1)(x_2 - a_2)^3$$

Let U be an open subset of R^n, let a $\in$ U, and let $f:U \to R$ a function whose partial derivatives of order $\leq$ (m+1) exist on U. *The Taylor Polynomial of f of degree* m *about* a is defined by

$$P_m(\vec{v}) = \sum_{|J| \leq m} \frac{1}{J!}(D_J f)(\vec{a})(x - \vec{a})^J$$

Here,

$$\sum_{|J| \leq m}$$

means that we sum over all sequences $J = (j_1, j_2, \ldots, j_a)$ of non-negative integers with $|J| = j_1 + j_2 + \cdots + j_n \leq m$.

It is easy to see that this definition agrees with that given in the previous section for n = 2.

As before, we define the remainder by

$$R_m(f)(\vec{v}) = f(\vec{v}) - P_m(\vec{v})$$

THEOREM 6.1. *Let* U *be an open subset of* R^n, $\vec{a} \in$ U, *and* $f:U \to R$ *a function whose partial derivatives of order* $\leq$ m + 1 *exist. Let* $\vec{v}$ *be a vector in* U *with the property that each point on the line segment between* $\vec{v}$ *and* $\vec{a}$ *is also in* U. *Then there is a vector* $\vec{c}$ *on the line segment from* $\vec{a}$ *to* $\vec{v}$ *such that*

$$R_m(\vec{v}) = \sum_{|J| = m+1} \frac{1}{J!}(D_J f)(\vec{c})(\vec{v} - \vec{a})^J$$

This is the *Taylor's Remainder Theorem* for functions of n-variables.

Proof. Define a function $g:[0,1] \to R^n$ by $g(t) = t\vec{v} + (1-t)\vec{a}$. The image of g, Im(g), is the line segment between $\vec{v}$ and $\vec{a}$, so Im(g) $\subseteq$ U. In fact, since U is open,

$$g(-\varepsilon, 1 + \varepsilon) \subset U$$

for some $\varepsilon > 0$.

The function g has derivatives of arbitrary order; in fact, $D_t g = \vec{v} - \vec{a}$ and those of order greater than one vanish. Hence the composite $F(t) = f(g(t))$ has derivatives of order $\leq m + 1$ on $(-\varepsilon, 1+\varepsilon)$. Applying Theorem 4.1, we have

$$F(t) = \sum_{j=0}^{m} \frac{F(0)^j}{j!} t^j + \frac{F^{(m+1)}(b)}{(m+1)!} t^{m+1}$$

for some b between 0 and t.

We now wish to express the $F^{(j)}(t)$ in terms of the partial derivatives of f. In principle, this is a matter of using the chain rule. In practice, it is a bit complicated.

To begin with,

$$F'(t) = f'(g(t))g'(t)$$

$$= (\frac{\partial f}{\partial x_1}, \ldots, \frac{\partial f}{\partial x_n}) \begin{pmatrix} x_1 - a_1 \\ \vdots \\ x_n - a_n \end{pmatrix} \tag{6.1}$$

$$= \sum_{j=1}^{n} \frac{\partial f}{\partial x_j}(g(t))(x_j - a_j)$$

At $t = 0$, $g(t) = \vec{a}$ and we have

$$F'(0) = \sum_{j=1}^{k} \frac{\partial f}{\partial x_j}(\vec{a})(x_i - a_i)$$

$$= \sum_{|J|=1} \frac{1}{J!} D_J f(\vec{a}) (\vec{v} - \vec{a})^J$$

since $J! = 1$ if $|J| = 1$.

To evaluate $F''(t)$, we differentiate $F'(t)$ as given in equation (6.1) above. The chain rule says that

$$\frac{d}{dt}\left(\frac{\partial f}{\partial x_j}(g(t))\right) = \left(\frac{\partial}{\partial x_i} \frac{\partial f}{\partial x_j}\right)(g(t))g'(t)$$

$$= \sum_{i=1}^{k} \frac{\partial^2 f}{\partial x_i \partial x_j}(g(t))(x_i - a_i)$$

There, using equation (6.1), we have

$$\frac{1}{2!}F''(t) = \frac{1}{2!} \sum_{j=1}^{k} \sum_{i=1}^{k} \frac{\partial^2 f}{\partial x_i \partial x_j}(g(t))(x_i - a_i)(x_j - a_j)$$

$$= \sum_{|J|=2} \frac{1}{J!}(D_J f)(g(t))(\vec{v} - \vec{a})^J$$

since if, say, $J = (2,0,0,\ldots)$, then the sum contributes only one $D_J f$ term (the $\frac{\partial^2}{\partial x_1^2}$ term), while if, say, $J = (1,0,1,\ldots)$ then the sum contributes two $D_J f$ terms (the $\frac{\partial^2 f}{\partial x_1 \partial x_2}$ and $\frac{\partial^2}{\partial x_2 \partial x_1}$ terms). It follows that

$$\frac{1}{2!}F''(0) = \sum_{|J|=2} \frac{1}{J!} (D_J f)(\vec{a})(\vec{v} - \vec{a})^J$$

Continuing in this way,[*] we get

$$F^{(j)}(t) = \sum_{i_1=1}^{k} \cdots \sum_{i_j=1}^{k} \frac{\partial^j}{\partial x_{i_1} \cdots \partial x_{i_j}} f(g(t))(x_{i_1} - a_{i_1}) \cdots (x_{i_j} - a_{i_j})$$

---

[*] As is usually the case, this phrase conceals a use of mathematical induction. See Exercise 7.

We need now to know how many terms in this sum correspond to each sequence J. This is a standard combinatorial problem, and we shall simply state the result:

$$F^{(j)}(t) = \sum_{|J|=j} \frac{|J|!}{J!}(D_J f)(g(t))(\vec{v} - \vec{a})^J \qquad (6.2)$$

Thus, if $\vec{v} = g(t)$, equations (6.1) and (6.2) imply that

$$R_m(\vec{v}) = F(t) - \sum_{j=0}^{m} \frac{F^{(j)}(0)}{j!} t^j$$

$$= \frac{F^{(m+1)}(b)}{(m+1)!} t^{m+1}$$

(for some b between 0 and 1)

$$= \sum_{|J|=m+1} \frac{1}{J!}(D_J f)(\vec{c})(\vec{v} - \vec{a})^J$$

where $\vec{c} = g(b)$.

Remark. We note that $F''(0)$ can be written in another way. Recall that the Hessian of f, $H(f)$, is the $n \times n$ matrix (a_{ij}) where each a_{ij} is the function

$$a_{ij} = \frac{\partial^2 f}{\partial x_i \partial x_j}$$

If f has continuous second partial derivative, $H(f)$ is a symmetric matrix (Theorem 3.3). Thus $H(f)(a)$ defines a linear transformation $T:R^n \rightarrow R^n$ and it is easy to check that $F''(0)$ is given by the formula

$$F''(0) = < T(\vec{v} - \vec{a}), \vec{v} - \vec{a} >$$

EXERCISES

1. Find the Taylor polynomial of the function $f: R^3 \to R$ of degree m about $\vec{a}$ where

    (a) $f(x_1, x_2, x_3) = x_1 x_2 x_3$, $\vec{a} = (0,0,0)$, $m = 3$

    (b) $f(x_1, x_2, x_3) = x_1 x_2 x_3$, $\vec{a} = (0,0,0)$, $m = 44$

    (c) $f(x_1, x_2, x_3) = x_1 x_2 x_3$, $\vec{a} = (1,2,-1)$, $m = 2$

    (d) $f(x_1, x_2, x_3) = \sin(x_1 x_2 x_3)$, $\vec{a} = (0,0,0)$, $m = 2$

    (e) $f(x_1, x_2, x_3) = \sin(x_1 x_2 x_3)$, $\vec{a} = (2,0,1)$, $m = 3$

    (f) $f(x_1, x_2, x_3) = \sin(x_1 x_2 x_3)$, $\vec{a} = (1,2,\frac{\pi}{6})$, $m = 2$

    (g) $f(x_1, x_2, x_3) = x_1 \cos(x_2 x_3 + x_1^2)$, $\vec{a} = (0,0,0)$, $m = 2$

    (h) $f(x_1, x_2, x_3) = x_1 \cos(x_2 x_3 + x_1^2)$, $\vec{a} = (0,2,\frac{\pi}{8})$, $m = 2$

2. Prove that $F''(0) = \langle T(\vec{v} - \vec{a}), \vec{v} - \vec{a} \rangle$ where $T: R^n \to R^n$ is the linear transformation defined by the matrix $H(f)(\vec{a})$. (This is the assertion of the last paragraph of this section.)

3. Let $f: R^3 \to R$ be a function having continuous partial derivatives of order ≤ 2 and let $g(x_1, x_2, x_3)$ be the polynomial of degree 2 such that

$$\frac{\partial^{i+j} f}{\partial x_1^i \partial x_2^j \partial x_3^k}(0,0) = \frac{\partial^{i+j} g}{\partial x_1^i \partial x_2^j \partial x_3^k}(0,0) \tag{6.3}$$

for $i + j + k \leq 2$. Prove that g is the Taylor polynomial of f of degree 2 about $\vec{a} = (0,0,0)$.

4. Let $f: R^3 \to R$ be a function having continuous partial derivatives of order $\leq m$ and let g be a polynomial of degree m such that (6.3) nolds for $i + j + k \leq m$. Prove that g is the Taylor polynomial of f of degree m about $\vec{a} = (0,0,0)$.

5. State and prove the analogue of Exercise 5 for arbitrary $\vec{a} = (a_1, a_2, a_3)$.

6.   State and prove the analogue of Exercise 5 for functions $f:R^n \to R$ for

(a) $\vec{a} = (0,0,\ldots,0)$

(b) $\vec{a} = (a_1,a_2,\ldots,a_n)$ arbitrary

*7.   Derive the formula for $F^{(j)}(t)$ of page 254 supplying the inductive step implied in the text.

*8,   Prove Equation (6.2). (Warning: the proof has no very direct connection with anything in the text.)

7.   A SUFFICIENT CONDITION FOR DIFFERENTIABILITY

Let U be an open set in R^n and let $f:U \to R^p$ be a function, $f(\vec{v}) = (f_1(\vec{v}),\ldots,f_p(\vec{v}))$. We saw in Section 3.7 that if f is differentiable at a point $\vec{v}_0 \in U$, then the first order partial derivatives of the functions $f_1,\ldots,f_p$ exist at $\vec{v}_0$, but we also saw that the existence of the partial derivatives does not guarantee the differentiability of f. We now give a condition on f which does insure that f is differentiable.

THEOREM 7.1. *Let U be an open subset of* R^n, *and let* $f:U \to R^p$ *be a function with* $f(\vec{v}) = (f_1(\vec{v}),\ldots,f_n(\vec{v}))$. *Suppose that all the partial derivatives* $\frac{\partial f_i}{\partial x_j}$ $(1 \le i \le p, 1 \le j \le n)$ *exist and are continuous on U. Then f is differentiable on U.*

Proof. We consider first the case where p = 1, so that $f:U \to R$. (Later we shall prove the general case from this special case.) Let $\vec{v}_0 = (a_1,\ldots,a_n)$, and let $L:R^n \to R$ be the linear transformation whose matrix is $J_f(\vec{v}_0)$. Thus

$$L(t_1,\ldots,t_n) = t_1 \frac{\partial f}{\partial x_1}(\vec{v}_0) + \cdots + t_n \frac{\partial f}{\partial x_n}(\vec{v}_0)$$

We shall prove that L is the derivative of f at $\vec{v}_0$; that is, that for any $\varepsilon > 0$ there is a $\delta > 0$ such that if $\|\vec{v}-\vec{v}_0\| < \delta$, then

$$|f(\vec{v}) - f(\vec{v}_0) - L(\vec{v}-\vec{v}_0)| < \varepsilon\|\vec{v}-\vec{v}_0\| \tag{7.1}$$

Since U is open, there is a number $r > 0$ such that $B_r(\vec{v}_0) \subset U$. Now let $s = r/\sqrt{n}$. If $\vec{v} = (x_1,\ldots,x_n)$ is any vector with

$$|x_j - a_j| < s \tag{7.2}$$

for $1 \le j \le n$, then

$$\|\vec{v}-\vec{v}_0\| = (\sum_{j=1}^{n} |x_j - a_j|^2)^{1/2}$$

$$< (ns^2)^{1/2} = r$$

and $\vec{v} \in S$. From now on, we shall work with vectors satisfying (7.2). (Such vectors lie in a "cube" centered about $\vec{v}_0$; the picture for $n = 2$ is shown in Figure 7.1).

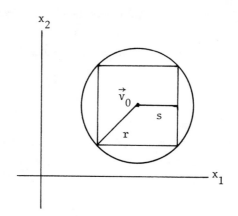

FIGURE 7.1. The cube centered at $\vec{v}_0$

Let $\vec{v} = (x_1,\ldots,x_n)$, and define

$$\vec{v}_j = (x_1,\ldots,x_j, a_{j+1},\ldots,a_n)$$

for $1 \le j \le n$. (For convenience, we let $\vec{v}_n = \vec{v}$.) Once $\vec{v}$ satisfies (7.2), so do $\vec{v}_1,\ldots,\vec{v}_{n-1}$, and

$$f(\vec{v}) - f(\vec{v}_0) = \sum_{j=1}^{n} f(\vec{v}_j) - f(\vec{v}_{j-1}) \tag{7.3}$$

We shall use this formula to estimate the left-hand side of equation (7.1).

Consider a typical expression on the right-hand side of (7.1), $f(\vec{v}_j) - f(\vec{v}_{j-1})$. We define a function $g_j:[a_j,x_j] \to R^*$ by

$$g_j(t) = f(x_1,\ldots,x_{j-1},t,a_{j+1},\ldots,a_n)$$

Then $g_j(a_j) = f(\vec{v}_{j-1})$, $g_j(x_j) = f(\vec{v}_j)$, and g_j is continuous on $[a_j,x_j]$ (because f is) and differentiable on (a_j,x_j) (because f has partial derivatives).

According to the Mean Value Theorem (see Exercise 8 of Section 3.5), there is a point $c_j \in (a_j,x_j)$ such that

$$f(\vec{v}_j) - f(\vec{v}_{j-1}) = g_j(x_j) - g_j(a_j)$$
$$= (x_j - a_j)g_j'(c_j)$$

Let $\vec{w}_j = (x_1,\ldots,x_{j-1},c_j,a_{j+1},\ldots,a_n)$. Then $g_j'(c_j) = \dfrac{\partial f}{\partial x_j}(\vec{w}_j)$, and so

$$f(\vec{v}_j) - f(\vec{v}_{j-1}) = (x_j - a_j)\frac{\partial f}{\partial x_j}(\vec{w}_j)$$

Thus, from (7.3),

$$f(\vec{v}) - f(\vec{v}_0) - L(\vec{v}-\vec{v}_0) = \sum_{j=1}^{n} (x_j - a_j)\frac{\partial f}{\partial x_j}(\vec{w}_j) - L(\vec{v}-\vec{v}_0)$$

$$= \sum_{j=1}^{n} (x_j - a_j)\frac{\partial f}{\partial x_j}(\vec{w}_j)$$

$$- \sum_{j=1}^{n} (x_j - a_j)\frac{\partial f}{\partial x_j}(\vec{v}_0)$$

---

* We assume, for convenience, that $x_j > a_j$. If $x_j < a_j$, we use the interval $[x_j,a_j]$, and the proof is essentially the same. If $x_j = a_j$, the term is 0.

$$= \sum_{j=1}^{n} (x_j - a_j)(\frac{\partial f}{\partial x_j}(\vec{w}_j) - \frac{\partial f}{\partial x_j}(\vec{v}_0))$$

and

$$|(f(\vec{v}) - f(\vec{v}_0) - L(\vec{v}-\vec{v}_0)| \leq \sum_{j=1}^{n} |x_j - a_j| |\frac{\partial f}{\partial x_j}(\vec{w}_j) - \frac{\partial f}{\partial x_j}(\vec{v}_0)|$$

$$\leq \sum_{j=1}^{n} \|\vec{v}-\vec{v}_0\| |\frac{\partial f}{\partial x_j}(\vec{w}_j) - \frac{\partial f}{\partial x_j}(\vec{v}_0)| \quad (7.4)$$

since $|x_j - a_j| < \|\vec{v}-\vec{v}_0\|$ for all j. Moreover,

$$\|\vec{w}_j - \vec{v}_0\| = (\sum_{i=1}^{j-1} |x_j - a_j|^2 + |c_j - a_j|^2)^{1/2}$$

$$\leq \|\vec{v}-\vec{v}_0\|$$

for all j.

Now we use the hypothesis that the partial derivatives are continuous: given $\varepsilon > 0$, there is a $\delta > 0$ such that if $\|\vec{w}-\vec{v}_0\| < \delta$, then

$$|\frac{\partial f}{\partial x_j}(\vec{w}) - \frac{\partial f}{\partial x_j}(\vec{v}_0)| < \frac{\varepsilon}{n}$$

for j = 1,2,...,n. So if $\|\vec{v}-\vec{v}_0\| < \delta$,

$$(f(\vec{v}) - f(\vec{v}_0) - L(\vec{v}-\vec{v}_0)| < \sum_{j=1}^{n} \|\vec{v}-\vec{v}_0\|\frac{\varepsilon}{n} \quad \text{(from (7.4))}$$

$$= \|\vec{v}-\vec{v}_0\|\varepsilon$$

and we have proved (7.1). Thus we have proved Theorem 7.1 in the special case p = 1.

We prove the general case by means of a proposition which is of some interest in its own right.

PROPOSITION 7.2. *Let U be an open subset of* R^n, *and let* $f:U \rightarrow R^p$ *be a function with* $f(\vec{v}) = (f_1(\vec{v}),...,f_n(\vec{v}))$. *Suppose that the function* $f_1,...,f_n$ *all are differentiable (as functions from U to R). Then f is differentiable on U.*

Proof. For $\vec{v}_0 \in U$, let $L_1,\ldots,L_n$ be the derivatives of $f_1,\ldots,f_n$ respectively at $\vec{v}_0$, and define $L:R^n \to R^p$ by

$$L(\vec{v}) = (L_1(\vec{v}),\ldots,L_n(\vec{v}))$$

It is easy to check that L is a linear transformation. We shall now check that $L = D_{\vec{v}_0} f$. For each $j = 1,\ldots,p$, define functions $r_j:U \to R$ by

$$r_j(\vec{v}) = f_j(\vec{v}) - f_j(\vec{v}_0) - L_j(\vec{v}-\vec{v}_0)$$

Then

$$f(\vec{v}) - f(\vec{v}_0) - L(\vec{v}-\vec{v}_0) = (r_1(\vec{v}),\ldots,r_p(\vec{v}))$$

and

$$\|f(\vec{v}) - f(\vec{v}_0) - L(\vec{v}-\vec{v}_0)\| = (\sum_{j=1}^{p} r_j(\vec{v})^2)^{1/2}$$

Choose $\varepsilon > 0$. We know that there is a $\delta > 0$ such that if $\|\vec{v}-\vec{v}_0\| < \delta$, then

$$|r_j(\vec{v})| < \frac{\varepsilon}{\sqrt{p}} \|\vec{v}-\vec{v}_0\|$$

for $j = 1,\ldots,n$. For this δ,

$$\|f(\vec{v}) - f(\vec{v}_0) - L(\vec{v}-\vec{v}_0)\| < (\sum_{j=1}^{p} \frac{\varepsilon^2}{p}\|\vec{v}-\vec{v}_0\|^2)^{1/2} = \varepsilon\|\vec{v}-\vec{v}_0\|$$

and the proposition is proved. Theorem 7.1 now follows immediately.

Remarks. 1. We have stated Theorem 7.1 for functions defined between Euclidean spaces. There is a similar theorem for inner product spaces; one chooses orthonormal bases and defines coordinate functions and partial derivatives accordingly. We leave the details to the reader.

2. Theorem 7.1 says that if the partial derivatives are continuous, f is differentiable. Proposition 3.7.1 says that if f is dif-

ferentiable, the partial derivatives exist. These statements are not
exact converses, and there is no easy way to fill the gap. We have
seen (in Exercise 3 of Section 3.7) that there are functions which
have all partial derivatives, but are not differentiable. (See also
Exercise 2.) On the other hand, differentiable functions need not
have continuous partial derivatives; Exercises 3 and 4 give examples.

EXERCISES

1.   (a) Verify that the function L in Proposition 7.2 is a linear
transformatión.

      (b) Suppose that the matrix for L_1 is $(a_{11}, \ldots, a_{1p})$, for L_2 is
$(a_{21}, \ldots, a_{2p})$, and so on. What is the matrix for L ?

2.   Define $f: R^2 \to R$ by

$$
f(x_1, x_2) = \begin{cases} \dfrac{x_1^2 x_2}{x_1^4 + x_2^2} & \text{if } (x_1, x_2) \neq (0,0) \\[2ex] 0 & \text{if } (x_1, x_2) = (0,0) \end{cases}
$$

      (a) Show that f has not only partial derivatives at 0, but di-
rectional derivatives in all directions there (see Exercise 5, Section 3.7).

      (b) Show that f is discontinuous at 0 (and hence not differen-
tiable there).

3.   Define $f: R \to R$ by

$$
f(x) = \begin{cases} x^2 \sin \dfrac{1}{x^2} & \text{if } x \neq 0 \\[2ex] 0 & \text{if } x = 0 \end{cases}
$$

Show that f is differentiable everywhere on R, but that f' is not continuous.

4.   (a) Let $g, h: R \to R$ be differentiable, and define $f: R^2 \to R$ by

$$f(x_1,x_2) = g(x_1) + h(x_2)$$

Show that f is differentiable.

(b)   Find a function $f:R^2 \to R$ such that f is differentiable, but the partial derivatives of f are discontinuous.

5.   State and prove a version of Theorem 7.1 for inner product spaces.

8.   THE EQUALITY OF MIXED PARTIAL DERIVATIVES

We now give the proof of Theorem 3.3.

We can clearly assume i = 1, j = 2 by reordering the variables if necessary. Since all other variables are held fixed, we can also assume n = 2. For convenience, we denote the points in R^2 by (x,y).

Let $\vec{v} = (a,b)$ and let $\vec{w} = (c,d)$ be a point in R^2 with a < c, b < d, and such that the rectangle $R(\vec{v},\vec{w})$

$$R(\vec{v},\vec{w}) = \{(x,y) \in R^2 | a \le x \le c, b \le y \le d\}$$

is entirely contained in U. Let $k(\vec{v},\vec{w}) \in R$ be defined by

$$k(\vec{v},\vec{w}) = f(c,d) - f(c,b) - f(a,d) + f(a,b)$$

We evaluate $k(\vec{v},\vec{w})$ in two different ways.

Define $g:R \to R$ by g(x) = f(x,d) - f(x,b). Then $k(\vec{v},\vec{w}) = g(c)$ - g(a) = (c-a)g'(s_1) for some s_1 between a and c, by the Mean Value Theorem for functions of one variable. (See Exercise 9, Section 3.5.) However,

$$g'(s_1) = \frac{\partial f}{\partial x}(s_1,d) - \frac{\partial f}{\partial x}(s_1,b)$$

$$= (d-b)\frac{\partial^2 f}{\partial y \partial x}(s_1,s_2)$$

for s_2 between b and d, again by the Mean Value Theorem. Thus

$$k(\vec{v},\vec{w}) = (c-a)(d-b)\frac{\partial^2 f}{\partial y \partial x}(s_1,s_2)$$

Similarly, we define $h:R \to R$ by

$$h(y) = f(c,y) - f(a,y)$$

Then, just as above,

$$k(\vec{v},\vec{w}) = (c-a)(d-b)\frac{\partial^2 h}{\partial x \partial y}(t_1,t_2)$$

for t_1 between a and c, t_2 between b and d. It follows that

$$\frac{\partial^2 f}{\partial y \partial x}(s_1,s_2) = \frac{\partial^2 f}{\partial x \partial y}(t_1,t_2) \qquad (8.1)$$

Let $\varphi,\psi:R^2 \to R$ be defined by

$$\varphi(\vec{z}) = \frac{\partial^2 f}{\partial y \partial x}(\vec{z}), \ \psi(\vec{z}) = \frac{\partial^2 f}{\partial x \partial y}(\vec{z})$$

Thus $\varphi(\vec{s}) = \psi(\vec{t})$, $\vec{s} = (s_1,s_2)$ and $\vec{t} = (t_1,t_2)$ by equation (8.1). We know that φ and ψ are continuous at $\vec{v}$; we need to prove that $\varphi(\vec{v}) = \psi(\vec{v})$. Let $\varepsilon > 0$ be any small number and choose $\delta > 0$ so that

$$\|\vec{v}-\vec{z}\| < \delta \Rightarrow |\varphi(\vec{v}) - \varphi(\vec{z})| < \varepsilon/2 \text{ and } |\psi(\vec{v}) - \psi(\vec{z})| < \varepsilon/2$$

Pick $\vec{w}$ above so that $\|\vec{v}-\vec{w}\| < \delta$. Then

$$\|\vec{v}-\vec{s}\| < \delta \text{ and } \|\vec{v}-\vec{t}\| < \delta$$

so that

$$|\varphi(\vec{v}) - \varphi(\vec{s})| < \varepsilon/2 \text{ and } |\psi(\vec{v}) -\psi(\vec{t})| < \varepsilon/2$$

Then

$$\begin{aligned}
|\varphi(\vec{v}) - \psi(\vec{v})| &= |\varphi(\vec{v}) - \varphi(\vec{s}) + \varphi(\vec{s}) - \psi(\vec{t}) + \psi(\vec{t}) - \psi(\vec{v})| \\
&\le |\varphi(\vec{v}) - \varphi(\vec{s})| + |\varphi(\vec{s}) - \psi(\vec{t})| + |\psi(\vec{t}) - \psi(\vec{v})| \\
&< \ \ \varepsilon/2 + \varepsilon/2 = \varepsilon
\end{aligned}$$

since $\varphi(\vec{s}) = \psi(\vec{t})$. Thus $|\varphi(\vec{v}) - \psi(\vec{v})| < \varepsilon$ for any $\varepsilon > 0$, and it follows that $\varphi(\vec{v}) = \psi(\vec{v})$.

Remark. 1. A somewhat stronger result than Theorem 3.3 is true: if f is differentiable and $\dfrac{\partial^2 f}{\partial y \partial x}$ exists in an open set about $\vec{v}$ and is continuous there, then $\dfrac{\partial^2 f}{\partial x \partial y}(\vec{v})$ exists and is equal to $\dfrac{\partial^2 f}{\partial y \partial x}(\vec{v})$. We shall not prove this result here. (See Exercise 1.) In most cases arising in practice, both mixed partials are known to exist and be continuous, and then, of course, they are equal.

EXERCISES

1. Modify the proof of Theorem 3.3 to prove the result mentioned in Remark 1: if f is continuously differentiable and $\dfrac{\partial^2 f}{\partial y \partial x}$ exists in an open set about v, and if $\dfrac{\partial^2 f}{\partial y \partial x}$ is continuous at $\vec{v}$, then $\dfrac{\partial^2 f}{\partial x \partial y}(\vec{v})$ exists and equals $\dfrac{\partial^2 f}{\partial y \partial x}(\vec{v})$.

2. Let $f : R^2 \to R$ be defined by

$$f(x,y) = |x|$$

Show that $\dfrac{\partial^2 f}{\partial x \partial y} = 0$ for all $(x,y) \in R^2$, but that $\dfrac{\partial^2 f}{\partial y \partial x}$ is not always defined. Does this contradict Theorem 3.3 ? Does it contradict Exercise 1?

CHAPTER 7

LINEAR TRANSFORMATIONS AND MATRICES

INTRODUCTION

The idea of associating a matrix to a linear transformation has already proved fruitful. We used it in Section 3.4 to prove that linear transformations between Euclidean spaces are continuous and in Section 4.7 to prove that linear transformations between finite dimensional inner product spaces are continuous. In this chapter, we shall again use matrices to study linear transformations.

We begin by deriving some of the elementary properties of the correspondence between linear transformations and matrices (once bases have been fixed) and then study the matrix of an isomorphism. We next determine the way in which the matrix of a linear transformation changes when we change bases, and then define the trace and adjoint of a linear transformation in terms of its matrix. The trace and adjoint are used to define the inner product on the space of linear transformations between two finite dimensional vector spaces referred to in Section 6.3.

We also define row and column operations. These operations can be used to simplify a matrix and to compute the inverse of an invertible matrix. In the final section, we study simultaneous linear equations in the spirit of these operations.

Unless explicitly stated to the contrary, *all vector spaces in this chapter will be finite dimensional.*

1. THE MATRIX OF A LINEAR TRANSFORMATION

We begin by reviewing the definition of the matrix of a linear transformation between finite dimensional vector spaces. (See Section 4.7.)

Let V, W, and W' be vector spaces with bases $\alpha = \{\vec{v}_1,\ldots,\vec{v}_n\}$ for
V, $\beta = \{\vec{w}_1,\ldots,\vec{w}_m\}$ for W, and $\gamma = \{\vec{w}_1',\ldots,\vec{w}_p'\}$ for W'. *These bases
are fixed for the remainder of this section.* We shall investigate
what happens when we change bases in Section 3.

Suppose $T:V \to W$ is a linear transformation. For each $j = 1,\ldots,n$,
we can express $T(\vec{v}_j)$ as a linear combination of $\vec{w}_1,\ldots,\vec{w}_m$:

$$T(\vec{v}_j) = \sum_{i=1}^{m} a_{ij}\vec{w}_i \tag{1.1}$$

The numbers a_{ij} (which depend on T and the bases α,β) define an m × n
matrix

$$A_T(\alpha,\beta) = (a_{ij})$$

This matrix is called the *matrix of T relative to the bases* α *and* β.
In cases where no confusion seems likely, we write A_T for $A_T(\alpha,\beta)$.

In particular, if $T:R^n \to R^m$, then the matrix associated to T in
Section 2.6 is the matrix of T relative to α and β where α and β are
the standard bases for R^n and R^m.

If $A = (a_{ij})$ is any m × n matrix, we can use formula (1.1) to
define a linear transformation $T_A:V \to W$. In this case, $A = A_{T_A}(\alpha,\beta)$.

Here are some examples of the matrices of certain linear transformations.

(1) Let $T:V \to W$ be the zero transformation: $T(\vec{v}) = \vec{0}$ for all
$\vec{v} \in V$. Then

$$T(\vec{v}_j) = \vec{0} = \sum_{i=1}^{m} 0\vec{w}_i$$

Therefore $a_{ij} = 0$ for all i and j, and A_T is the matrix of all whose
entries are 0. Unsurprisingly, this matrix is called the *zero matrix*.

(2) Let $I:V \to V$ be the identity transformation: $I(\vec{v}) = \vec{v}$, for
all $\vec{v} \in V$. Then $I(\vec{v}_j) = \vec{v}_j$, and so

$$a_{ij} = \delta_{ij} = \begin{cases} 1 & \text{if } i = j \\ 0 & \text{if } i \neq j \end{cases}$$

The matrix A_I is called the *identity matrix*, and is denoted by I_n (where n is the dimension of V) or simply by I. Explicitly, I_n is the n × n matrix

$$I_n = \begin{pmatrix} 1 & 0 & 0 & \cdots & 0 \\ 0 & 1 & 0 & \cdots & 0 \\ 0 & 0 & 1 & \cdots & 0 \\ & & \cdots & & \\ 0 & \cdots & & 0 & 1 \end{pmatrix}$$

(3)   Recall (Example 6 of Section 2.2) that P_n is the space of polynomials of degree $\leq$ n. Define $T:P_3 \to P_2$ by $Tf = f'$. Let $\alpha = \{1,x,x^2,x^3\}$ be the basis of P_3, and let $\beta = \{1,x,x^2\}$ be the basis of P_2. Then it is not hard to check that

$$A_T(\alpha,\beta) = \begin{pmatrix} 0 & 1 & 0 & 0 \\ 0 & 0 & 2 & 0 \\ 0 & 0 & 0 & 3 \end{pmatrix}$$

In earlier sections, we have defined sums, scalar multiples, and products for matrices and sums, scalar multiples, and composites of linear transformations. The next results show that these operations are preserved under the correspondence $T \to A_T$.

PROPOSITION 1.1.   *Let* $T,T':V \to W$ *be linear transformations and* r *a real number. Then*

$$A_{T+T'}(\alpha,\beta) = A_T(\alpha,\beta) + A_{T'}(\alpha,\beta)$$
$$A_{rT}(\alpha,\beta) = rA_T(\alpha,\beta)$$

PROPOSITION 1.2.   *Let* $T:V \to W$, $S:W \to W'$ *be linear transformations. Then*

$$A_{S \circ T}(\alpha,\gamma) = A_S(\beta,\gamma)A_T(\alpha,\beta)$$

The proofs of these two propositions proceed just as in the case of the corresponding results in Section 2.6. We leave the details as Exercises 7 and 8.

The next result shows that we can pick bases so that the matrix of a linear transformation takes a particularly simple form.

PROPOSITION 1.3. *Let* T:V $\to$ W *be a linear transformation. Then we can find bases* α *for* V *and* β *for* W *such that*

$$A_T(\alpha,\beta) = \begin{pmatrix} A_1 & A_2 \\ A_3 & A_4 \end{pmatrix}$$

where A_1 *is a* k $\times$ k *identity matrix and* A_2, A_3, *and* A_4 *are zero matrices of the appropriate size. (This size may be* 0 $\times$ m *or* m $\times$ 0, *in which case some of those matrices do not appear.)*

Remark. Whereas this result will be useful to us in what follows, our main interest will be in studying linear transformations T:V $\to$ V of a vector space into itself. In this case, we will be concerned with finding a basis α for V so that $A_T(\alpha,\alpha)$ is as simple as possible. This is the object of study in Chapter 10.

Proof of Proposition 1.3. Let $\vec{v}_1,\ldots,\vec{v}_n$ be a basis for V so that $\vec{v}_{k+1},\ldots,\vec{v}_n$ is a basis for h(T). (We use Proposition 4.4.5 here.) Set $\vec{w}_j = T(\vec{v}_j)$, $1 \leq j \leq k$. Then, just as in the proof of Proposition 4.4.6, $\vec{w}_1,\ldots,\vec{w}_k$ are independent. We now use Proposition 4.2.3 to find vectors $\vec{w}_{k+1},\ldots,\vec{w}_m$ so that $\vec{w}_1,\ldots,\vec{w}_m$ is a basis for W. It follows that

$$T(\vec{v}_j) = \begin{cases} \vec{w}_j, & 1 \leq j \leq k \\ \vec{0}, & k < j \leq n \end{cases}$$

Therefore, if we set $\alpha = \{\vec{v}_1,\ldots,\vec{v}_n\}$ and $\beta = \{\vec{w}_1,\ldots,\vec{w}_m\}$, $A_T(\alpha,\beta)$ has the required form.

EXERCISES

1.   Determine the matrix of the linear transformation $T: P_3 \to P_2$ given by $T(f) = f' + 3f''$ relative to the usual bases for P_3 and P_2.   (See Example 3 above.)

2.   Determine the matrix of the linear transformation $T: V \to V$ given by $T(\vec{v}) = 7\vec{v}$.

3.   Let $T: R^2 \to R^2$ be the linear transformation which rotates the plane through $30°$. Find the matrix of T relative to the standard bases.

4.   Find the matrix of the linear transformation T of Exercise 3 relative to the basis $\{(1,1),(1,-2)\}$ for R^2.

5.   Let $T: R^2 \to R^2$ be reflection about the line $x - 2y = 0$. Find the matrix of T relative to the standard basis for R^2.

6.   Let $T: R^2 \to R^2$ be as in Exercise 5. Find its matrix relative to the basis $\{(2,1),(1,-2)\}$ for R^2.

7.   Prove Proposition 1.1.

8.   Prove Proposition 1.2.

9.   Let I be the $n \times n$ identity matrix and 0 the $n \times n$ zero matrix. Prove that

   (a) $IA = A$ and $BI = B$ whenever these products are defined.

   (b) $0A = 0$ and $B0 = 0$ whenever these products are defined.

What are the dimensions of the matrices for which the products are defined?

10.   Let α be the standard basis for R^2. Find a basis $\beta = (\vec{v}_1, \vec{v}_2)$ such that $I(\alpha, \beta) = \begin{pmatrix} 2 & 1 \\ 1 & 1 \end{pmatrix}$.

11.   Let $T: R^2 \to R^3$ be given by

$$T(x,y) = (x-y, x+y, x+y)$$

Find bases α for R^2, β for R^3 so that $A_T(\alpha,\beta)$ has the form described in Proposition 1.3.

12. Let $T:R^3 \to R^3$ be given by

$$T(x,y,z) = (2x+y, y+z, 2x-z)$$

Find bases α,β for R^3 so that $A_T(\alpha,\beta)$ has the form described in Proposition 1.3.

13.   Let $A = (a_{ij})$ be an m × n matrix. The *transpose* of A, A^t, is the n × m matrix whose (i,j) component is a_{ji}. (If, for example, $A = \begin{pmatrix} 1 & 2 & 3 \\ 4 & 5 & 6 \end{pmatrix}$, then

$$A^t = \begin{pmatrix} 1 & 4 \\ 2 & 5 \\ 3 & 6 \end{pmatrix}$$

Show that $(AB)^t = B^t A^t$ (if both products are defined).

*14. If $T:V \to W$, define $\widetilde{T}:W^* \to V^*$ by

$$(\widetilde{T}(f))(\vec{v}) = f(T(\vec{v}))$$

for $f \in W^*$ and $\vec{v} \in V$. (Here V^* and W^* are the dual spaces of V and W respectively; see Exercise 5, Section 4.3). Let α be a basis for V, β a basis for W, and α^*, β^* the dual bases for V^* and W^* respectively. Prove that

$$A_{\widetilde{T}}(\beta^*,\alpha^*) = (A_T(\alpha,\beta))^t$$

2.   ISOMORPHISMS AND INVERTIBLE MATRICES

We now discuss the matrix of an isomorphism between two vector spaces. We begin with a useful characterization.

PROPOSITION 2.1. *A linear transformation* $T:V \to W$ *is an isomorphism if and only if there is a linear transformation* $S:W \to V$ *such that*

$$SoT = I_V, \quad ToS = I_W$$

where I_V *and* I_W *are the identity mappings of* V *and* W *respectively.*

Proof. If T is an isomorphism, then $S = T^{-1}$ satisfies the conditions of the proposition. (See Proposition 4.3.3.)

If $S:W \to V$ satisfies the conditions of the proposition, then T is surjective since, for any $\vec{w} \in W$,

$$\vec{w} = I_W(\vec{w})$$
$$= (ToS)(\vec{w})$$
$$= T(S(\vec{w}))$$

If $T(\vec{v}) = \vec{0}$, then

$$\vec{0} = S(T(\vec{v}))$$
$$= (SoT)(\vec{v})$$
$$= I_V(\vec{v}) = \vec{v}$$

Thus T is injective , and the proof is complete.

Suppose now that $T:V \to W$ is an isomorphism. Let $S:W \to V$ be as in Proposition 2.1, and let $\alpha = \{\vec{v}_1,\ldots,\vec{v}_n\}$, $\beta = \{\vec{w}_1,\ldots,\vec{w}_n\}$ be bases for V and W respectively. Set $A = A_T(\alpha,\beta)$, $B = A_S(\beta,\alpha)$. Then

$$AB = A_T(\alpha,\beta)A_S(\beta,\alpha)$$
$$= A_{ToS}(\beta,\beta) \qquad \text{(by Proposition 1.2)}$$
$$= A_{I_W}(\beta,\beta)$$
$$= I_n$$

and

$$BA = A_S(\beta,\alpha)A_T(\alpha,\beta)$$

$$= A_{S \circ T}(\alpha,\alpha) \qquad \text{(by Proposition 1.2)}$$

$$= A_{I_V}(\alpha,\alpha)$$

$$= I_n$$

We say an $n \times n$ matrix A is *invertible* if there is an $n \times n$ matrix B such that

$$AB = BA = I_n$$

We call B an *inverse* for A.

Here are some examples.

(1) A square matrix $D = (d_{ij})$ is called a *diagonal matrix* if $d_{ij} = 0$ for $i \neq j$. We sometimes write $D = \text{diag}(d_1,\ldots,d_n)$ for the diagonal matrix with $d_{ii} = d_i$, $1 \leq i \leq n$. The matrix $D = \text{diag}(d_1,\ldots, d_n)$ is invertible if and only if each $d_i \neq 0$, $1 \leq i \leq n$. (See Exercise 5.) In this case,

$$D^{-1} = \text{diag}(d_1^{-1},\ldots,d_n^{-1})$$

(2) Let e_{ij} be the $n \times n$ matrix with a single nonzero entry, that being a one in the (i,j) position. An *elementary matrix* is one of the form $E_{ij}(r) = I_n + re_{ij}$, $i \neq j$, $r \in R$. For instance, if $n = 3$,

$$E_{1,3}(4) = \begin{pmatrix} 1 & 0 & 4 \\ 0 & 1 & 0 \\ 0 & 0 & 1 \end{pmatrix} \qquad E_{2,1}(-2) = \begin{pmatrix} 1 & 0 & 0 \\ -2 & 1 & 0 \\ 0 & 0 & 1 \end{pmatrix}$$

Any elementary matrix is invertible, with

$$(E_{ij}(r))^{-1} = E_{ij}(-r)$$

To see this, we compute

$$E_{ij}(r)E_{ij}(-r) = (I_n + re_{ij})(I_n - re_{ij})$$

$$= I_n I_n + re_{ij}I_n - I_n re_{ij} - r^2 e_{ij}e_{ij}$$

$$= I_n + re_{ij} - re_{ij} + r^2 e_{ij} e_{ij}$$
$$= I_n + r^2 e_{ij} e_{ij}$$

However, $e_{ij} e_{ij}$ is the zero matrix since $i \neq j$. (See Exercise 8).

(3) A *permutation matrix* is a square matrix with a single non-zero entry in each row and in each column, that entry being a one. For example,

$$P = \begin{pmatrix} 0 & 1 & 0 \\ 0 & 0 & 1 \\ 1 & 0 & 0 \end{pmatrix}$$

is a permutation matrix. If P is a permutation matrix, then P is invertible and $P^{-1} = P^t$, the transpose of P. (See Exercise 7.)

Note. The reason for calling the matrices of Example 3 permutation matrices is that, for any matrix A, PA is obtained by permuting the rows of A and AP is obtained by permuting the columns of A (provided these products are defined).

We now prove some results about invertible matrices.

PROPOSITION 2.2. *If A is an invertible* n × n *matrix, then the* n × n *matrix* B *satisfying* AB = BA = I_n *is unique.*

Proof. Suppose C is another n × n matrix with

$$AC = CA = I_n$$

Then

$$C = CI_n$$
$$= C(AB)$$
$$= (CA)B$$
$$= I_n B$$
$$= B$$

Thus, an invertible matrix A has a unique inverse, which we denote by A^{-1}.

PROPOSITION 2.3. *A linear transformation* $T:V \to W$ *is an isomor-*
phism if and only if $A_T(\alpha,\beta)$ *is invertible for any bases* α *for* V *and*
β *for* W.

Proof. If T is an isomorphism, then $A_{T^{-1}}(\beta,\alpha)$ is an inverse
matrix for $A_T(\alpha,\beta)$, (from Proposition 1.2).

If $A = A_T(\alpha,\beta)$ is invertible, we use the inverse matrix, A^{-1}, to
define a linear transformation $S:W \to V$ (using the analogue of equation (1.1) for A^{-1}). Then

$$T \circ S = T_A \circ T_{A^{-1}}$$

$$= T_{AA^{-1}}$$

$$= T_{I_n}$$

$$= I_W$$

Similarly, $S \circ T = I_V$ so, T is an isomorphism.

EXERCISES

1.   Prove that if A and B are invertible n × n matrices, then so is
AB, and $(AB)^{-1} = B^{-1}A^{-1}$.

2.   If A and B are invertible, is it true that A + B is invertible?

3.   Prove that the matrix

$$A = \begin{pmatrix} a & b \\ c & d \end{pmatrix}$$

is invertible if and only if $\Delta = ad - bc \neq 0$. Show that, in this case,
A^{-1} is given by

$$A^{-1} = \begin{pmatrix} d/\Delta & -b/\Delta \\ -c/\Delta & a/\Delta \end{pmatrix}$$

4.   Recall (Exercise 13 of Section 1). The transpose A^t of a matrix $A = (a_{ij})$ is the matrix (c_{ij}) defined by $c_{ij} = a_{ji}$. (One gets A^t by interchanging the rows and columns of A.) Show that if A is invertible, then A^t is invertible, and $(A^t)^{-1} = (A^{-1})^t$.

5.   Prove that the diagonal matrix $D = \text{diag}(d_1,\ldots,d_n)$ (defined in Example 1) is invertible if and only if $d_i \neq 0$, $i = 1,\ldots,n$ and that $D^{-1} = \text{diag}(d_1^{-1},\ldots,d_n^{-1})$ in this case.

6.   Prove that the matrices e_{ij} defined in Example 2 satisfy the following:

$$e_{ij}e_{k\ell} = \begin{cases} 0, & \text{if } j \neq k \\ e_{i\ell}, & \text{if } j = k \end{cases}$$

7.   Let P be a permutation matrix (see Example 3). Prove that $P^{-1} = P^t$.

8.   Find the inverse of the matrix

$$\begin{pmatrix} 1 & a & b \\ 0 & 1 & c \\ 0 & 0 & 1 \end{pmatrix}$$

9.   Show that if $T:V \to V$ is injective, then T is also surjective (and therefore an isomorphism).

10.   Show that if $T:V \to V$ is surjective, then T is also injective (and therefore an isomorphism).

*11.   Give conditions under which the matrix

$$\begin{pmatrix} a & b & c \\ 0 & d & e \\ 0 & 0 & f \end{pmatrix}$$

is invertible and determine its inverse when it is.

*12.   Show that if $S,T:V \to V$ are linear transformations such that $S \circ T$ is invertible, then S and T are invertible.   (See Exercises 9 and 10.)

3.   CHANGE OF BASIS

We now investigate the way the matrix of a linear transformation changes when we change the bases of the vector spaces involved.

Suppose that $\alpha = \{\vec{v}_1,\ldots,\vec{v}_n\}$ and $\alpha' = \{\vec{v}_1',\ldots,\vec{v}_n'\}$ are two bases for V.   We define an $n \times n$ matrix

$$B(\alpha,\alpha') = (b_{ij})$$

by the condition

$$\vec{v}_j = \sum_{i=1}^{n} b_{ij}\vec{v}_i' \tag{3.1}$$

for $1 \leq j \leq n$.

LEMMA 3.1.   *Let* $I = I_V$ *be the identity mapping of* V.   *Then*

$$B(\alpha,\alpha') = A_I(\alpha,\alpha')$$

Furthermore, $B(\alpha',\alpha) = B(\alpha,\alpha')^{-1}$.

Proof.   Set $A_I(\alpha,\alpha') = (a_{ij})$ so that, by definition,

$$I\vec{v}_j = \sum_{i=1}^{n} a_{ij}\vec{v}_i'$$

Since $I\vec{v}_j = \vec{v}_j$, this is exactly the defining equation (3.1) for the matrix $B(\alpha,\alpha')$.   Thus $B(\alpha,\alpha') = A_I(\alpha,\alpha')$.   It follows that

$$B(\alpha,\alpha')B(\alpha',\alpha) = A_I(\alpha,\alpha')A_I(\alpha',\alpha)$$

$$= A_{II}(\alpha',\alpha') \qquad \text{(by Proposition 1.2)}$$

$$= I$$

Similarly,

$$B(\alpha',\alpha)B(\alpha,\alpha') = I$$

and $B(\alpha',\alpha) = B(\alpha,\alpha')^{-1}$.

The next result gives the relationship between the matrices of a linear transformation relative to different bases.

PROPOSITION 3.2. *Let* $T:V \to W$ *be a linear transformation,* α,α' *bases for* V, *and* β,β' *bases for* W. *Then*

$$A_T(\alpha',\beta') = B(\beta,\beta')A_T(\alpha,\beta)B(\alpha',\alpha)$$

Proof. Let I_V be the identity mapping of V so that $T = TI_V$. If we consider I_V to map V with basis α' into V with basis α and T to map V with basis α to W with basis β, we see that

$$A_T(\alpha',\beta) = A_{TI_V}(\alpha',\beta)$$

$$= A_T(\alpha,\beta)A_{I_V}(\alpha',\alpha) \qquad \text{(by Proposition 1.2)}$$

$$= A_T(\alpha,\beta)B(\alpha',\alpha) \qquad \text{(by Proposition 3.1)}$$

Similarly, if I_W is the identity mapping on W, we have

$$A_T(\alpha',\beta') = A_{I_W T}(\alpha',\beta')$$

$$= A_{I_W}(\beta,\beta')A_T(\alpha',\beta) \qquad \text{(by Proposition 1.2)}$$

$$= B(\beta,\beta')A_T(\alpha',\beta) \qquad \text{(by Proposition 3.1)}$$

Combining these two equations, we have

$$A_T(\alpha',\beta') = B(\beta,\beta')A_T(\alpha',\beta)$$

$$= B(\beta,\beta')A_T(\alpha,\beta)B(\alpha',\alpha)$$

For example, suppose $T:R^3 \to R^2$ is defined by the matrix

$$A = \begin{pmatrix} 2 & 0 & 1 \\ -1 & 1 & 3 \end{pmatrix}$$

relative to the standard bases α for R^3 and β for R^2. Suppose $\alpha' = \{\vec{v}_1', \vec{v}_2', \vec{v}_3'\}$ is the new basis for R^3 defined by

$$\vec{v}_1' = 2\vec{e}_1 - \vec{e}_2 + 3\vec{e}_3$$
$$\vec{v}_2' = \vec{e}_1 + \vec{e}_2$$
$$\vec{v}_3' = \vec{e}_1 - \vec{e}_2 - \vec{e}_3$$

and $\beta' = \{\vec{w}_1', \vec{w}_2'\}$ is the base for R^2 defined by

$$\vec{w}_1' = 2\vec{e}_1 - \vec{e}_2$$
$$\vec{w}_2' = \vec{e}_1$$

Then

$$B(\beta, \beta') = \begin{pmatrix} 0 & -1 \\ 1 & 2 \end{pmatrix}$$

since

$$\vec{e}_1 = \vec{w}_2'$$
$$\vec{e}_2 = -\vec{w}_1' + 2\vec{w}_2'$$

Similarly,

$$B(\alpha', \alpha) = \begin{pmatrix} 2 & 1 & 1 \\ -1 & 1 & -1 \\ 3 & 0 & -1 \end{pmatrix}$$

Thus

$$A_T(\alpha', \beta') \quad \begin{pmatrix} 0 & -1 \\ 1 & 2 \end{pmatrix} \begin{pmatrix} 2 & 0 & 1 \\ -1 & 1 & 3 \end{pmatrix} \begin{pmatrix} 2 & 1 & 1 \\ -1 & 1 & -1 \\ 3 & 0 & -1 \end{pmatrix}$$

$$= \begin{pmatrix} -6 & 0 & 5 \\ 19 & 2 & -9 \end{pmatrix}$$

We can check this calculation by computing T directly on $\vec{v}_i'$, $1 \le i \le 3$. For example, the matrix $A_T(\alpha',\beta')$ above tells us that

$$\begin{aligned}
T\vec{v}_1' &= -6\vec{w}_1' + 19\vec{w}_2' \\
&= -6(2\vec{e}_1 - \vec{e}_2) + 19\vec{e}_1 \\
&= 7\vec{e}_1 + 6\vec{e}_2
\end{aligned}$$

However, from the original definition of T we see that

$$\begin{aligned}
T\vec{v}_1' &= T(2\vec{e}_1 - \vec{e}_2 + 3\vec{e}_3) \\
&= 2T(\vec{e}_1) - T(\vec{e}_2) + 3T(\vec{e}_3) \\
&= 2(2\vec{e}_1 - \vec{e}_2) - \vec{e}_2 + 3(\vec{e}_1 + 3\vec{e}_2) \\
&= 7\vec{e}_1 + 6\vec{e}_2
\end{aligned}$$

Similar computations can be carried out for $T\vec{v}_2'$ and $T\vec{v}_3'$.

Let A and B be m × n matrices. We say that A is *equivalent* to B if there is an invertible m × m matrix P and an invertible n × n matrix Q such that

$$B = PAQ$$

For example, we have just seen that the matrices $A_T(\alpha,\beta)$ and $A_T(\alpha',\beta')$ are equivalent; $A_T(\alpha',\beta') = PA_T(\alpha,\beta)Q$ where $P = B(\beta,\beta')$ and $Q = B(\alpha',\alpha)$.

PROPOSITION 3.3. *The relation "A is equivalent to B" between n × n matrices is an equivalence relation.*

Proof. First of all, A is equivalent to A; set $P = Q = I_n$. If A is equivalent to B, so that $B = PAQ$, then $A = P^{-1}BQ^{-1}$ and B is equivalent to A. Finally, if A is equivalent to B ($B = PAQ$) and B is equivalent to C ($C = P'BQ'$), then

$$\begin{aligned}
C &= P'(PAQ)Q' \\
&= (P'P)A(QQ')
\end{aligned}$$

Thus A is equivalent to C, since P'P and Q'Q are invertible (Exercise 1 of Section 2).

Suppose now that $T:V \to V$ and α is a basis for V. We write $A_T(\alpha)$ for $A_T(\alpha,\alpha)$. If α' is a second basis for V, then

$$A_T(\alpha') = B(\alpha,\alpha')A_T(\alpha)B(\alpha',\alpha)$$
$$= B(\alpha,\alpha')A_T(\alpha)B(\alpha,\alpha')^{-1}$$

This motivates the following definition.

Let A and B be $n \times n$ matrices. We say A is *similar* to B if there is an invertible $n \times n$ matrix P such that

$$B = P^{-1}AP$$

In particular, $A_T(\alpha)$ and $A_T(\alpha')$ are similar matrices.

PROPOSITION 3.4. *The relation on* $n \times n$ *matrices of being similar is an equivalence relation.*

We leave the proof of this result as Exercise 8.

We conclude this section with a useful restatement of Proposition 1.3. The proof is left to the reader as Exercise 9.

PROPOSITION 3.5. *Let A be an* $m \times n$ *matrix. Then A is equivalent to a matrix of the form*

$$B = \begin{pmatrix} A_1 & A_2 \\ A_3 & A_4 \end{pmatrix}$$

where A_1 *is the* $k \times k$ *identity matrix,* A_2 *is the* $k \times (n-k)$ *zero matrix,* A_3 *is the* $(m-k) \times k$ *zero matrix, and* A_4 *is the* $(m-k) \times (n-k)$ *zero matrix.*

EXERCISES

1. Suppose $T:R^2 \to R^3$ has a matrix

$$\begin{pmatrix} 2 & 1 \\ 0 & -3 \\ 1 & 4 \end{pmatrix}$$

relative to the standard bases for R^2 and R^3. Use Proposition 3.2 to find its matrix relative to the bases $\alpha' = \{\vec{v}_1', \vec{v}_2'\}$ for R^2, $\beta' = \{\vec{w}_1', \vec{w}_2', \vec{w}_3'\}$ for R^3 where

$$\vec{v}_1' = 7\vec{e}_1 - 11\vec{e}_2$$
$$\vec{v}_2' = 2\vec{e}_1 + 5\vec{e}_2$$

and

$$\vec{w}_1' = \vec{e}_2 - \vec{e}_3$$
$$\vec{w}_2' = \vec{e}_3$$
$$\vec{w}_3' = \vec{e}_1 - \vec{e}_2 + \vec{e}_3$$

Check the answer by computing $T\vec{v}_2'$ in two different ways.

2.   Suppose $T : R^2 \to R^2$ has the matrix

$$\begin{pmatrix} 2 & 0 \\ 1 & 1 \end{pmatrix}$$

relative to the standard bases for R^2. Determine the matrix $A_T(\alpha')$, where $\alpha' = \{(1,-1),(1,1)\}$.

3.   Let A_1, B_1, A_2, B_2 be $n \times m$ matrices. Suppose A_1 is equivalent to B_1 and A_2 is equivalent to B_2. Prove (or disprove by counterexample) that

(a) $A_1 + A_2$ is equivalent to $B_1 + B_2$

(b) rA_1 is equivalent to rB_1 for any real number r

4.   If $n = m$ in Exercise 3, is it true that $A_1 A_2$ is equivalent to $B_1 B_2$?

5.   With the notation of Exercise 3, if A_1 is similar to B_1 and A_2 similar to B_2, prove (or disprove by counterexample) that

(a) $A_1 + A_2$ is similar to $B_1 + B_2$

(b) rA_1 is similar to rB_1 for any real number r

6.   In Exercise 5, does it follow that A_1A_2 is similar to B_1B_2?

7.   (a) Suppose A is an invertible matrix and A is equivalent to B. Prove that B is invertible.

(b) Suppose A is an invertible matrix and A is equivalent to B. Prove that A^{-1} is equivalent to B^{-1}.

(c) Prove that statements (a) and (b) hold if "equivalent" is replaced by "similar."

(d) Show that the matrices

$$A = \begin{pmatrix} 0 & 1 \\ 0 & 0 \end{pmatrix} , \quad B = \begin{pmatrix} 1 & 2 \\ 0 & 1 \end{pmatrix}$$

cannot be equivalent.

8.   Prove Proposition 3.4.

9.   Prove Proposition 3.5.

10.   Show that similar $n \times n$ matrices are equivalent, but that the converse need not hold.

*11.   Show that the matrix $\begin{pmatrix} a & b \\ c & d \end{pmatrix}$ is similar to $\begin{pmatrix} 1 & 0 \\ 0 & 2 \end{pmatrix}$ if and only if $a + d = 3$ and $ad - bc = 2$.

12.   (a) Show that A is similar to I_n if and only if $A = I_n$.

(b) Show that A is equivalent to I_n if and only if A is an $n \times n$ invertible matrix.

*13.   Are $\begin{pmatrix} 1 & 1 \\ 0 & 0 \end{pmatrix}$ and $\begin{pmatrix} 0 & 1 \\ 0 & 0 \end{pmatrix}$ equivalent? Are they similar?

*14. (a) Let $\begin{pmatrix} a & 0 \\ 0 & b \end{pmatrix}$ and $\begin{pmatrix} a' & 0 \\ 0 & b' \end{pmatrix}$ be diagonal 2 × 2 matrices. Show
that they are similar if and only if they have the same diagonal terms
(perhaps in a different order).

(b) Prove the corresponding result for n × n matrices.

4. THE RANK OF A MATRIX

We continue the study of the relationship between a linear transformation and its matrix.

Let $A = (a_{ij}$ be an m × n matrix. We consider the m rows of A
as elements of R^n and the n columns of A as elements of R^m. The *row
rank* of A is the dimension of the span of the rows of A in R^n and the
column rank of A the dimension of the span of the columns of A in R^m.

For example, if A is given by

$$A = \begin{pmatrix} 1 & 0 & 1 \\ 0 & 2 & 1 \\ 1 & 2 & 2 \\ 1 & -2 & 0 \end{pmatrix}$$

then the row rank is two, since

$$(1,2,2) = (1,0,1) + (0,2,1)$$
$$(1,-2,0) = (1,0,1) - (0,2,1)$$

and (1,0,1), (0,2,1) are clearly independent. The column rank of A
is also two since

$$2(1,0,1,1) + (0,2,2,-2) - 2(1,1,2,0) = (0,0,0,0)$$

and the first two columns are independent. We shall see later that
the equality of these two numbers is no accident.

PROPOSITION 4.1. *Let* $\alpha = \{\vec{v}_1, \ldots, \vec{v}_n\}$ *be a basis for the vector
space V,* $\beta = \{\vec{w}_1, \ldots, \vec{w}_m\}$ *a basis for the vector space W, and let
T:V → W be a linear transformation. Then the rank of T is equal to
the column rank of the matrix* $A_T(\alpha, \beta)$.

Proof. By the definition of the rank of T,

$$\text{rank } T = \dim(\text{image } T)$$
$$= \dim \text{span}(T(\vec{v}_1),\ldots,T(\vec{v}_n))$$

We need to show, therefore, that $\dim \text{span}(T(\vec{v}_1),\ldots,T(\vec{v}_n))$ is the column rank of T. Let $S:W \to R^m$ be the linear transformation defined by

$$S(\vec{w}_i) = \vec{e}_i$$

$1 \leq i \leq m$, where $\vec{e}_1,\ldots,\vec{e}_m$ is the standard basis for R^m. Then S is easily seen to be an isomorphism, and if $A_T(\alpha,\beta) = (a_{ij})$, we have

$$S(T(\vec{v}_j)) = S(\sum_{i=1}^{m} a_{ij}\vec{w}_i)$$

$$= \sum_{i=1}^{m} a_{ij}\vec{e}_i$$

It follows that S takes $T(\vec{v}_j)$ onto the element of R^m determined by the jth column of $A_T(\alpha,\beta)$. Thus S defines an isomorphism between the image of T and the subspace of R^m spanned by the column of $A_T(\alpha,\beta)$. Since an isomorphism preserves dimension (by the corollary to Proposition 4.3.5), the proposition follows.

Recall (Exercise 13 of Section 1) that if $A = (a_{ij})$ is an $n \times n$ matrix, the transpose of A, A^t, is defined to be the $n \times m$ matrix obtained from A by interchanging the rows and columns of A. That is, $A^t = (c_{ij})$ where $A = (a_{ij})$ and $c_{ij} = a_{ji}$.

PROPOSITION 4.2. *Let A and B be m × n matrices and let C be an n × p matrix. Then*

$$(A+B)^t = A^t + B^t$$
$$(AC)^t = C^t A^t$$

Furthermore, if n = m and A is invertible, then

$$(A^{-1})^t = (A^t)^{-1}$$

Proof. This result is easy. (See Exercise 13 of Section 1.)
We prove only the last assertion.

$$A^t(A^{-1})^t = (A^{-1}A)^t$$
$$= I^t = I$$
$$(A^{-1})^t A^t = (AA^{-1})^t$$
$$= I^t = I$$

Note, that, if A is any matrix,

row rank A = column rank A^t
column rank A = row rank A^t

In addition, if A is equivalent to B, we have

$$B = PAQ$$

for invertible matrices P, Q. Taking the transpose of both sides of
this equation, we have

$$B^t = Q^t A^t P^t$$

so A^t is equivalent to B^t. We use these observations in the proof of
the following.

PROPOSITION 4.3. *If A and B are equivalent* m × n *matrices, then
the row rank of A is equal to the row rank of B and the column rank
of A is equal to the column rank of B.*

Proof. Let $T:R^n \to R^m$ be the linear transformation defined by

$$T\vec{e}_j = \sum_{i=1}^{m} a_{ij} \vec{e}_i$$

$1 \le j \le n$, where $A = (a_{ij})$. Then, if α and β are the standard bases
for R^n and R^m, $A = A_T(\alpha,\beta)$.

Let P be an invertible m × m matrix, Q an invertible n × n matrix with B = PAQ. Suppose that $P^{-1} = (p_{ij})$ and $Q = (q_{ij})$ and define vectors

$$\vec{v}_j' = \sum_{i=1}^{n} q_{ij}\vec{e}_i$$

$1 \le j \le n$,

$$\vec{w}_\ell' = \sum_{k=1}^{m} p_{k\ell}\vec{e}_k$$

$1 \le \ell \le m$. Equivalently, $\vec{v}_j' = T_Q(\vec{e}_j)$ and $\vec{w}_\ell' = T_{P^{-1}}(\vec{e}_\ell)$. Since P^{-1} and Q are invertible, T_Q and $T_{P^{-1}}$ are isomorphisms and it follows that $\alpha' = \{\vec{v}_1', \ldots, \vec{v}_n'\}$ is a basis for R^n and $\beta' = \{\vec{w}_1', \ldots, \vec{w}_m'\}$ a basis for R^m. Furthermore, by definition, $B(\alpha', \alpha) = Q$ and $B(\beta', \beta) = P^{-1}$ so $B(\beta, \beta') = P$. Thus

$$\begin{aligned} A_T(\alpha', \beta') &= B(\beta, \beta')A_T(\alpha, \beta)B(\alpha', \alpha) \\ &= PAQ \\ &= B \end{aligned}$$

so that A and B are matrices of T relative to different bases for R^n and R^m. In particular,

$$\begin{aligned} \text{column rank A} &= \text{rank T} && \text{(by Proposition 4.1)} \\ &= \text{column rank B} && \text{(by Proposition 4.1)} \end{aligned}$$

Applying the same argument to A^t and B^t, we see that

$$\begin{aligned} \text{row rank A} &= \text{column rank } A^t \\ &= \text{column rank } B^t \\ &= \text{row rank B} \end{aligned}$$

This proposition has an interesting consequence.

THEOREM 4.4. *For any matrix A, the row rank of A is equal to the column rank of A.*

Proof. According to Proposition 3.5, A is equivalent to a matrix B of a very simple form:

$$B = \begin{pmatrix} I_k & 0 \\ 0 & 0 \end{pmatrix}$$

for some k. It is clear that row rank B = column rank B = k. On the other hand,

row rank A = row rank B

column rank A = column rank B

by Proposition 4.2. Putting these equalities together, we get the theorem.

We can now define the *rank* of a matrix A by

rank A = row rank A

= column rank A

It follows from Proposition 4.1 that, for any linear transformation T:V → W and bases α for V, β for W, we have

rank $A_T(\alpha,\beta)$ = rank T

Another useful consequence is the following:

PROPOSITION 4.5. *An* n × n *matrix* A *is invertible if and only if* rank A = n.

This follows easily from Theorem 4.4, Proposition 3.5 and Proposition 4.2. We leave the details as an exercise.

EXERCISES

1.   Compute the row rank of each of the following matrices.

(a) $\begin{pmatrix} 1 & 0 \\ 2 & 3 \end{pmatrix}$ (b) $\begin{pmatrix} 1 & 2 \\ 2 & 7 \\ -2 & 3 \end{pmatrix}$

(c) $\begin{pmatrix} 2 & 0 & 4 & -8 \\ 1 & 0 & 2 & -4 \\ -4 & 0 & -8 & 16 \end{pmatrix}$ (d) $\begin{pmatrix} 1 & 0 & -1 & 2 \\ 1 & 0 & 1 & 2 \\ 5 & 1 & 1 & -7 \end{pmatrix}$

(e) $\begin{pmatrix} 2 & 0 & 1 \\ -2 & 1 & 7 \\ 5 & 0 & 5 \\ -1 & -1 & -1 \\ 0 & 7 & 0 \end{pmatrix}$ (f) $\begin{pmatrix} 0 & 1 & 1 & 7 & 4 \\ -1 & 5 & 1 & 2 & 1 \\ 1 & -4 & 0 & 5 & 3 \\ -2 & 9 & 1 & -3 & -2 \end{pmatrix}$

2.   Determine, by direct computation, the column rank of the matrices in Exercise 1.

3.   Prove that a 2 × 2 matrix is invertible if and only if its rank is two.

4.   Prove Proposition 4.5.

5.   Prove that if A is equivalent to B, the rank of A is equal to the rank of B.

*6.   Let A and B be m × n matrices. Prove that if rank A = rank B, then A and B are equivalent.

7.   Show by example that the conclusion of Exercise 6 does not remain true if we replace equivalent by similar.

*8.   (a) Let A be an m × n matrix (m ≤ n) of rank m. Show that there is an n × m matrix B with AB = I_m.

     (b) Show, conversely, that if A is an m × n matrix and B is an n × m matrix such that AB = I_m, then rank(A) = m.

5.   THE TRACE AND ADJOINT OF A LINEAR TRANSFORMATION

The principal aim of this section is to define a natural inner product on the vector space of all linear transformations between two finite

dimensional inner product spaces. This is the inner product referred
to when we discussed higher derivatives in Chapter 6. In the course
of defining this inner product, we introduce two important notions,
the *trace* and the *adjoint* of a linear transformation.

If $A = (a_{ij})$ is an $n \times n$ matrix, we define the *trace* of A, tr(A),
to be the sum of the diagonal entries in A. Explicitly,

$$\text{tr}(A) = \sum_{i=1}^{n} a_{ii}$$

For instance, the trace of the zero matrix is zero and the trace of
the $n \times n$ identity matrix is n.

It is easy to see that

$$\text{tr}(A+B) = \text{tr}(A) + \text{tr}(B)$$

and

$$\text{tr}(aA) = a(\text{tr}(A))$$

for any two $n \times n$ matrices A, B and real number a. However, if we set

$$A = \begin{pmatrix} 1 & 0 \\ 0 & 0 \end{pmatrix} \quad , \qquad B = \begin{pmatrix} 0 & 0 \\ 0 & 1 \end{pmatrix}$$

then

$$\text{tr}(AB) \neq \text{tr}(A)\text{tr}(B)$$

since AB is the zero matrix. Nevertheless, we can prove the following.

PROPOSITION 5.1.   *For any two* $n \times n$ *matrices A and B, we have*

$$\text{tr}(AB) = \text{tr}(BA)$$

Proof.   Let $AB = (c_{ij})$ and $BA = (d_{ij})$ so that

$$c_{ij} = \sum_{k=1}^{n} a_{ik}b_{k\ell}$$

$$d_{ij} = \sum_{k=1}^{n} b_{ik} a_{kj}$$

Thus

$$tr(AB) = \sum_{i=1}^{n} c_{ii}$$

$$= \sum_{i=1}^{n} \sum_{k=1}^{n} a_{ik} b_{ki}$$

$$= \sum_{k=1}^{n} \sum_{i=1}^{n} b_{ki} a_{ik}$$

$$= \sum_{k=1}^{n} d_{kk} = tr(BA)$$

Suppose now that V is a vector space with basis $\alpha = \{\vec{v}_1,\ldots,\vec{v}_n\}$, $T:V \to V$ is a linear transformation, and $A = A_T(\alpha,\alpha)$ is the matrix of T relative to the basis α. We define the *trace of* T, tr(T), to be the trace of the matrix A. In order for this definition to make sense, we must show that it does not depend on which basis we choose for V.

PROPOSITION 5.2. *Let* $\alpha = \{\vec{v}_1,\ldots,\vec{v}_n\}$ *and* $\alpha' = \{\vec{v}_1',\ldots,\vec{v}_n'\}$ *be bases for V, T:V $\to$ V a linear transformation, A the matrix of T relative to the basis* α*, and A' the matrix of T relative to the basis* α'*. Then*

$$Tr(A) = tr(A')$$

Proof. Let $S:V \to V$ be the linear transformation defined by $S(\vec{v}_i) = \vec{v}_i'$, $1 \le i \le n$ (using Proposition 4.3.1). Then S is an isomorphism and its matrix P relative to the basis α is invertible. According to the discussion in Section 4, we see that $A' = PAP^{-1}$ so that

$$tr(A') = tr(PAP^{-1})$$

$$= tr((PA)P^{-1})$$

$$= tr(P^{-1}(PA)) \qquad \text{(by Proposition 5.1)}$$

$$= \text{tr}((P^{-1}P)A)$$

$$= \text{tr}A$$

This proves Proposition 5.2.

We now define the adjoint of a linear transformation between inner product spaces.

PROPOSITION 5.3. *Let V and W be finite dimensional inner product spaces and T:V → W a linear transformation.* Then there is a unique linear transformation T^* :W → V satisfying the condition

$$< \vec{Tv}, \vec{w} > = < \vec{v}, T^* \vec{w} >$$

for all $\vec{v}$ *in V and* $\vec{w}$ *in W.*

The linear transformation T^* is called the *adjoint* of T.

Proof. We first prove uniqueness. Suppose T_1 and T_2 are two linear transformations from W to V satisfying

$$< \vec{Tv}, \vec{w} > = < \vec{v}, T_1\vec{w} > = < \vec{v}, T_2\vec{w} >$$

for all $\vec{v}$ in V and $\vec{w}$ in W. Then

$$< \vec{v}, T_1\vec{w} - T_2\vec{w} > = 0$$

for all $\vec{v}$ in V and $\vec{w}$ in W. In particular, we can set $\vec{v} = T_1\vec{w} - T_2\vec{w}$ and we have

$$< T_1\vec{w} - T_2\vec{w}, T_1\vec{w} - T_2\vec{w} > = 0$$

It follows from the defining properties of the inner product that $T_1\vec{w} - T_2\vec{w} = \vec{0}$, so that $T_1\vec{w} = T_2\vec{w}$ for all $\vec{w}$ in W.

To prove the existence of T^*, we choose orthonormal bases $\alpha = \{\vec{v}_1, \ldots, \vec{v}_n\}$ for V and $\beta = \{\vec{w}_1, \ldots, \vec{w}_m\}$ for W. Let $A = (a_{ij})$ be the $m \times n$ matrix of T relative to these bases. Then the transpose of A; A^t, is an $n \times m$ matrix which defines a linear transformation T^* :W → V. To prove that

$< \vec{Tv}, \vec{w} > = < \vec{v}, T^* \vec{w} >$ for all $\vec{v}$ in V and $\vec{w}$ in W, it is clearly sufficient to check it on vectors in the bases α and β. Thus

$$< \vec{Tv_i}, \vec{w_j} > = < \sum_{k=1}^{m} a_{ki} \vec{w_k}, \vec{w_j} >$$

$$= \sum_{k=1}^{m} a_{ki} < \vec{w_k}, \vec{w_j} >$$

$$= a_{ji}$$

Similarly,

$$< \vec{v_i}, T^* \vec{w_j} > = < \vec{v_i}, \sum_{\ell=1}^{n} a_{j\ell} \vec{v_\ell} >$$

$$= a_{ji}$$

so the Proposition is proved.

We note that, in the course of proving Proposition 5.3, we have also proved the following.

PROPOSITION 5.4. *Let V and W be inner product spaces with orthonormal bases* α *and* β. *Let* $T:V \rightarrow W$ *be a linear transformation with* $A_T(\alpha,\beta) = A$. *Then* $A_{T^*}(\beta,\alpha) = A^t$.

The next result gives several of the elementary properties of the adjoint of a linear transformation.

PROPOSITION 5.5. *Let V and W be finite dimensional inner product spaces,* $S,T:V \rightarrow V$ *linear transformation, and* c *a real number. Then*

   (i) $(T^*)^* = T$
   (ii) $(S+T)^* = S^* + T^*$
   (iii) $(cT)^* = cT^*$
   (iv) If $V = W$, $tr(T^*) = tr(T)$

Furthermore, if W' is a third inner product space and $S:W \rightarrow W'$ *a linear transformation, then*

(v) $(S \circ T)^* = T^* \circ S^*$

(vi) $I^* = I$

If $T:V \to V$ is invertible, then T^* is also invertible and

(vii) $(T^*)^{-1} = (T^{-1})^*$

Proof. For (i), we must show that

$$< T^* \vec{w}, \vec{v} > = < \vec{w}, T\vec{v} >$$

for all $\vec{v}$ in V and $\vec{w}$ in W. This follows from the symmetry of the inner product and the definition of T^*.

Statements (ii) and (iii) are straightforward, while (iv) is an immediate consequence of Proposition 5.4. We leave the details as Exercise 5.

To prove (v), we note that

$$
\begin{aligned}
< (S \circ T)(\vec{v}), \vec{w} > &= < S(T(\vec{v})), \vec{w} > \\
&= < T(\vec{v}), S^* \vec{w} > \\
&= < \vec{v}, T^*(S^*(\vec{w})) > \\
&= < \vec{v}, (T^* \circ S^*)(\vec{w}) >
\end{aligned}
$$

Thus $T^* \circ S^* = (S \circ T)^*$.

(vi) is an easy exercise. For (vii), we use (v): since $T \circ T^{-1} = I = T^{-1} \circ T$, we have

$$
\begin{aligned}
(T^{-1})^* \circ T^* = (T \circ T^{-1})^* \\
= I^*
\end{aligned}
$$

and

$$
\begin{aligned}
T^* \circ (T^{-1})^* = (T^{-1} \circ T)^* \\
= I^*
\end{aligned}
$$

Since $I^* = I$, $(T^{-1})^*$ is the inverse of T^*, and the proposition is proved.

We define here for future reference two interesting classes of
linear transformations. Let V be an inner product space and $T:V \to V$
a linear transformation. We say that T is *self-adjoint* if $T = T^*$ and
orthogonal if $T^* \circ T = T \circ T^* = I$. (See Exercise 8 for a characterization
of the matrices of these transformations.)

Let V and W be inner product spaces and $\mathcal{L}(V,W)$ the vector space
of all linear transformations from V to W. The next result defines
a natural inner product on this space.

PROPOSITION 5.6. *For* $S,T \in \mathcal{L}(V,W)$, *the function* $< S,T >$ *given*
by

$$< S,T > = tr(S \circ T^*)$$

defines an inner product on $\mathcal{L}(V,W)$.

Proof. Note first of all that

$$< T,S > = tr(T \circ S^*) = tr((S \circ T^*)^*)$$

by (i) and (v) of Proposition 5.5. According to statement (iv) of
this same Proposition,

$$< T,S > = tr(T \circ S^*) = tr((S \circ T^*)^*) = tr(S \circ T^*) = < S,T >$$

so $< S,T > = < T,S^* >$ for all S,T in $\mathcal{L}(V,W)$.

If S_1, S_2, T are elements of $\mathcal{L}(V,W)$ and b_1, b_2 are real numbers,
then the equation

$$< b_1 S_1 + b_2 S_2, T > = b_1 < S_1, T > + b_2 < S_2, T >$$

follows from statements (ii) and (iii) of Proposition 5.5.

To prove that $< S,S > \geq 0$ for all S in $\mathcal{L}(V,W)$ and $< S,S > = 0$ if
and only if $S = 0$, we let $A = (a_{ij})$ be the m × n matrix of S relative
to orthonormal bases for V and W. Then

$$< S,S > = tr(S \circ S^*) = tr(AA^t) = \sum_{i=1}^{m} \sum_{j=1}^{n} a_{ij}^2$$

by direct computation. Thus $< S,S > = 0$ and $< S,S > = 0$ if and only if $a_{ij} = 0$ for all i,j, $1 \le i \le m$, $1 \le j \le n$; that is, if and only if S is the zero transformation.

We conclude this section by showing how to construct orthonormal bases in $\mathcal{L}(V,W)$ relative to the inner product defined above.

PROPOSITION 5.7. *Let* $\alpha = \{\vec{v}_1,\ldots,\vec{v}_n\}$, $\beta = \{\vec{w}_1,\ldots,\vec{w}_m\}$ *be orthonormal bases for the inner product spaces* V *and* W. *For each* i,j, $1 \le i \le n$, $1 \le j \le m$, *define linear transformations* $T_{ij}:V \to W$ *by*

$$T_{ij}(\vec{v}_k) = \begin{cases} \vec{0} & \text{if } i \ne k \\ \vec{w}_j & \text{if } i = k \end{cases}$$

Then the elements T_{ij} *form an orthonormal basis for the inner product space* $\mathcal{L}(V,W)$ *with the inner product defined in Proposition 5.6.*

Proof. We give a sketch of the proof, leaving the details to the reader.

We first compute T_{ij}^*. Note that

$$< T_{ij}^*\vec{w}_\ell,\vec{v}_k > = < \vec{w}_\ell,T_{ij}\vec{v}_k >$$
$$= 0$$

unless $k = i$ and $\ell = j$. It follows that

$$T_{ij}^*\vec{w}_\ell = \begin{cases} \vec{0} & \text{if } j \ne \ell \\ \vec{v}_i & \text{if } j = \ell \end{cases}$$

Next, one easily proves that

$$\text{trS} = \sum_{j=1}^{m} < S\vec{w}_j,\vec{w}_j > \tag{5.1}$$

for any linear transformation $S:W \to W$. Thus

$$< T_{ij}, T_{pq} > = tr(T_{ij} \circ T_{pq}^{*})$$

$$= \sum_{\ell=1}^{m} < T_{ij} \circ T_{pq}^{*} \vec{w}_{\ell}, \vec{w}_{\ell} >$$

$$= \sum_{\ell=1}^{m} < T_{pq}^{*} \vec{w}_{\ell}, T_{ij}^{*} \vec{w}_{\ell} >$$

$$= \begin{cases} < \vec{v}_j, \vec{v}_j > & \text{if } i = p, \ j = q, \\ 0 & \text{otherwise} \end{cases}$$

Proposition 5.7 now follows, since $< \vec{v}_j, \vec{v}_j > = 1$.

EXERCISES

1.   Prove that for any $n \times n$ matrices A,B and real number a, we have
   (i)   tr(A+B) = tr(A) + tr(B),
   (ii) tr(aA) = atr(A).

2.   Give a detailed proof of Proposition 5.7.

3.   Let V and W be finite dimensional inner product spaces and
S,T:V $\rightarrow$ W linear transformation. Let α be an orthonormal basis for
V and β an orthonormal basis for W. Then, if $A = (a_{ij}) = A_S(\alpha, \beta)$ and
$B = (b_{ij}) = A_T(\alpha, \beta)$, use Proposition 5.7 to show that

$$< S, T > = \sum_{i,j} a_{ij} b_{ij}$$

4.   Let $\mathcal{L}_0$ be the subset of all T in $\mathcal{L}(V,V)$ with tr(T) = 0. Prove
that $\mathcal{L}_0$ is a subspace of $\mathcal{L}(V,V)$ and determine its dimension.

5.   Prove in detail statements (i) through (iv) of Proposition 5.5.

6.   Prove that the sum of two self-adjoint transformations is self
adjoint, and determine conditions under which the product of two self-
adjoint transformations is self-adjoint.

7.   Prove that the product of two orthogonal transformations is orthogonal.

8.   Let $A = (a_{ij})$ be the matrix of a linear transformation $T:V \to V$ relative to an orthonormal basis for V. Prove that

(i) T is self-adjoint if and only if A is symmetric; that is $a_{ij} = a_{ji}$ for all $1 \leq i,\ j \leq n$

(ii) T is orthogonal if and only if its rows are orthonormal; that is, $\Sigma_{k=1}^{n} a_{ik} a_{jk} = \delta_{ij},\ 1 \leq i,\ j \leq n$

9.   Prove that if V is a finite dimensional inner product space, then $T:V \to V$ is orthogonal if and only if $< \vec{Tv}, \vec{Tv'} > = < \vec{v}, \vec{v'} >$ for all $\vec{v}, \vec{v'}$ in V.

10.   Let $T:V \to V$ be a self-adjoint linear transformation and prove the following:

(i) $T^2 = 0$ implies $T = 0$

(ii) $T^4 = 0$ implies $T = 0$

(iii) $T^3 = 0$ implies $T = 0$

11.   Let $T:V \to W$ be a linear transformation between two inner product spaces. Show that

$$\mathcal{I}(T) = h(T^*)^{\perp}$$

Use this result to show that

$$h(T) = \mathcal{I}(T^*)^{\perp}$$

(See Exercise 8, Section 4.7.)

12.   Verify equation (5.1).

13.   Let V and W be normed vector spaces and $T:V \to W$ a linear transformation. Define the *norm of* T, $\|T\|$, by

$$\|T\| = \text{lub} \ \|T(\vec{v})\|$$
$$\|\vec{v}\| = 1$$

when this upper bound exists. In this case, T is said to be *bounded*.

(a) Prove that

$$\|T\| = \text{lub} \ \|T(\vec{v})\|$$
$$\|\vec{v}\| \leq 1$$

$$= \text{lub} \ \frac{\|T(\vec{v})\|}{\vec{v} \neq \vec{0} \ \ \|\vec{v}\|}$$

(b) Assume that V and W are finite dimensional inner product spaces and show that $\|T\|$ defined here is *not* the same as the norm obtained from the inner product on $\mathcal{L}(V,W)$ defined in this section.

14. Let $T: R^n \to R^m$ have matrix (a_{ij}) relative to the standard bases.

(a) Suppose that both R^n and R^m have norms defined by

$$\|(x_i)\| = \sum_i |x_i|$$

Prove that, if the norm is given as in Exercise 13,

$$\|T\| = \max_j \ \sum_i |a_{ij}|$$

(b) Suppose that both R^n and R^m have norms defined by

$$\|(x_i)\| = \max_i |x_i|$$

Prove that

$$\|T\| = \max_i \ \sum_j |a_{ij}|$$

6. ROW AND COLUMN OPERATIONS

In this section, we introduce elementary row and column operations on matrices. Suitably interpreted, these operations give a method for computing the inverse of an invertible matrix.

We begin with the elementary row operations. The first elementary row operation interchanges two rows of a matrix. For example, if A is the matrix

$$A = \begin{pmatrix} 1 & 2 & 3 \\ 4 & 5 & 6 \\ 7 & 8 & 9 \end{pmatrix}$$

we can interchange the first and third row of A obtaining

$$\begin{pmatrix} 7 & 8 & 9 \\ 4 & 5 & 6 \\ 1 & 2 & 3 \end{pmatrix}$$

The second row operation multiplies any row of a matrix by a non-zero real number and the third operation adds a multiple of one row to another row. For example, we can multiply the second row of the matrix A above by - 1/2 obtaining

$$\begin{pmatrix} 1 & 2 & 3 \\ -2 & -\frac{5}{2} & -3 \\ 7 & 8 & 9 \end{pmatrix}$$

or we can multiply the first row of A by -2 and add it (componentwise) to the second row, obtaining

$$\begin{pmatrix} 1 & 2 & 3 \\ 2 & 1 & 0 \\ 7 & 8 & 9 \end{pmatrix}$$

We can define three elementary column operations in an analogous way. Our first result of this section states that these elementary row and column operations can be used to transform any matrix into a particularly simple form (see Proposition 1.3).

PROPOSITION 6.1. *Let* A *be any* n × m *matrix. Using the elementary row and column operations, the matrix* A *can be transformed into a matrix of the form*

$$\begin{pmatrix} A_1 & A_2 \\ A_3 & A_4 \end{pmatrix}$$

where A_1 is a $k \times k$ *identity matrix*, k = rank A , and A_2, A_3, A_4 are zero matrices.

Before proving this proposition, we illustrate it in a special case. Let A be the matrix

$$A = \begin{pmatrix} 0 & 1 & 2 \\ 1 & 3 & -2 \end{pmatrix}$$

Interchanging the first and second column, we obtain

$$\begin{pmatrix} 1 & 0 & 2 \\ 3 & 1 & -2 \end{pmatrix}$$

If we now multiply the first row by -3 and add it to the second row, we obtain

$$\begin{pmatrix} 1 & 0 & 2 \\ 0 & 1 & -8 \end{pmatrix}$$

Multiplying the first column by -2 and adding it to the third column gives us

$$\begin{pmatrix} 1 & 0 & 0 \\ 0 & 1 & -8 \end{pmatrix}$$

Finally, we multiply the second column by 8 and add it to the third column, obtaining

$$\begin{pmatrix} 1 & 0 & 0 \\ 0 & 1 & 0 \end{pmatrix}$$

which has the required form with

$$A_1 = \begin{pmatrix} 1 & 0 \\ 0 & 1 \end{pmatrix} \qquad A_2 = \begin{pmatrix} 0 \\ 0 \end{pmatrix}$$

and A_3, A_4 not needed.

Proof of Proposition 6.1. Let $A = (a_{ij})$ be an n × m matrix. We proceed by induction of the integer $N = n + m - 1$.

If $N = 1$, it follows that $n = m = 1$ and the proposition is trivial in this case. Suppose $N > 1$ and assume the proposition true when $n + m - 1 < N$. If A is the zero matrix, it has the required form, so we may as well assume some $a_{ij} \neq 0$. By interchanging rows and columns if necessary, we can make the first entry in the first row non zero. In fact, multiplying the first row by the inverse of this first entry, we can make the first entry in the first row equal to one.

Suppose now that the second entry in the first row (or column) is non zero. Then, by adding the appropriate multiple of the first column (or row) to the second column (or row), we can make this entry zero. Continuing in this way, we can bring A into the form

$$\begin{pmatrix} 1 & 0 & \cdots & 0 \\ 0 & & & \\ \vdots & & A' & \\ 0 & & & \end{pmatrix}$$

where A' is an (n-1) by (m-1) matrix. Since

$$(n-1) + (m-1) - 1 = n + m - 3 < N$$

it follows by induction that A' can be put into the required form by elementary row and column operations. Therefore, the same is true for A, and the proposition is proved.

COROLLARY. *If A is an invertible* n × n *matrix, then A can be transformed into the identity matrix by elementary row and column operations.*

This corollary follows from Proposition 6.1 and the fact that A
is invertible if and only if rank A = n. (See Proposition 4.5.)

In order to apply the elementary row and column operations to
the problem of finding inverses we need to reinterpret them in terms
of matrix multiplication. First of all, interchanging two rows of a
matrix A corresponds to multiplying A on the left by a permutation
matrix (see Example 2 of Section 3.) For example, interchanging the
first and third row of the matrix.

$$A = \begin{pmatrix} a & b & c \\ d & e & f \\ g & h & j \end{pmatrix}$$

is accomplished by multiplying A on the left by the permutation matrix

$$P = \begin{pmatrix} 0 & 0 & 1 \\ 0 & 1 & 0 \\ 1 & 0 & 0 \end{pmatrix}$$

as an easy calculation shows. More generally, interchanging the kth
and ℓ^{th} rows of an n × m matrix A corresponds to multiplying A on the
left by the matrix n × n, $P_{k,\ell} = (c_{ij})$ where

$$c_{ij} = \begin{cases} 1 & \text{if } i = k \text{ and } j = \ell \text{ or } i = \ell \text{ and } j = k \\ 1 & \text{if } i = j \text{ and } i \neq k \text{ or } \ell, j \neq k \text{ or } \ell \\ 0 & \text{otherwise} \end{cases}$$

Multiplying the ith row of an n × m matrix A by a real number $r \neq 0$
corresponds to multiplying A on the left by the matrix $\text{diag}(d_1,\ldots,d_n)$
where $d_j = 1$ for $j \neq i$ and $d_i = r$, (see Example 1 of Section 2). For
example,

$$\begin{pmatrix} 1 & 0 \\ 0 & r \end{pmatrix} \begin{pmatrix} a & b \\ c & d \end{pmatrix} = \begin{pmatrix} a & b \\ rc & rd \end{pmatrix}$$

Finally, multiplying a matrix A on the left by the elementary
matrix $E_{ij}(r)$ corresponds to adding r times the jth row of A to the ith
row of A. For instance,

$$\begin{pmatrix} 1 & 0 & 0 \\ 0 & 1 & r \\ 0 & 0 & 1 \end{pmatrix} \begin{pmatrix} a & b \\ c & d \\ e & f \end{pmatrix} = \begin{pmatrix} a & b \\ c+re & d+rf \\ e & f \end{pmatrix}$$

In a similar way, the elementary column operations correspond to multiplying a matrix on the right by a permutation matrix, a diagonal matrix, or an elementary matrix. With this interpretation, we can restate Proposition 6.1 as follows:

PROPOSITION 6.2. *Let A be any n × m matrix. Then there are matrices P and Q, each of which is the product of permutation matrices, diagonal matrices, and elementary matrices such that PAQ has the form described in Proposition 6.1.*

In particular, according to the corollary to Proposition 6.1, if A is an invertible n × n matrix, we can find matrices P and Q, each of which is the product of permutation matrices, diagonal matrices, and elementary matrices such that

$$PAQ = I_n$$

It follows that

$$A = P^{-1}IQ^{-1} = P^{-1}Q^{-1}$$

so that

$$A^{-1} = QP$$

This gives a method of computing the inverse of a matrix. For example, if

$$A = \begin{pmatrix} 2 & 3 \\ 3 & 5 \end{pmatrix}$$

we multiply the first row by 1/2, obtaining

$$\begin{pmatrix} 1 & 3/2 \\ 3 & 5 \end{pmatrix}$$

then add -3 times the first row to the second row to get

$$\begin{pmatrix} 1 & 3/2 \\ 0 & 1/2 \end{pmatrix}$$

We then add - 3/2 the first column to the second column and multiply the second column by 2 to obtain I_2. In other words,

$$\begin{pmatrix} 1 & 0 \\ 0 & 1 \end{pmatrix} = \begin{pmatrix} 1 & 0 \\ -3 & 1 \end{pmatrix} \begin{pmatrix} 1/2 & 0 \\ 0 & 1 \end{pmatrix} \begin{pmatrix} 2 & 3 \\ 3 & 5 \end{pmatrix} \begin{pmatrix} 1 & -3/2 \\ 0 & 1 \end{pmatrix} \begin{pmatrix} 1 & 0 \\ 0 & 2 \end{pmatrix}$$

so that I_2 = PAQ where

$$P = \begin{pmatrix} 1 & 0 \\ -3 & 1 \end{pmatrix} \begin{pmatrix} 1/2 & 0 \\ 0 & 1 \end{pmatrix} = \begin{pmatrix} 1/2 & 0 \\ -3/2 & 1 \end{pmatrix}$$

$$Q = \begin{pmatrix} 1 & -3/2 \\ 0 & 1 \end{pmatrix} \begin{pmatrix} 1 & 0 \\ 0 & 2 \end{pmatrix} = \begin{pmatrix} 1 & -3 \\ 0 & 2 \end{pmatrix}$$

Thus

$$A^{-1} = QP = \begin{pmatrix} 1 & -3 \\ 0 & 2 \end{pmatrix} \begin{pmatrix} 1/2 & 0 \\ -3/2 & 1 \end{pmatrix}$$

$$= \begin{pmatrix} 5 & -3 \\ -3 & 2 \end{pmatrix}$$

EXERCISES

1.   Use elementary row and column operations to transform the matrices of Exercise 1 of Section 4 into the form described in Proposition 6.1.

2.   For each of the matrices in Exercise 1, find the matrices P and Q of Proposition 6.2.

3.   Using the method described above, find the inverse of the matrix

$$\begin{pmatrix} 0 & 3 & 1 \\ 2 & 1 & 0 \\ 3 & 5 & 1 \end{pmatrix}$$

4.   A square matrix $A = (a_{ij})$ is *upper triangular* if $a_{ij} = 0$ for $i > j$ and *lower triangular* if $a_{ij} = 0$ for $i < j$.

   (a)   Show that any $n \times n$ matrix can be transformed into an upper triangular matrix using elementary row operations.

   (b)   Show that any $n \times n$ matrix can be transformed into a lower triangular matrix using elementary column operations.

*5.   Suppose that A is invertible. Show that A can be transformed into the identity matrix by using only row operations (or only column operations).

7.   GAUSSIAN ELIMINATION

We saw in the previous section that row and column operations can be used to invert matrices. Actually, as we shall see, row operations alone will suffice, and this procedure provides an efficient method for solving simultaneous linear equations.

   We begin with an example: solving the equation

$$\begin{pmatrix} 0 & 2 & 6 \\ 1 & 2 & 1 \\ 3 & 5 & 1 \end{pmatrix} \begin{pmatrix} x_1 \\ x_2 \\ x_3 \end{pmatrix} = \begin{pmatrix} 1 \\ 2 \\ 3 \end{pmatrix}$$

In solving this equation, we usually omit the column vector $\vec{x}$, and simply write

$$\begin{pmatrix} 0 & 2 & 6 \\ 1 & 2 & 1 \\ 3 & 5 & 1 \end{pmatrix} \qquad \begin{pmatrix} 1 \\ 2 \\ 3 \end{pmatrix}$$

We now perform row operations on the matrix to turn it into the identity, while performing the *same* operations on the column vector. The first step is to make sure that there is a nonzero entry in the upper left-hand corner. In this case, we can put one there by interchanging the top two rows, getting

$$\begin{pmatrix} 1 & 2 & 1 \\ 0 & 2 & 6 \\ 3 & 5 & 1 \end{pmatrix} \qquad \begin{pmatrix} 2 \\ 1 \\ 3 \end{pmatrix}$$

Ordinarily, we would now divide to get this first entry equal to 1, but in this case there is no need. Now we make every term except the top one in the first column equal to 0. In our case, we can arrange this by subtracting three times the first row from the third:

$$\begin{pmatrix} 1 & 2 & 1 \\ 0 & 2 & 6 \\ 0 & -1 & -2 \end{pmatrix} \qquad \begin{pmatrix} 2 \\ 1 \\ -3 \end{pmatrix}$$

The next step is to insure that the second entry in the second column is not zero; in this case, it is 2, and there is no problem. Now we make it equal to 1, by dividing through by 2:

$$\begin{pmatrix} 1 & 2 & 1 \\ 0 & 1 & 3 \\ 0 & -1 & -2 \end{pmatrix} \qquad \begin{pmatrix} 2 \\ \frac{1}{2} \\ -3 \end{pmatrix}$$

We could now make all the other terms in the second column equal to 0. Instead, we temporarily worry only about the terms below the diagonal. Adding the second row to the third, we get

$$\begin{pmatrix} 1 & 2 & 1 \\ 0 & 1 & 3 \\ 0 & 0 & 1 \end{pmatrix} \qquad \begin{pmatrix} 2 \\ \dfrac{1}{2} \\ -\dfrac{5}{2} \end{pmatrix}$$

The lowest diagonal element is now 1; if it had not been, we would have divided to make it 1. Next, we get rid of the terms above the diagonal. Subtract twice the second row from the first:

$$\begin{pmatrix} 1 & 0 & -5 \\ 0 & 1 & 3 \\ 0 & 0 & 1 \end{pmatrix} \qquad \begin{pmatrix} 1 \\ \dfrac{1}{2} \\ -\dfrac{5}{2} \end{pmatrix}$$

Now subtract three times the third row from the second.

$$\begin{pmatrix} 1 & 0 & -5 \\ 0 & 1 & 0 \\ 0 & 0 & 1 \end{pmatrix} \qquad \begin{pmatrix} 1 \\ 8 \\ -\dfrac{5}{2} \end{pmatrix}$$

Finally, add five times the third row to the first:

$$\begin{pmatrix} 1 & 0 & 0 \\ 0 & 1 & 0 \\ 0 & 0 & 1 \end{pmatrix} \qquad \begin{pmatrix} -\dfrac{23}{2} \\ 8 \\ -\dfrac{5}{2} \end{pmatrix}$$

The matrix is now the identity - brought there from the original matrix solely by means of row operations. The vector is now the desired solution,

$$\begin{pmatrix} x_1 \\ x_2 \\ x_3 \end{pmatrix} = \begin{pmatrix} \dfrac{-23}{2} \\ 8 \\ -\dfrac{5}{2} \end{pmatrix}$$

(The skeptical reader should check this.)

Why does this work? The easiest way to see what we did is to write the equation as

$$\vec{Ax} = \vec{v}$$

where (in general) A is an n × n matrix. If B is any other n × n matrix, then

$$\vec{BAx} = \vec{Bv}$$

We saw in the last section that row operations correspond to multiplication on the left by n × n matrices. Performing successive row operations therefore, corresponds to multiplying on the left by a matrix B. We arranged the row operations so that BA = I. Therefore

$$\vec{x} = (BA)\vec{x} = \vec{Bv}$$

and we can evaluate $\vec{Bv}$ by performing the successive row operations on $\vec{v}$.

This way of looking at row operations has another advantage: often we can solve the equations without changing A to I. For instance, in the example we had

$$\begin{pmatrix} 1 & 2 & 1 \\ 0 & 1 & 3 \\ 0 & 0 & 1 \end{pmatrix} \quad \begin{pmatrix} 2 \\ \frac{1}{2} \\ -\frac{5}{2} \end{pmatrix}$$

at one point. This means that we have transformed the original equation, $\vec{Ax} = \vec{v}$, into a new equation,

$$B_1 \vec{Ax} = B_1 \vec{v}$$

where

$$B_1 A = \begin{pmatrix} 1 & 2 & 1 \\ 0 & 1 & 3 \\ 0 & 0 & 1 \end{pmatrix}$$

and

$$B_1 \vec{v} = \begin{pmatrix} 2 \\ \frac{1}{2} \\ -\frac{5}{2} \end{pmatrix}$$

Any solution of the original equation solves this new one; since B_1 is invertible (Why?), every solution of this new equation solves the original one. But the new equation is easy to solve: it reads

$$x_1 + 2x_2 + x_3 = 2$$
$$x_2 + 3x_3 = \frac{1}{2}$$
$$x_3 = -\frac{5}{2}$$

which means that $x_3 = -5/2$, $x_2 = 1/2 - 3x_3 = 8$, and $x_1 = 2 - 2x_2 - x_3 = -23/2$. Notice that because of the form of the matrix $B_1 A$, we can read off the values of x_3, x_2 and x_1 successively.

This procedure for solving equations is called *Gaussian elimination*. The directions, once again, are as follows: to solve $A\vec{x} = \vec{v}$, where $A = (a_{ij})$, interchange rows (if necessary) to make $a_{11} \neq 0$. Then divide the first row to make $a_{11} = 1$, and add multiples of the first row to the others so that the first column becomes

$$\begin{pmatrix} 1 \\ 0 \\ \vdots \\ 0 \end{pmatrix}$$

Next, arrange to make $a_{22} \neq 0$ (interchanging the second row with some row *other than the first*, if necessary). Divide the second row to make $a_{22} = 1$; then add multiples of the second row to the others so that the second column becomes

$$\begin{pmatrix} 0 \\ 1 \\ 0 \\ \vdots \\ 0 \end{pmatrix}$$

Keep going in this way until A has become I. At each stage, perform the same operations on $\vec{v}$, and when A becomes I, $\vec{v}$ will be the solution $\vec{x}$. If we begin with A and I (instead of A and $\vec{v}$), then, as we saw in the previous section, changing A to I changes I to A^{-1}.

The reader may have wondered why we dealt only with row operations in this section. The reason is that only row operations make sense for the vector $\vec{v}$. (Another way of saying the same thing is that $B\vec{v}$ makes sense if B is an $n \times n$ matrix, but $\vec{v}B$ does not.) Another question is: "What happens when A is not invertible?" The answer is that at some stage it will be impossible to make $a_{ii} \neq 0$ for some i. For example, if

$$A = \begin{pmatrix} 1 & 2 \\ 2 & 4 \end{pmatrix}$$

then we get

$$\begin{pmatrix} 1 & 2 \\ 0 & 0 \end{pmatrix}$$

(subtract twice the top row from the second), and now we cannot put a diagonal element in the a_{22} position without disturbing the first column. One can get information about solutions of $A\vec{x} = \vec{v}$ even in this case - for instance, one can tell whether any solutions exist - but we shall not pursue the matter here.

Guassian elimination is a very efficient and very mechanical procedure. These facts make it ideal for computers. (There are some minor differences in the procedure as it is actually performed by the computer; these are the result of trying to minimize certain problems, like round-off errors, which arise in part because of the nature of computers. For instance, most programs try to avoid dividing by small numbers.) It is easy to appreciate the usefulness of the computer for such problems, especially if one has a 60×60 matrix to invert. Such matrices - and, indeed, bigger ones - actually arise in economics.

EXERCISES

1. Use the method of Gaussian Elimination to solve the following systems of linear equations when possible.

(a) $2x_1 + x_2 + x_3 = 0$

$x_1 - x_2 + x_3 = -1$

$3x_1 + 2x_2 + 2x_3 = 1$

(b) $3x_1 + x_2 + 9x_3 = 10$

$3x_1 + 2x_2 + 10x_3 = 13$

$2x_1 + x_2 + 7x_3 = 9$

(c) $x_2 + x_4 = 1$

$x_1 - x_2 + x_3 + 2x_4 = 1$

$2x_1 - x_2 + 3x_4 = 0$

$-x_1 + x_2 + x_3 + x_4 = -1$

(d) $x_1 - x_2 + 2x_3 = 1$

$3x_1 - x_2 + 2x_3 = 3$

$2x_1 - x_2 + 2x_3 = 2$

(This is an example where the solution is not unique.)

(e) $x_1 + 2x_2 + 3x_3 = 3$

$2x_1 + x_2 + 3x_3 = 3$

$x_1 + x_2 + 2x_3 = 0$

(This is an example where no solution exists.)

(f) $x_1 + x_2 + x_3 + x_4 = 3$

$x_1 - x_2 - 2x_3 + 3x_4 = 4$

$2x_1 + x_2 + x_3 + 3x_4 = 7$

$x_1 + 2x_2 - x_3 - 2x_4 = -5$

(g) $x_1 - x_2 = 0$

   $x_2 + x_3 = 1$

   $2x_1 + 3x_2 + 5x_3 = 4$

2.   Let $\vec{b} = (b_1, b_2, b_3)$ be the ith basis vector $\vec{e}_i$ for R^3, $1 \le i \le 3$. Show that the solution (x_1, x_2, x_3) to the linear equations

   $a_{11}x_1 + a_{12}x_2 + a_{13}x_3 = b_1$

   $a_{21}x_1 + a_{22}x_2 + a_{23}x_3 = b_2$

   $a_{31}x_1 + a_{32}x_2 + a_{33}x_3 = b_3$

gives the ith column for the inverse matrix A^{-1} to the matrix $A = (a_{ij})$ (when the inverse exists).

3.   Use Exercise 2 to find the inverse to the matrix

$$A = \begin{pmatrix} 2 & 1 & 1 \\ 1 & -1 & 1 \\ 3 & 2 & 2 \end{pmatrix}$$

(Hint: Find all three columns to A^{-1} at the same time.)

CHAPTER 8

MAXIMA AND MINIMA

INTRODUCTION

In many applications of mathematics, one wishes to maximize (or minimize) some function. In economics, for example, it is usual to try to maximize profit or minimize cost. Finding the extreme values of a function of one variable is one of the applications of elementary calculus. In this chapter, we develop criteria for finding the points at which a function of several variables takes on its maximum or minimum.

Just as in the case of functions of one variable, two cases need to be considered. To illustrate, let $f:[-1,1] \to R$ be defined by $f(x) = x^2$. The minimum value of f is at $x = 0$, where $f'(x) = 2x = 0$. On the other hand, f has a maximum at $x = 1$ and $x = -1$, but $f'(x) \neq 0$ at these points. Thus we can use the vanishing of the derivative to find maxima and minima on the "interior" of $[-1,1]$, but not at the "boundary" points $\{-1,1\}$ of $[-1,1]$. In this case, of course, there are only two boundary points, while for a function of several variables there are infinitely many. As a result, we need to develop a special procedure for dealing with maxima and minima at boundary points. This procedure, called the method of *Lagrange multipliers*, is one of the major topics of this chapter. The other is the analysis of maxima and minima at interior points; as we shall see, this analysis for finite dimensional inner product spaces is similar to the case of R^1.

315

1.   MAXIMA AND MINIMA AT INTERIOR POINTS

Let S be a subset of the finite dimensional inner product space V,
and let f:S → R be a differentiable function. If S is closed and
bounded, we know from Theorems 5.5.3 and 5.6.1 that S has a maximum
and a minimum somewhere in S. These theorems however, give us no
information about where the maximum or minimum is. In this section,
we discuss a method of locating the maximum and minimum values.

We shall solve only a part of the problem here, and we need some
definitions to describe what we shall do. We say that a point $\vec{v} \in S$
is an *interior* point of S if some open ball of positive radius containing $\vec{v}$ is contained in S. A point $\vec{v} \in S$ is a *boundary* point if
each open ball containing $\vec{v}$ meets both S and the complement of S.
(See Exercises 12-21 of Section 3.2.) Here are some examples:
1.   Let S be the closed unit disc in R^2:$S = \{\vec{v} = (x_1,x_2):x_1^2 + x_2^2 \le 1\}$.
(See Figure 1.1.) A point $\vec{v}_0 \in S$ with $\|\vec{v}_0\| < 1$ is an interior point,

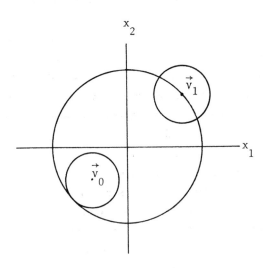

FIGURE 1.1. An interior point and a boundary
point of a disc in R^2

since the open ball of radius $1 - \|\vec{v}_0\|$ about $\vec{v}_0$ is in S. On the other hand, a point $\vec{v}_1 \in$ S with $\|\vec{v}_1\| = 1$ is a boundary point; any ball about $\vec{v}_1$ contains points $\vec{w}$ with $\|\vec{w}\| > 1$, and therefore no ball about $\vec{v}_1$ is contained in S.

2. Let S be the closed square in R^2 shown in Figure 1.2:

$$S = \{\vec{v} = (x_1, x_2) : |x_1| \leq 1 \text{ and } |x_2| \leq 1\}$$

Then any point $\vec{v}_0 = (x_1, x_2)$ with $|x_1| < 1$ and $|x_2| < 1$ is in the interior of S, since if $|x_1| \leq 1 - \varepsilon$ and $|x_2| \leq 1 - \varepsilon$, then $B_\varepsilon(\vec{v}_0) \subset$ S.

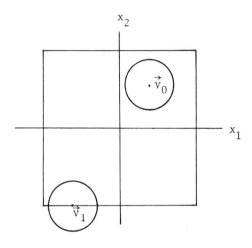

FIGURE 1.2. An interior point and a boundary
point of a rectangle in R^2

(See Exercise 2.) On the other hand, if $\vec{v}_1 = (x_1, x_2)$ with, say, $x_2 = -1$, then every ball about $\vec{v}_1$ contains vectors $\vec{w} = (y_1, y_2)$ with $|y_2| > 1$. Therefore, any such point $\vec{v}_1$ is a boundary point.

3. Let S be the closed unit ball in R^n: S = $\{\vec{v} \in R^n : \|\vec{v}\| \leq 1\}$. Then, just as in Example 1, the interior points are the points $\vec{v}$ with $\|\vec{v}\| < 1$ and the boundary points are those with $\|\vec{v}\| = 1$. (See Exercise 3.)

4.   If S is open, then every point $\vec{v}$ of S is in an open ball contained in S; hence, every point of S is an interior point.

5.   If S = $\{\vec{v} \in R^2 : \vec{v} = (x,x)\}$, then no point of S is in an open ball
contained in S (see Figure 1.3), and so every point of S is a boundary
point.

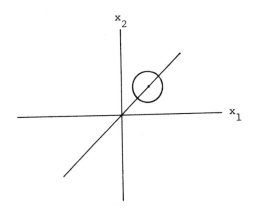

FIGURE 1.3.   A line in R^2 has no interior points

We shall be concerned here with maxima and minima at interior
points; we study the corresponding question for boundary points in
Sections 4, 5, and 6.

As in the case of functions of one variable, we say that an in-
terior point $\vec{v}_0 \in S$ is a *local maximum for* f if $f(\vec{v}) \leq f(\vec{v}_0)$ for all
$\vec{v}$ in some ball about $\vec{v}_0$; similarly, $\vec{v}_1$ is a *local minimum for* f if
$f(\vec{v}) \geq f(\vec{v}_1)$ for all $\vec{v}$ in some ball about $\vec{v}_1$. (See Figure 1.4.) We
call $\vec{v}_0$ an *extremum for* f if it is either a local maximum or a local
minimum.

Just as in the case of functions of one variable, we say that
the interior point $\vec{v}_0$ is a *critical point* of f if $f'(\vec{v}_0) = D_{\vec{v}_0} f = 0$
(the zero transformation). One reason for this definition is the
following theorem.

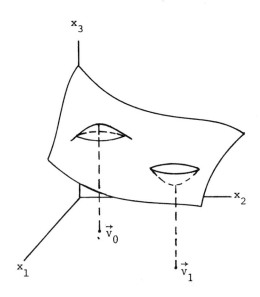

FIGURE 1.4.   The graph of a function with a
local maximum at $\vec{v}_0$ and a local
minimum at $\vec{v}_1$

THEOREM 1.1. *If $\vec{v}_0$ is an interior local maximum or local minimum point of $f:S \to R$, and if f is differentiable near $\vec{v}_0$, then $\vec{v}_0$ is
a critical point of f.*

Proof. We can, by choosing an orthonormal basis for V, identify
V with R^n. Let $\vec{v}_0 = (a_1,\ldots,a_n)$ be an interior point of S which is
a local maximum for f, and set $\vec{v} = (x_1,a_2,\ldots,a_n)$. Then $f(\vec{v}_0) \geq f(v)$
for all x_1 near a_1; that is, f, regarded as a function of its first
variable only, has a local maximum at $\vec{v}_0$. A standard calculus result
states that

$$\frac{\partial f}{\partial x_1}(\vec{v}_0) = 0$$

(See Exercise 7, Section 3.5.) The same reasoning, of course, applies
to the other components; thus

$$\frac{\partial f}{\partial x_1}(\vec{v}_0) = \cdots = \frac{\partial f}{\partial x_n}(\vec{v}_0) = 0$$

and it follows from Proposition 3.7.1 that $f'(\vec{v}_0) = D_{\vec{v}_0} f = 0$. Therefore $\vec{v}_0$ is a critical point of f. The same reasoning works when $\vec{v}_0$ is a local minimum.

For some purposes, Theorem 1.1 suffices to find the maximum and minimum of a function f. We often know in practice that S is a compact set, that $f(\vec{v}) = 0$ if $\vec{v}$ is a boundary point for S, and that $f > 0$ somewhere in S. (If, for instance, S is determined by physical constraints, these conditions are often met. See Exercises 6 through 10 for other examples.) Now we know that the maximum value of f is attained at an interior point of S. As the maximum is a local maximum, we know that it is attained at a critical point of S. We need only evaluate f at every critical point and see where it is biggest.

We do not always have such information at hand, however, and it is useful to have a method of distinguishing among maxima, minima, and critical points which are neither. Undoubtedly the reader has seen examples of critical points which are not extrema. For instance, the function $f(x) = x^3$ has a critical point at $x = 0$ but neither a maximum or a minimum. With more variables, more complicated phenomena can occur. For example, the function $f(x,y) = x^2 - y^2$ has a critical point at $(0,0)$ which is neither a maximum or minimum, since f has a maximum along the y-axis and a minimum along the x-axis. (This surface is called a saddle, for obvious reasons; see Figure 1.5.) In the next sections, we shall develop methods for describing the behavior of a function near a critical point.

EXERCISES

1. Determine the set of interior points and the set of boundary points of the following sets.

(a) $[a,b) \subset R^1$

(b) The set of rational numbers in R^1

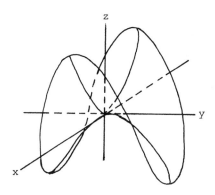

FIGURE 1.5.   A saddle surface

(c) $\{(x,y) \in R^2 : xy = 1\}$

(d) $\{(x,y) \in R^2 : xy < 1\}$

(e) $\{(x,y) \in R^2 : x > 0 \text{ and } y = \sin \frac{1}{x}\}$

(f) $\{(x,y,z) \in R^3 : x + y + z = 0\}$

(g) $\{(x,y,z) \in R^3 : x + y + z < 0\}$

(h) $\{(x,y,z) \in R^3 : xy \leq 1\}$

2.   Let S be the cube in R^n defined by

$$S = \{(x_1,\ldots,x_n) : |x_i| \leq 1, \ 1 \leq i \leq n\}$$

Prove that the interior of S is the set

$$\text{Int } S = \{(x_1,\ldots,x_n) : |x_i| < 1, \ 1 \leq i \leq n\}$$

and that all other points of S are boundary points.   (See Example 2 of this section.)

3.   Prove that the interior of the closed ball $\bar{B}_r(\vec{v}_0)$ in R^n is the open ball $B_r(\vec{v}_0)$ and that all other points of $\bar{B}_r(\vec{v}_0)$ are boundary points.

4.   Find all critical points of the following functions.

(a) $f:R \rightarrow R, \ f(x) = 2x^3 - 9x^2 + 12x$

(b) $f:R \rightarrow R, \ f(x) = e^{2x}$

(c) $f:R \rightarrow R, \ f(x) = \sin x$

(d) $f:R^2 \rightarrow R, \ f(x,y) = 2x + 3y - 6$

(e) $f:R^2 \rightarrow R, \ f(x,y) = 3 - x^2 - y^2$

(f) $f:R^2 \rightarrow R, \ f(x,y) = xy$

(g) $f:R^2 \rightarrow R, \ f(x,y) = e^{x(y-1)}$

(h) $f:R^2 \rightarrow R, \ f(x,y) = 2x^3y - 10x^2y + 12xy$

(i) $f:R^2 \rightarrow R, \ f(x,y) = \cos(x + y)$

5.   For each of the following functions, $(0,0)$ is a critical point.
Determine whether it is a local maximum, a local minimum, or neither.

(a) $f(x,y) = x^4 + y^4$

(b) $f(x,y) = x^4 - y^4$

(c) $f(x,y) = x^4 + y^3$

(d) $f(x,y) = -x^4 - y^4$

6.   Find the shortest distance from the point $(0,1)$ to the graph of
the parabola $y = x^2$.

7.   Find the shortest distance from the point $(0,5,1)$ to the graph
of the circular paraboloid $z = 2x^2 + 2y^2$.

8.   Find the point on a given line such that the sum of the distances from two given fixed points is a minimum (See Figure 1.6.) The
solution to this exercise is related to the optical law of reflection.
According to *Fermat's principle*, a light ray travels along a path that
minimizes time.

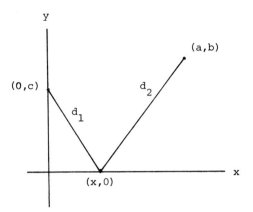

FIGURE 1.6. Minimize $d_1 + d_2$

*9. (The method of least squares.) (a) Suppose that $(x_1,y_1),\ldots,$
(x_n,y_n) are n points in the plane. We wish to find the line L, given
by the equation $y = ax + b$ which minimizes the sum of the squares of
the distances between the points and the line, measured along the
y-axis. That is, we want to minimize

$$F(a,b) = \sum_{j=1}^{n} (ax_j + b - y_j)^2$$

Show that the minimum of F is attained at

$$a = (\overline{xy} - \frac{1}{n} \sum_{j=1}^{n} x_j y_j)(\overline{x}^2 - \frac{1}{n} \sum_{j=1}^{n} x_j^2)^{-1}$$

$$b = \overline{y} - a\overline{x}$$

where $\overline{x}$, $\overline{y}$ are the *means* of the x_j, y_j respectively:

$$\overline{x} = \frac{1}{n} \sum_{j=1}^{n} x_j$$

$$\overline{y} = \frac{1}{n} \sum_{j=1}^{n} y_j$$

The line $y = ax + b$, with a and b chosen as above, is generally regarded in statistics as the line which comes closest to fitting the
data points $(x_1,y_1),\ldots,(x_n,y_n)$.

(b) Consider the corresponding problem in R^3, with points

$$(x_1,y_1,z_1),\ldots,(x_n,y_n,z_n)$$

and the plane $P:z = ax + by + c$, with

$$F(a,b,c) = \sum_{j=1}^{n} (z_j - ax_j - by_j - c_j)^2$$

*10. (a) Let $S \subset R^n$ be defined by

$$S = \{(x_1,\ldots,x_n):x_i > 0, \ 1 \le i \le n\}$$

and define $f:S \to R$ by

$$f(x_1,\ldots,x_n) = \frac{1}{n}(x_1 +\cdots+ x_n)(x_1x_2\ldots x_n)^{-1/n}$$

Find the critical points of f.

(b) Assuming that the critical point of f is a minimum, prove
that

$$(x_1x_2\ldots x_n)^{1/n} \le \frac{x_1 +\cdots+ x_n}{n}$$

11. Find the minimum distance from $(6,3,4)$ to the paraboloid
$x_3 = x_1^2 + x_2^2$.

2. QUADRATIC FORMS

Because this section may seem pointless without some explanation of
purpose, we begin with some motivational material which will not be
used in the rest of the section.

Suppose that $f:R \to R$ is a differentiable function and that a is
a critical point of f. Assuming that f has a continuous second derivative, we can expand f near a by Taylor's Theorem (Theorem 6.4.1):

$$f(x) = f(a) + f(x - a)f'(a) + \frac{1}{2}(x - a)^2 f''(c) \qquad (2.1)$$

where c is between a and x. Since a is a critical point, $f'(a) = 0$; thus we may rewrite (2.1) as

$$f(x) - f(a) = (x - a)^2 f''(c) \qquad (2.2)$$

Now suppose that $f''(a) > 0$. Then $f''(c) > 0$ if c is sufficiently near a (since f'' is continuous), and so $f(x) - f(a) > 0$ if $x \neq a$ is sufficiently near a (by (2.2)). Thus a is a local minimum if $f''(a) > 0$. Similarly, a is a local maximum if $f''(a) < 0$. This result is usually referred to as the "second derivative test" in elementary calculus. (If $f''(a) = 0$, then the test fails; the point a can be a local maximum, a local minimum, or neither. See Exercise 10.)

We wish to use a similar procedure for critical points in R^n. Now, however, we need to use Taylor's theorem for functions of several variables (Theorem 6.6.1), and the expression for the second derivative is more complicated. To analyze it, we first discuss quadratic forms.

A *quadratic form* in n variables $x_1, \ldots, x_n$ is an expression of the form

$$Q(x_1, \ldots, x_n) = \sum_{i=1}^{n} \sum_{j=1}^{n} a_{ij} x_i x_j$$

For instance, if $n = 2$, then

$$Q(x_1, x_2) = a_{11} x_1^2 + (a_{12} + a_{21}) x_1 x_2 + a_{22} x_2^2$$

If $i \neq j$, there are two terms involving $x_i x_j$, $a_{ij} x_i x_j$ and $a_{ji} x_i x_j$, *and we always assume that* $a_{ij} = a_{ji}$. (This assumption does not restrict the form Q at all, of course. It simply makes expressions more symmetric.)

The quadratic form Q is called *positive definite* if $Q(x_1, \ldots, x_n) > 0$ unless $x_1 = \cdots = x_n = 0$, *negative definite* if $Q(x_1, \ldots, x_n) < 0$ unless $x_1 = \cdots = x_n = 0$, and *indefinite* if $Q(x_1, \ldots, x_n)$ takes on positive and negative values.

EXAMPLES

1. The quadratic form $Q_1(x_1,x_2) = x_1^2 + x_2^2$ is positive definite.

2. The quadratic form $Q_2(x_1,x_2) = -x_1^2 - 2x_2^2$ is negative definite.

3. The quadratic form $Q_3(x_1,x_2) = x_1^2 - 3x_2^2$ is indefinite.

4. An $n \times n$ matrix $A = (a_{ij})$ is *symmetric* if $a_{ij} = a_{ji}$ for all i,j. If A is any symmetric $n \times n$ matrix, we denote by Q_A the quadratic form

$$Q_A(x_1,\ldots,x_n) = \sum_{i=1}^{n} \sum_{j=1}^{n} a_{ij}x_i x_j$$

In this context, we say that the matrix A is positive definite when Q_A is positive definite, and similarly for the other kinds of definiteness. For instance, if

$$A = \begin{pmatrix} -1 & 0 \\ 0 & -2 \end{pmatrix}$$

then Q_A is the same as in Example 2 above, and we say that A is negative definite.

Note that if $T:R^n \to R^n$ is the linear transformation defined by A and $\vec{v} = (x_1,\ldots,x_n)$, then

$$Q_A(\vec{v}) = \, < T(\vec{v}),\vec{v} > \tag{2.3}$$

We should also remark that there are quadratic forms which are not positive definite, negative definite, or indefinite. For instance, the quadratic form

$$Q_4(x_2,x_2) = x_1^2 + 2x_1x_2 + x_2^2 = (x_1 + x_2)^2$$

is always non negative but it is not positive definite, since $Q_4(1,-1) = 0$. We say that the quadratic form Q is *positive semidefinite* if $Q(\vec{v}) \geq 0$ for all $\vec{v}$; thus Q_4 is positive semidefinite, but not positive definite. Similarly, Q is called *negative semidefinite* if $Q(\vec{v}) \leq 0$ for all $\vec{v}$. We shall not be concerned with such quadratic forms in the rest of this section.

Deciding whether a symmetric matrix A is positive definite or not can be difficult. In the case of 2×2 matrices, however, there is a fairly simple criterion.

THEOREM 2.1. *The symmetric 2×2 matrix $A = (a_{ij})$ is*

(a) *positive definite if $a_{11} > 0$ and $a_{11}a_{22} - a_{12}^2 > 0$*

(b) *negative definite if $a_{11} < 0$ and $a_{11}a_{22} - a_{12}^2 > 0$*

(c) *indefinite if $a_{11}a_{22} - a_{12}^2 < 0$*

Proof. We begin by writing out Q_A, assuming that $a_{11} \neq 0$:

$$Q_A(x_1, x_2) = a_{11}x_1^2 + 2a_{12}x_1x_2 + a_{22}x_2^2$$

$$= a_{11}(x_1^2 + \frac{2a_{12}}{a_{11}}x_1x_2) + a_{22}x_2^2$$

$$= a_{11}(x_1^2 + \frac{2a_{12}}{a_{11}}x_1x_2 + (\frac{a_{12}}{a_{11}})^2x_2^2) + (a_{22} - \frac{a_{12}^2}{a_{11}})x_2^2$$

$$= a_{11}[(x_1 + \frac{a_{12}}{a_{11}}x_2)^2 + \frac{1}{a_{11}^2}(a_{11}a_{22} - a_{12}^2)x_2^2] \qquad (2.4)$$

Suppose now that $a_{11} > 0$ and $a_{11}a_{22} - a_{12}^2 > 0$. Then (2.4) is a sum of positive multiples of squares, and therefore $Q_A(x_1, x_2) \geq 0$ for all (x_1, x_2). If $Q_A(x_1, x_2) = 0$, then both terms of (2.4) must be 0; in particular, $(1/a_{11})(a_{11}a_{22} - a_{12}^2)x_2^2 = 0$. It follows that $x_2 = 0$. Going back to (2.4), we see that $a_{11}x_1^2 = 0$ so that $x_1 = 0$. It follows that $Q_A(x_1, x_2) > 0$ if either x_1 or x_2 is non-zero, and (a) is proved. Part (b) is proved in the same way.

For (c), there are a number of cases. If, for instance, $a_{11} > 0$, then (2.2) shows that $Q_A(1, 0) > 0$ and $Q_A(-a_{12}/a_{11}, 1) < 0$. The case where $a_{11} < 0$ is similar. If $a_{11} = 0$, then we cannot use (2.2), and we must go back to the original expression for Q_A:

$$Q_A(x_1, x_2) = 2a_{12}x_1x_2 + a_{22}x_2^2$$

$$= x_2(2a_{12}x_1 + a_{22}x_2)$$

In this expression, $a_{12} \neq 0$, since our assumption is that $- a_{12}^2 < 0$.
It is easy to check (Exercise 12) that we can choose x_1 and x_2 to make
$Q_A(x_1, x_2) > 0$ or < 0. Thus Q_A is indefinite in this case, too, and
the theorem is proved.

The quantity $a_{11}a_{22} - a_{12}^2$ is called the *determinant* of the symmetric 2×2 matrix A and denoted Det A. We shall define the determinant of an arbitrary $n \times n$ matrix in Chapter 10; for a general 2×2
matrix $A = (a_{ij})$, the formula is

$$\text{Det } A = a_{11}a_{22} - a_{12}a_{21}$$

(This reduces to $a_{11}a_{22} - a_{12}^2$ if $a_{12} = a_{21}$.) For a 3×3 matrix A,

$$\text{Det } A = a_{11}(a_{22}a_{33} - a_{23}a_{32}) - a_{12}(a_{21}a_{33} - a_{23}a_{31})$$
$$+ a_{13}(a_{21}a_{32} - a_{22}a_{31})$$

The extension of Theorem 2.1 to $n \times n$ matrices involves determinants
and is given in Section 8 of Chapter 10. We state it here for 3×3
matrices, leaving the proof as Exercise 13.

THEOREM 2.2. *Let $A = (a_{ij})$ be a symmetric 3×3 matrix and define matrices A_1, A_2 and A_3 by*

$$A_1 = (a_{11}), \quad A_2 = \begin{pmatrix} a_{11} & a_{12} \\ a_{21} & a_{22} \end{pmatrix}, \quad A_3 = A$$

*(a) If Det $A_1 = a_{11} > 0$, Det $A_2 > 0$, and Det $A_3 > 0$, then A is
positive definite.*

*(b) If Det $A_1 = a_{11} < 0$, Det $A_2 > 0$, and Det $A_3 < 0$, then A is
negative definite.*

*(c) If Det $A_j \neq 0$ for $j = 1,2,3$ and neither (a) nor (b) holds,
then A is indefinite.*

For example, consider the 3×3 matrix

$$A = \begin{pmatrix} 3 & 2 & 5 \\ 2 & 3 & 4 \\ 5 & 4 & 9 \end{pmatrix}$$

Then $A_1 = (3)$ and $A_2 = \begin{pmatrix} 3 & 2 \\ 2 & 3 \end{pmatrix}$, while $A_3 = A$. We have

Det $A_1 = 3$, Det $A_2 = 5$, Det $A_3 = 2$

Since all determinants are positive, A is positive definite. In fact, some algebra shows that

$$Q_A(x_1, x_2, x_3) = 2(x_1 + x_2 + 2x_3) + (x_1 + x_3)^2 + x_2^2$$

In this form, it is easy to check that Q is positive definite.

When $n = 1$, $A = (a)$ is given by a number, and the quadratic form is $Q_A(x) = ax^2$. It is positive definite when $a > 0$, negative definite when $a < 0$. When $a = 0$, the form is only semi-definite, and Theorem 2.2 does not apply.

We shall need another result about quadratic forms in the next section; it states, for instance, that if A and B are symmetric matrices, A is positive definite, and B is "close" to A, then B is positive definite. We shall merely state the result here, postponing the proof until Section 7. First, however, we need to define what "close" means for matrices. This is not difficult. We have defined an inner product on the space $\mathcal{L}(R^n, R^n)$ (in Section 7.5) and we may regard the n × n matrices as linear transformations using the standard basis. If $A = (a_{ij})$, then the formula for $\|A\|$ is

$$\|A\| = \left(\sum_{i=1}^{n} \sum_{j=1}^{n} a_{ij}^2 \right)^{\frac{1}{2}}$$

Now we can state the result.

PROPOSITION 2.3. *Let* A *be a symmetric* n × n *matrix.*

(a) If A *is positive definite, then there is an* $\varepsilon > 0$ *such that any symmetric* n × n *matrix* B *with* $\|A - B\| < \varepsilon$ *is positive definite.*

(b) If A is negative definite, then there is an $\varepsilon > 0$ such that any symmetric n × n matrix B with $\|A - B\| < \varepsilon$ is negative definite.

(c) If A is indefinite and $\vec{v}_1, \vec{v}_2$ are unit vectors with $Q_A(\vec{v}_1)$ > 0 and $Q_A(\vec{v}_2) < 0$, then there is an $\varepsilon > 0$ such that if B is any symmetric n × n matrix with $\|A - B\| < \varepsilon$, then $Q_B(\vec{v}_1) > 0$ and $Q_B(\vec{v}_2) < 0$. In particular, B is indefinite.

EXERCISES

1.   Find the quadratic forms associated to these matrices:

(a) $\begin{pmatrix} 1 & 2 \\ 2 & 1 \end{pmatrix}$ (d) $\begin{pmatrix} 3 & -1 \\ -1 & -2 \end{pmatrix}$

(b) $\begin{pmatrix} 3 & 2 \\ 2 & 4 \end{pmatrix}$ (e) $\begin{pmatrix} -2 & 1 \\ 1 & -3 \end{pmatrix}$

(c) $\begin{pmatrix} 3 & 2 \\ 2 & 7 \end{pmatrix}$ (f) $\begin{pmatrix} 4 & 3 \\ 3 & 2 \end{pmatrix}$

2.   Which of the above forms are positive definite, which are negative definite, and which indefinite?

3.   Find the quadratic forms associated to these matrices:

(a) $\begin{pmatrix} 1 & 0 & 0 \\ 0 & 1 & 0 \\ 0 & 0 & 1 \end{pmatrix}$ (d) $\begin{pmatrix} 3 & 1 & 2 \\ 1 & 5 & 3 \\ 2 & 3 & 7 \end{pmatrix}$

(b) $\begin{pmatrix} -4 & 0 & 1 \\ 0 & -3 & 2 \\ 1 & 2 & -5 \end{pmatrix}$ (e) $\begin{pmatrix} -3 & 1 & 0 \\ 1 & -6 & 1 \\ 0 & 1 & 7 \end{pmatrix}$

(c) $\begin{pmatrix} 4 & 2 & 0 \\ 2 & 5 & 3 \\ 0 & 3 & 5 \end{pmatrix}$ (f) $\begin{pmatrix} 1 & -1 & 0 \\ -1 & 2 & 0 \\ 0 & 0 & 3 \end{pmatrix}$

4. Determine which of the matrices in Exercise 3 are positive definite, which are negative definite, and which are indefinite.

5. Find the matrix A corresponding to each of the quadratic forms Q_A given below:

    (a) $Q_A(x_1,x_2) = x_1^2 + 6x_1x_2 - 3x_2^2$

    (b) $Q_A(x_1,x_2) = 4x_1x_2 + x_2^2$

    (c) $Q_A(x_1,x_2) = x_1^2 + x_1x_2 + x_2^2$

    (d) $Q_A(x_1,x_2) = 4x_1x_2 - 7x_1^2 - 3x_2^2$

    (e) $Q_A(x_1,x_2) = 6(x_1^2 + x_2^2) - 5x_1x_2$

    (f) $Q_A(x_1,x_2) = x_1x_2$

6. Determine which of the above forms are positive definite, which are negative definite, and which are indefinite.

7. Find the matrix A corresponding to each of the following quadratic forms:

    (a) $Q_A(x_1,x_2,x_3) = x_1^2 - 2x_2^2 + 4x_3^2$

    (b) $Q_A(x_1,x_2,x_3) = 3x_1^2 - 2x_1x_2 + x_2^2 + 4x_2x_3 + 8x_3^2$

    (c) $Q_A(x_1,x_2,x_3) = 2x_1x_2 + 4x_1x_3 + 6x_2x_3$

    (d) $Q_A(x_1,x_2,x_3) = x_1x_2 + x_2x_3 - 4(x_1^2 + x_2^2 + x_3^2)$

    (e) $Q_A(x_1,x_2,x_3) = x_1^2 + x_2^2 - x_3^2 - 6x_1x_2 + x_2x_3$

8. Determine which of the above forms are positive definite, which are negative definite, and which are indefinite.

9. Give examples of symmetric 3 × 3 matrices which are

    (a) Positive definite

    (b) Positive semidefinite but not positive definite

    (c) Indefinite

    (d) Negative definite

    (e) Negative semidefinite but not negative definite

10. Construct an example of a function $f:R \to R$ with $f'(0) = f''(0) = 0$ such that 0 is

    (a) a local maximum

    (b) a local minimum

    (c) neither

11. Let Q be a quadratic form on R^n. Prove that $Q(\lambda \vec{v}) = \lambda^2 Q(\vec{v})$ for any $\vec{v} \in R^n$, $\lambda \in R$.

12. Complete the proof of Theorem 2.1. (Hint: choose $x_2 > 0$; then let x_1 be alternately > 0 and < 0, with $|x_1|$ large.)

*13. Prove Theorem 2.2.

3. CRITERIA FOR LOCAL MAXIMA AND MINIMA

Since any finite dimensional inner product space is isometrically isomorphic to a Euclidean space, we will deal only with functions defined on subsets of Euclidean space for the remainder of this chapter. The extension to arbitrary finite dimensional inner product spaces is straightforward.

Let S be a subset of R^n, let $f:S \to R$ be a function with a continuous second derivative, and let $\vec{v}_0$ be an interior point of S which is a critical point for f. We would like to have a method of determining whether $\vec{v}_0$ is a local maximum, a local minimum, or neither. Just as in the case of functions on R^1, we look at the second derivative for a criterion.

Recall (from Section 6.3) that the Hessian of f at a point $\vec{v}$, $H(f)(\vec{v})$, is a matrix formed from the second partial derivatives of f. If the partial derivatives are continuous, as we are assuming, then $H(f)(\vec{v})$ is a symmetric matrix. (See Proposition 6.3.3.)

THEOREM 3.1. *Let U be an open set in R^n and $f:U \to R$ a function having continuous second order partial derivatives. Let $\vec{v}_0$ be a critical point of f and let $H(f)(\vec{v}_0)$ be the Hessian of f at $\vec{v}_0$.*

(a) If $H(f)(\vec{v}_0)$ is positive definite, then $\vec{v}_0$ is a local minimum.

(b) If $H(f)(\vec{v}_0)$ is negative definite, then $\vec{v}_0$ is a local maximum.

(c) If $H(f)(\vec{v}_0)$ is indefinite, then $\vec{v}_0$ is neither a local maximum nor a local minimum.

Proof. The proof of Theorem 3.1 is like that for functions on R, sketched at the beginning of Section 2. At a few places, however, we need Proposition 2.3.

We begin with Taylor's Theorem (Theorem 6.6.1), with m = 1. We expand f about the critical point $\vec{v}_0$. Since $\frac{\partial f}{\partial x_i}(\vec{v}_0) = 0$, $1 \leq i \leq n$, we find that

$$f(\vec{v}) - f(\vec{v}_0) = \frac{1}{2} \sum_{i=1}^{n} \sum_{j=1}^{n} (x_i - a_i)(x_j - a_j) \frac{\partial^2 f}{\partial x_i \partial x_j}(\vec{v}_1) \qquad (3.1)$$

where $\vec{v}_1$ is on the line segment from $\vec{v}$ to $\vec{v}_0$. The right side of (3.1) is, except for the factor 1/2, just the quadratic form associated with the Hessian at $\vec{v}_1$ (evaluated at $\vec{v} - \vec{v}_0$). Let us call this quadratic form $Q(\vec{v}_1)$; then we may rewrite (3.1) as

$$f(\vec{v}) - f(\vec{v}_0) = \frac{1}{2}Q(\vec{v}_1)(\vec{v} - \vec{v}_0) \qquad (3.2)$$

(a) Suppose $H(f)(\vec{v}_0)$ is positive definite. Then, by Proposition 2.3, there is an $\varepsilon > 0$ such that if $\|A - H(f)(\vec{v}_0)\| < \varepsilon$, then A is positive definite. Since the second derivative of f is continuous, we may pick $\delta > 0$ such that if $\|\vec{v} - \vec{v}_0\| < \delta$, then $\vec{v} \in U$ and $\|H(f)(\vec{v}) - H(f)(\vec{v}_0)\| < \varepsilon$. Since $\vec{v}_1$ is on the line segment from $\vec{v}$ to $\vec{v}_0$, we have

$$\|\vec{v} - \vec{v}_0\| < \delta \Rightarrow \|\vec{v}_1 - \vec{v}_0\| < \delta$$

Therefore $H(f)(\vec{v}_1)$ is positive definite, and $Q(\vec{v}_1)(\vec{v} - \vec{v}_0) > 0$ when $0 < \|\vec{v} - \vec{v}_0\| < \delta$. Now (3.2) tells us that $f(\vec{v}) - f(\vec{v}_0) > 0$ when $0 < \|\vec{v} - \vec{v}_0\| < \delta$, so that $\vec{v}_0$ is a local minimum for f.

(b) This is essentially the same as (a), with appropriate sign changes.

(c) We know that $Q(\vec{v}_0)$ is indefinite; therefore we can find vec-
tors $\vec{w}_1$ and $\vec{w}_2$ with $Q(\vec{v}_0)(\vec{w}_1) > 0$ and $Q(\vec{v}_0)(\vec{w}_2) < 0$. We shall show,
essentially, that f increases in the direction of $\vec{w}_1$ and decreases in
the direction of $\vec{w}_2$. Thus $\vec{v}_0$ will not be an extremum.

Using Proposition 2.3, we may choose $\varepsilon > 0$ such that if A is any
symmetric matrix with

$$\|A - H(f)(\vec{v}_0)\| < \varepsilon$$

then $Q_A(\vec{w}_1) > 0$ and $Q_A(\vec{w}_2) < 0$. Choose $\delta > 0$ so that if $\|\vec{v} - \vec{v}_0\| < \delta$,
then $\vec{v} \in U$ and

$$\|H(f)(\vec{v}) - H(f)(\vec{v}_0)\| < \varepsilon$$

We now let $\vec{u}_1$ be a vector in the direction $\vec{w}_1$ from $\vec{v}_0$, but within δ of $\vec{v}_0$: that is, we let

$$\vec{u}_1 = \vec{v}_0 + \lambda\vec{w}_1$$

where $\lambda > 0$ and

$$\|\vec{u}_1 - \vec{v}_0\| = \|\lambda\vec{w}_1\| < \delta$$

(See Figure 3.1.) Then, according to formula (3.2),

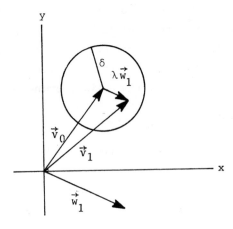

FIGURE 3.1.

$$f(\vec{u}_1) - f(\vec{v}_0) = \frac{1}{2}Q(\vec{v}_1)(\vec{u}_1 - \vec{v}_0)$$

where $\vec{v}_1$ is on the line segment from $\vec{v}_0$ to $\vec{u}_1$. Thus,

$$f(\vec{u}_1) - f(\vec{v}_0) = \frac{1}{2}Q(\vec{v}_1)(\lambda\vec{w}_1)$$

$$= \frac{\lambda^2}{2}Q(\vec{v}_1)(\vec{w}_1) > 0$$

since

$$\|\vec{v}_1 - \vec{v}_0\| < \|\vec{u}_1 - \vec{v}_0\| < \delta$$

Hence $\vec{v}_0$ is not a local maximum.

On the other hand, if $\mu > 0$ is chosen such that the vector $\vec{u}_2 = \vec{v}_0 + \mu\vec{w}_2$ satisfies

$$\|\vec{u}_2 - \vec{v}_0\| = \|\mu\vec{w}_2\| < \delta$$

then

$$f(\vec{u}_2) - f(\vec{v}_0) = \frac{1}{2}Q(\vec{v}_2)(\vec{u}_2 - \vec{v}_0)$$

where $\vec{v}_2$ is on the line segment from $\vec{v}_0$ to $\vec{u}_2$. Therefore,

$$f(\vec{u}_2) - f(\vec{v}_0) = \frac{1}{2}(\mu\vec{w}_2)$$

$$= \frac{\mu^2}{2}Q(\vec{v}_2)(\vec{w}_2) < 0$$

since

$$\|\vec{v}_2 - \vec{v}_0\| < \|\vec{u}_2 - \vec{v}_0\| < \delta$$

It follows that $\vec{v}_0$ is not a local minimum either, and part (c) of Theorem 3.2 is proved.

Here are some examples using Theorem 3.1.

1. Let $f : R^2 \rightarrow R$ be defined by

$$f(x,y) = x^3 - 6xy + 3y^2 - 24x + 4$$

Since

$$\frac{\partial f}{\partial x} = 3x^2 - 6y - 24, \quad \frac{\partial f}{\partial y} = -6x + 6y$$

$f'(x,y) = 0$ when

$$x = y, \quad 3x^2 - 6y - 24 = 0$$

The critical points are therefore $(4,4)$ and $(-2,-2)$.

Next, we compute the second partial derivatives of f:

$$\frac{\partial^2 f}{\partial x^2} = 6x, \quad \frac{\partial^2 f}{\partial x \partial y} = \frac{\partial^2 f}{\partial y \partial x} = -6, \quad \frac{\partial^2 f}{\partial y^2} = 6$$

Thus

$$H(f)(x,y) = \begin{pmatrix} 6x & -6 \\ -6 & 6 \end{pmatrix}$$

In particular,

$$H(f)(4,4) = \begin{pmatrix} 24 & -6 \\ -6 & 6 \end{pmatrix}, \quad H(f)(-2,-2) = \begin{pmatrix} -12 & -6 \\ -6 & 6 \end{pmatrix}$$

Theorem 2.1 tells us that $H(f)(4,4)$ is positive definite and $H(f)(-2,-2)$ is indefinite; hence $(4,4)$ is a local minimum and $(-2,-2)$ is not a local extremum.

2.   Next, consider the function $g:R^2 \to R$,

$$g(x,y) = x^2 - 8xy + 2y^2 - 6x - 4y + 9$$

Since

$$\frac{\partial g}{\partial x} = 2x - 8y - 6, \quad \frac{\partial g}{\partial y} = -8x + 4y + 4$$

the critical points of g occur where

$$2x - 8y - 6 = 0, \quad -8x + 4y + 4 = 0$$

or at $(1/7, - 5/7)$. Furthermore,

$$\frac{\partial^2 g}{\partial x^2} = 2, \quad \frac{\partial^2 g}{\partial x \partial y} = \frac{\partial^2 g}{\partial y \partial x} = - 8, \quad \frac{\partial^2 f}{\partial y^2} = 4$$

so

$$H(g)(x,y) = \begin{pmatrix} 2 & -8 \\ -8 & 4 \end{pmatrix} \tag{3.3}$$

In particular, $H(g)(1/7,-5/7)$ is given by (3.3). Applying Theorem 2.1, we see that $H(g)(1/7,-5/7)$ is indefinite. Thus $(1/7,-5/7)$ is not an extremum of g.

The point $(1/7,-5/7)$ is often called a *saddle point* of g, and $(-2,-2)$ is called a saddle point of f. (See Figure 1.5 for a similar surface.) In dimensions higher than 2, there is a considerable variety of types of saddle points possible. Unfortunately, it is difficult to sketch them.

Theorem 3.1 does not describe the nature of all critical points, since not all quadratic forms are positive definite, negative definite, or indefinite. A few simple examples show why. The functions

$$f_1(x_1,x_2) = x_1^2 + x_2^4$$

$$f_2(x_1,x_2) = x_1^2 - x_2^4$$

$$f_3(x_1,x_2) = x_1^2 + x_2^3$$

all have critical points at $(0,0)$, and in each case the Hessian is $\begin{pmatrix} 1 & 0 \\ 0 & 0 \end{pmatrix}$. However, f_1 has a local minimum at the origin, f_2 has a saddle point there, and the origin is neither a saddle point nor a local minimum or maximum for f_3. When the test fails, one is usually forced to test nearby points to determine the nature of the critical points. (There are more mathematically sophisticated procedures, but they are cumbersome.) In practice, however, Theorem 3.1 takes care of most cases.

We should say a few words more about the nature of critical points. Suppose for the moment that A is a diagonal $n \times n$ matrix (see Section 7.2 for the definition), with all diagonal terms nonzero:

$$A = \begin{pmatrix} a_{11} & 0 & \cdots & 0 \\ 0 & a_{22} & \cdots & 0 \\ & \cdots & & \\ 0 & \cdots & 0 & a_{nn} \end{pmatrix}$$

$= \mathrm{diag}(a_{11}, \ldots, a_{nn})$, for short. It is easy to see that

A is positive definite $\iff a_{ii} > 0$, all i
A is negative definite $\iff a_{ii} < 0$, all i
A is indefinite $\iff \exists i,j \ (1 \le i,j \le n)$ with $a_{ii} > 0 > a_{jj}$

In the case $n = 2$, there are exactly three possibilities: both diagonal terms are positive, both are negative, or one is positive and one is negative.

Now let $f: R^n \to R$ be twice differentiable, let $\vec{v}$ be a critical point of f, and let $A = H(f)(\vec{v})$. Suppose again that A is diagonal and that all diagonal entries are nonzero. If $n = 2$, the three possibilities for A correspond to the three geometric possibilities for f near $\vec{v}$: $\vec{v}$ can be a local minimum for f, a local maximum for f, or a saddle point. When $n > 2$, however, there are more possibilities for A; if $n = 3$, for instance, A could have two positive entries and one negative one, or two negative entries and one positive one. In both of these cases $\vec{v}$ is not an extremum, but the two look different geometrically. (Unfortunately, it is almost impossible to draw the graph of such a function.) It is possible, however, to get more information on the behavior of the function f by using this more detailed analysis of the diagonal entries of A. One simple example may show how this analysis works. Suppose f has local maxima at $\vec{v}_1$ and $\vec{v}_2$. Then f must have another critical point, v_3, because it is impossible to have two mountains without some sort of valley in between. The other critical point can be a saddle point (a pass between the mountains) or a local minimum (a true valley). In either case, we learn something of the "geography" of the graph of f from the critical points alone. The branch of mathematics called Morse theory (after the late Marston Morse, who developed it) gives a systematic treatment of this procedure.

It may seem that we have considered a rather special case in restricting ourselves to diagonal matrices. In fact, that is no real restriction at all. The Spectral Theorem, which we shall prove in Chapter 10, implies that if $T:R^n \to R^n$ is given by a symmetric matrix A (with respect to the standard basis), then there is an orthonormal bases α for R^n with respect to which the matrix for T is diagonal. As a result, we can give a similar analysis for any critical point $\vec{v}$ such that $H(f)(\vec{v})$ is invertible. However, we shall not pursue this subject further.

EXERCISES

1. Find and classify the critical points of the following.

   (a) $f(x,y) = x^2 - 4x + 2y^2 + 7$

   (b) $f(x,y) = 3xy - x^2 - y^2$

   (c) $f(x,y) = e^{-(x^2 + y^2)}$

   (d) $f(x,y) = x^2 - 2xy + \frac{1}{3}y^3 - 8$

   (e) $f(x,y) = xy + \frac{4}{x} + \frac{2}{y}$

   (f) $f(x,y) = (\cosh x)(\sin y)$

   (g) $f(x,y) = 12x^3 - 36xy - 2y^3 + 9y^2 - 72x + 60y + 5$

   (h) $f(x,y,z) = x^2 + y^2 + z^2 + xz + yz + 2x - 2y - 4z + 10$

2. Determine the critical points and the local maxima and minima of the function

   $$f(x,y) = x^3 - 3axy + y^3$$

   for all values of a.

3. Show that for all the functions of Section 1, Exercise 5, the test of Theorem 3.1 fails at the critical point (0,0).

4.   Show that $f(x,y) = x^3 - xy^2$ has a critical point at $(0,0)$ which is not an extremum. Sketch the graph of f near $(0,0)$.

5.   Let $g:R^2 \to R$ be given by $g(x,y) = (x - y^2)(x - 2y^2)$.

(a) Sketch the regions in the plane where g is positive, where g is negative, and where g is zero.

(b) Show that $(0,0)$ is a critical point for g and that, restricted to any line through $(0,0)$, g has a local minimum at $(0,0)$.

(c) Show that $(0,0)$ is not a local minimum for g.

6.   Find the dimensions of the rectangular box of maximum volume whose surface area is 150 inches.

7.   A rectangular parallelepiped has three of its faces on the coordinate planes. If one vertex is to be on the plane $3x + 2y + z = 4$, find the dimensions for maximum volume.

8.   A somewhat fancy rectangular box with a volume of 1000 cubic inches has sides and bottom made out of tin and a top made out of platinum. If sheet tin costs \$.50 a square inch and sheet platinum \$500 a square inch, what dimensions minimize the cost?

9.   Find three positive numbers x,y,z whose sum is 24 which maximize xy^2z^3.

10.   Find three positive numbers x,y, and z whose sum is 33 which minimize $x^2 + 2y^2 + 3z^2$.

11.   Let $f:R^2 \to R$ be a function whose Laplacian Δf is everywhere negative. (See Exercise 8 of Section 6.3). Show that f has no local minima.

*12.   Bob and Larry are engaged in a game. The rules of the game specify that Bob picks a number x, then Larry picks a number y and pays Bob $f(x,y)$ dollars where $f(x,y) = y^2 - x^2 + 6y + 8x - 2$. Describe the best strategy for each player. What happens?

*13. Let $(x_1,y_1),\ldots,(x_n,y_{n_2})$ be points in the plane. Find the equation of the parabola $y = ax^2 + bx + c$ such that the quantity

$$f(a,b,c) = \sum_{j=1}^{n} (y_j - ax_j^2 - bx_j - c)^2$$

is minimized. (This parabola is the best fit to the data in the sense of least squares; compare Exercise 9 of Section 1.)

4. CONSTRAINED MAXIMA AND MINIMA: I

In our discussion of maxima and minima in the previous sections, we had to defer any investigation of points on the boundary of a set which might be extreme points. In this section, we see how to deal with these points.

Let $g:R^k \rightarrow R^m$ be a continuously differentiable function and define $S \subset R^k$ by

$$S = \{\vec{v} \in R^k : g(\vec{v}) = \vec{0}\}$$

Since $S = g^{-1}(\{0\})$, S is closed. (See Proposition 5.2.5.) Let U be an open set containing S, and let $f:U \rightarrow R$ be a continuously differentiable function on U. The question we shall study here is the following: how can we find the maximum and minimum values (if any) of f restricted to S?

To see the relationship of this question to the question of extreme points on the boundary, consider an example mentioned in Section 1: we want to maximize a function on the compact set $S = \{\vec{v} \in R^n : \|\vec{v}\| \leq 1\}$. S is not a set of the sort described above. We know, however, how to find all extreme points on the interior of S, the set $\{\vec{v} \in R^n : \|\vec{v}\| < 1\}$. That leaves only the set

$$\partial S = \{\vec{v} \in R^n : \|\vec{v}\| = 1\}$$

(customarily the boundary of a set S is denoted by ∂S), and ∂S is a set of the appropriate kind. (We can rewrite the criterion for mem-

bership in ∂S as $< \vec{v},\vec{v} > - 1 = 0$, and $g(\vec{v}) = < \vec{v},\vec{v} > - 1$ is continuously differentiable.)

Not all boundaries can be expressed in the appropriate form; for
instance, the boundary of the unit square in R^2 cannot be. In many
cases, however, the methods we shall now develop for finding extreme
points can be applied anyway. We shall discuss this matter later.

To explain how the method works, we consider a special case.
First of all, we assume that $g:R^3 \rightarrow R^1$. (Later, we shall generalize
this somewhat.) Secondly, we assume that g is of the special form

$$g(x_1,x_2,x_3) = h(x_1,x_2) - x_3 \qquad\qquad (4.1)$$

where h is a continuously differentiable function. While this assumption may seem quite drastic, we shall see later that it really is not.

Under these assumptions, we need to maximize $f(x_1,x_2,x_3)$ subject
to the constraint

$$x_3 = h(x_1,x_2)$$

That is, we effectively need to maximize the function

$$F(x_1,x_2) = f(x_1,x_2, h(x_1,x_2))$$

We can do this with the help of Theorem 1.1 and the chain rule. Let
$H:R^2 \rightarrow R^3$ be defined by

$$H(x_1,x_2) = (x_1,x_2, h(x_1,x_2))$$

Then $F = f \circ H$ and therefore

$$F'(x_1,x_2) = f'(H(x_1,x_2)) \circ H'(x_1,x_2)$$

Since f' has the matrix

$$\left(\frac{\partial f}{\partial x_1} \quad \frac{\partial f}{\partial x_2} \quad \frac{\partial f}{\partial x_3}\right)$$

and H' the matrix

$$\begin{pmatrix} 1 & 0 \\ 0 & 1 \\ \dfrac{\partial h}{\partial x_1} & \dfrac{\partial h}{\partial x_2} \end{pmatrix}$$

we see that

$$F'(\vec{v}) = \left(\frac{\partial f}{\partial x_1} + \frac{\partial f}{\partial x_3} \frac{\partial h}{\partial x_1} \qquad \frac{\partial f}{\partial x_2} + \frac{\partial f}{\partial x_3} \frac{\partial h}{\partial x_2} \right)$$

where derivatives of h are evaluated at $\vec{v}$, and derivatives of f are evaluated at $H(\vec{v})$. Hence at a critical point of F,

$$\frac{\partial f}{\partial x_j} + \frac{\partial f}{\partial x_3}(H(\vec{v})) \frac{\partial h}{\partial x_j}(\vec{v}) = 0 \qquad (4.2)$$

$j = 1,2$. We can rewrite these equations more symmetrically by recalling the function g. Using (4.1), we have

$$\frac{\partial g}{\partial x_1} = \frac{\partial h}{\partial x_1}, \ \frac{\partial g}{\partial x_2} = \frac{\partial h}{\partial x_2}, \ \frac{\partial g}{\partial x_3} = -1$$

and (4.2) becomes

$$\frac{\partial f}{\partial x_j} \frac{\partial g}{\partial x_3} = \frac{\partial f}{\partial x_3} \frac{\partial g}{\partial x_j} \qquad (4.3)$$

$j = 1,2$. We add on one trivial equation,

$$\frac{\partial f}{\partial x_3} \frac{\partial g}{\partial x_3} = \frac{\partial f}{\partial x_3} \frac{\partial g}{\partial x_3}$$

Now (4.3), with this last equation, becomes

$$\frac{\partial g}{\partial x_3} \text{ grad } f = \frac{\partial f}{\partial x_3} \text{ grad } g \qquad (4.4)$$

(See Exercise 7 of Section 3.7 for the definition of the gradient.) We may summarize this result by saying that if $\vec{v}$ *is an extreme point of f on S, then* $(\text{grad } f)(\vec{v})$ *and* $(\text{grad } g)(\vec{v})$ *are proportional.*

Let us pause for an example. We shall maximize the function $x + y + z$ on the subset of R^3 given by $z = 1 - 7x^2 - 3y^2$. Then

$$f(x,y,z) = x + y + z, \quad g(x,y,z) = 1 - 7x^2 - 3y^2 - z$$

and

$$\text{grad } f = (1,1,1), \quad \text{grad } g = (-14x, -6y, -1)$$

If grad f and grad g are proportional, then their ratio must be -1, since that is the ratio of their third components. Thus we must have

$$(-14x, -6y, -1) = (-1)(1,1,1)$$

or

$$14x = 1, \quad 6y = 1$$

It follows that $x = 1/14$ and $y = 1/6$. From the equation defining the constraint, we have

$$z = 1 - \frac{7}{14^2} - \frac{3}{6^2} = \frac{37}{42}$$

Therefore, the point $\vec{v}_0 = (1/14, 1/6, 37/42)$ is a "critical point" for f given the constraint imposed by g; it could be a maximum, a minimum, or neither. In fact, $\vec{v}_0$ is the point where f is largest (given the constraint imposed by g), as one can see by evaluating f at nearby points.

We assumed in our previous work that g was of the form $g(x_1, x_2, x_3) = h(x_1, x_2) - x_3$; put differently, we have assumed that our constraint expressed x_3 as a function of x_1 and x_2. If we had assumed instead that x_2 was a function of x_1 and x_3, for instance, we would arrive at the same result. The reason is that our conclusion (grad f and grad g are proportional) does not treat any one coordinate differently from any other. (The skeptical reader may go through the calculation; see Exercise 11).

This remark, that the criterion for a constrained maximum holds if any coordinate is a function of the rest, may not seem to make the conclusion hold much more generally, but it does. As we shall see in the next chapter, a result called the Implicit Function Theorem states that if $g:R^3 \to R$ is continuously differentiable and if $g(\vec{v}_0)$ = 0, grad $g(\vec{v}_0) \neq \vec{0}$, then the constraint $g(\vec{v})$ = 0 may be regarded as expressing one of the coordinates of $\vec{v}$ as a function of the rest (at least when $\vec{v}$ is near $\vec{v}_0$). This theorem, therefore, says that our reasoning applies wherever (grad g)$(\vec{v}_0) \neq \vec{0}$.

Assuming for the moment the Implicit Function Theorem, we may sum up our discussion as follows:

THEOREM 4.1. *Let* $g:R^3 \to R$ *be a continuously differentiable function, and let* $S = \{\vec{v} \in R^3 : g(\vec{v}) = 0\}$. *Let* U *be an open set in* R^3 *containing* S, *and let* $f:U \to R$ *be continuously differentiable. Suppose that* grad g *is nonzero everywhere on* S. *Then, if* $\vec{v}$ *is a point of* S *where* f *has a maximum or minimum,* (grad f)$(\vec{v})$ *and* (grad g)$(\vec{v})$ *are proportional.*

We discussed the situation above where we wanted a maximum; of course, the discussion for a minimum is similar. There is a test like Theorem 3.1 to distinguish maxima, minima, and critical points which are neither; it is fairly cumbersome, and it is almost never used. Instead, one determines the nature of the critical point by testing nearby points or by using other information.

Here is an example of Theorem 4.1: we maximize $2x + 3y + 6z$ subject to the constraint $x^2 + y^2 + z^2 = 1$. Then

$f(x,y,z) = 2x + 3y + 6z$
grad $f = (2,3,6)$

and

$g(x,y,z) = x^2 + y^2 + z^2 - 1$
grad $g = (2x,2y,2z)$

Notice that grad $g = \vec{0}$ only at $(0,0,0)$, and $g(0,0,0) \neq 0$. Thus the conditions of Theorem 4.1 are met. We conclude that at a maximum or

minimum of f (under the constraint), $(2,3,6)$ and $(2x,2y,2z)$ are proportional. That is

$$(x,y,z) = \lambda(2,3,6)$$

where λ is some real number. We determine λ by using the constraint:
since $g(x,y,z) = 0$, we have

$$\lambda^2(2^2 + 3^2 + 6^2) = 1$$

or $\lambda^2 = 1/49$, $\lambda = \pm 1/7$. We need, therefore, to consider the points

$$\vec{v}_1 = (\tfrac{2}{7}, \tfrac{3}{7}, \tfrac{6}{7}), \quad \vec{v}_2 = (\tfrac{-2}{7}, \tfrac{-3}{7}, \tfrac{-6}{7})$$

As is easily checked, $f(\vec{v}_1) = 7$, $f(\vec{v}_2) = -7$. We know that f has a
maximum and a minimum on the set of points satisfying $g(\vec{v}) = 0$, by
Theorem 5.5.3, since this set is closed and bounded and therefore
compact (by Theorem 5.6.1). Since $\vec{v}_1$ and $\vec{v}_2$ are the only possible
extreme points, $\vec{v}_1$ is a maximum and $\vec{v}_2$ is a minimum.

We have assumed throughout this discussion that f and g were
functions from R^3 to R^1. Exactly the same reasoning applies, however,
for functions from R^{n+1} to R^1. (Exercise 12 gives a sketch of the
steps.) Furthermore, the Implicit Function Theorem still applies to
generalize the situation. Thus we may state an analogue of Theorem
4.1 for R^{n+1}:

THEOREM 4.2. *Let* $g:R^{n+1} \to R$ *be a continuously differentiable
function, and let* $S = \{\vec{v} \in R^{n+1} : g(\vec{v}) = 0\}$. *Let U be an open set in*
R^{n+1} *containing S, and let* $f:U \to R$ *be continuously differentiable.
Suppose that* grad g *is nonzero everywhere on S. Then if* $\vec{v}$ *is a point
of S where f has a maximum or minimum,* (grad f)$\vec{v}$ *and* (grad g)$\vec{v}$ *are
proportional.*

We shall give a geometric interpretation of this result in the
next section.

We note finally that g need not be defined on all of R^{n+1}; the
theorem also applies if g is defined on an open subset of R^{n+1}.

EXERCISES

1.   Maximize $f(x,y,z) = 6z - x^2 - y^2$ subject to the condition $x + y + z = 3$. (There are two ways to do this problem: by using Theorem 4.1 or by solving for z and reducing to a 2-variable problem. Try it both ways.)

2.   Find the points at which the function $f(x,y) = xy$ takes its maximum and minimum values subject to the condition $x^2 + y^2 = 4$.

3.   Find the point on the unit sphere $x^2 + y^2 + z^2 = 1$ which is furthest from the point $(1,2,-2)$.

4.   Do Exercise 6 of the previous section by the methods of this section.

5.   Do Exercise 7 of the previous section by the methods of this section.

6.   Do Exercise 8 of the previous section by the methods of this section.

7.   Do Exercise 9 of the previous section by the methods of this section.

8.   Do Exercise 10 of the previous section by the methods of this section.

9.   A rectangular box has three of its faces on the coordinate planes. Find the dimensions of the box of largest volume if the vertex of the box in the first octant lies on the ellipsoid

$$\frac{x^2}{4} + \frac{y^2}{9} + \frac{z^2}{25} = 1$$

10.   A rectangular box has three of its faces on the coordinate planes. Find the dimensions of the box of largest volume if the vertex of the box in the first octant lies on the surface

$$x^3 + y^3 + z^3 = 1$$

11.   Give the proof of Theorem 4.1 for the case where g has the form

$$g(x_1, x_2, x_3) = h(x_2, x_3) - x_1$$

*12.   Prove Theorem 4.2.   (Hint: Assume first that

$$g(x_1, \ldots, x_{n+1}) = h(x_1, \ldots, x_n) - x_{n+1}$$

Thus the problem is to maximize

$$f(x_1, \ldots, x_n, h(x_1, \ldots, x_n))$$

Define

$$H(x_1, \ldots, x_n) = (x_1, \ldots, x_n, h(x_1, \ldots, x_n))$$

Now find the critical points of $F = f \circ H$.   These give equations like
(4.2).   Use the expression for g to get equations like (4.3).   Add on
the equation

$$\frac{\partial f}{\partial x_{n+1}} \frac{\partial g}{\partial x_{n+1}} = \frac{\partial f}{\partial x_{n+1}} \frac{\partial g}{\partial x_{n+1}}$$

and conclude that an equation like (4.4) holds.   Now apply the Implicit Function Theorem, as in the proof of Theorem 4.1, to complete the
proof; see Section 9.5 for a discussion of the Implicit Function
Theorem.)

5.   THE METHOD OF LAGRANGE MULTIPLIERS

There is a standard way of arranging the calculations which arise
when Theorem 4.1 or Theorem 4.2 is used: the method of Lagrange multipliers.

Let U be an open subset of R^n, let $f, g : U \to R$ be continuously differentiable and let $S = g^{-1}(0)$, as in Theorem 4.2.   At a critical

point $\vec{v}$ of f on S, (grad f)$(\vec{v})$ = λ(grad g)$(\vec{v})$, where λ is some constant. Thus

$$\frac{\partial f}{\partial x_j}(\vec{v}) = \lambda\frac{\partial g}{\partial x_j}(\vec{v}) \qquad\qquad (5.1)$$

$1 \le j \le n + 1$. Moreover,

$$g(\vec{v}) = 0 \qquad\qquad (5.2)$$

since $\vec{v} \in S$.

Now consider the function F, where

$$F(x_1,\ldots,x_{n+1},\lambda) = f(x_1,\ldots,x_{n+1}) - \lambda g(x_1,\ldots,x_{n+1})$$

Then

$$\frac{\partial F}{\partial x_j} = \frac{\partial f}{\partial x_j} - \lambda\frac{\partial g}{\partial x_j}$$

$1 \le j \le n + 1$, and

$$\frac{\partial F}{\partial \lambda} = - g$$

Thus we may summarize (5.1) and (5.2) as

$$\frac{\partial F}{\partial x_1} = \frac{\partial F}{\partial x_2} = \ldots = \frac{\partial F}{\partial x_{n+1}} = \frac{\partial F}{\partial \lambda} = 0 \qquad\qquad (5.3)$$

That is, we solve the problem of finding extreme of f subject to the
constraint g by setting $F = f - \lambda g$ and setting all derivatives of F
equal to 0. This procedure is called the *method of Lagrange multipliers*.

Here is an example. We determine the volume of the largest rectangular solid with faces parallel to the coordinate planes inscribed
in the ellipsoid $x^2 + 4y^2 + 9z^2 = 1$. If the rectangle touches the
ellipse at (x,y,z) with $x > 0$, $y > 0$, and $z > 0$, then its length, width,
and height are 2x, 2y, and 2z respectively; thus we need to maximize
8xyz subject to the constraint $x^2 + 4y^2 + 9z^2 = 1$. It is easy to check
that Theorem 4.1 applies.

Let $F(x,y,z,\lambda) = 8xyz - \lambda(x^2 + 4y^2 + 9z^2 - 1)$; then

$8yz - 2\lambda x = 0$
$8xz - 8\lambda y = 0$
$8xy - 18\lambda z = 0$

and

$$x^2 + 4y^2 + 9z^2 - 1 = 0$$

Solving these equations may seem difficult, but it is not so hard if we exploit their symmetry. Multiply the first equation by $x/2$, the second by $y/2$, and the third by $z/2$; we then have

$$4xyz = \lambda x^2 = 4\lambda y^2 = 9\lambda z^2$$

Hence $x^2 = 4y^2 = 9z^2$, and the last equation shows that they are all $1/3$. Thus $x = 1/\sqrt{3}$, $y = 1/2\sqrt{3}$, $z = 1/3\sqrt{3}$, and the volume is

$$8xyz = \frac{4}{9\sqrt{3}}$$

(We know that there is a maximum, by Theorems 5.3.3 and 5.6.1).

Notice that we never found λ. There is no need to, of course, since λ does not appear in the eventual answer. When possible, one uses λ to find symmetries in the equations, as above.

For a second example, we maximize $x^2 - xy + 2y^2$ subject to the constraint $x^2 + y^2 = 1$. Thus

$f(x,y) = x^2 - xy + 2y^2$
$g(x,y) = x^2 + y^2 - 1$

Now $(\text{grad } g)(x,y) = (2x,2y)$, which is nonzero whenever $g(x,y) = 0$. Thus Theorem 4.1 applies.

Let $F(x,y,\lambda) = x^2 - xy + 2y^2 - \lambda(x^2 + y^2 - 1)$. Then

$$\frac{\partial F}{\partial x} = 2x - y - 2\lambda x = 0$$

$$\frac{\partial F}{\partial y} = - x + 4y - 2\lambda y = 0$$

$$\frac{\partial F}{\partial \lambda} = - (x^2 + y^2 - 1) = 0$$

at a critical point of f subject to the constraint. From the first two equations, we have

$$2\lambda xy = 2xy - y^2$$

$$2\lambda xy = 4xy - x^2$$

Thus $2xy - y^2 = 4xy - x^2$ or, equivalently

$$x^2 - 2xy - y^2 = 0$$

Solving this equation (with the help of the quadratic formula) we obtain

$$x = (1 \pm \sqrt{2})y$$

In addition, we know that $x^2 + y^2 = 1$, so we have

$$((1 \pm \sqrt{2})^2 + 1)y^2 = 1$$

or

$$y^2 = \frac{1}{4 \pm 2\sqrt{2}}$$

Hence

$$x^2 = (1 \pm \sqrt{2})^2 y^2 = \frac{3 \pm \sqrt{2}}{4 \pm 2\sqrt{2}}$$

and

$$xy = (1 \pm \sqrt{2})y^2 = \frac{1 \pm \sqrt{2}}{4 \pm 2\sqrt{2}}$$

It follows that

$$f(x,y) = \frac{3 \pm \sqrt{2} - (1 \pm \sqrt{2}) + 2}{4 \pm 2\sqrt{2}} = \frac{2}{2 \pm \sqrt{2}}$$

at the critical points. Since f must attain its maximum and minimum
on S (again by Theorems 5.5.3 and 5.6.1) we know that f has a maximum
on the set, and clearly it is $4/(4 - 2\sqrt{2}) = 2 + \sqrt{2}$; it is reached when
$y^2 = 1/(4 - 2\sqrt{2}) = (2 + \sqrt{2})/4$. Similarly, the minimum value of f is
$4/(4 - 2\sqrt{2}) = 2 - \sqrt{2}$.

The reader may wonder what Theorem 4.1 (or 4.2) means geometrically. Here is one interpretation. First of all, consider the relationship between grad $g(\vec{v})$ and S, the set where $g(\vec{v}) = 0$. The directional derivative of g at $\vec{v}$ in the direction $\vec{w}$ is equal to $(D_{\vec{v}}g)(\vec{w})$.
(See Exercises 5 and 6 of Section 3.7.) However, it is easily seen
that

$$(D_{\vec{v}}g)(\vec{w}) = < (\text{grad } g)\vec{v},\vec{w} > \tag{5.4}$$

The function g is identically 0 on S; therefore the directional derivatives in directions along S should be 0. That is, $(\text{grad } g)(\vec{v})$ should
be orthogonal to S at $\vec{v}$, or, equivalently, $(\text{grad } g)(\vec{v})$ *points in the*
direction of the normal to S *at* $\vec{v}$. (See Figure 5.1.)

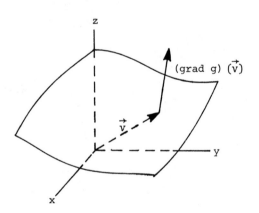

FIGURE 5.1.

Suppose now that $\vec{v}$ is a local maximum (or minimum for f on S.
Then the same sort of argument as in Section 1 shows that the directional derivatives of f in directions along S should be 0. Therefore

(grad f)$(\vec{v})$ too, must point in the direction of the normal to S at $\vec{v}$.
This means that (grad f)$(\vec{v})$ is a multiple of (grad g)$(\vec{v})$, as Theorem
4.1 states.

EXERCISES

1-10. Do each of the exercises of the previous section by the method
of Lagrange multipliers.

11. Find the minimum value of the function

$$f(x,y) = x^2 + 16xy - 11y^2$$

subject to the constraint $x^2 + y^2 = 25$.

12. Find the maximum value of the function

$$f(x,y,z) = x^2 + xy + y^2 + xz + z^2$$

subject to the constraint $x^2 + y^2 + z^2 = 4$.

13. Maximize the function $g(x,y,z) = xyz$ subject to the constraints
$x + y + z = 3$, $x > 0$, $y > 0$, $z > 0$.

14. A rectangular box has three of its faces on the coordinate planes.
Find the dimension of the box of largest volume if the vertex of the
box in the first octant lies on the ellipsoid

$$\frac{x^2}{a^2} + \frac{y^2}{b^2} + \frac{z^2}{c^2} = 9$$

15. A cylindrical can is to contain one liter (= 1000 cm^3) of liquid.
Find the dimensions which minimize the surface area if
   (a)   the can has a top
   (b)   the can has no top

16. A box has a volume of 4 liters. Find the dimension of the box
which minimize surface area if

(a)   the box has a top

(b)   the box has no top

(c)   the box has no top and is twice as long as it is wide

17.   A box with volume 3 liters is to be made with tin sides, a copper bottom, and no top. If sheet tin costs $.50 a square inch and sheet copper costs $2.00 a square inch, find the dimensions of the least expensive box.

18.   French wine sells at wholesale for $5 a bottle, German wine for $4 a bottle, and California wine for $3 a bottle. Assume that the profit on x bottles of French wine is $2x - x^2/120$, on y bottles of German wine $3y - y^2/60$, and on z bottles of California wine is z. How much of each wine should a merchant buy to maximize his profit if he has $600?

19.   The load that a rectangular beam of given length can support is proportional to the width and the square of the thickness. Suppose that a cylindrical log with a circular cross-section of radius r is to be pared down until it has a rectangular cross-section. What dimensions make it strongest?   (See Figure 5.2.)

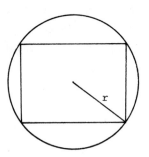

FIGURE 5.2.   Cutting a beam from a log

20.   Maximize $(x_1 + \cdots + x_n)^2$, subject to $\sum_{j=1}^{n} x_j^2 = 1$, and show that the result implies that the geometric mean of n positive numbers is no greater than their arithmetic mean.

21.   If x,y,z,u,v,w are positive and $x/u + y/v + z/w = 1$, show that $27xyz \leq uvw$.

22.   Let p and q be positive.

(a)   Find the minimum of the function

$$\frac{x^p}{p} + \frac{y^q}{q}$$

subject to the constraint $xy = 1$.

*(b)   Use (a) to prove that, if $(1/p) + (1/q) = 1$ and $x > 0$, $y > 0$, then

$$xy \leq \frac{x^p}{p} + \frac{y^q}{q}$$

23.   Verify Equation (5.4.)

6.   CONSTRAINED MAXIMA AND MINIMA: II

In the previous section, we investigated extrema of functions f subject to the condition $g(\vec{v}) = 0$, where $g = R^{n+1} \to R^1$; that is, subject to one constraint. In this section, we extend this discussion to constraints $G:R^{n+k} \to R^k$. For simplicity, we shall deal with the case $G:R^3 \to R^2$, but the reasoning holds more generally.

We assume, then, that $G:R^3 \to R^2$ is a continuously differentiable function, and we let $S = \{\vec{v} \in R^3 : G(\vec{v}) = \vec{0}\}$. We wish to maximize (or minimize) the continuously differentiable function f on S. Ordinarily S will be a curve in R^3. For instance, if

$$G(x_1,x_2,x_3) = (x_2 - x_1^2, \; x_3 - x_1^3)$$

then

$$S = \{(x_1,x_1^2,x_1^3):x_1 \in R\}$$

Let us write $G(\vec{v}) = (g_1(\vec{v}),g_2(\vec{v}))$; then S is the set of points satisfying $g_1(\vec{v}) = g_2(\vec{v}) = 0$. We can use the geometric explanation at the end of the previous section to see what the procedure should be. At a typical point $\vec{v} \in S$, there is a plane of vectors normal to S. By the same reasoning as in Section S, grad $g_1(\vec{v})$ should be normal to S (g_1 is identically 0 on S, and therefore the directional derivative of g_1 in the direction tangent to S should be 0). Similarly, grad $g_2(\vec{v})$ should be normal to S. If grad $g_1(\vec{v})$ and grad $g_2(\vec{v})$ are linearly independent, therefore, they should span the plane normal to S. At an extremum of f on S, the directional derivative of f along the tangent to S will be 0. Therefore (grad f)$(\vec{v})$ will be orthogonal to S, and therefore (grad f)$(\vec{v})$ should be a linear combination of (grad g_1)$(\vec{v})$ and (grad g_2)$(\vec{v})$. The next theorem makes this result precise. We state it for general n and k.

THEOREM 6.1. *Let* $G:R^{n+k} \to R^k$ *be continuously differentiable, and let* $g_1,\ldots,g_k$ *be the coordinate functions of G. Assume that* (grad g_1)$(\vec{v}),\ldots,$(grad g_k)$(\vec{v})$ *are linearly independent whenever* $G(\vec{v})$ $= \vec{0}$. *Let* $S = \{\vec{v}:G(\vec{v}) = \vec{0}\}$, *and let* $f:U \to R$ *be a continuously differentiable function defined on the open set* U *containing S. If* $\vec{v}_0$ *is a point of S where f (restricted to S) has a maximum or minimum, then* (grad f)$(\vec{v}_0)$ *is a linear combination of* (grad g_1)$(\vec{v}_0),\ldots,$(grad g_k)$(\vec{v}_0)$.

We defer the proof of Theorem 6.1 briefly, and first illustrate how to apply it. Again, the most efficient method is to use Lagrange multipliers. Suppose, for instance, that $G:R^3 \to R^2$. At an extreme point $\vec{v}_0$, we know that there are real numbers λ_1 and λ_2 such that

$$(\text{grad } f)(\vec{v}_0) = \lambda_1(\text{grad } g_1)(\vec{v}_0) + \lambda_2(\text{grad } g_2)(\vec{v}_0) \qquad (6.1)$$

Furthermore,

$$g_1(\vec{v}_0) = g_2(\vec{v}_0) = 0 \qquad (6.2)$$

Set

$$F(\vec{v},\lambda_1,\lambda_2) = f(\vec{v}) - \lambda_1 g_1(\vec{v}) - \lambda_2 g_2(\vec{v})$$

Then

$$\frac{\partial F}{\partial x_j} = \frac{\partial f}{\partial x_j} - \lambda_1 \frac{\partial g_1}{\partial x_j} - \lambda_2 \frac{\partial g_2}{\partial x_j}$$

$j = 1,2,3,$ and

$$\frac{\partial F}{\partial \lambda_1} = - g_1(\vec{v}), \quad \frac{\partial F}{\partial \lambda_2} = - g_2(\vec{v})$$

Hence (6.1) and (6.2) become

$$\frac{\partial F}{\partial x_1} = \frac{\partial F}{\partial x_2} = \frac{\partial F}{\partial x_3} = \frac{\partial F}{\partial \lambda_1} = \frac{\partial F}{\partial \lambda_2} = 0 \qquad (6.3)$$

This gives the conditions for an extreme point. In practice, solving the resulting equations is tedious at best.

Here is a relatively simple example of a problem with two con- straints. We wish to find the highest point on the curve given by the intersection of $x^2 + y^2 + z^2 = 1$ with $4z = 3xy$. That is, we wish to maximize

$$f(x,y,z) = z$$

with the constraints

$$g_1(x,y,z) = x^2 + y^2 + z^2 - 1 = 0$$

and

$$g_2(x,y,z) = 3xy - 4z = 0$$

It is not hard to check that the conditions of Theorem 6.1 are met. We set

$$F(x,y,z,\lambda_1,\lambda_2) = z - \lambda_1(x^2 + y^2 + z^2 - 1) - \lambda_2(3xy - 4z)$$

The condition for an extreme point is that all derivatives of F are 0. Thus

$$- 2\lambda_1 x - 3\lambda_2 y = 0$$

$$- 2\lambda_1 y - 3\lambda_2 x = 0$$

$$1 - 2\lambda_1 z - 4\lambda_2 = 0$$

$$- (x^2 + y^2 + z^2 - 1) = 0$$

$$- (3xy - 4z) = 0$$

The first two equations say that

$$\frac{2\lambda_1}{3\lambda_2} = \frac{-y}{x} = - \frac{x}{y}$$

Thus

$$x^2 = y^2, \ y = \pm \ x$$

The last equation then asserts that

$$z = \frac{3x^2}{4}$$

and the fourth equation gives

$$2x^2 + \frac{9x^4}{16} - 1 = 0$$

or

$$9x^4 + 32x^2 - 16 = 0$$

This is a quadratic equation in x^2. Solving, we find that

$$x^2 = - 4 \text{ or } \frac{4}{9}$$

The first root does not give a solution for x; the second gives

$$x = \pm \ 2/3$$

Then, since $y = \pm \ x$ and $z = 3xy/4$, we find that the critical points
are

$(\frac{2}{3}, \frac{2}{3}, \frac{1}{3})$, $(\frac{2}{3}, \frac{-2}{3}, \frac{-1}{3})$, $(\frac{-2}{3}, \frac{-2}{3}, \frac{1}{3})$, $(\frac{-2}{3}, \frac{2}{3}, \frac{-1}{3})$

Since the set S satisfying the constraints is a closed subset of the unit sphere, S is compact (by Theorem 5.6.1) and f has a maximum on S (by Theorem 5.5.3). It is now clear that the maximum is 1/3, attained at (2/3,2/3,1/3) and (-2/3,-2/3,1/3).

We mentioned at the beginning of the previous section that not all boundaries of sets S are given by constraints like $G(\vec{v}) = \vec{0}$, where G is continuously differentiable. For instance, if S is the unit cube in R^3,

$$S = \{(x,y,z) \in R^3 : 0 \le x,y,z \le 1\}$$

then ∂S consists of all $(x,y,z) \in S$ such that at least one of x,y,z is 0 or 1. ∂S cannot be described in the way needed to apply Theorem 4.1 or 6.1. (For one thing, ∂S has corners, and is therefore not smooth enough.) However, S is made up of pieces (like the square

$$\{(x,y,z) \in R^3 : x = 0,\ 0 < y,\ z < 1\}$$

or the line

$$\{(x,y,z) \in R^3 : x = y = 1,\ 0 \le z \le 1\})$$

for each of which Theorem 4.2 or Theorem 6.1 is applicable. Thus we can apply our theorems by cutting ∂S into enough smaller pieces. In practice, most problems involving constrained maxima and minima can be solved in this way.

We conclude this section by proving Theorem 6.1 in the special case where $G:R^3 \to R^2$; the proof we give generalizes, however. (See Exercise 10.) We assume first that $G = (g_1, g_2)$ where

$$g_1(x_1, x_2, x_3) = h_1(x_1) - x_2$$
$$g_2(x_1, x_2, x_3) = h_2(x_1) - x_3$$

Then S is the set of points of the form

$$(x, h_1(x_1), h_2(x_1))$$

Define $H: R^1 \to R^3$ by

$$H(x) = (x, h_1(x), h_2(x))$$

Then, as in the proof of Theorem 4.1, finding the extreme points of f subject to the conditions $G(\vec{v}) = \vec{0}$ is equivalent to finding the extreme points of foH.

We apply the chain rule: at a critical point of foH,

$$\frac{d}{dx}(foH) = \frac{\partial f}{\partial x_1} + \frac{\partial f}{\partial x_2}\frac{dh_1}{dx_1} + \frac{\partial f}{\partial x_3}\frac{dh_2}{dx_1} = 0 \qquad (6.4)$$

We add two more equations,

$$\frac{\partial f}{\partial x_2} - \frac{\partial f}{\partial x_2} + 0 = 0$$

$$\frac{\partial f}{\partial x_3} + 0 - \frac{\partial f}{\partial x_3} = 0 \qquad (6.5)$$

Since

$$\frac{\partial g_1}{\partial x_1} = \frac{dh_1}{dx_1}, \; \frac{\partial g_1}{\partial x_2} = -1, \; \frac{\partial g_1}{\partial x_3} = 0$$

and

$$\frac{\partial g_2}{\partial x_1} = \frac{dh_1}{dx_1}, \; \frac{\partial g_2}{\partial x_2} = 0, \; \frac{\partial g_2}{\partial x_3} = 1$$

we can rewrite (6.4) and (6.5) together as

$$(\text{grad } f)(\vec{v}) + \frac{\partial f}{\partial x_2}(\text{grad } g_1)(\vec{v}) + \frac{\partial f}{\partial x_3}\text{grad } g_2(\vec{v}) = \vec{0}$$

if $\vec{v}$ is a critical point of f. That is, at a critical point $\vec{v}$, $(\text{grad } f)(\vec{v})$ is a linear combination of grad $g_1(\vec{v})$ and grad $g_2(\vec{v})$.

The theorem will follow in general if we can show that the constraint $G(\vec{v}) = 0$ can be regarded as defining two of the coordinates of $\vec{v}$ as functions of the third. We shall see in the next chapter that the Implicit Function theorem permits this. Thus the proof is complete except for this one step, which we take care of in the next chapter.

EXERCISES

1. Find the maximum and minimum of $f(x,y,z) = z$ subject to the constraints $x^2 + y^2 + z^2 = 1$ and $x + y + z = -1$.

2. Find the maxima and minima of $f(x,y,z) = x^2yz$ subject to the constraints $x + y + z = 3$ and $x - 2y + 2z = 1$.

3. Find the minimum value of $xy + yz$ given that $x^2 + y^2 = 2$ and $x^2 + z^2 = 2$.

4. Find the point on the line $x = y = 2z$ closest to $(3,4,5)$.

5. Find the point on the line $x = y = z$ closest to the line $x = 0$, $y + z = 4$.

6. Find the point on the curve in R^3 given by

$$x^2 + y^2 = 1 \text{ and } z^2 - x^2 = 1$$

nearest the origin.

7. Find the distance between the closest points on the curves $y = x^2$ and $x^2 - y^2 = 27/16$. (Hint: Regard the square of the distance as a function of four variables.)

8. What point on the intersection of the unit sphere with $x + y + z = 1$ is closest to $(2,3,1/3)$?

9. Maximize $x_1y_1 + x_2y_2$, given that $x_1^2 + x_2^2 = y_1^2 + y_2^2 = 1$. What does the answer mean?

*10. Give the proof of Theorem 6.1 in the general case, assuming the Implicit Function Theorem, Theorem 9.5.1. (Hint: the Implicit Func-

tion Theorem says, in effect, that we may assume that the constraints
are given by functions of the form

$$g_1(\vec{v}) = h_1(x_1,\ldots,x_n) - x_{n+1}$$
$$g_2(\vec{v}) = h_2(x_1,\ldots,x_n) - x_{n+2}$$
$$\cdot \quad \cdot \quad \cdots \quad \cdot \quad \cdot \qquad\qquad (6.6)$$
$$g_k(\vec{v}) = h_k(x_1,\ldots,x_n) - x_{n+k}$$

Define

$$H(x_1,\ldots,x_n) = (x_1,\ldots,x_n,h_1(x_1,\ldots,x_n),\ldots,h_k(x_1,\ldots,x_n))$$

Show that we need to maximize (or minimize)

$$F(x_1,\ldots,x_n) = f \circ H(x_1,\ldots,x_n)$$

Find the condition for grad F to be $\vec{0}$. Then show, as in the proof of
Theorems 4.1, 4.2, and the special case of Theorem 6.1, that it can
be rewritten as

$$\text{grad } f + \frac{\partial f}{\partial x_{n+1}} \text{ grad } g_1 + \cdots + \frac{\partial f}{\partial x_{n+k}} \text{ grad } g_k = \vec{0}$$

Show that this proves the theorem for the constraints in the form
(6.6). As mentioned above, the Implicit Function Theorem then gives
the result in general.)

*11. Try to maximize $f(x,y,z) = xyz$ subject to the constraints $xy = 1$,
$x^2 + z^2 = 1$. What goes wrong and why?

7 THE PROOF OF PROPOSITION 2.3

Proposition 2.3 depends on two lemmas. The first gives a criterion
for a matrix to be positive (and negative) definite, and the second
gives an inequality concerning $\|A\|$.

LEMMA 7.1.   (a) *The symmetric matrix* A *is positive definite if there is an* $\varepsilon > 0$ *such that* $Q_A(\vec{v}) > 0$ *for all* $\vec{v}$ *with* $0 < \|\vec{v}\| < \varepsilon$.

(b) *The symmetric matrix* A *is positive definite if* $Q_A(\vec{v}) > 0$ *for all* $\vec{v}$ *with* $\|\vec{v}\| = 1$.

Similar statements hold for negative definite and indefinite A. The proofs are easy; we shall do (a), leaving (b) and the other statements as exercises.

Let $\vec{v} \in R^n$ be a non-zero vector. Then, if

$$\vec{w} = \frac{\varepsilon \vec{v}}{2\|\vec{v}\|}$$

$\|\vec{w}\| < \varepsilon$, so that $Q_A(\vec{w}) > 0$. However,

$$Q_A(\vec{v}) = < T_A\vec{v}, \vec{v} >$$

$$= < T_A(\frac{2\|\vec{v}\|}{\varepsilon} \vec{w}), \frac{2\|\vec{v}\|}{\varepsilon} \vec{w} >$$

$$= \frac{4\|\vec{v}\|^2}{\varepsilon^2} < T_A\vec{w}, \vec{w} >$$

$$= \frac{4\|\vec{v}\|^2}{\varepsilon^2} Q_A(\vec{w}) > 0$$

where $T_A : R^n \to R^n$ is the linear transformation defined by A. Thus A is positive definite.

LEMMA 7.2.   *If* $T = T_A : R^n \to R^n$ *is the linear transformation with matrix* $A = (a_{ij})$, *then*

$$\|T_A(\vec{v})\| \leq \|A\|$$

for any unit vector $\vec{v} \in R^n$.

Proof. Let $\vec{v} = \Sigma_{i=1}^n b_i \vec{e}_i$ be a unit vector in R^n, where $\vec{e}_1, \ldots, \vec{e}_n$ is the standard basis for R^n. Then

$$\|T\vec{v}\|^2 = < T\vec{v}, T\vec{v} >$$

$$= \sum_{i=1}^n \sum_{j=1}^n b_i b_j < T\vec{e}_i, T\vec{e}_j >$$

$$\leq \sum_{i=1}^{n} \sum_{j=1}^{n} b_i b_j \|\vec{Te_i}\| \cdot \|\vec{Te_j}\|$$

by the Schwarz Inequality (Theorem 4.6.2.) This last expression can be rewritten as

$$\left(\sum_{j=1}^{n} b_j \|\vec{Te_j}\| \right)^2 = \; < \vec{v}, \sum_{j=1}^{n} \|\vec{Te_j}\|\vec{e_j} >^2$$

$$\leq \|\vec{v}\|^2 \left(\sum_{j=1}^{n} \|\vec{Te_j}\|^2 \right)^2$$

$$= \left(\sum_{j=1}^{n} \|\vec{Te_j}\|^2 \right)^2$$

again by the Schwarz Inequality (recall that $\vec{v}$ is a unit vector). Thus

$$\|\vec{Tv}\|^2 \leq \sum_{j=1}^{n} \|\vec{Te_j}\|^2$$

Since $\vec{Te_j} = \sum_{i=1}^{n} a_{ij}\vec{e_i}$,

$$\|\vec{Te_j}\|^2 = \sum_{i=1}^{n} (a_{ij})^2$$

and

$$\sum_{j=1}^{n} \|\vec{Te_j}\|^2 = \sum_{i=1}^{n} \sum_{j=1}^{n} (a_{ij})^2$$

Thus $\|\vec{Tv}\|^2 \leq \|T\|^2$, which proves the lemma.

We now use these lemmas to prove Proposition 2.3.

Statements (a) and (b) have essentially identical proofs; we prove only (a). Let $C = \{\vec{v} \in R^n : \|\vec{v}\| = 1\}$. Then C is compact, by Theorem 5.6.1 and the function $Q_A : R^n \to R$ is continuous, so, by Theorem 5.5.3, Q_A has a minimum value, ε, on C. We know that $\varepsilon > 0$, since A is positive definite.

Now suppose that B is a symmetric matrix with $\|B - A\| < \varepsilon$, and let $\vec{v}$ be any unit vector. Then

$$
\begin{aligned}
|Q_B(\vec{v}) - Q_A(\vec{v})| &= |< B\vec{v}, \vec{v} > - < A\vec{v}, \vec{v} >| \\
&= |< (B - A)\vec{v}, \vec{v} >| \\
&\leq \|\vec{v}\| \|(B - A)\vec{v}\| \quad \text{(by the Schwarz inequality)} \\
&= \|(B - A)\vec{v}\| \\
&\leq \|B - A\| \quad \text{(by Lemma 7.2)} \\
&< \varepsilon
\end{aligned}
$$

Thus $Q_B(\vec{v}) - Q_A(\vec{v}) > -\varepsilon$, and $Q_B(\vec{v}) > Q_A(\vec{v}) - \varepsilon > 0$ for all $\vec{v} \in C$. By Lemma 7.1, B is positive definite.

To prove (c), pick ε such that $Q_A(\vec{v}_1) \geq \varepsilon$, $Q_A(\vec{v}_2) \leq -\varepsilon$. The same calculation as in the previous proof shows that

$$
|Q_A(\vec{v}_1) - Q_B(\vec{v}_1)| < \varepsilon, \quad |Q_A(\vec{v}_2) - Q_B(\vec{v}_2)| < \varepsilon
$$

which means that

$$
Q_B(\vec{v}_1) - Q_B(\vec{v}_1) > -\varepsilon, \quad Q_B(\vec{v}_2) - Q_A(\vec{v}_2) < \varepsilon
$$

Therefore

$$
Q_B(\vec{v}_1) > 0, \quad Q_B(\vec{v}_2) < 0
$$

as claimed.

Note. For semidefinite transformations (positive or negative), the corresponding result is false. For instance,

$$
A = \begin{pmatrix} 1 & 0 \\ 0 & 0 \end{pmatrix}
$$

is positive semidefinite, but if ε is positive, then

$$A_\varepsilon = \begin{pmatrix} 1 & 0 \\ 0 & \varepsilon \end{pmatrix}$$

is positive definite, and if ε is negative, then A_ε is indefinite. The most that can be proved is that if A is nonzero and positive semi-definite, then no sufficiently close B is negative definite. (See Exercise 3.)

EXERCISES

1. Complete the proof of Lemma 7.1.

2. State and prove the analogue of Lemma 7.1 for indefinite symmetric matrices.

3. Prove that, if A is nonzero and positive semidefinite, then there is an $\varepsilon > 0$ such that whenever B is negative definite, $\|A - B\| \geq \varepsilon$.

4. Use Exercise 3 to show that if $H(f)(\vec{v})$ is positive semidefinite and $H(f)(\vec{v}) \neq 0$, then $\vec{v}$ is not a local maximum for f. State and prove a similar result about nonzero negative semidefinite transformations.

*5. In Exercise 13 of Section 7.5, we defined a different norm on linear transformations $T:V \to W$, where V, W are normed vector spaces:

$$\|T\|_0 = \operatorname*{lub}_{\|\vec{v}\|=1} \|T\vec{v}\|$$

(The subscript $_0$ is to distinguish this norm from the one used earlier in this section.) If A is an $n \times n$ matrix, it defines a linear transformation $T_A:R^n \to R^n$ (one uses the standard basis); define

$$\|A\|_0 = \|T_A\|_0$$

(where R^n has the standard norm). Prove that Lemma 7.2 and Proposition 2.3 still hold if we use the $\|\ \|_0$ norm on the $n \times n$ matrices.

CHAPTER 9

THE INVERSE AND IMPLICIT FUNCTION THEOREMS

INTRODUCTION

We saw in Chapter 3 that the derivative of a function at a point provides a good linear approximation to the function near the point.
This fact suggests that properties of the derivative at a point should
be reflected in properties of the function near the point. The Inverse
Function Theorem and Implicit Function Theorem are both instances of
this phenomenon. The Inverse Function Theorem states that if the derivative of a function $f:V \to V$ is invertible at $\vec{v}_0$, then so is f, at
least near $\vec{v}_0$. The Implicit Function Theorem gives conditions (on
the derivative of a function) under which a collection of equations
can be solved for some of the variables in terms of the rest. We note
that the Implicit Function Theorem is exactly what is needed to complete the discussion of constrained maxima and minima of the previous
chapter.

Both these theorems are quite important in mathematics, and the
reader should, at the least, learn what they say. The proofs, however, are quite difficult, and the reader may wish to skip them on
first reading.

1. THE INVERSE FUNCTION THEOREM

In this section, we state the Inverse Function Theorem for functions
between open subsets of finite dimensional inner product spaces.

Since the result is stronger when the dimension of these spaces is one, we give a separate statement in this case; the proof is given in the next section. The proof of the general case of the inverse function theorem is given in Section 3.

In order to state the Inverse Function Theorem, we need to know what an inverse function is. Let V and W be finite dimensional inner product spaces, U an open subset of V, U' an open subset of W, and $f:U \rightarrow U'$ a function. We say that a function $g:U' \rightarrow U$ is an *inverse* for f if $(g \circ f)(\vec{v}) = \vec{v}$ for all $\vec{v} \in U$ and $(f \circ g)(\vec{w}) = \vec{w}$ for all $\vec{w} \in U'$. Clearly g is an inverse for f if and only if f is an inverse for g. It is also immediate that $f:U \rightarrow U'$ has an inverse if and only if f is bijective (that is, both 1 - 1 and onto); see Exercise 1.

For example, the function $f:(0,1) \rightarrow (0,1)$ given by $f(x) = x^2$ has an inverse $g(y) = \sqrt{y}$, since

$$(g \circ f)(x) = g(x^2)$$
$$= \sqrt{x^2} = x$$

and

$$(f \circ g)(y) = f(\sqrt{y})$$
$$= (\sqrt{y})^2 = y$$

However, the function $h:(-1,1) \rightarrow (0,1)$ defined by $h(x) = x^2$ does not have an inverse since, h is not injective: if k is an inverse, then $k(1/4) = k(h(1/2)) = 1/2$ and $k(1/4) = k(h(-1/2)) = -1/2$. Thus we have two values for $k(1/4)$, which is impossible.

If $T:V \rightarrow W$ is a linear transformation, then T has an inverse if and only if T is an isomorphism. If T is an isomorphism, we know from Proposition 4.3.3 that $T^{-1}:W \rightarrow V$ is also an isomorphism. Finally, we know that $T:V \rightarrow W$ is an isomorphism if and only if its matrix (relative to any bases for V and W) is invertible. (See Proposition 7.2.3.)

Our first result is an analogue of Proposition 7.2.1.

PROPOSITION 1.1. *Let* $g_1, g_2:U' \rightarrow U$ *be inverses for the function* $f:U \rightarrow U'$. *Then* $g_1 = g_2$.

Note. If $f:U \to U'$ has an inverse, the inverse is usually de-noted by $f^{-1}:U' \to U$.

Proof of Proposition 1.1. We use the definition of inverse: For any $\vec{v} \in V$, we have

$$g_1(\vec{v}) = g_1 \circ (f \circ g_2)(\vec{v})$$
$$= (g_1 \circ f) \circ g_2(\vec{v})$$
$$= g_2(\vec{v})$$

The next result gives a necessary condition for a differentiable function to have a differentiable inverse.

PROPOSITION 1.2. Let V and W be finite dimensional inner product spaces, U an open subset of V, and U' an open subset of W. Let $g:U' \to U$ be an inverse for the function $f:U \to U'$. If f is differentiable at $\vec{v}$ and g is differentiable at $\vec{w} = f(\vec{v})$, then both $D_{\vec{v}}f$ and $D_{\vec{w}}g$ are invertible, and

$$D_{\vec{w}}g = (D_{\vec{v}}f)^{-1}$$

Proof. Let $I:U \to U$ be the identity mapping; $I(\vec{v}) = \vec{v}$. Then $D_{\vec{v}}I = I_V$, the identity linear transformation in V. On the other hand, we have

$$D_{\vec{v}}I = D_{\vec{v}}(g \circ f)$$
$$= (D_{\vec{w}}g) \circ (D_{\vec{v}}f)$$
$$= I_V$$

from the chain rule. In the same way, we can show that $(D_{\vec{v}}f) \circ (D_{\vec{w}}g) = I_W$, and the proposition follows.

Remark. Note that, under the hypothesis of Proposition 1.2, dim V = dim W (since the isomorphism $D_{\vec{v}}f:V \to W$ preserves dimension by Theorem 4.4.2). In fact, the hypothesis that f and f^{-1} are differentiable is convenient but unnecessary. It can be shown that if $f:U \to W$

(where $U \subseteq V$ is open) is continuous and injective, and if $f(U)$ is an open set in W, then dim V = dim W. This result is fundamental in the subject of dimension theory; its proof is difficult and will not be given here.

It may not seem too surprising that there is no continuous injective function mapping an open interval in R^1 onto an open disc in R^2, for instance. However, if one removes the requirement that f be injective, then such maps exist. Exercise 7 gives an example. (The sets involved there are closed, for convenience, but similar examples exist for open sets.)

We now state the one dimensional version of the Inverse Function Theorem.

THEOREM 1.3. *Let* $f:(a,b) \to R$ *be a differentiable function and suppose that* $f'(x) \neq 0$ *for all* $x \in (a,b)$. *Then*

(a) The function f *is injective on* (a,b).

(b) The image of f *is an open interval* (c,d).

(c) The inverse $f^{-1}:(c,d) \to (a,b)$ *to* f *is differentiable with derivative*

$$(f^{-1})'(y) = \frac{1}{f'(x)}$$

where $y = f(x)$.

The proof will be given in the next section.

In view of Proposition 1.2, Theorem 1.3 is the best result that could be expected; if f is to have a differentiable inverse, then $f'(x)$ must be nonzero at every point in (a,b).

We now consider a few examples from the point of view of Theorem 1.3.

1.   Let $f:(0,1) \to R$ be given by $f(x) = x^2$. Then f has a differentiable inverse g given by $g(y) = \sqrt{y}$, $y \in (0,1)$. This is consistent with Theorem 1.3, since $f'(x) = 2x$ is nonzero on $(0,1)$.

2.   Let $f:(-1,1) \to R$ be given by $f(x) = x^2$. As we saw at the beginning of the last section, f is not invertible. This is again consistent with Theorem 1.3, since $f'(x) = 2x = 0$ when $x = 0$.

3.   Let $f: R \to R$ be given by $f(x) = x^3$. Then f has an inverse $g: R \to R$, $g(y) = y^{1/3}$. However,

$$g'(y) = \frac{1}{3y^{2/3}}$$

for $y \neq 0$ and $g'(y)$ does not exist at $y = 0$. This is not surprising since $f'(x) = 3x^2 = 0$ when $x = 0$.

4.   Theorem 1.3 can be used to define interesting functions. For example, let $f: (-\pi/2, \pi/2) \to R$ be given by $f(x) = \sin x$. Then $f'(x) = \cos x \neq 0$ on $(-\pi/2, \pi/2)$, so we know from Theorem 1.3 that f has a differentiable inverse $g(y) = $ arc sin y defined for $y \in (-1,1)$ (since $\sin(\pi/2) = 1$ and $\sin(-\pi/2) = -1$). Furthermore, we know that

$$g'(y) = \frac{1}{f'(x)}$$

$$= \frac{1}{\cos x}$$

where $\sin x = y$. However,

$$\cos x = (1 - \sin^2 x)^{1/2}$$

$$= (1 - y^2)^{1/2}$$

so we have

$$\frac{d}{dy} \text{ arc sin } y = (1 - y^2)^{-1/2}$$

This is the usual formula from elementary calculus.

The analogue of Theorem 1.3 in dimensions greater than one might seem to be that if f is differentiable on a connected open set U and $f'(\vec{v})$ is invertible for all $\vec{v} \in U$, then f has an inverse. This statement is false, however. Figure 1.1 should illustrate what can go wrong. (An explicit example is given in Exercise 3.) The problem is that in higher dimensions, functions have too many directions to move around in. The reader should convince himself that no such problem exists

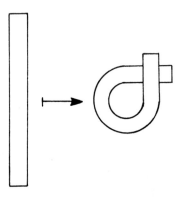

FIGURE 1.1.

in R^1. The difference is that a point moving in R^1 cannot return to
its starting point without stopping somewhere, while in higher dimensions there is the possibility of going around in circles.

For higher dimensions, then, we need to lower our sights. We do
this by asking only for a "local inverse." That is, if $\vec{v}_0 \in U$, then
we ask that the restriction of f to some small open set contained in
U be invertible. In the case of Figure 1.1, for instance, the function f does indeed have such "local inverses."

The content of the general Inverse Function Theorem is that local
inverses exist when f' is invertible.

THEOREM 1.4. (Inverse Function Theorem). *Let V and W be finite
dimensional inner product spaces, U_1 be an open set in V, and let
$f:U_1 \to W$ be a function with a continuous differential. Suppose that
$\vec{v}_0$ is a point in U_1 such that $D_{\vec{v}_0} f$ is an isomorphism. Then there is
an open subset U of U_1 containing $\vec{v}_0$ such that*

   (a)   the function f is injective on U

   (b)   the set $U' = f(U)$ is open

   *(c)   the inverse $f^{-1}:U' \to U$ to f is differentiable with deriva-
tive*

$$D_{\vec{w}}f^{-1} = (D_{\vec{v}}f)^{-1}$$

where $\vec{w} = f(\vec{v})$.

We give the proof of this result in Section 3.

Remark. The Inverse Function Theorem applies only to *some* open set U about $\vec{v}_0$. Since U is not specified, the theorem is hard to apply to specific examples. It is most useful as a theoretical tool, in cases where it is simply important that *some* inverse should exist.

EXERCISES

1. (a) Prove that a function $f:U \to U'$ has an inverse if and only if f is bijective.

(b) Suppose that f is bijective and that $g:U' \to U$ satisfies $(g \circ f)(\vec{v}) = \vec{v}$, all $\vec{v} \in U$. Show that $(f \circ g)(\vec{w}) = \vec{w}$, all $\vec{w} \in U'$.

2. Which of the following functions have inverses? Determine the inverse function when it exists and obtain a formula for the derivative of the inverse function.

(a) $f:(2,6) \to R$, $f(x) = x^2 + 6$

(b) $f:(-3,-1) \to R$, $f(x) = x^2 + 6$

(c) $f:(-3,1) \to R$, $f(x) = x^2 + 6$

(d) $f:(0,3) \to R$, $f(x) = x^4$

(e) $f:(-7,10) \to R$, $f(x) = x^3 - 1$

(f) $f:R \to R$, $f(x) = x^m$, m odd

(g) $f:(0,\pi) \to R$, $f(x) = \cos x$

(h) $f:(-\pi/2,\pi/2) \to R$, $f(x) = \tan x$

(i) $f:(0,\pi) \to R$, $f(x) = \sin x$

(j) $f:R \to R$, $f(x) = e^x$

(k) $f:[-1/2,1/2] \to R$, $f(x) = x^3 - x$

(*ℓ*) $f:(0,\infty) \rightarrow R$, $f(x) = \cosh x = (e^x + e^{-x})/2$

(m) $f:R \rightarrow R$, $f(x) = \sinh x = (e^x - e^{-x})/2$

3.   Show that the function $f(r,\theta) = (r \cos \theta, r \sin \theta)$, defined for $r > 0$ and $\theta \in R$, has a local inverse near every point in its image, but no global inverse.

4.   The functions f below have differentiable inverses when restricted to some interval including x_0. Find the largest such open interval.

   (a) $f(x) = x^2 - 7x + 6 : x_0 = 1$

   (b) $f(x) = x^3 - 5x^2 + 3x + 3 : x_0 = 0$

   (c) $f(x) = \sin x : x_0 = 3$

   (d) $f(x) = \sin x : x_0 = -1$

   (e) $f(x) = \tan x : x_0 = 0$

   (f) $f(x) = e^x : x_0 = 2$

   (g) $f(x) = 16 + e^{x^2} : x_0 = -1$

   (h) $f(x) = \sin x + \cos x : x_0 = 0$

5.   Compute $(f^{-1})'(y_0)$, for each of the functions in Exercise 4. Here $y_0 = f(x_0)$.

6.   Let V and W be vector spaces, $T:V \rightarrow W$ a linear transformation, and let $\vec{w}_0$ a fixed vector in V. Define $f:V \rightarrow W$ by $f(\vec{v}) = T(\vec{v}) + \vec{w}_0$, and prove that f is invertible if and only if T is.

*7.   (The Polya space-filling curve.) Let S be a non-isoceles right triangle (including the inside) in R^2. We shall define a continuous function P mapping the interval $[0,1]$ onto S.

   Any point in the interval may be written as a binary "decimal" (where everything is in the base 2 instead of 10). For instance, $1/3 = .01010101...,1/2 = .10000... = .01111...$, and so on. Given a number $x = .a_1 a_2 a_3 ...$, we define P(x) as follows: the altitude to the hypotenuse of S divides S into two triangles, similar to S and of

different sizes. If $a_1 = 1$, we work with the bigger of the two, if
$a_1 = 0$, with the smaller. Next, look at a_2 and repeat the process,
dividing the appropriate triangle in the same way. Keep repeating
the process. (Figure 1.2 shows an early stage in the construction.)
The triangles get smaller and smaller, and their intersection will be
a single point. This point is $P(x)$.

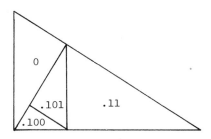

FIGURE 1.2

(a) Prove that the intersection of the triangles is a single
point. (Hint: use Exercise 5 of Section 5.3.)

(b) As noted earlier, 1/2 has two different binary expansions.
Show that the two expansions lead to the same value for $P(1/2)$. (If
$x = a/2^n$, where a and n are positive integers and $a/2^n < 1$, then x
has two binary expansions, but $P(x)$ is well-defined for essentially
the same reason that $P(1/2)$ is.)

(c) Show that P is continuous. (Hint: if x_0 and x_1 are sufficiently close, they have binary expansions which agree to a large
number of places.)

(d) Show that P is onto S.

(e) (For the especially brave). Show that P is nowhere differentiable if the smaller acute angle of the triangle is $> 30°$ (and $< 45°$,
of course).

*8. Let V and W be finite dimensional inner product spaces; let $U \subset V$,
$U' \subset W$ be open subsets. We say U and U' are *diffeomorphic* if there

is a differentiable function f taking U onto U' with differentiable
inverse. We call the mapping f a *diffeomorphism*.

(a) Show that any two non-empty open balls in V are diffeomorphic.

(b) Suppose U and U' are as above and that f:U → U' is differentiable and surjective with $D_{\vec{v}}f$ invertible for all $\vec{v} \in U$. Is f a
diffeomorphism?

(c) If we assume in addition that f is injective, can we conclude f is a diffeomorphism?

2. THE PROOF OF THEOREM 1.3

This section is devoted to proving the Inverse Function Theorem for
functions from R^1 to R^1. As we shall see, part (a) is fairly easy,
but (b) and (c) will take more work. This work will, however, be useful preparation for the proof of Theorem 1.4 given in the next section.

Let x_1 and x_2 be two distinct points in (a,b). According to the
Mean Value Theorem (Exercise 9, Section 3.5), there is a point ξ between x_1 and x_2 such that

$$f(x_1) - f(x_2) = f'(\xi)(x_1 - x_2)$$

Since $f'(\xi) \neq 0$, $f(x_1) - f(x_2) \neq 0$, which means that $f(x_1) \neq f(x_2)$.
Thus f is injective and (a) is proved.

To prove part (b), we use the Intermediate Value Theorem (Theorem
5.8.6). Let x_1 and x_2 be points in (a,b), as above, and let $c = f(x_1)$
and $d = f(x_2)$. Suppose that y is any number between x_1 and x_2. Then
Theorem 5.8.6 states that for some x between x_1 and x_2, $y = f(x)$.
That is, the set of points $I = \{f(x):x \in (a,b)\}$ has the property that
if c and d are any two points in I, then every point between c and d
is in I. We saw in Section 5.8 that I is therefore an interval.
(See Exercise 9 of Section 5.8.)

To finish the proof of (b), we need to show that I is an open
interval. If not, then I is of the form [c,d), [c,d), (c,d], (-∞,d],

or $[c,-\infty)$. In every one of these cases, I has either a largest or smallest element. Thus f takes on either a maximum or a minimum at some point $x_0 \in (a,b)$. We know (Exercise 7, Section 3.5) that $f'(x_0)$ = 0. This contradicts the hypothesis about f; therefore I is open, and (b) is proved.

In order to prove (c), we need the following.

LEMMA 2.1. *Let V and W be normed vector spaces, C a compact subset of V, and $D \subset W$. Suppose $f:C \to D$ is a continuous bijective function. Then $f^{-1}:D \to C$ is continuous.*

Proof. We use the characterization of continuity given in Exercise 11, Section 5.2. Thus we need to show that $(f^{-1})^{-1}(B)$ is a closed set whenever B is a closed subset of V. However,

$$(f^{-1})^{-1}(B) = \{\vec{v} \in D: f^{-1}(\vec{v}) \in B\}$$

$$= \{\vec{v} \in D: f^{-1}(\vec{v}) \in B \cap C\}$$

$$= D \cap f(B \cap C)$$

since f is defined only for points in C. Now B is closed and C is compact; thus $B \cap C$ is compact. Using Theorem 5.5.2, we see that $f(B \cap C)$ is compact and therefore closed.

The set D, being compact (Proposition 5.5.2), is also closed, and therefore $D \cap f(B \cap C)$ is closed (by Proposition 3.2.2). We have thus shown that $(f^{-1})^{-1}(B)$ is closed wherever B is closed and therefore that f^{-1} is continuous. (See also Exercise 3.)

We can now prove statement (c) of Theorem 1.3. First of all, let y be any point in (c,d) and choose $x \in (a,b)$ with $f(x) = y$. Let $a',b' \in (a,b)$ be such that $a < a' < x < b' < b$. Then $[a',b']$ is compact and f is continuous and injective on $[a',b']$. Thus, f^{-1} is continuous on the image of $[a',b']$. Furthermore, the image under f of the open interval (a',b') is an open interval containing y. It follows easily that f^{-1} is continuous at y. Since y was arbitrarily chosen in (c,d), f^{-1} is continuous on (c,d).

Let y_0 be a point in (c,d) with $f^{-1}(y_0) = x_0$ (so that $f(x_0) = y_0$). We now show that

$$\lim_{y \to y_0} \left| \frac{f^{-1}(y) - f^{-1}(y_0)}{y - y_0} - \frac{1}{f'(x_0)} \right| = 0$$

This will prove that $(f^{-1})'(y_0) = 1/f'(x_0)$, which is statement (c).

Let $\varepsilon > 0$ be any small number, $\varepsilon < 1/2$, and let $x = f^{-1}(y)$ (so that $f(x) = y$). To begin with,

$$\left| f^{-1}(y) - f^{-1}(y_0) - \frac{y - y_0}{f'(x_0)} \right| = \left| x - x_0 - \frac{y - y_0}{f'(x_0)} \right| \qquad (2.1)$$

$$= \frac{1}{|f'(x_0)|} \left| f'(x_0)(x - x_0) \right.$$
$$\left. - (f(x) - f(x_0)) \right|$$

Since $f'(x_0)$ is the derivative of f at x_0, we can choose $\delta_1 > 0$ so that

$$\left| f(x) - f(x_0) - f'(x_0)(x - x_0) \right| < \varepsilon |f'(x_0)| |x - x_0| \qquad (2.2)$$

when $|x - x_0| < \delta_1$. (We apply the definition of differentiability with $\varepsilon |f'(x_0)|$ replacing ε.) Furthermore, since f^{-1} is continuous at y_0, we can find $\delta > 0$ such that

$$|x - x_0| = |f^{-1}(y) - f^{-1}(y_0)| < \delta_1$$

whenever $|y - y_0| < \delta$. It follows from (2.1) and (2.2) that

$$\left| f^{-1}(y) - f^{-1}(y_0) - \frac{1}{f'(x_0)}(y - y_0) \right| < \varepsilon |x - x_0| \qquad (2.3)$$

whenever $|y - y_0| < \delta$. Now

$$|x - x_0| = |f^{-1}(y) - f^{-1}(y_0)|$$

$$= \left| f^{-1}(y) - f^{-1}(y_0) - \frac{1}{f'(x_0)}(y - y_0) + \frac{1}{f'(x_0)}(y - y_0) \right|$$

$$\leq \left| f^{-1}(y) - f^{-1}(y_0) - \frac{1}{f'(x_0)}(y - y_0) \right| + \frac{1}{|f'(x_0)|}|y - y_0|$$

$$< \varepsilon |x - x_0| + \frac{1}{|f'(x_0)|}|y - y_0|$$

when $|y - y_0| < \delta$. (We use (2.3) here.) Thus

$$(1 - \varepsilon)|x - x_0| < \frac{1}{|f'(x_0)|}|y - y_0|$$

when $|y - y_0| < \delta$. Since $\varepsilon < 1/2$, we have

$$|x - x_0| < \frac{1}{1-\varepsilon} \frac{|y - y_0|}{|f'(x_0)|} < \frac{2}{|f'(x_0)|}|y - y_0| \qquad (2.4)$$

when $|y - y_0| < \delta$. Combining (2.3) and (2.4), we obtain the inequality

$$\left| f^{-1}(y) - f^{-1}(y_0) - \frac{1}{|f'(x_0)|}(y - y_0) \right| < \frac{2\varepsilon}{|f'(x_0)|}|y - y_0| \qquad (2.5)$$

when $|y - y_0| < \delta$. Therefore

$$\left| \frac{f^{-1}(y) - f^{-1}(y_0)}{y - y_0} - \frac{1}{f'(x_0)} \right| < \frac{2\varepsilon}{|f'(x_0)|}$$

whenever $|y - y_0| < \delta$. Since ε was arbitrarily small, it follows that

$$\lim_{y \to y_0} \left| \frac{f^{-1}(y) - f^{-1}(y_0)}{y - y_0} - \frac{1}{f'(x_0)} \right| = 0$$

and statement (c) is proved. This completes the proof of Theorem 1.3.

EXERCISES

*1. Show that if U is an open interval in R and $f:U \to R$ is a continuous function with the property that $f(a) < f(b)$ if $a < b$ (and $a,b \in U$), then f has a continuous inverse on U.

2.   Use Exercise 1 to show that if n is odd, then $f(x) = x^n$ has a continuous inverse on all of R.

3.   Use the definition of compactness directly to prove Lemma 2.1.
(Hint: if f^{-1} is not continuous at $\vec{w}_0$, show that there is an $\varepsilon > 0$
and a sequence $\{\vec{w}_n\}$ such that $\vec{w}_n \to \vec{w}_0$ and $\|f^{-1}(\vec{w}_n) - f^{-1}(\vec{w}_0)\| \geq \varepsilon$ for
all n. Let $\vec{v}_n = f^{-1}(\vec{w}_n)$, $\vec{v}_0 = f^{-1}(\vec{w}_0)$. Pick a convergent subsequence
of the $\vec{v}_n$'s with limit $\vec{v}_0'$, say, and show that $f(\vec{v}_0') = \vec{w}_0 = f(\vec{v}_0)$. Deduce a contradiction.)

3.   THE PROOF OF THE GENERAL INVERSE FUNCTION THEOREM

We now prove Theorem 1.4. The proofs of statements (a) and (c) are
analogous to the proofs of the corresponding statements of Theorem
1.3. However, the proof of statement (b) is quite different; this is
due to the fact that the Intermediate Value Theorem has no analogue
in higher dimensions. As usual, we assume that the inner product
spaces dealt with here are Euclidean spaces.

We begin with another extension of the Mean Value Theorem.

PROPOSITION 3.1. *Let* $B = B_r(\vec{v}_0)$ *be the open ball of radius r
about* $\vec{v}_0$, *and let* $f:B \to R^m$ *be a function with a continuous partial derivative on B with component functions* $f_1,\ldots,f_m:B \to R$. *Then, for any*
$\vec{v}_1,\vec{v}_2 \in B$, *there are points* $\vec{c}_1,\ldots,\vec{c}_m$ *on the line segment from* $\vec{v}_1$ *to*
$\vec{v}_2$ *such that*

$$f(\vec{v}_1) - f(\vec{v}_2) = L(\vec{v}_1 - \vec{v}_2)$$

where $L:R^n \to R^m$ *is the linear transformation with matrix* (a_{ij}) *given
by*

$$a_{ij} = \frac{\partial f_i}{\partial x_j}(\vec{c}_i)$$

This result is proved by applying Proposition 6.6.1 to each of
the component functions $f_i:U \to R$, $1 \leq i \leq m$. We leave the details
to the reader.

We can now prove statement (a) of Theorem 1.4. (Compare this
proof with the proof of statement (a) of Theorem 1.3.)

PROPOSITION 3.2. *Let U_1 be an open subset in R^n and $f:U_1 \to R^n$ a function with a continuous differential. Suppose $\vec{v}_0$ is a point in U_1 such that $D_{\vec{v}_0} f$ is an isomorphism. Then f is injective on some open set U with $\vec{v}_0 \in U \subset U_1$.*

Proof. In order to prove this proposition, we need the notion of the determinant of an n × n matrix and two of its elementary properties. We state the result that we need here, leaving the detailed development to the next chapter.

For any n × n matrix $A = (a_{ij})$, the determinant of A is a real number Det A. As we shall see, if n = 1, then Det $A = a_{11}$, and if n = 2, then

$$\text{Det } A = a_{11}a_{22} - a_{12}a_{21}$$

We need two results here:

Det A is a continuous function of the matrix
entries a_{ij} (3.1)

A is invertible if and only if Det $A \neq 0$ (3.2)

(As we shall see, Det A is a polynomial in the matrix coefficients. We have seen an example of the second result in Exercise 16 of Section 2.6.)

Now let $W = R^n \times \cdots \times R^n$ be the n-fold Cartesian product of R^n with itself, with inner product

$$< (\vec{v}_1,\ldots,\vec{v}_n),(\vec{w}_1,\ldots,\vec{w}_n) > = \sum_{j=1}^{n} < \vec{v}_j,\vec{w}_j >$$

Let $\tilde{U}_1 = U_1 \times \cdots \times U_1 \subset W$ and define $G:\tilde{U}_1 \to R$ by

$$G(\vec{v}_1,\ldots,\vec{v}_n) = \text{Det}(b_{ij})$$

where

$$b_{ij} = \frac{\partial f_i}{\partial x_j}(\vec{v}_i)$$

G is a continuous function, from property (3.1) above and the fact that the b_{ij} are all continuous functions. Furthermore, $G(\vec{v}_0,\ldots,\vec{v}_0)$ = $\text{Det}(D_{\vec{v}_0}f)$, which is non zero using (3.2) (since $D_{\vec{v}_0}(f)$ is invertible, by assumption). Thus we can find a $\delta > 0$ such that

$$\left| G(\vec{v}_1,\ldots,\vec{v}_n) - G(\vec{v}_0,\ldots,\vec{v}_0) \right| < \left| G(\vec{v}_0,\ldots,\vec{v}_0) \right|$$

whenever $\|\vec{v}_i - \vec{v}_0\| < \delta$, for all $i = 1,\ldots,n$. As a result, $G(\vec{v}_1,\ldots,\vec{v}_n) \neq 0$ whenever $\vec{v}_1,\ldots,\vec{v}_n < B_\delta(\vec{v}_0)$.

Suppose now that $\vec{v}_1,\vec{v}_2 \in B_\delta(\vec{v}_0)$ with $f(\vec{v}_1) = f(\vec{v}_2)$. According to Proposition 3.1, we can find points $\vec{c}_1,\ldots,\vec{c}_n$ on the line segment from $\vec{v}_1$ to $\vec{v}_2$ such that

$$\vec{0} = f(\vec{v}_1) - f(\vec{v}_2) = L(\vec{v}_1 - \vec{v}_2)$$

where L is the linear transformation with matrix (b_{ij}),

$$b_{ij} = \frac{\partial f_i}{\partial x_j}(\vec{c}_i)$$

However, since $\vec{v}_1,\vec{v}_2 \in B_\delta(\vec{v}_0)$, it follows that $\vec{c}_1,\ldots,\vec{c}_n \in B_\delta(\vec{v}_0)$, so that $G(\vec{c}_1,\ldots,\vec{c}_n) = \text{Det } L \neq 0$. Thus L is an isomorphism and $L(\vec{v}_1 - \vec{v}_2) = \vec{0}$ can only happen if $\vec{v}_1 - \vec{v}_2 = \vec{0}$. Therefore $\vec{v}_1 = \vec{v}_2$ whenever $f(\vec{v}_1) = f(\vec{v}_2)$ on $B_\delta(\vec{v}_0)$ and Proposition 3.2 is proved with $U = B_\delta(v_0)$.

It follows that the restriction of f to U has an inverse. In fact, we now see that this inverse is continuous.

PROPOSITION 3.3. *Let U be an open set in* R^n *and* $f:U \to R^n$ *a continuous injective function. Then the inverse to f,* $f^{-1}:f(U) \to R^n$ *is continuous.*

To prove f^{-1} continuous at $\vec{w}_0$, we use Proposition 3.1 to show that $\|f(\vec{w}) - f(\vec{w}_0)\| \geq M\|\vec{w} - \vec{w}_0\|$ for some $M > 0$ and all $\vec{w}$ near $\vec{w}_0$. We leave the details to the reader.

The next result gives statement (b) of Theorem 1.4.

PROPOSITION 3.4. *Let U be an open set in* R^n, *and let* $f:U \to R^n$ *be a function. Suppose that the differential of f is continuous and* that $D_{\vec{v}}f:R^n \to R^n$ is an isomorphism for all $\vec{v} \in U$. Then the image V of U under f is an open set.

Proof. Let $\vec{v}_0$ be an arbitrary point in U and set $\vec{w}_0 = f(\vec{v}_0)$. We must show that there is some open ball about $\vec{w}_0$ contained entirely in V.

Since U is open, $\vec{v}_0 \in U$, and $D_{\vec{v}_0}f$ is an isomorphism, we can find an $r > 0$ such that the closed ball $\bar{B} = \bar{B}_r(\vec{v}_0)$ is contained in U and f is injective on $\bar{B}$. Let C be the boundary of $\bar{B}$,

$$C = \{\vec{v} \in U : \|\vec{v} - \vec{v}_0\| = r\}$$

Then, both $\bar{B}$ and C are compact (since they are both closed and bounded; see Theorem 5.6.1) so that both $f(\bar{B})$ and $f(C)$ are also compact (by Proposition 5.5.2). Let $d > 0$ be the number defined by

$$d = glb\{\|\vec{w} - \vec{w}_0\| : \vec{w} \in f(C)\}$$

We now need the following:

LEMMA 3.5. The number d defined above is positive.

Proof. Note, first of all, that $\vec{w}_0 \notin f(C)$. For if $f(\vec{v}_1) = \vec{w}_0$ for $\vec{w}_1 \in C$, then we would have $\vec{v}_0 = \vec{v}_1$, since f is injective on $\bar{B}$ and $f(\vec{v}_0) = \vec{w}_0$. However, $\vec{v}_0 \notin C$, so this is impossible.

Now define a function $h:C \to R$ by

$$h(\vec{v}) = \|f(\vec{v}) - \vec{w}_0\|$$

As h is continuous and C is compact, h has a minimum (Theorem 5.5.3). This minimum is nonnegative (since h is), and is not 0 (otherwise $\vec{w} \in f(c)$). But it is easy to see that this minimum value is d, and the lemma is proved.

Proposition 3.4 is now a consequence of the next result.

LEMMA 3.6. Let $s = 1/3\ d$. Then $B_s(\vec{w}_0) \subset f(U)$.

Proof. Let $\vec{w}_1$ be any point in $B_s(\vec{w}_0)$, and define $\varphi : \bar{B} \to R$ by

$$\varphi(\vec{v}) = \| f(\vec{v}) - \vec{w}_1 \|^2$$

Since $\bar{B}$ is compact and φ is continuous, φ attains its maximum and minimum on $\bar{B}$ (by Proposition 5.5.3). Let $\vec{v}_1 \in \bar{B}$ be the point at which φ attains its minimum. We shall show that $f(\vec{v}_1) = \vec{w}_1$, so that $\vec{w}_1 \in f(\bar{B}) \subseteq f(U)$ and the lemma follows.

To prove that $f(\vec{v}_1) = \vec{w}_1$, we first note that $\vec{v}_1 \notin C$. For if $\vec{v}_1 \in C$, we have

$$\| f(\vec{v}_1) - \vec{w}_1 \| = \| f(\vec{v}_1) - \vec{w}_0 + \vec{w}_0 - \vec{w}_1 \|$$

$$\geq \| f(\vec{v}_1) - \vec{w}_0 \| - \| \vec{w}_0 - \vec{w}_1 \|$$

$$\geq d - \frac{d}{3} = \frac{2d}{3}$$

Thus $\varphi(\vec{v}_1) \geq (4/9)d$. However $\varphi(\vec{v}_0) \leq (1/9)d$ (since $\vec{w}_1 \in B_s(\vec{w}_0)$), so that $\varphi(\vec{v}_1)$ cannot be a minimum value for φ.

Therefore $\vec{v}_1$ is in the interior of $\bar{B}$, and it follows from Theorem 8.1.1 that $D_{\vec{v}_1} \varphi$ is the zero transformation. To complete the proof, we need to compute $D_{\vec{v}_1} \varphi$.

LEMMA 3.7. *If φ and f are as above, then*

$$(D_{\vec{v}} \varphi)(\vec{w}) = 2 < (D_{\vec{v}} f)\vec{w}, f(\vec{v}) - \vec{w}_1 >$$

Proof. Let $\vec{w}_1 = (a_1, \ldots, a_n)$ and $f(\vec{v}) = (f_1(\vec{v}), \ldots, f_n(\vec{v}))$. Then

$$\varphi(\vec{v}) = \sum_{j=1}^{n} (f_j(\vec{v}) - a_j)^2$$

Using the chain rule, we see that

$$\frac{\partial \varphi}{\partial x_i}(\vec{v}) = 2 \sum_{j=1}^{n} \frac{\partial f_j}{\partial x_i}(\vec{v}) (f_j(\vec{v}) - a_j)$$

Thus, if $\vec{w} = (b_1, \ldots, b_n)$,

$$(D_{\vec{v}}\varphi)(\vec{w}) = 2 \sum_{i=1}^{n} \sum_{j=1}^{n} b_i \frac{\partial f_j}{\partial x_i}(\vec{v})(f_j(\vec{v}) - a_j)$$

$$= 2 < (D_{\vec{v}}f)\vec{w}, \; f(\vec{v}) - \vec{w}_1 >$$

We may rewrite this formula as

$$(D_{\vec{v}}\varphi)(\vec{w}) = 2 < \vec{w}, (D_{\vec{v}}f)^*(f(\vec{v}) - \vec{w}_1) > \qquad (3.3)$$

(See Section 7.5).

Since $D_{\vec{v}_1}\varphi$ is the zero transformation, we must have $(D_{\vec{v}_1}\varphi)(\vec{w}) = \vec{0}$ for all $\vec{w} \in R^n$. It follows from Lemma 3.7 (or equation (3.1)) that

$$< (D_{\vec{v}_1} f)^*(f(\vec{v}_1) - \vec{w}_1), \vec{w} > = 0$$

for all $\vec{w} \in R^n$. Hence

$$(D_{\vec{v}_1} f)^*(f(\vec{v}_1) - \vec{w}_1) = \vec{0}$$

However, $D_{\vec{v}_1} f$ is an isomorphism for all $\vec{v}_1 \in U_1$, and therefore $(D_{\vec{v}_1} f)^*$ is also (Proposition 7.5.5), so that $f(\vec{v}_1) - \vec{w}_1$ must be the zero vector. This completes the proof of Lemma 3.6 and of Proposition 3.4.

It remains to prove that the local inverse to f is differentiable.

PROPOSITION 3.8. *Let U be an open subset of* R^n *and* $f:U \to R^n$ *an injective function with a continuous differential. Suppose further that* $D_{\vec{v}}f$ *is an isomorphism for all* $\vec{v} \in U$ *and let V be the image of U under f (which is open by Proposition 3.4). Then* $f^{-1}:V \to U$ *is differentiable.*

Proof. Let $\vec{v}_0$ be a fixed point of U, $\vec{w}_0 = f(\vec{v}_0)$, and $L = D_{\vec{v}_0} f$. We must show that

$$\lim_{\vec{w} \to \vec{w}_0} \frac{\|f^{-1}(\vec{w}) - f^{-1}(\vec{w}_0) - L^{-1}(\vec{w} - \vec{w}_0)\|}{\|\vec{w} - \vec{w}_0\|} = 0$$

Let ε be any small number, $0 < \varepsilon < 1/2$. If $\vec{v} = f^{-1}(\vec{w})$ (so that $f(\vec{v}) = \vec{w}$), we have

$$\|f^{-1}(\vec{w}) - f^{-1}(\vec{w}_0) - L^{-1}(\vec{w} - \vec{w}_0)\| = \|L^{-1}(f(\vec{v}) - f(\vec{v}_0) - L(\vec{v} - \vec{v}_0)\|$$

$$\leq M\|f(\vec{v}) - f(\vec{v}_0) - L(\vec{v} - \vec{v}_0)\|$$

where M is chosen so that

$$\|L^{-1}\vec{w}\| \leq M\|\vec{w}\|$$

for all $\vec{w} \in R^n$. (See Exercise 18 of Section 3.3.) Since $L = D_{\vec{v}_0} f$, we can find $\delta_1 > 0$ such that

$$\|f(\vec{v}) - f(\vec{v}_0) - L(\vec{v} - \vec{v}_0)\| < \varepsilon M\|\vec{v} - \vec{v}_0\|$$

whenever $\|\vec{v} - \vec{v}_0\| < \delta_1$. Furthermore,

$$\|\vec{v} - \vec{v}_0\| = \|f^{-1}(\vec{w}) - f^{-1}(\vec{w}_0)\|$$

and, since f^{-1} is continuous (by Proposition 3.3), we can find $\delta > 0$ such that

$$\|f^{-1}(\vec{w}) - f^{-1}(\vec{w}_0)\| < \delta_1$$

whenever $\|\vec{w} - \vec{w}_0\| < \delta$. It follows that

$$\|f^{-1}(\vec{w}) - f^{-1}(\vec{w}_0) - L^{-1}(\vec{w} - \vec{w}_0)\| < \varepsilon\|\vec{v} - \vec{v}_0\| \qquad (3.4)$$

whenever $\|\vec{w} - \vec{w}_0\| < \delta$.

Now

$$\|\vec{v} - \vec{v}_0\| = \|f^{-1}(\vec{w}) - f^{-1}(\vec{w}_0)\|$$

$$= \|f^{-1}(\vec{w}) - f^{-1}(\vec{w}_0) - L^{-1}(\vec{w} - \vec{w}_0) + L^{-1}(\vec{w} - \vec{w}_0)\|$$

$$\leq \|f^{-1}(\vec{w}) - f^{-1}(\vec{w}_0) - L^{-1}(\vec{w} - \vec{w}_0)\| + \|L^{-1}(\vec{w} - \vec{w}_0)\|$$

$$\leq \varepsilon\|\vec{v} - \vec{v}_0\| + M\|\vec{w} - \vec{w}_0\|$$

when $\|\vec{w} - \vec{w}_0\| < \delta$. (We use (3.4) here). Therefore

$$(1 - \varepsilon)\|\vec{v} - \vec{v}_0\| < M\|\vec{w} - \vec{w}_0\|$$

when $\|\vec{w} - \vec{w}_0\| < \delta$. Since $\varepsilon < 1/2$, we have

$$\|\vec{v} - \vec{v}_0\| < 2M\|\vec{w} - \vec{w}_0\| \tag{3.5}$$

when $\|\vec{w} - \vec{w}_0\| < \delta$. Combining (3.4) and (3.5), we obtain the inequality

$$\|f^{-1}(\vec{w}) - f^{-1}(\vec{w}_0) - L^{-1}(\vec{w} - \vec{w}_0)\| < 2\varepsilon M\|\vec{w} - \vec{w}_0\|$$

when $\|\vec{w} - \vec{w}_0\| < \delta$. Therefore

$$\frac{\|f^{-1}(\vec{w}) - f^{-1}(\vec{w}_0) - L^{-1}(\vec{w} - \vec{w}_0)\|}{\|\vec{w} - \vec{w}_0\|} < 2\varepsilon M$$

whenever $\|\vec{w} - \vec{w}_0\| < \delta$. Since ε was arbitrary, it follows that

$$\lim_{\vec{w} \to \vec{w}_0} \frac{\|f^{-1}(\vec{w}) - f^{-1}(\vec{w}_0) - L^{-1}(\vec{w} - \vec{w}_0)\|}{\|\vec{w} - \vec{w}_0\|} = 0$$

and statement (c) of Theorem 1.4 is proved.

EXERCISES

1.   Prove Proposition 3.1.

2.   Let $W = R^n \times \cdots \times R^n$ be the cartesian product of n copies of R^n with inner product

$$< (\vec{w}_1,\ldots,\vec{w}_n),(\vec{v}_1,\ldots,\vec{v}_n) > = \sum_{i=1}^{n} < \vec{v}_i,\vec{w}_i >$$

Let U be an open subset of R^n and $\tilde{U} = U \times \cdots \times U \subset W$. Prove that $\tilde{U}$ is open.

3.   Prove the analogue of Exercise 2 for closed sets.

4.   Prove Proposition 3.3.

5.   Let C, D be disjoint subsets of R^n, C compact and D closed.   Let
d $\in$ R be defined by

$$d = \text{glb}\{\|\vec{v} - \vec{w}\| : \vec{v} \in C, \vec{w} \in D\}$$

Prove that d > 0.   (See the proof of Lemma 3.5.)

6.   Give an example to show Exercise 5 is not true if we only assume
C and D closed.

7.   Verify that the result of Lemma 3.7. is in accord with that of
Exercise 10, Section 6.1.

4.   THE IMPLICIT FUNCTION THEOREM: I

It is often necessary to deal with relations between variables which
do not define functions.   Consider, for instance, the set of points
$(x,y) \in R^2$ satisfying

$$x^2 + y^2 - 1 = 0 \qquad\qquad\qquad (4.1)$$

These points lie on the unit circle about the origin (See Figure 4.1).

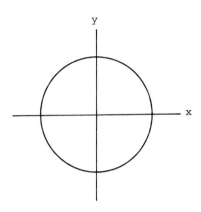

FIGURE 4.1.

Notice that (4.1) does not define y as a function of x, since there are two values of y satisfying (4.1) for every x with $-1 < x < 1$. Similarly, x is not a function of y.

We can use (4.1) to define y as a function of x (for $-1 < x < 1$) by the simple expedient of choosing one of the values of y for each x. If, as is reasonable, we want the function to be continuous, then we have two ways of defining the function, one corresponding to the upper half of the circle and one corresponding to the lower half: $y = (1 - x^2)^{1/2}$ or $y = - (1 - x^2)^{1/2} (-1 < x < 1)$. There is no parti- cular reason for picking one of these or the other. If, however, we start out with a point on the curve $((- 3/5, - 4/5)$, say), then there is a natural choice: we pick the function whose graph contains the given point (in our case, we let $y = g(x) = - (1 - x^2)^{1/2}$).

The points $(1,0)$ and $(-1,0)$ are special: we do not have a good way of defining y as a function of x on any open interval containing one of these points. At $(1,0)$, for instance, we have two problems:

(a) for $x > 1$, there is no value of y such that (x,y) satisfies (4.1);

(b) for $x < 1$, we have a choice of functions (as above), and no way to decide which to take.

Now we consider a slightly more general situation. Let $f:R^2 \to R$ be a continuous differentiable function; consider the set of points (x,y) such that

$$f(x,y) = 0 \qquad\qquad\qquad (4.2)$$

We would like to use (4.2) to define y explicitly as a function of x. When is that possible?

The first point to notice is that generally we do not have pre- cisely one y satisfying (4.2) for each x. Thus we cannot simply use (4.2) to define the function. What we really hope to have is a con- tinuous function $g(x)$ such that

$$f(x,g(x)) = 0 \qquad\qquad\qquad (4.3)$$

Then we can say that (4.2) gives y implicitly as a function of x, the function being g.

As the example showed, we often have a choice of functions g.
To restrict our choice, we assume that we know a point (x_0, y_0) satisfying (4.2), and we require this point to be in the graph of g:

$$g(x_0) = y_0$$

We have not yet specified the domain of g, but, since $g(x_0)$ is defined,
g should be defined at least on some open interval containing x_0.

Does g exist? The example shows that we need some further condition. (We cannot find g for (4.1) if $(x_0, y_0) = (1,0)$, for instance.)
The condition is fairly simple: a certain partial derivative must not
vanish.

THEOREM 4.1. *Let* $f : R^2 \to R^1$ *be continuously differentiable, let*
$f(x_0, y_0) = 0$, *and suppose that* $\partial f / \partial y(x_0, y_0) \neq 0$. *Then there is an*
open interval U *about* y_0 *with the following property: there exists*
a continuous function $g : U \to R^1$ *such that* $g(x_0) = y_0$ **and**

$$f(x, g(x)) = 0$$

These properties determine g *uniquely. Moreover,* g *is differentiable,*
and

$$g'(x) = - \left(\frac{\partial f}{\partial x}\right)\left(\frac{\partial f}{\partial y}\right)^{-1} \tag{4.4}$$

(where the partial derivatives are evaluated at $(x, g(x))$.

Theorem 4.1 is a special case of the Implicit Function Theorem,
which we shall soon state more generally. Its proof, like that of the
general Implicit Function Theorem, is based on the Inverse Function
Theorem. We shall give the proof of Theorem 4.1 separately, however,
because it is somewhat more explicit and easier to follow.

First, however, we note something about formula (4.4). The reader may have learned about differentiation of implicit functions in
elementary calculus. For instance, suppose that y is defined implicitly as a function of x by

$$f(x,y) = x^2 + xy + y^2 - 7 = 0$$

The usual rule for computing $\frac{dy}{dx}$ is to differentiate both sides of this equation according to the usual rules, remembering that y is a function of x:

$$2x + y + x\frac{dy}{dx} + 2y\frac{dy}{dx} = 0$$

or

$$(2x + y) + (x + 2y)\frac{dy}{dx} = 0 \tag{4.5}$$

Now it is easy to compute $\frac{dy}{dx}$ by solving:

$$\frac{dy}{dx} = -\frac{2x + y}{x + 2y}$$

This procedure is equivalent to (4.4). Notice that in (4.5), the terms involving $\frac{dy}{dx}$ always come from differentiating some function of y. In fact, the coefficient of $\frac{dy}{dx}$ in (4.5) is just $\frac{\partial f}{\partial y}$. Similarly, the terms involving $\frac{dy}{dx}$ are just $\frac{\partial f}{\partial x}$. Thus (4.5) amounts to

$$\frac{\partial f}{\partial x} + \frac{\partial f}{\partial y}\frac{dy}{dx} = 0$$

and this equation gives (4.4).

Proof of Theorem 4.1. Define $F:R^2 \to R^2$ by

$$F(x,y) = (x,f(x,y))$$

Then F' has matrix

$$\begin{pmatrix} 1 & 0 \\ \frac{\partial f}{\partial x} & \frac{\partial f}{\partial y} \end{pmatrix}$$

which is invertible at (x_0,y_0) since $\frac{\partial f}{\partial y}(x_0,y_0) \neq 0$. (Verify this!). The Inverse Function Theorem says that there is a ball B of radius $r > 0$ about $(x_0,0) = F(x_0,y_0)$ in R^2 and a function $G:B \to R^2$ which is a differentiable inverse for F. Let

$$G(x,y) = (g_1(x,y), g_2(x,y))$$

Then

$$(x,y) = F(G(x,y)) = F(g_1(x,y), g_2(x,y)) \qquad (4.6)$$
$$= (g_1(x,y), f(g_1(x,y), g_2(x,y)))$$

Equating coordinates, we find that

$$x = g_1(x,y)$$

and

$$y = f(g_1(x,y), g_2(x,y))$$
$$= f(x, g_2(x,y))$$

Let $U = \{x: |x - x_0| < r\}$, so that $(x,0) \in B$ if $x \in U$.   Set

$$g(x) = g_2(x,0)$$

Then g is differentiable, and

$$0 = f(x, g_2(x,0))$$
$$= f(x, g(x))$$

Furthermore, since

$$G(x_0, 0) = (x_0, y_0)$$

we have

$$g(x_0) = g_2(x_0, 0) = y_0$$

Therefore g is the required function.

To finish the proof, we compute $g'(x)$.   We know that

$$\begin{pmatrix} 1 & 0 \\ 0 & 1 \end{pmatrix} = F'(G(x,y))G'(x,y)$$

$$= \begin{pmatrix} 1 & 0 \\ \frac{\partial f}{\partial x}(G(x,y)) & \frac{\partial f}{\partial y}(G(x,y)) \end{pmatrix} \begin{pmatrix} \frac{\partial g_1}{\partial x} & \frac{\partial g_1}{\partial y} \\ \frac{\partial g_2}{\partial x} & \frac{\partial g_2}{\partial y} \end{pmatrix}$$

by the chain rule. Since $g_1(x,y) = x$ we have

$$\frac{\partial g_1}{\partial x} = 1, \quad \frac{\partial g_1}{\partial y} = 0$$

Now compute the lower left-hand term of the product:

$$0 = \frac{\partial f}{\partial x}(G(x,y)) + \frac{\partial f}{\partial y}(G(x,y))\frac{\partial g_2}{\partial x}$$

Let $y = 0$, so that $g_2(x,0) = g(x)$; we get

$$0 = \frac{\partial f}{\partial x}(x,g(x)) + \frac{\partial f}{\partial y}(x,g(x))\frac{dg}{dx}$$

or

$$\frac{dg}{dx} = - \left(\frac{\partial f}{\partial x}\right)\left(\frac{\partial f}{\partial y}\right)^{-1}$$

(where the derivatives are evaluated at $(x,g(x))$). This completes the proof.

Notice that in the example at the beginning of the section, there are two points, $(1,0)$ and $(-1,0)$, near which y is not defined implicitly as a function of x; at those points, $\frac{\partial f}{\partial y} = 0$. However, $\frac{\partial f}{\partial x} \neq 0$ there, so that near those points x is defined implicitly as a function of y.

Notice also that the function f need not be defined on all of R^2; the theorem holds even if f is defined only on an open set of R^2.

EXERCISES

1. Each of the following equations define y as a function of x on an interval including the point given. Compute $\frac{dy}{dx}$ at that point.

(a) $x^2 + xy + y^3 - 11 = 0$: $(1,2)$

(b) $\sin x + \cos y - 1 = 0$: $(\pi/2, \pi/2)$

(c) $x^2 + y^3 - 12 = 0$: $(-2,2)$

(d) $e^x + \tan y - 1 = 0$: $(0,0)$

(e) $xy - x^2 + y^6 + 1 = 0$: $(2,1)$

2.   Regard x as a function of y and compute $\frac{dx}{dy}$ for each of the equations of Exercise 1 at the points indicated.

3.   Let $f(x,y) = x^4 + y^4 - 1$. At what points (x,y) on the set $S = \{(x,y):f(x,y) = 0\}$ does the condition $f(x,y) = 0$ fail to define y implicitly as a function of x? Where is x not implicitly a function of y?

4.   Let $f(x,y) = y - x \cos y + (x^2 + \pi + 2)/4$. At what points (x,y) on the set $\{(x,y):f(x,y) = 0\}$ does the condition $f(x,y) = 0$ fail to define either of x or y as a function of the other?

5.   (a) For what points on the lemniscate $(x^2 + y^2)^2 - 2a^2(x^2 - y^2) = 0$ $(a > 0)$ is it impossible to express either x or y as a function of the other?

(b) Find the points on the lemniscate where y can be regarded as a function of x and compute $\frac{dy}{dx}$ at these points.

*6.   Let $f:R^2 \to R$ have a continuous derivative, and let $S = \{(x,y): f(x,y) = 0\}$. Suppose there exists $\varepsilon > 0$, $N > 0$ such that $\frac{\partial f}{\partial y}(x,y) \geq \varepsilon$ and $\frac{\partial f}{\partial x}(x,y) \leq N$ for all $(x,y) \in S$ with $a < x < b$, and let $f(x_0,y_0) = 0$, $x_0 \in (a,b)$. Show that there is a differentiable function $g:(a,b) \to R$ such that $g(x_0) = y_0$ and $f(x,g(x)) = 0$. (The point is that the interval on which g is defined is specified.)

5.   THE IMPLICIT FUNCTION THEOREM: II

The general Implicit Function Theorem deals with functions from R^{n+k} to R^n. To see what it should say, consider a linear transformation

$T:R^{n+k} \to R^n$. We regard R^{n+k} as $R^k \times R^n$, think of T as a function of two variables $\vec{x} \in R^k$, $\vec{y} \in R^n$, and write $T(\vec{x},\vec{y}) = T_1(\vec{x}) + T_2(\vec{y})$, where $T_1:R^k \to R^n$ and $T_2:R^n \to R^n$ are linear transformations. For instance, if $T:R^3 \to R^2$ is defined by the matrix

$$\begin{pmatrix} 1 & 3 & 5 \\ 2 & 4 & 6 \end{pmatrix}$$

then we regard a point (x_1,x_2,x_3) in R^3 as $(x_1,(x_2,x_3))$, where $\vec{x} = x_1 \in R$ and $\vec{y} = (x_2,x_3) \in R^2$; T_1 is given by the matrix

$$\begin{pmatrix} 1 \\ 2 \end{pmatrix}$$

and T_2 by the matrix

$$\begin{pmatrix} 3 & 5 \\ 4 & 6 \end{pmatrix}$$

We would like to know when the equation $T(\vec{x},\vec{y}) = \vec{0}$ defines $\vec{y}$ implicitly as a function of $\vec{x}$. To begin, we consider the example above. We know that

$$T(x_1,x_2,x_3) = T_1(x_1) + T_2(x_2,x_3)$$

$$= \begin{pmatrix} 1 \\ 2 \end{pmatrix} x_1 + \begin{pmatrix} 3 & 5 \\ 4 & 6 \end{pmatrix} \begin{pmatrix} x_2 \\ x_3 \end{pmatrix}$$

Given x_1, we want to find x_2 and x_3 so that

$$T(x_1,x_2,x_3) = \vec{0}$$

that is, so that

$$\begin{pmatrix} 3 & 5 \\ 4 & 6 \end{pmatrix} \begin{pmatrix} x_2 \\ x_3 \end{pmatrix} = - \begin{pmatrix} 1 \\ 2 \end{pmatrix} x_1 = \begin{pmatrix} -x_1 \\ -2x_1 \end{pmatrix} \tag{5.1}$$

The matrix $\begin{vmatrix} 3 & 5 \\ 4 & 6 \end{vmatrix}$ is invertible; in fact, its inverse is

$$\begin{pmatrix} -3 & \dfrac{5}{2} \\ 2 & -\dfrac{3}{2} \end{pmatrix}$$

as is easily checked. Multiplying both sides of (5.1) by this matrix, we obtain

$$\begin{pmatrix} x_2 \\ x_3 \end{pmatrix} = \begin{pmatrix} -3 & \dfrac{5}{2} \\ 2 & -\dfrac{3}{2} \end{pmatrix} \begin{pmatrix} -x_1 \\ -2x_1 \end{pmatrix} = \begin{pmatrix} -2x_1 \\ x_1 \end{pmatrix} \tag{5.2}$$

This gives $\vec{y} = (x_2, x_3)$ as a function of $x_1 = \vec{x}$; this function is defined implicitly by $T(\vec{x}, \vec{y}) = \vec{0}$. The critical step in deriving (5.2) was inverting the linear transformation T_2.

Now we return to the general case, where $T : R^{n+k} \to R^n$. Suppose, as was the case in the example, that T_2 is invertible. Then if $\vec{x}$ is any vector in R^k, $T_1(\vec{x}) \in R^n$. We want to find $\vec{y} \in R^n$ such that

$$\vec{0} = T(\vec{x}, \vec{y}) = T_1(\vec{x}) + T_2(\vec{y}) \tag{5.3}$$

or

$$T_2(\vec{y}) = -T_1(\vec{x})$$

or

$$y = T_2^{-1}(-T_1(\vec{x})) = -T_2^{-1}T_1(\vec{x}) \tag{5.4}$$

Equation (5.4) defines $\vec{y}$ uniquely, and thus makes $\vec{y}$ a function of $\vec{x}$. Working backwards, we see that this choice of $\vec{y}$ lets us satisfy (5.3). That is, (5.3) defines $\vec{y}$ implicitly as a function of $\vec{x}$, and this function is given explicitly by (5.4).

We know from Theorem 7.4.4 and Proposition 7.4.5 that an $n \times n$ matrix is invertible if and only if its columns are independent (as vectors in R^n). Using this fact, we can summarize the discussion above as follows. Let $T : R^{n+k} \to R^n$ be a linear transformation considered as a function $T(\vec{x}, \vec{y})$, $\vec{x} \in R^k$, $\vec{y} \in R^n$. Then the equation $T(\vec{x}, \vec{y}) = \vec{0}$ can be solved for $\vec{y}$ as a function of $\vec{x}$ if the last n columns of the matrix of T are linearly independent.

Now let $f:R^{n+k} \to R^n$ be a function with continuous partial derivatives near a point $\vec{v}_0$ and suppose that $f(\vec{v}_0) = \vec{0}$. Since f can be approximated near $\vec{v}_0$ by the linear transformation $f'(\vec{v}_0)$, we might hope that the above analysis applies to f whenever it applies to $f'(\vec{v}_0)$. The statement that it does (at least near $\vec{v}_0$) is the Implicit Function Theorem.

THEOREM 5.1. (The Implicit Function Theorem). *Let U be an open subset of* R^{n+k} *and* $f:R^{n+k} \to R^n$ *a function with continuous partial derivatives on U.* (We write $f(\vec{x},\vec{y})$ where $\vec{x} \in R^k$, $\vec{y} \in R^n$.) *Suppose that* $f(\vec{x}_0,\vec{y}_0) = \vec{0}$ *and that the last n columns of the Jacobian of f at* $(\vec{x}_0,\vec{y}_0)$ *are independent. Then there is an open set* U_1 *in* R^k *containing* $\vec{x}_0$ *and a unique continuous function* $h:U_1 \to R^n$ *with* $h(\vec{x}_0) = \vec{y}_0$ *and* $f(\vec{x},h(\vec{x})) = \vec{0}$ *for all* $\vec{x}$ *in* U_1. *Furthermore, the function h is differentiable. In fact, if one writes the matrix for* $f'(\vec{x}_0,\vec{y}_0)$ *as* (A,B), *where A is the matrix defined by the first k columns of the Jacobian of f at* $(\vec{x}_0,\vec{y}_0)$ *and B is the matrix defined by the last n columns, then the Jacobian matrix for h at* $\vec{x}_0$ *is* $-B^{-1}A$.

We shall give the proof later in this section.

While we have stated the theorem in general, we shall be concerned primarily with the case where $n = 1$. Here are some examples.

1. Suppose that $f:R^3 \to R$ is given by

$$f(x_1,x_2,x_3) = x_1^2 + x_2^2 + x_3^2 - 1$$

We let $\vec{x} = (x_1,x_2)$ and $y = x_3$. The point $((1/3,2/3),-2/3) = (\vec{x}_0,\vec{y}_0)$ satisfies

$$f(\vec{x}_0,\vec{y}_0) = 0$$

Moreover, $\frac{\partial f}{\partial x_3} = 2x_3 \neq 0$ at $(\vec{x}_0,\vec{y}_0)$. (In this example, $n = 1$, and the condition on the Jacobian matrix is that the last entry should be non-zero.) Thus the conditions of Theorem 5.1 are met, and we can conclude that the equation $f(x_1,x_2,x_3) = 0$ defines x_3 implicitly as a function of x_1 and x_2 near the point $(1/3,2/3,-2/3)$. In this case, of course, we can easily write the function explicitly:

$$x_3 = - (1 - x_1^2 - x_2^2)^{1/2}$$

2. More generally, let $f:R^3 \to R$ be a continuously differentiable function, and suppose that

$$f(a_1,a_2,a_3) = 0, \quad \frac{\partial f}{\partial x_3}(a_1,a_2,a_3) \neq 0$$

Then the conditions of Theorem 5.1 are met at $(\vec{x}_0,y_0)$, where $\vec{x}_0 = (a_1,a_2)$ and $y_0 = a_3$. We conclude that $f(x_1,x_2,x_3) = 0$ defines x_3 as an implicit function of x_1 and x_2 near the point (a_1,a_2,a_3).

3. Similarly, if $f:R^{k+1} \to R$ is continuously differentiable and $f(\vec{x}_0,y_0) = 0$, $\frac{\partial f}{\partial x_{k+1}}(\vec{x}_0,y_0) \neq 0$, then Theorem 5.1 says that the equation $f(\vec{x},y) = 0$ implicitly defines y as a function of $\vec{x}$ near $(\vec{x}_0,y_0)$.

In the above examples, we always expressed the last variable as a function of the others. There is, of course, nothing special about the last variable. If, for instance, we had $\frac{\partial f}{\partial x_1}(a_1,a_2,a_3) \neq 0$ in Example 2, we could then use $f(x_1,x_2,x_3) = 0$ to express x_1 implicitly as a function of x_2 and x_3.

This fact is often quite useful. For example, let $f:R^{k+1} \to R^1$ be a continuously differentiable function with the property that if $f(\vec{v}) = 0$, then $(\text{grad } f)(\vec{v}) \neq \vec{0}$. Let $S = \{\vec{v}:f(\vec{v}) = 0\}$. If $\vec{v}_0 \in S$, then one of the partial derivatives of f at $\vec{v}_0$ is nonzero (since grad $f \neq \vec{0}$ on S), and it follows that the set S can be regarded near $\vec{v}_0$ as defining one coordinate as a function of the rest. Thus S can be regarded as made up of pieces of the graphs of functions from R^k to R^1. This is exactly what is needed to complete the proof of Theorem 8.4.1.

Similarly, we have used the general Implicit Function Theorem in Theorem 8.6.1. We had a function $G:R^{n+k} \to R^n$; we assumed that at each point $\vec{v}_0 \in S = \{\vec{v}:f(\vec{v}) = \vec{0}\}$, the gradients of the coordinate functions were independent. That assumption means that at each $\vec{v}_0 \in S$, the row rank of $J_G(\vec{v}_0)$ is n. By Theorem 7.4.4, the column rank of $J_G(\vec{v}_0)$ is also n, and the Implicit Function Theorem says that we can therefore

express some n coordinates as a function of the other k, at least
near $\vec{v}_0$. This is the fact we cited in Section 8.6.

Proof of Theorem 5.1. The proof follows the lines of the proof
of Theorem 4.1; the major difference is that the notation is more
complicated.

Define $F: R^{n+k} \to R^{n+k}$ by

$$F(\vec{x}, \vec{y}) = (\vec{x}, f(\vec{x}, \vec{y}))$$

Then F is differentiable, and it is not hard to check that the Jacobian matrix of F is

$$\begin{pmatrix} I_k & 0 \\ & J_f \end{pmatrix}$$

where I_k is the k × k identity matrix, 0 is a k × n matrix of zeroes,
and J_f is the Jacobian matrix of f. Moreover, our condition on
$D_{(\vec{x}_0, \vec{y}_0)}(f)$ makes $D_{(\vec{x}_0, \vec{y}_0)}(F)$ invertible; see Exercise 5 for a proof.
We may thus apply Theorem 1.4 to F, and conclude that there is an in-
verse G for F defined in some ball U_0 about $F(\vec{x}_0, \vec{y}_0) = (\vec{x}_0, \vec{0})$. Write

$$G(\vec{x}, \vec{y}) = (g_1(\vec{x}, \vec{y}), g_2(\vec{x}, \vec{y}))$$

where $g_1(\vec{x}, \vec{y}) \in R^k$ and $g_2(\vec{x}, \vec{y}) \in R^n$. We have

$$\begin{aligned}
(\vec{x}, \vec{y}) &= F(G(\vec{x}, \vec{y})) \\
&= F(g_1(\vec{x}, \vec{y}), g_2(\vec{x}, \vec{y})) \\
&= (g_1(\vec{x}, \vec{y}), f(g_1(\vec{x}, \vec{y}), g_2(\vec{x}, \vec{y})))
\end{aligned} \tag{5.5}$$

Hence

$$g_1(\vec{x}, \vec{y}) = \vec{x} \tag{5.6}$$

and

$$f(\vec{x}, g_2(\vec{x}, \vec{y})) = \vec{y} \tag{5.7}$$

(from (5.5) and (5.6)). We now define (for all $\vec{x}$ with $(\vec{x},0) \in U_0$)

$$h(\vec{x}) = g_2(\vec{x},\vec{0})$$

Then h is defined on an open set in R^k, h is differentiable, and (5.7) implies that

$$f(\vec{x},h(\vec{x})) = \vec{0}$$

Also, $G(\vec{x}_0,\vec{0}) = (\vec{x}_0,\vec{y}_0)$, and therefore $h(\vec{x}_0) = \vec{y}_0$. Thus h meets the conditions of the theorem.

To complete the proof, we need to compute $J_h(\vec{x}_0)$. Define a function $H:R^k \rightarrow R^{n+k}$ by

$$H(\vec{x}) = (\vec{x},h(\vec{x}))$$

Then $f(H(\vec{x})) = 0$, and the Jacobian matrix of H is

$$J_H(\vec{x}) = \begin{pmatrix} I_k \\ J_h(\vec{x}) \end{pmatrix}$$

where $J_h(x)$ is the Jacobian matrix of h. Moreover, the chain rule says that

$$J_f(H(\vec{x}))J_H(\vec{x}) = 0$$

We apply this formula at $\vec{x}_0$. Then $H(\vec{x}_0) = (\vec{x}_0,\vec{y}_0)$, and

$$0 = (A \quad B) \begin{pmatrix} I_k \\ J_h(\vec{x}_0) \end{pmatrix}$$

$$= A + BJ_h(\vec{x}_0)$$

(the reader can easily verify that the usual rule for matrix multiplication holds even when we deal with blocks of matrices; here, 0 is the n × k matrix containing entirely of zeroes). Now it is clear that

$$-B^{-1}A = J_h(\vec{x}_0)$$

and the theorem is proved.

EXERCISES

1.   Which of the following equations implicitly define z as a function of x and y near the given point? When the equation does, compute $\frac{\partial z}{\partial x}$ and $\frac{\partial z}{\partial y}$ at the point

   (a) $x^2 + y^2 + z^2 = 49$: $(6,-3,-2)$

   (b) $xy + yz + xz - xyz - 4 = 0$: $(2,2,-3)$

   (c) $xye^z + z \cos(x^2 + y^2) = 0$: $(0,0,0)$

   (d) $xy^6 - yz^3 + z - xy - 11 = 0$: $(1,2,3)$

   (e) $2xy + e^{xz} - z \log y - 1 = 0$: $(0,1,1)$

   (f) $x + y + z + \cos xyz - 1 = 0$: $(0,0,0)$

2.   Which of the following equations define z and w as functions of
x and y near the given point? When the equation does, compute $\frac{\partial z}{\partial x}$,
$\frac{\partial z}{\partial y}$, $\frac{\partial w}{\partial x}$, and $\frac{\partial^2 z}{\partial x \partial y}$ at the point. (The coordinates are (x,y,z,w).)

   (a) $z^2 + w^2 = x^2 + y$, $z + w = x^2 - y$: $(2,1,1,2)$

   (b) $x - y^2 + z^2 + 2w^2 = 6$, $x^2 + y^2 = z^2 + w^2$: $(1,2,-1,2)$

   (c) $x^2 + y^2 - z + w = 3$, $xy + zw + 1 = 0$: $(1,0,-1,1)$

   (d) $xe^w + yz - z^2 = 0$, $y \cos w + x^2 - z^2 = 1$: $(2,1,2,0)$

3.   The point $(0,2,4)$ is on the graph of

$$x^2 + y^2 + z^2 = 20$$

$$x - xy + z = 4$$

Can you use these equations to define y and z as functions of x near
$(0,2,4)$? How about x and z as functions of y, or (x,y) as functions
of z?

*4.   Let U be an open subset of R^n and $f:U \to R^m$ a function whose com-
ponent functions have continuous partial derivatives. We say that f
is an *immersion* if $D_{\vec{v}} f$ is injective for all $\vec{v}$ in U and a *submersion*
if $D_{\vec{v}} f$ is surjective for all $\vec{v}$ in U.

(a) Suppose that $f:U \to R^m$ is an immersion. Prove that, for each $\vec{v}$ in U, we can find an open set V of U containing $\vec{v}$, an open set W of R^m containing $f(\vec{v})$, and a diffeomorphism h of W onto an open subset of R^m such that

$$hf(x_1,\ldots,x_n) = (x_1,\ldots,x_n,0,\ldots,0)$$

(b) Suppose that $f:U \to R^m$ is a submersion. Prove that, for each $\vec{v}$ in U, we can find an open subset V of U containing $\vec{v}$, an open subset W of R^m containing $f(\vec{v})$, and a diffeomorphism h of W onto an open subset of R^m such that

$$hf(x_1,\ldots,x_n) = (x_1,\ldots,x_m)$$

5.   Show that if we have an $(n + k) \times (n + k)$ matrix of the form

$$C = \begin{pmatrix} I_k & 0 \\ A & B \end{pmatrix}$$

where I_k is the $k \times k$ identity matrix, 0 is a $k \times n$ matrix of zeroes, A is an $n \times k$ matrix, and B is an invertible $n \times n$ matrix, then C is invertible, and

$$C^{-1} = \begin{pmatrix} I_k & 0 \\ -B^{-1}A & B^{-1} \end{pmatrix}$$

CHAPTER 10

THE SPECTRAL THEOREM

INTRODUCTION

Among the easiest matrices to deal with are the diagonal matrices,
those in which all terms not on the main diagonal are 0. It is reasonable to ask which linear transformations T from a vector space V
to itself have diagonal matrices with respect to some basis. If T
has such a basis and $\vec{v}$ is a basis vector, then $T\vec{v}$ is a multiple of $\vec{v}$.
Thus we are led to study vectors v which are multiplied by some con-
stant λ under the action of T. These vectors, called *eigenvectors,*
and the constants that arise, called *eigenvalues,* are of considerable
importance in many situations.

In studying eigenvectors and eigenvalues, it is necessary to introduce complex vector spaces. This is because the eigenvalues of a
linear transformation T are the roots of a polynomial associated with
T, its *characteristic polynomial,* and polynomials always have complex
roots but need not have real roots.

The first two sections of this chapter deal with complex numbers
and complex vector spaces. In the next two sections, we introduce
the determinant and develop some of its properties. In Section 5, we
begin the study of eigenvectors and eigenvalues, and in Sections 6 and
7, we show that certain linear transformations always have a diagonal
matrix relative to some basis; this result is called the Spectral Theorem.

403

In the final two sections of the chapter, we apply the Spectral Theorem to the study of quadratic forms and give a proof of the Spectral Theorem using the results of Chapter 8.

1. COMPLEX NUMBERS

This section contains a review of complex numbers. It can be omitted by the reader already familiar with them.

A complex number c is an expression of the form a + bi where a and b are real numbers and i is a symbol whose "square" i^2 is -1. The number a is called the *real part* of c, Re(c), and b is the *imaginary part* of c, Im(c). (Note that Re(c) and Im(c) are real numbers by this definition.) Two complex numbers c and c' are equal if and only if their real and imaginary parts are equal.

The standard way of representing complex numbers uses the complex plane. The complex number a + bi is identified with the point (a,b) in R^2. (See Figure 1.1.)

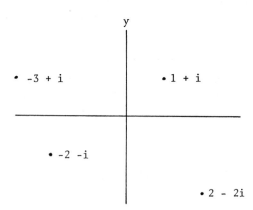

FIGURE 1.1. The complex plane

If c = a + bi and c' = a' + b'i are two complex numbers, we can add and multiply them in the obvious way, remembering that i^2 = - 1.

Explicitly,

$$(a + bi) + (a' + b'i) = (a + a') + (b + b')i$$

$$(a + bi)(a' + b'i) = aa' + ab'i + a'bi + bb'i^2$$

$$= (aa' - bb') + (ab' + a'b)i$$

For example,

$$(2 - 3i) + (7 + 4i) = 9 + i$$

$$(2 - 3i)(7 + 4i) = 14 + 8i - 21i - 12i^2$$

$$= 26 - 13i$$

All of the usual properties hold for these operations. Both are commutative and associative, and the distributive laws hold.

The *complex conjugate* of $c = a + bi$ is defined to be $a - bi$, and is usually written $\bar{c}$. It is easy to check that $\overline{(c_1 + c_2)} = \bar{c}_1 + \bar{c}_2$ and $\overline{c_1 c_2} = \bar{c}_1 \bar{c}_2$. Notice that

$$\text{Re}(c) = \frac{1}{2}(c + \bar{c}), \quad \text{Im}(c) = \frac{1}{2i}(c - \bar{c})$$

We often regard the real numbers as a subset of the complex numbers: a real number is a complex number whose imaginary part is 0. (Equivalently, c is real if and only if $c = \bar{c}$). A number whose real part is 0 is sometimes called *pure imaginary*.

The product $c\bar{c} = a^2 + b^2$ is always a nonnegative real number, and $c\bar{c} > 0$ unless $c = 0$. The *norm* of c, $|c|$, is defined to be $\sqrt{c\bar{c}}$. Therefore $c\bar{c}/|c|^2 = 1$ if $c \neq 0$. The norm function satisfies the triangle inequality:

$$|c_1 + c_2| \leq |c_1| + |c_2|$$

Using the fact that $c\bar{c}$ is always real, we can divide any complex number $c' = a' + b'i$ by $c = a + bi$ if $c \neq 0$ and obtain a complex number:

$$\frac{c'}{c} = \frac{c'\bar{c}}{c\bar{c}} = \frac{(a'+b'i)(a-bi)}{a^2+b^2},$$

$$= \left(\frac{a'a+b'b}{a^2+b^2}\right) + \left(\frac{ab'-a'b}{a^2+b^2}\right) i$$

For example,

$$\frac{2 + 3i}{4 - 5i} = \frac{(2+3i)(4+5i)}{(4-5i)(4+5i)}$$

$$= \frac{8+10i+12i+15i^2}{16+25}$$

$$= -\frac{7}{41} + \frac{22}{41} i$$

In particular,

$$\frac{1}{c} = \frac{a}{a^2+b^2} - \frac{b}{a^2+b^2} i$$

PROPOSITION 1.1.   *Let* $P(x) = a_n x^n + a_{n-1}x^{n-1} + \cdots + a_0$ *be any polynimial with real coefficients.   If* c *is a complex number with* $P(c) = 0$ *then* $P(\bar{c}) = 0$.

That is, if P is a polynomial with real coefficients, the complex roots occur in conjugate pairs.

Proof.   By hypothesis, $P(c) = \sum_{j=0}^{n} a_j c^j = 0$.   Taking complex conjugates, we obtain

$$0 = \left(\overline{\sum_{j=0}^{n} a_j c^j}\right) = \sum_{j=0}^{n} \bar{a}_j \bar{c}^j$$

$$= \sum_{j=0}^{n} a_j \bar{c}^j \qquad \text{(since } a_1,\ldots,a_n \text{ are real)}$$

$$= P(\bar{c})$$

The important property which makes complex numbers so useful is given in the following theorem.

THEOREM 1.2.   (Fundamental Theorem of Algebra). *Let* $P(x) =$ $\sum_{j=0}^{n} a_j x^j$, $a_n \neq 0$, *be any polynomial of degree at least 1 with complex coefficients. Then there is a complex number* c *with* $P(c) = 0$. *(Every non-constant polynomial with complex coefficients has a complex root.)*

In fact, a polynomial of degree n has n roots (counting multiplicities), as is easily proved by induction.

The proof of this theorem is not beyond the scope of this book, but including it here would be a considerable digression. We give it in Appendix 2.

EXERCISES

1.   Let $c_1 = 2 + 3i$, $c_2 = 4 - 7i$, and $c_3 = -2 + 5i$ and evaluate the following. (That is, write each of the following in the form a + bi, a,b $\in$ R.)

(a) $c_1 + c_2$ (f) $1/c_1$

(b) $c_1 c_2$ (g) c_2/c_3

(c) $c_1 (c_2 + c_3)$ (h) $c_1/(c_2 + \bar{c}_3)$

(d) $c_2 (c_1 + \bar{c}_2)$ (i) $c_3^2 c_1/(c_1^2 - c_2 c_3)$

(e) $c_2^2 c_3 (1 - c_2)$ (j) $(c_2 - c_1^2)/(c_1 c_3 + \bar{c}_2^2)$

2.   Prove that $\overline{c_1 + c_2} = \bar{c}_1 + \bar{c}_2$ and $\overline{c_1 c_2} = \bar{c}_1 \bar{c}_2$ for any two complex numbers c_1 and c_2.

3.   Show that for any complex number c, there is a complex number z with $|z| = 1$ and $cz = |c|$.

4.   (Polar representation of complex numbers) If $(x,y) \in R^2$, then (x,y) is given in polar coordinates by (r,θ), where $r = (x^2 + y^2)^{1/2}$ and $\theta = $ arc tan y/x; also, $x = r \cos \theta$ and $y = r \sin \theta$. This description of (x,y) carries over to complex numbers.

If $\theta \in R$, define

$$e^{i\theta} = \cos \theta + i \sin \theta$$

(a) Show that $|e^{i\theta}| = 1$ for all real θ.

(b) Show that if $z = x + iy$ is any complex number, then there are real numbers r, θ, with $r \geq 0$, such that $z = re^{i\theta}$. (This is the *polar representation* of z.)

(c) Show that z determines r uniquely. (However, θ is not uniquely determined; for instance, $e^{2\pi i} = 1$.)

(d) Show that $(r_1 e^{i\theta_1})(r_2 e^{i\theta_2}) = r_1 r_2 e^{i(\theta_1 + \theta_2)}$. In particular,

$$(re^{i\theta})^n = r^n e^{in\theta}$$

(This last result is known as De Moivre's Theorem.)

5. Define a function $f:R \to C$ by $f(t) = e^{it} = \cos t + i \sin t$. Show that

$$f'(t) = ie^{it}$$

(This is one justification for the definition in Exercise 4; the complex exponential satisfies the right formula for differentiation.)

6. Use Theorem 1.2 to prove that any real polynomial has n roots (counting multiplicities).

7. Use Theorem 1.2 to prove that any real polynomial of odd degree has a real root.

8. Use Theorem 5.8.6 to prove that any real polynomial of odd degree has a real root.

*9. Let $f:C \to C$ be a polynomial in z:

$$f(z) = a_0 + a_1 z + \cdots + a_n z^n$$

Set $u(z) = \text{Re } f(z)$, $v(z) = \text{Im } f(z)$. We may also regard f as a function of real variables x,y if we let $z = x + iy$. Thus we may regard f as a function from R^2 to R^2, with

$$f(x,y) = (u(x,y),v(x,y))$$

Show that

$$\frac{\partial u}{\partial x} = \frac{\partial v}{\partial y}$$

$$\frac{\partial v}{\partial x} = -\frac{\partial u}{\partial y}$$

(These are the *Cauchy-Riemann* equations; they are basic in the theory of functions of a complex variable.)

2. COMPLEX VECTOR SPACES

The axioms for a vector space given in Section 2 of Chapter 2 dealt with two operations: addition and multiplication by real numbers. For this reason, the vector spaces we dealt with are called real vector spaces (though we have generally suppressed the work "real"). For many purposes, however, it is more convenient to deal with vector spaces in which scalar multiplication is defined for complex numbers. We shall discuss the properties of these vector spaces in this section.

A *complex vector space* is a set, V, on which two operations are defined: addition, and scalar multiplication by complex numbers. These operations satisfy the axioms (2.1) through (2.8) of Section 2.2, except that now the number r is in C, not R. Here are some examples:

(1) Corresponding to R^2, we have C^2, the set of all pairs of complex numbers, with operations defined as follows:

$$(x_1,x_2) + (y_1,y_2) = (x_1 + y_1, x_2 + y_2)$$
$$c(x_1 x_2) = (cx_1, cx_2)$$

if c is a complex number. This example looks very much like R^2; the
only difference is that the components are complex numbers.

(2) Similarly, we have C^n, the set of n-tuples of complex numbers, with operations defined in the usual way:

$$(x_1,\ldots,x_n) + (y_1,\ldots,y_n) = (x_1 + y_1,\ldots,x_n + y_n)$$
$$c(x_1,\ldots,x_n) = (cx_1,\ldots,cx_n)$$

(3) If X is a subset of a real normed vector space V, we let
$C^0(X;C)$ be the set of all complex valued continuous functions defined
on X. (Here the norm on C defined in the previous section is used
to define continuity.) We define addition and scalar multiplication
in $C^0(X;C)$ by

$$(f_1 + f_2)(\vec{v}) = f_1(\vec{v}) + f_2(\vec{v})$$
$$(cf_1)(\vec{v}) = c(f_1(\vec{v}))$$

where f_1, f_2 are elements of $C^0(X;C)$ and c is a complex number. It is
easy to check that $C^0(X;C)$ is a complex vector space.

(4) The set $\{\vec{0}\}$ consisting of a single element $\vec{0}$ is the smallest
complex vector space. Addition and scalar multiplication are defined
in the only way possible:

$$\vec{0} + \vec{0} = \vec{0}$$
$$c\vec{0} = \vec{0}$$

for all c $\in$ C.

Other examples are given in Exercises 2-5.

Notice that any complex vector space can be regarded as a real
vector space; just restrict scalar multiplication to the reals. However, the procedure cannot be reversed; not every real vector space
is a complex vector space. See Exercise 8.

Most of the results proved earlier for real vector spaces also
hold for complex vector spaces. All of the results about bases and
dimension, for instance, still hold. In particular, any complex

n-dimensional vector space is isomorphic to C^n. There is one subtlety we should note: dimension may not mean what one might expect. For instance, C^1 is a 1-dimensional *complex* vector space, though the complex plane may seem to have two dimensions. If one regards C^1 as a *real* vector space, it is 2-dimensional; one basis consists of 1 and i. Similarly, C^n is n-dimensional as a complex vector space, but 2n-dimensional as a real vector space. See Exercises 7 and 8.

All our results about linear transformations also continue to hold; now, of course, all matrices may have complex coefficients.

The major change when dealing with complex vector spaces concerns inner product spaces. The axioms in Section 4.5 cannot be satisfied in a complex vector space (see Exercise 9); instead, we use the following properties:

$$< \vec{v},\vec{v} > \geq 0 \text{ for all } \vec{v} \in V, \text{ and } < \vec{v},\vec{v} > = 0$$
$$\text{if and only if } \vec{v} = \vec{0} \tag{2.1}$$

$$< \vec{v},\vec{w} > = \overline{< \vec{w},\vec{v} >} \text{ if } \vec{v},\vec{w} \in W \tag{2.2}$$

$$< c_1\vec{v}_1 + c_2\vec{v}_2,\vec{w} > = c_1 < \vec{v}_1,\vec{w} > + c_2 < \vec{v}_2,\vec{w} >$$
$$\text{for all } \vec{v}_1,\vec{v}_2,\vec{w} \in V \text{ and } c_1,c_2 \in C \tag{2.3}$$

The change in the definition is the complex conjugate in (2.2). As a consequence

$$< \vec{v},c_1\vec{w}_1 + c_2\vec{w}_2 > = \bar{c}_1 < \vec{v},\vec{w}_1 > + \bar{c}_2 < \vec{v},\vec{w}_2 > \tag{2.4}$$

(See Exercise 11.)

An inner product satisfying (2.1) through (2.3) above is called a *Hermitian inner product*. The standard example of a complex inner product space is C^n, with

$$< (x_j),(y_j) > = \sum_{j=1}^{n} x_j\bar{y}_j \tag{2.5}$$

(See Exercise 10.)

We call a complex vector space with a Hermitian inner product a *Hermitian inner product space*.

As in the case of real vector spaces, we define $\|\vec{v}\|$ by

$$\|\vec{v}\| = (< \vec{v},\vec{v} >)^{1/2}$$

Then $\|c\vec{v}\| = |c| \|\vec{v}\|$ for any complex number c, $\|\vec{v}\| \geq 0$ for all $\vec{v} \in V$, and $\|\vec{v}\| = 0$ if and only if $\vec{v} = \vec{0}$. (See Exercise 12.)

As in the case of real vector spaces, the Schwarz inequality also holds in a Hermitian inner product space. The proof is quite similar but a bit more complicated. We leave it to the reader.

PROPOSITION 2.1. *If $\vec{v},\vec{w}$ are vectors in the Hermitian inner product space V, then*

$$|< \vec{v},\vec{w} >| \leq \|\vec{v}\| \|\vec{w}\|$$

The triangle inequality

$$\|\vec{v} + \vec{w}\| \leq \|\vec{v}\| + \|\vec{w}\|$$

follows just as in the proof of Proposition 4.6.3. We omit the details.

If V is a finite dimensional Hermitian inner product space, the Gram-Schmidt process (Proposition 4.7.2) applies, exactly as in the real case, to give an orthonormal basis. It follows that any n-dimensional Hermitian inner product space is isometrically isomorphic to C^n with inner product defined in (2.5).

The trace of a linear transformation T on a complex vector space is again defined to be the sum of the diagonal elements of a matrix for T. All of the properties of the trace of a real linear transformation proved earlier hold equally well for complex linear transformation.

Suppose now that V and W are Hermitian inner product spaces and $T:V \rightarrow W$ is a linear transformation. As in Proposition 7.5.3, we define the *adjoint* of T, $T^*:W \rightarrow V$ by the equation

$$< T\vec{v},\vec{w} > = < \vec{v},T^*\vec{w} >$$

$$(2.6)$$

The uniqueness of the transformation satisfying this equation follows just as in the proof of Proposition 7.5.3. To prove existence, we suppose that $\alpha = \{\vec{v}_1,\ldots,\vec{v}_n\}$ is an orthonormal basis for V, $\beta = \{\vec{w}_1,\ldots,\vec{w}_m\}$ an orthonormal basis for W, and

$$\vec{Tv}_i = \sum_{j=1}^{m} a_{ji}\vec{w}_j$$

$1 \le i \le n$. Define $T^*:W \to V$ by

$$T^*\vec{w}_j = \sum_{i=1}^{n} \bar{a}_{ji}\vec{v}_i$$

$1 \le j \le m$. Then

$$< \vec{Tv}_i,\vec{w}_k > = < \sum_{j=1}^{m} a_{ji}\vec{w}_j,\vec{w}_k >$$

$$= \sum_{j=1}^{m} a_{ji}< \vec{w}_j,\vec{w}_k >$$

$$= a_{ki}$$

and

$$< \vec{v}_i,T^*\vec{w}_k > = < \vec{v}_i, \sum_{j=1}^{n} \bar{a}_{kj}\vec{v}_j >$$

$$= \sum_{j=1}^{n} a_{kj}< \vec{v}_i,\vec{v}_j >$$

$$= a_{ki}$$

Therefore equation (2.6) holds for basis vectors, which clearly implies that it holds for all vectors.

As a consequence of the discussion above, we see that, if α is an orthonormal basis for V, β an orthonormal basis for W, and $T:V \to W$ a linear transformation, then $A_{T^*}(\alpha,\beta)$ is the *conjugate transpose* $\bar{A}^t$

of the matrix, $A = A_T(\alpha,\beta)$. Explicitly, if $A_T(\alpha,\beta) = (a_{ij})$, then
$A_{T^*}(\beta,\alpha) = (b_{ij})$ where $b_{ij} = \bar{a}_{ji}$. For example, if

$$A = \begin{pmatrix} 2+i & 2-i \\ 3+i & 4-i \end{pmatrix}$$

then

$$\bar{A}^t = \begin{pmatrix} 2-i & 3-i \\ 2+i & 4+i \end{pmatrix}$$

Two parts of Proposition 7.5.5 need changing:

If $T:V \rightarrow W$ is a linear transformation and $c \in C$, then

$$(iii)' \quad (cT)^* = \bar{c}T^*$$

and

$$(iv)' \quad \text{If } V = W, \ tr(T^*) = \overline{tr(T)}$$

Again, the proofs are straightforward and left as exercises. The other parts of the proposition need no modification.

Propositions 7.5.6 and 7.5.7 remain true for complex vector spaces, but we shall not need them later. The proof of Proposition 7.5.6 needs some modification, and we leave the details as an exercise.

Recall that a linear transformation $T:V \rightarrow V$ is called *self-adjoint* if $T^* = T$. If V is a Hermitian inner product space, a self-adjoint linear transformation is sometimes called *Hermitian* and a linear transformation $T:V \rightarrow V$ satisfying $TT^* = T^*T = I$ is usually called *unitary* (rather than orthogonal). The linear transformation T is Hermitian if and only if its matrix $A = (a_{ij})$ (with respect to some orthonormal basis) satisfies $a_{ij} = \overline{a_{ji}}$. (This is an immediate consequence of the modified form of Proposition 7.5.4.) There is also a criterion for unitary matrices, like the one for orthogonal matrices given in Exercise 9 of Section 7.5; we leave it as an exercise.

EXERCISES

1.   Prove that $C^0(X;C)$ of Example 3 is a complex vector space.

2.   Let $P_n(C)$ be the set of complex polynomials $p(x)$ of degree $\leq n$;

$$p(x) = a_0 + a_1 x + \cdots + a_n x^n$$

where $a_0, a_1, \ldots, a_n \in C$. Define addition and scalar multiplication as in Example 6 of Section 2.2 and prove that $P_n(C)$ is a complex vector space of dimension $n + 1$.

3.   Let $m(n,m;C)$ be the set of $n \times m$ complex matrices (a_{ij}), $a_{ij} \in C$, and define addition and scalar multiplication as in Example 5 of Section 2.2. Prove that $m(n,m;C)$ is a complex vector space of dimension nm.

4.   Let V be the set of A in $m(n,m;C)$ with $A^t = A$. Prove that V is a complex subspace of $m(n,m;C)$ and determine its dimension.

5.   Let V be the set of $A \in m(n,m;C)$ with $A^t = -A$. Prove that V is a complex subspace of $m(n,m;C)$ and determine its dimension.

6.   Let V be the set of all $A \in m(n,m;C)$ with $\bar{A}^t = -A$. Prove that V is a real vector space and determine its dimension. Is V a complex subspace of $m(n,m;C)$? Why or why not?

7.   Consider C^n as a real vector space by restricting the scalars to the real numbers. Prove that the dimension of C^n as a real vector space is $2n$.

*8.   More generally, show that if V is any n-dimensional complex vector space, then as a real vector space V is $2n$-dimensional. (It follows that any odd-dimensional real vector space cannot be a complex vector space as well.)

9.   Show that a complex vector space cannot have an inner product satisfying the axioms in Section 4.5.

10.   Prove that $< (x_i),(y_i) > = \Sigma_{i=1}^{n} x_i \bar{y}_i$ defines a Hermitian inner product on C^n.

11.   Prove Equation (2.4).

12.   If V is a Hermitian inner product space, define

$$\|\vec{v}\| = (< \vec{v},\vec{v} >)^{1/2}$$

for $\vec{v}$ in V.   Prove that $\|c\vec{v}\| = |c|\|\vec{v}\|$ if c is a complex number, $\|\vec{v}\| \geq 0$, and $\|\vec{v}\| = 0$ if and only if $\vec{v} = \vec{0}$.

13.   Prove Proposition 2.1 and the triangle inequality for the norm defined in Exercise 12.   (Exercise 3 of Section 1 may help.)

14.   State and prove the analogue of Proposition 7.5.4 for complex vector spaces.

15.   State and prove the analogue of Proposition 7.5.5 for complex vector spaces.

16.   State and prove the analogue of Proposition 7.5.6 for complex vector spaces.

17.   Let V be a finite dimensional Hermitian inner product space and $T:V \rightarrow V$ a linear transformation with matrix $A = (a_{ij})$ with respect to some orthonormal basis.

(a)   Prove that T is self adjoint if and only if $a_{ij} = \overline{a_{ji}}$ for all $1 \leq i, j \leq n$.

(b)   Prove that T is unitary if and only if, for all $1 \leq i, j \leq n$,

$$\sum_{k=1}^{n} a_{ik}\bar{a}_{jk} = \delta_{ij}$$

18.   Let V be a finite dimensional Hermitian inner product space. Prove that a linear transformation $T:V \rightarrow V$ is unitary if and only if

$$< T\vec{v},T\vec{w} > = < \vec{v},\vec{w} >$$

for all $\vec{v},\vec{w}$ in V.

19. Let V be a Hermitian inner product space. Prove the *complex polarization identity*

$$< \vec{v},\vec{w} > = \frac{1}{4}(\|\vec{v} + \vec{v}\|^2 - \|\vec{v} - \vec{w}\|^2 + i\|\vec{v} + i\vec{w}\|^2 - i\|\vec{v} - i\vec{w}\|^2)$$

(Compare Exercise 8 of Section 4.5.)

20. Let V be a complex vector space and $T:V \to C$ a linear transformation. Define mappings $T_1, T_2 : V \to R$ by $T_1(\vec{v}) = \text{Re}(T(\vec{v}))$, $T_2(\vec{v}) = \text{Im}(T(\vec{v}))$ so that $T(\vec{v}) = T_1(\vec{v}) + iT_2(\vec{v})$. Prove that T_1 and T_2 are real linear transformations (if one regards V as a real vector space).

21. If V is a finite dimensional Hermitian inner product space and $T:V \to C$ is linear, prove that there is a unique vector $\vec{w} \in V$ with

$$T(\vec{v}) = < \vec{v},\vec{w} >$$

for all $\vec{v} \in C$. (This is the *Riesz Representation Theorem*. (Compare Exercise 16 of Section 4.7.)

3.   DETERMINANTS

We now introduce the very important notion of the determinant of an n × n matrix. The determinant is a function which assigns to each n × n matrix A a number, Det A. We shall see in the next section that A is invertible if and only if Det A $\neq$ 0. This fact will be useful to us when we study eigenvectors and eigenvalues in Section 5.

To begin with, it will be convenient to think of the determinant somewhat differently.

Definition. The *determinant* is a function, Det, from n-tuples of vectors in R^n to R,

$$\text{Det}:R^n \times R^n \times...\times R^n \to R$$

with the following properties:

If all of the vectors but one are fixed, Det is linear
in the remaining variable.   (3.1)

If any two of the vectors $\vec{v}_1,\ldots,\vec{v}_n$ are equal, then
$\text{Det}(\vec{v}_1,\ldots,\vec{v}_n) = 0$.   (3.2)

If $\vec{e}_1,\vec{e}_2,\ldots,\vec{e}_n$ is the standard basis for R^n, then   (3.3)

$\text{Det}(\vec{e}_1,\vec{e}_2,\ldots,\vec{e}_n) = 1$.

As we shall see below, these properties determine the function
Det uniquely.

Instead of using R^n, we could have worked with C^n. Then the
determinant would be a function from n-tuples of vectors in C^n to C
satisfying properties (3.1) - (3.3) above, and the results which follow would apply to it as well.

For most of this discussion, we shall not need to use property
(3.3). We deal, therefore, with a function Δ which has properties
(3.1) and (3.2), but not necessarily (3.3). Whatever properties we
prove for Δ, of course, apply to Det as well.

PROPOSITION 3.1. *Let* $\Delta: R^n \times \cdots \times R^n \to R$ *be a function satisfying*
(3.1) and (3.2). Then interchanging any two of $\vec{v}_1,\ldots,\vec{v}_n$ *reverses*
the sign of Δ.

For instance, if n = 4, then

$$\Delta(\vec{v}_1,\vec{v}_4,\vec{v}_3,\vec{v}_2) = -\Delta(\vec{v}_1,\vec{v}_2,\vec{v}_3,\vec{v}_4)$$

Proof. For convenience, we interchange $\vec{v}_1$ and $\vec{v}_2$; the same proof
works for any two vectors.

We know, by property (3.2) that

$$\Delta(\vec{v}_1 + \vec{v}_2, \vec{v}_1 + \vec{v}_2, \vec{v}_3, \ldots, \vec{v}_n) = 0$$

However, using property (3.1) twice, we have

$$\Delta(\vec{v}_1 + \vec{v}_2, \vec{v}_1 + \vec{v}_2, \vec{v}_3, \ldots, \vec{v}_n) = \Delta(\vec{v}_1 + \vec{v}_2, \vec{v}_1, \vec{v}_3, \ldots, \vec{v}_n)$$
$$+ \Delta(\vec{v}_1 + \vec{v}_2, \vec{v}_2, \vec{v}_3, \ldots, \vec{v}_n)$$

$$= \Delta(\vec{v}_1,\vec{v}_1,\vec{v}_3,\ldots,\vec{v}_n) + \Delta(\vec{v}_2,\vec{v}_1,\vec{v}_3,\ldots,\vec{v}_n)$$
$$+ \Delta(\vec{v}_1,\vec{v}_2,\vec{v}_3,\ldots,\vec{v}_n) + \Delta(\vec{v}_2,\vec{v}_2,\vec{v}_3,\ldots,\vec{v}_n)$$
$$= \Delta(\vec{v}_2,\vec{v}_1,\vec{v}_3,\ldots,\vec{v}_n) + \Delta(\vec{v}_1,\vec{v}_2,\vec{v}_3,\ldots,\vec{v}_n)$$

by property (3.2). Thus

$$\Delta(\vec{v}_2,\vec{v}_1,\vec{v}_3,\ldots,\vec{v}_n) = -\Delta(\vec{v}_1,\vec{v}_2,\vec{v}_3,\ldots,\vec{v}_n)$$

as claimed.

THEOREM 3.2. *Properties (3.1), (3.2), and (3.3) uniquely deter-mine the function* Det. *In fact, if* Δ *satisfies properties (3.1), (3.2) and*

$$\Delta(\vec{e}_1,\ldots,\vec{e}_n) = c \qquad\qquad (3.3)'$$

then $\Delta = c\cdot$Det.

Proof. First of all, suppose that Δ satisfies (3.1), (3.2), and (3.3)'. Then we can determine $\Delta(\vec{v}_1,\ldots,\vec{v}_n)$ whenever $\vec{v}_1,\ldots,\vec{v}_n$ are all in the standard basis. For if two of the vectors $\vec{v}_1,\ldots,\vec{v}_n$ are iden-tical, then $\Delta(\vec{v}_1,\ldots,\vec{v}_n) = 0$. If $\vec{v}_1,\ldots,\vec{v}_n$ are all different, then they are the vectors $\vec{e}_1,\ldots,\vec{e}_n$ in some other order. We can use pro-perty (3.1) to interchange pairs of vectors until they are in the or-der $\vec{e}_1,\ldots,\vec{e}_n$, and then use (3.3)' to evaluate Δ.

Next, we can evaluate $\Delta(\vec{v}_1,\ldots,\vec{v}_n)$ when $\vec{v}_2,\ldots,\vec{v}_n$ are in the stan-dard basis and $\vec{v}_1$ is any vector in R^n. For if $\vec{v}_1 = \Sigma_{i=1}^n x_i\vec{e}_i$, then property (3.1) states that

$$\Delta(\vec{v}_1,\vec{v}_2,\ldots,\vec{v}_n) = \sum_{i=1}^n x_i\Delta(\vec{e}_i,\vec{v}_2,\vec{v}_3,\ldots,\vec{v}_n)$$

We can evaluate each of the terms in the sum, since all the vectors are in the standard basis.

We continue the same way: we can evaluate $\Delta(\vec{v}_1,\ldots,\vec{v}_n)$ when $\vec{v}_3,\ldots,\vec{v}_n$ are in the standard basis and $\vec{v}_1$, $\vec{v}_2$ are any vectors in R^n; if $\vec{v}_2 = \Sigma_{i=1}^n y_i\vec{e}_i$, then

$$\Delta(\vec{v}_1, \vec{v}_2, \ldots, \vec{v}_n) = \sum_{i=1}^{n} y_i \Delta(\vec{v}_1, \vec{e}_i, \vec{v}_3, \ldots, \vec{v}_n)$$

and we already know how to evaluate each term in the sum. By induction (the induction step is clear, but complicated to write out), we see that $\Delta(\vec{v}_1, \ldots, \vec{v}_n)$ is uniquely determined for all vectors $\vec{v}_1, \ldots, \vec{v}_n$ in R^n.

In particular, Det is uniquely determined. Moreover, c·Det satisfies properties (3.1), (3.2), and (3.3)', and therefore (since Δ is uniquely determined) Δ = c·Det.

This proof also provides a method of evaluating Det. It turns out to be a very inefficient method.

We should emphasize one fact about Theorem 3.2. It does *not* show that any function satisfying (3.1), (3.2), and (3.3) exists; it just shows that at most one such function does. It is conceivable that properties (3.1), (3.2), and (3.3) are somehow inconsistent. (Of course, if they were, we would not have written the section this way.)

We now introduce a slight notational change. From now on, we shall also think of Det as a function of n × n matrices. If A is an n × n matrix, we regard it as a row of n column vectors:

$$A = \begin{pmatrix} a_{11} & \cdots & a_{1n} \\ & \cdots & \\ a_{n1} & \cdots & a_{nn} \end{pmatrix} = (A_1, \ldots, A_n)$$

where

$$A_1 = \begin{pmatrix} a_{11} \\ \vdots \\ a_{n1} \end{pmatrix}, \qquad A_2 = \begin{pmatrix} a_{12} \\ \vdots \\ a_{n2} \end{pmatrix}, \ldots, A_n = \begin{pmatrix} a_{1n} \\ \vdots \\ a_{nn} \end{pmatrix}$$

We then identify each of these column vectors with a vector in R^n and define Det A to be $\text{Det}(A_1, \ldots, A_n)$. Property (3.3) now states that Det I = 1 where I is the identity matrix.

THEOREM 3.3. *The function* Det *really does exist.*

Here is a sketch of the proof; the details are given in Appendix 1.

If $n = 1$, $A = (a_{11})$ and we define Det A to be a_{11}. This definition clearly satisfies the properties (3.1), (3.2), and (3.3).

To continue, we use induction. Suppose that we have defined determinants for $k \times k$ matrices. Let A be a $(k + 1) \times (k + 1)$-matrix and let A_{ij} ($1 \leq i, j \leq k + 1$) be the matrix obtained by deleting the ith row and the jth column from A. Equivalently, one obtains A_{ij} from A by deleting the row and column containing a_{ij}. For example, if

$$A = \begin{pmatrix} a_{11} & a_{12} & a_{13} \\ a_{21} & a_{22} & a_{23} \\ a_{31} & a_{32} & a_{33} \end{pmatrix}$$

then

$$A_{23} = \begin{pmatrix} a_{11} & a_{12} \\ a_{31} & a_{32} \end{pmatrix}$$

To define Det A, we pick any i with $1 \leq i \leq k + 1$ and define

$$\text{Det } A = \sum_{j=1}^{k+1} (-1)^{i+j} a_{ij} \text{Det}(A_{ij})$$

Note that A_{ij} is a $k \times k$ matrix, so $\text{Det}(A_{ij})$ is defined by induction. If $k + 1 = 2$ and $i = 1$, this formula gives

$$\text{Det } A = a_{11}a_{22} - a_{12}a_{21}$$

If $k + 1 = 3$, $A = (a_{ij})$, and $i = 1$, we have

$$\text{Det } A = a_{11}\text{Det } A_{11} - a_{12}\text{Det } A_{12} + a_{13}\text{Det } A_{13}$$
$$= a_{11}(a_{22}a_{33} - a_{23}a_{32}) - a_{12}(a_{21}a_{33} - a_{23}a_{31})$$
$$+ a_{13}(a_{21}a_{32} - a_{22}a_{31})$$

This definition does not depend on the choice of i, and it can be checked that the function so defined satisfies properties (3.1), (3.2), and (3.3). The details are somewhat complicated, and we leave them to Appendix 1.

The above procedure for computing the determinant is sometimes called "expanding by minors."

EXERCISES

1. Compute the determinant of A when

(a) $A = \begin{pmatrix} 1 & 7 \\ 3 & 4 \end{pmatrix}$

(b) $A = \begin{pmatrix} 2 & 2 \\ -7 & 4 \end{pmatrix}$

(c) $A = \begin{pmatrix} 1 & 0 & 2 \\ -1 & 3 & 0 \\ 2 & 1 & 1 \end{pmatrix}$

(d) $A = \begin{pmatrix} -1 & 2 & 2 \\ 0 & 3 & -1 \\ 4 & 2 & -4 \end{pmatrix}$

(e) $A = \begin{pmatrix} 2 & 1 & -1 \\ 3 & -2 & 2 \\ 1 & -1 & -1 \end{pmatrix}$

(f) $A = \begin{pmatrix} 3 & 1 & 1 & -1 \\ 0 & 2 & 0 & 0 \\ 1 & 6 & -1 & -2 \\ 2 & 5 & 0 & 1 \end{pmatrix}$

(g) $A = \begin{pmatrix} 2 & 4 & -2 & 0 \\ -2 & 0 & 0 & 9 \\ -1 & -2 & 1 & 1 \\ 7 & 0 & 1 & -5 \end{pmatrix}$

(h) $A = \begin{pmatrix} 3 & 6 & 0 & 1 \\ 1 & 4 & -1 & 0 \\ 1 & 1 & -2 & 6 \\ 2 & 1 & 0 & -2 \end{pmatrix}$

2.   Prove that $\text{Det}(\lambda A) = \lambda^n \text{Det } A$ for any n × n matrix A and any real number λ.

3.   Show by example that $\text{Det}(A + B)$ need not equal $\text{Det } A + \text{Det } B$.

*4.   Let $f:R \rightarrow M_n(R)$ be a differentiable function from the reals to the n × n real matrices, and let $g(x) = \text{Det } f(x)$. We can write f(x) as $(A_1(x), \ldots, A_n(x))$, where $A_i(x)$ is the ith column of $f(x)$. Show that

$$g'(x) = \sum_{i=1}^{n} \text{Det}(A_1(x), \ldots, A_{i-1}(x), A_i'(x), A_{i+1}(x), \ldots, A_n(x))$$

(Hint: check the formula for $n = 1$ and 2 by brute force. Then use induction, plus the description of Det given in the proof of Theorem 3.3.)

5. The matrix $A = (a_{ij})$ is said to be *upper triangular* if $a_{ij} = 0$ whenever $i > j$. Show that $\text{Det } A = a_{11}a_{22} \cdots a_{nn}$ for any upper triangular matrix.

6. Show that the absolute value of the determinant of any permutation matrix is 1. (See Section 7.2 for the definition of a permutation matrix.)

*7. Let M_n be the number of computations required to compute the determinant of an $n \times n$ matrix using the expansion by minors method. (Then $M_1 = 0$, $M_2 = 3$ - two multiplications and a subtraction - and so on.) Prove that

$$\lim_{n \to \infty} \left(\frac{M_n}{n!}\right) = e$$

(See "The Ubiquitous e" by Peter G. Sawtelle, *Mathematics Magazine*, 49 (1976), p. 244-245.)

8. Let $A = (a_{ij})$ be a 3×3 matrix and define vectors $\vec{u}_1, \vec{u}_2, \vec{u}_3 \in R^3$ by $\vec{u}_i = (a_{i1}, a_{i2}, a_{i3})$, $1 \le i \le 3$. Prove that $< \vec{u}_1, \vec{u}_2 \times \vec{u}_3 > = \text{Det } A$. (See Section 4.8 for the definition of the cross product.)

9. Write out in detail the induction step in the proof of Theorem 3.2.

10. (a) Show that $\text{Det}(\vec{v}_1, \ldots, \vec{v}_n)$ is not changed if one adds a linear combination of $\vec{v}_2, \ldots, \vec{v}_n$ to $\vec{v}_1$. (A similar statement holds for $\vec{v}_2, \ldots, \vec{v}_n$ instead of $\vec{v}_1$.)

(b) Show that if $\vec{v}_1,\ldots,\vec{v}_n$ are linearly dependent, then
$\text{Det}(\vec{v}_1,\ldots,\vec{v}_n) = 0$.

*11. Show that Det is a polynomial function of the coordinates of a
matrix - that is, if $A = (a_{ij})$, Det A is a polynomial in the a_{ij}.
(We used this fact in Section 9.3, when proving the Inverse Function
Theorem; see also Exercise 11 of the next section.)

4. PROPERTIES OF THE DETERMINANT

We develop here some of the elementary properties of the determinant
of a matrix. We also show how to define the determinant of a linear
transformation.

PROPOSITION 4.1. *If* A *and* B *are* n × n *matrices, then*

Det(AB) = (Det A)(Det B)

Proof. Define a function Δ on n × n matrices by

$\Delta(B) = \text{Det}(AB)$

We shall show that Δ satisfies (3.1), (3.2), and (3.3)' with c = Det(A).
Proposition 4.1 will then follow from Theorem 3.2.

Suppose $B = (B_1,\ldots,B_n)$ is a row of column vectors, or, equivalently, n × 1 matrices. Then

$AB = (AB_1,\ldots,AB_n)$

so that

$\Delta(B) = \text{Det}(AB_1,\ldots,AB_n)$

The fact that Δ satisfies property (3.1) follows immediately from the
fact that Det does; if two of the B_i are the same, so are two of the
AB_i.

To see that Δ satisfies (3.2), we proceed directly. For instance,

$$\Delta(B_1 + C_1, B_2, \ldots, B_n) = \text{Det}(A(B_1 + C_1), AB_2, \ldots, AB_n)$$
$$= \text{Det}(AB_1 + AC_1, AB_2, \ldots, AB_n)$$
$$= \text{Det}(AB_1, AB_2, \ldots, AB_n)$$
$$+ \text{Det}(AC_1, AB_2, \ldots, AB_n)$$
$$= \Delta(B_1, B_2, \ldots, B_n) + \Delta(C_1, B_2, \ldots, B_n)$$

Theorem 3.2 now states that $\Delta(B) = c \, \text{Det} \, B$ with $c = \Delta(I)$, since $\text{Det} \, I = 1$. However,

$$\Delta(I) = \text{Det} \, AI$$
$$= \text{Det} \, A$$

Thus

$$\text{Det}(AB) = \Delta(B)$$
$$= c \, \text{Det} \, B$$
$$= (\text{Det} \, A)(\text{Det} \, B)$$

as claimed.

An immediate and useful corollary is:

PROPOSITION 4.2. *If* A *is invertible, then* $\text{Det} \, A \neq 0$ *and* $\text{Det} \, A^{-1}$ $= (\text{Det} \, A)^{-1}$.

Proof. If A is invertible, let B be its inverse. Then

$$1 = \text{Det} \, I$$
$$= \text{Det}(AB)$$
$$= (\text{Det} \, A)(\text{Det} \, B)$$

Thus, $\text{Det} \, A \neq 0$ and

$$\text{Det} \, A^{-1} = \text{Det} \, B$$
$$= (\text{Det} \, A)^{-1}$$

This proposition has a converse.

PROPOSITION 4.3. *If A is not invertible,* Det A = 0.

Thus A is invertible if and only if Det A $\neq$ 0.

Proof. Let A = $(A_1, \ldots, A_n)$. Proposition 7.4.5 says that if A is not invertible, then $A_1, \ldots, A_n$ are linearly dependent. Hence one of these vectors is a linear combination of the others; suppose

$$A_n = \sum_{j=1}^{n-1} c_j A_j$$

Then

$$\text{Det } A = \sum_{j=1}^{n-1} c_j \text{Det}(A_1, \ldots, A_{n-1}, A_j) = 0$$

by property (3.2).

Thus the determinant provides (at least in principle) a way of finding which matrices are invertible. As we shall soon see, it also provides a way of computing the inverse. (However, the method is *not* computationally practical.)

Another useful fact is the following.

PROPOSITION 4.4. *Let A be an* n × n *matrix. Adding a multiple of one column of A to a different column does not affect the value of* Det A.

For instance,

$$\text{Det}(A_1, A_2, \ldots, A_n) = \text{Det}(A_1 + cA_2, A_2, \ldots, A_n)$$

Proof.

$$\text{Det}(A_1 + cA_2, A_2, \ldots, A_n)$$
$$= \text{Det}(A_1, A_2, \ldots, A_n) + \text{Det}(cA_2, A_2, \ldots, A_n)$$
$$= \text{Det}(A_1, \ldots, A_n) + c\text{Det}(A_2, A_2, \ldots, A_n)$$

and the second term is 0, by property (3.2). The general case follows in the same way.

THEOREM 4.5. (Cramer's rule). *Suppose that* $A_1, \ldots, A_n$, *C are
vectors in* R^n *with* $x_1 A_1 + \cdots + x_n A_n = C$. *Then, for any i, $1 \leq i \leq n$,
we have*

$$x_i \text{Det}(A_1, \ldots, A_n) = \text{Det}(A_1, \ldots, A_{i-1}, C, A_{i+1}, \ldots, A_n)$$

Proof. $\text{Det}(A_1, \ldots, A_{i-1}, C, A_{i+1}, \ldots, A_n)$

$$= \text{Det}(A_1, \ldots, A_{i-1}, \sum_{j=1}^{n} x_j A_j, A_{i+1}, \ldots, A_n)$$

$$= \sum_{j=1}^{n} x_j \text{Det}(A_1, \ldots, A_{i-1}, A_j, A_{i+1}, \ldots, A_n)$$

However, if $j \neq i$, then two vectors in $(A_1, \ldots, A_{i-1}, A_j, A_{i+1}, \ldots, A_n)$
are identical and the determinant is 0. Thus only the term $j = i$
remains, and

$$\text{Det}(A_1, \ldots, A_{i-1}, C, A_{i+1}, \ldots, A_n) = x_i \text{Det}(A_1, \ldots, A_n)$$

as claimed.

Remarks.

1. If $\text{Det}(A_1, \ldots, A_n) \neq 0$, this theorem provides a method of solving the set of simultaneous linear equations

$$A_1 x_1 + \cdots + A_n x_n = C$$

Once n is greater than 2 or 3, it usually involves more work than
Gaussian elimination (Section 7.7).

Even if $\text{Det}(A_1, \ldots, A_n) = 0$, the theorem says something: the equation $A_1 x_1 + \cdots + A_n x_n = C$ then cannot have a solution unless

$$\text{Det}(A_1, \ldots, A_{i-1}, C, A_{i+1}, \ldots, A_n) = 0$$

for all i. Even then, it need not have a solution; consider the case
$n = 2$,

$$A_1 = A_2 = \begin{pmatrix} 0 \\ 0 \end{pmatrix}, \ C = \begin{pmatrix} 0 \\ 1 \end{pmatrix}$$

2. Suppose that $A = (A_1, \ldots, A_n) = (a_{ij})$ is invertible. If $B = (b_{ij})$, then the condition $AB = I$ can be rewritten as

$$\sum_{i=1}^{n} b_{ij} A_i = e_j$$

the standard basis vector. Cramer's rule now states that

$$b_{ij} = (\text{Det } A)^{-1} \text{Det}(A_1, \ldots, A_{i-1}, e_j, A_{i+1}, \ldots, A_n)$$

$$= (\text{Det } A)^{-1} \text{Det} \begin{pmatrix} a_{11} & \cdots & a_{1,i-1} & 0 & a_{1,i+1} & \cdots & a_{1n} \\ & \cdots & & 1 & & \cdots & \\ a_{n1} & \cdots & a_{n,i-1} & 0 & a_{n,i+1} & \cdots & a_{nn} \end{pmatrix}$$

(the 1 is in the jth row and i^{th} column). If we expand the second determinant by minors (as in Theorem 3.3) along the ith column, all terms except the jth are 0, and the remaining term is $(-1)^{i+j} \text{Det } A_{ji}$. (Recall: A_{ji} is formed from A by deleting the jth row and i^{th} column.) Thus

$$b_{ij} = (-1)^{i+j} (\text{Det } A)^{-1} \text{Det } A_{ji}$$

This gives a method of computing A^{-1}. Once again, it is very cumbersome once n is greater than 2 or 3. However, it is important, because it shows explicitly how A^{-1} depends on A.

We still need to see how to compute Det A efficiently. First, we note another useful result.

PROPOSITION 4.6. *Let* A^t *be the transpose of* A. *Then* Det A = Det A^t.

Because the proof follows easily from the proof of the existence of the determinant, we give it in Appendix 1.

Note. Proposition 4.6 states, in effect, that whatever is true about determinants for the columns of a matrix is also true for the rows. (For instance, Det is linear as a function of any row of A if the other rows are fixed.) The reason is that the rows of A are the columns of A^t.

To compute the determinant of a matrix one uses the various properties of determinants to reduce the computation to one for a matrix with lots of zeroes. Perhaps the best explanation is an example. Let A be the 4 × 4 matrix

$$A = \begin{pmatrix} 1 & 1 & 2 & 3 \\ 2 & 3 & 4 & 0 \\ 3 & 2 & 3 & 1 \\ 4 & 1 & 1 & -1 \end{pmatrix}$$

Then, adding column 4 to column 3 and using Proposition 4.4, we have

$$\text{Det } A = \text{Det} \begin{pmatrix} 1 & 1 & 5 & 3 \\ 2 & 3 & 4 & 0 \\ 3 & 2 & 4 & 1 \\ 4 & 1 & 0 & -1 \end{pmatrix}$$

$$= \text{Det} \begin{pmatrix} 1 & 4 & 5 & 3 \\ 2 & 3 & 4 & 0 \\ 3 & 3 & 4 & 1 \\ 4 & 0 & 0 & -1 \end{pmatrix}$$

(adding column 4 to column 2)

$$= \text{Det} \begin{pmatrix} 13 & 4 & 5 & 3 \\ 2 & 3 & 4 & 0 \\ 7 & 3 & 4 & 1 \\ 0 & 0 & 0 & -1 \end{pmatrix}$$

(adding 4 times column 4 to column 1). Thus, expanding by minors along row 4, we have

$$\text{Det } A = (-1)^{4+4}(-1)\text{Det}\begin{pmatrix} 13 & 4 & 5 \\ 2 & 3 & 4 \\ 7 & 3 & 4 \end{pmatrix}$$

$$= -\text{Det}\begin{pmatrix} 13 & 4 & 1 \\ 2 & 3 & 1 \\ 7 & 3 & 1 \end{pmatrix}$$

(subtracting column 2 from column 3). Now, subtracting 3 times column 3 from column 2, we have

$$\text{Det } A = -\text{Det}\begin{pmatrix} 13 & 1 & 1 \\ 2 & 0 & 1 \\ 7 & 0 & 1 \end{pmatrix}$$

$$= -(-1)^{1+2}\text{Det}\begin{pmatrix} 2 & 1 \\ 7 & 1 \end{pmatrix}$$

$$= -5$$

expanding by minors along column 2 and using the formula in the 2×2 case. This procedure is never pleasant, but with practice it gets better than it looks at first.

To conclude this section, we define the determinant of a linear transformation $T:V \to V$, where V is a finite-dimensional vector space. Let $\alpha = \{\vec{v}_1, \ldots, \vec{v}_n\}$ be any basis for V; let $A_T(\alpha,\alpha)$ be the matrix for T in this basis. We define the determinant of T to be the determinant of the matrix $A_T(\alpha,\alpha)$,

$$\text{Det } T = \text{Det } A_T(\alpha,\alpha)$$

PROPOSITION 4.7. *The determinant of* T *is well defined; it does not depend on the basis chosen. Furthermore,* T *is an isomorphism if and only if* $\text{Det } T \neq 0$.

Proof. Let β be another basis for V. Then

$$A_T(\beta,\beta) = PA_T(\alpha,\alpha)P^{-1}$$

where $P = B(\alpha,\beta)$. (See Proposition 7.3.2.) Hence, using Proposition 4.1, we have

$$
\begin{aligned}
\text{Det } A_T(\beta,\beta) &= \text{Det}(PA_T(\alpha,\alpha)P^{-1}) \\
&= \text{Det}(P)\text{Det}(A_T(\alpha,\alpha))\text{Det}(P^{-1}) \\
&= \text{Det}(P)\text{Det}(A_T(\alpha,\alpha))(\text{Det}(P))^{-1} \\
&= \text{Det } A_T(\alpha,\alpha)
\end{aligned}
$$

This proves the first part of Proposition 4.7. The second part follows from Proposition 4.3 and the fact that T is an isomorphism if and only if $A_T(\alpha,\alpha)$ is invertible (by Proposition 7.2.3).

EXERCISES

1. Use Cramer's Rule to solve the systems of equations

(a)
$$
\begin{aligned}
2x + y &= 4 \\
3x + 2y &= 7
\end{aligned}
$$

(b)
$$
\begin{aligned}
8x + 7y &= 10 \\
5x + 4y &= 7
\end{aligned}
$$

(c)
$$
\begin{aligned}
x + 2y + 3z &= 6 \\
3x - y + 6z &= 1 \\
x + 4y - 2z &= 5
\end{aligned}
$$

(d)
$$
\begin{aligned}
2x + 2y - z &= 1 \\
-x + 2y + 2z &= 0 \\
2x - y + 2z &= 0
\end{aligned}
$$

2. Determine the inverse of the matrices

(a) $\begin{pmatrix} 2 & 1 \\ 3 & 2 \end{pmatrix}$

(b) $\begin{pmatrix} 8 & 7 \\ 5 & 4 \end{pmatrix}$

(c) $\begin{pmatrix} 1 & 3 \\ -3 & 5 \end{pmatrix}$

3.   Determine the inverse of the matrix

$A = \begin{pmatrix} a & b \\ c & d \end{pmatrix}$

when it exists.

4.   Let A and B be the 2 × 2 matrices

$A = \begin{pmatrix} 1 & 2 \\ 3 & 4 \end{pmatrix}$, $B = \begin{pmatrix} 0 & 7 \\ 5 & 1 \end{pmatrix}$

Compute AB and verify Det(AB) = (Det A)(Det B) in this case.

5.   (a) Prove that the product of k n × n matrices, $A_1 \cdots A_k$, is invertible if and only if each term of the product is.

   (b) Let A and B be n × n matrices. Prove that AB is invertible if and only if BA is invertible.

6.   Let A be a complex n × n matrix and $T:C^n \to C^n$ the linear transformation with matrix A. Prove that

$Det(T^*) = \overline{Det\ A}$

where T^* is the adjoint of T.

7.   Let V be a finite dimensional real inner product space and let $T:V \to V$ be an orthogonal transformation. Prove that Det T = ± 1. (See Section 7.5 for the definition of an orthogonal linear transformation.)

8.   Let V be a finite dimensional Hermitian inner product space (see Section 2 for the definition) and $T:V \to V$ a unitary linear transformation. Prove that |Det T| = 1.

9.   Use the method described after Proposition 4.6 to compute the determinants of Exercise 1 of Section 3.

10. Let $\vec{v}_1 = (a_{11}, a_{12})$, $\vec{v}_2 = (a_{21}, a_{22})$ be independent vectors in R^2 and let S be the parallelogram spanned by $\vec{v}_1$ and $\vec{v}_2$. (See Figure 4.1.) Prove that the area of S is given by

$$\left| \mathrm{Det} \begin{pmatrix} a_{11} & a_{12} \\ a_{21} & a_{22} \end{pmatrix} \right|$$

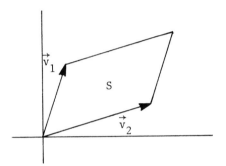

FIGURE 4.1. The parallelogram spanned by $\vec{v}_1$ and $\vec{v}_2$

*11. Let $A = (a_{ij})$ be an invertible n × n real matrix, and give $\mathrm{m}(n,n)$ the inner product of Section 7.5. Show that:

(a) There is an open set U of $\mathrm{m}(n,n)$ containing A such that if B ∈ U, then B is invertible.

(b) The function $F:U \rightarrow \mathrm{m}(n,n)$ defined by

$$F(B) = B^{-1}$$

is continuous.

*12. Prove the result corresponding to Exercise 11 for arbitrary finite-dimensional inner product spaces. (The result is true for finite-dimensional complex Hermitian inner product spaces, too.)

  Note. Exercise 11 supplies a fact needed in Section 9.3 for the proof of the Inverse Function Theorem.

5.   EIGENVECTORS AND EIGENVALUES

Let V be a finite-dimensional real or complex vector space and let T:V → V be a linear transformation. A non-zero vector $\vec{v} \in V$ is called an *eigenvector* of T, with *eigenvalue* λ, if $T(\vec{v}) = \lambda\vec{v}$; that is, if the linear transformation T simply multiplies the vector $\vec{v}$ by a constant.

Here are some examples.

1.   If V is one dimensional, then any non-zero vector $\vec{v}$ is an eigenvector for T. This follows from the fact that any vector in V is a multiple of $\vec{v}$.

2.   Suppose $T:R^2 \to R^2$ is represented relative to the standard basis for R^2 by the matrix

$$A = \begin{pmatrix} 2 & 1 \\ 0 & 2 \end{pmatrix}$$

Then $\vec{v} = (1,0)$ is an eigenvector for A with eigenvalue 2.

3.   If $T:R^2 \to R^2$ has matrix

$$A = \begin{pmatrix} 0 & 1 \\ -1 & 0 \end{pmatrix}$$

relative to the standard basis for R^2, then T has no eigenvectors. (T is rotation by 90°.) However, if we consider T as a transformation from C^2 to C^2, then T has an eigenvector $\vec{v} = (1,i)$ with eigenvalue i and an eigenvector $\vec{w} = (1,-i)$ with eigenvalue -i.

4.   Let $T:R^2 \to R^2$ be defined by the matrix

$$A = \begin{pmatrix} 2 & 1 \\ 1 & 2 \end{pmatrix}$$

Then $(1,1)$ and $(1,-1)$ are eigenvectors with eigenvalues 3 and 1 respectively.

We shall say that the number λ is an eigenvalue of $T:V \to V$ if it is the eigenvalue for some nonzero eigenvector of T. For instance,

3 and 1 are eigenvalues of T in the fourth example above, and we shall
soon see that there are no others.

Remark. The vector $\vec{0}$ is always a problem when one discusses
eigenvectors and eigenvalues. Since $T\vec{0} = \vec{0} = \lambda\vec{0}$ for all linear transformations T and all numbers λ, $\vec{0}$ is an eigenvector for all linear
transformations with all eigenvalues. This state of affairs is often
inconvenient. To avoid the worst problems, we make a convention:
the vector $\vec{0}$ is always considered an eigenvector of a linear transformation. However, a number λ is an eigenvalue for the eigenvector $\vec{0}$
only if λ is also an eigenvalue for a nonzero eigenvector of T.

We should also mention that eigenvectors are sometimes called
characteristic vectors or *proper vectors,* and eigenvalues *characteristic values or proper values.*

We begin our study of eigenvectors and eigenvalues with some
simple results.

PROPOSITION 5.1. *Let* λ *be an eigenvalue of* T:V $\rightarrow$ V. *The set*
V_λ *of all eigenvectors with eigenvalue* λ *is a subspace of* V.

Proof. Since λ is an eigenvalue of T, V_λ is not empty, and
$\vec{0} \in V_\lambda$ by our convention. Furthermore, if $\vec{v}_1$ and $\vec{v}_2$ are in V_λ and
a_1, a_2 are scalars, then

$$T(a_1\vec{v}_1 + a_2\vec{v}_2) = a_1 T(\vec{v}_1) + a_2 T(\vec{v}_2)$$
$$= a_1\lambda\vec{v}_1 + a_2\lambda\vec{v}_2$$
$$= \lambda(a_1\vec{v}_1 + a_2\vec{v}_2)$$

so that $a_1\vec{v}_1 + a_2\vec{v}_2$ is in V_λ. It follows from Proposition 7.4.3 that
V_λ is a subspace.

PROPOSITION 5.2. *Let* $\vec{v}_1,\ldots,\vec{v}_k$ *be nonzero eigenvectors for the
linear transformation* T:V $\rightarrow$ V *with eigenvalues* $\lambda_1,\ldots,\lambda_k$. *Assume*
$\lambda_i \neq \lambda_j$ *unless* i = j. *Then* $\vec{v}_1,\ldots,\vec{v}_k$ *are linearly independent.*

In other words, nonzero eigenvectors with distinct eigenvalues are linearly independent. In particular, a given nonzero eigenvector has only one eigenvalue.

Proof. Suppose that the $\vec{v}_i$ are linearly dependent. Then we can find a relation of the form

$$\sum_{i=1}^{k} \alpha_i \vec{v}_i = \vec{0} \tag{5.1}$$

in which not all of the α_i are 0. It follows that at least two of the α_i are nonzero. For if, say, $\alpha_2 = \alpha_3 = \ldots = \alpha_k = 0$ and $\alpha_1 \neq 0$, then we would have $\vec{v}_1 = \vec{0}$; this contradicts the assumption that all $\vec{v}_i$ are nonzero.

Multiplying Equation (5.1) by $-\lambda_k$, we have

$$\sum_{i=1}^{k} -\alpha_i \lambda_k \vec{v}_i = \vec{0}$$

Applying T to Equation (5.1), we obtain

$$\sum_{i=1}^{k} \alpha_i \lambda_i \vec{v}_i = \vec{0}$$

Adding the last two equations, we have

$$\sum_{i=1}^{k} \alpha_i (\lambda_i - \lambda_k) \vec{v}_i = \vec{0}$$

Since the coefficient of $\vec{v}_k$ is 0, this equation becomes

$$\sum_{i=1}^{k-1} \alpha_i (\lambda_i - \lambda_k) \vec{v}_i = \vec{0} \tag{5.2}$$

and $\lambda_i - \lambda_k \neq 0$ if $1 \leq i \leq k - 1$, by hypothesis. Moreover, at least one of the α_i in (5.2) is nonzero, so that the vectors $\vec{v}_1, \ldots, \vec{v}_{k-1}$ are linearly dependent.

Repeating this procedure, we see that $\vec{v}_1, \ldots, \vec{v}_{k-2}$ are linearly dependent, and so on until we find that $\vec{v}_1$ is linearly dependent. This means that $\vec{v}_1 = \vec{0}$, which contradicts our assumption. It follows that $\vec{v}_1, \ldots, \vec{v}_k$ are linearly independent, and the proposition is proved.

How does one find the eigenvectors and eigenvalues of a linear transformation? The next theorem provides the crucial information.

THEOREM 5.3. *The scalar* λ *is an eigenvalue for* T *if and only if* $\text{Det}(T - \lambda I) = 0$.

Proof. Suppose that λ is an eigenvalue for T. Then there is a nonzero vector $\vec{v} \in V$ such that $T\vec{v} = \lambda\vec{v}$, or $(T - \lambda I)\vec{v} = \vec{0}$. Therefore the linear transformation $(T - \lambda I)$ is not an isomorphism, and Proposition 4.3 implies that $\text{Det}(T - \lambda I) = 0$.

Conversely, suppose that $\text{Det}(T - \lambda I) = 0$. Then $T - \lambda I$ is not an isomorphism, and $h(T - \lambda I)$ is non-trivial. Therefore there is a nonzero vector $\vec{v}$ such that $(T - \lambda I)\vec{v} = \vec{0}$, or $T\vec{v} = \lambda\vec{v}$, and λ is the eigenvalue for the eigenvector $\vec{v}$.

The equation $\text{Det}(T - \lambda I) = 0$ is a polynomial equation in λ of degree n, where n = dimension of V. It is called the *characteristic equation* of T. For example, if $T:R^2 \to R^2$ has matrix A, where

$$A = \begin{pmatrix} a & b \\ c & d \end{pmatrix}$$

then

$$\text{Det}(T - \lambda I) = \text{Det}\begin{pmatrix} a-\lambda & b \\ c & d-\lambda \end{pmatrix}$$

$$= (a - \lambda)(d - \lambda) - bc$$

$$= \lambda^2 - (a + d)\lambda + ad - bc$$

We can now see an important advantage of complex vector spaces: if V is complex, the Fundamental Theorem of Algebra (Theorem 1.2) tells us that the characteristic equation of T necessarily has a complex root, so that T has an eigenvalue; if V is real, the characteristic equation need not have a real root and no eigenvalues need exist. (See Example 3 above for one example of this phenomenon. For the linear transformation given there, $\text{Det}(T - \lambda I) = \lambda^2 + 1$.)

COROLLARY 1. *Suppose that* V *is an* n-*dimensional complex vector space and that the characteristic equation of* $T:V \to V$ *has* n *distinct roots. Then* V *has a basis of eigenvectors of* T. *If* V *is a real vector space and the characteristic equation of* T *has* n *distinct real roots, then* V *has a basis of eigenvectors of* T.

Proof. Let $\lambda_1, \ldots, \lambda_n$ be the eigenvalues of T. We can find corresponding nonzero eigenvectors $\vec{v}_1, \ldots, \vec{v}_n$ in V. The vectors $\vec{v}_1, \ldots, \vec{v}_n$ are linearly independent, by Proposition 4.3.2, and thus form a basis.

COROLLARY 2. *If* V *is* n-*dimensional, then* $T:V \to V$ *has at most* n *eigenvalues.*

Proof. An equation of degree n has at most n roots.

We now reconsider the examples given at the beginning of the section.

1. In Example 4 at the beginning of this section, $\text{Det}(T - \lambda I) = \lambda^2 - 4\lambda + 3$. The roots of the equation $\lambda^2 - 4\lambda + 3 = 0$ are 3 and 1; thus 3 and 1 are the eigenvalues of T. The corresponding eigenvectors are $(1,1)$ and $(1,-1)$ (or any nonzero multiple), and they form a basis for R^2. (Of course, any multiples of the given eigenvectors are eigenvectors. We shall not repeat this remark in the following examples.)

2. Let T be the linear transformation of Example 3. The characteristic equation of T is $\lambda^2 + 1 = 0$, and its roots are $\pm i$. Thus if T is regarded as operating on C^2, it has $\pm i$ as eigenvalues, with corresponding eigenvectors $(1, \pm i)$.

3. Let T be the linear transformation of Example 2. The characteristic equation of T is $\lambda^2 - 4\lambda + 4 = 0$, and its only root is 2. Thus 2 is the only eigenvalue of T. If $\vec{v} = (x_1, x_2)$ is an eigenvector of T, then $T\vec{v} = 2\vec{v}$, or

$$(2x_1 + x_2, 2x_2) = (2x_1, 2x_2)$$

which implies that $x_2 = 0$. The only eigenvectors of T, therefore, are multiples of $(1,0)$, and T does not have a basis of eigenvectors. This has nothing to do with the fact that T acts on a real vector space; the same thing happens if we regard T as acting on C^2. Even in complex vector spaces, therefore, not every linear transformation yields a basis of eigenvectors.

We have seen how to find the eigenvalues of a linear transformation. Finding the eigenvectors for these eigenvalues is a matter of solving simultaneous linear equations. For example, let $T: R^2 \to R^2$ be the linear transformation with matrix

$$\begin{pmatrix} 1/2 & 1 \\ 1 & 2 \end{pmatrix}$$

The characteristic equation for T,

$$\text{Det}(T - \lambda I) = \lambda^2 - (5/2)\lambda = 0$$

has roots $\lambda = 0$ and $\lambda = 5/2$. To find a (nonzero) eigenvector for $\lambda = 0$, we need a vector $(x,y) \in R^2$, with $T(x,y) = 0$. Now

$$\begin{aligned} T(x,y) &= ((1/2)x + y, \ x + 2y) \\ &= (0,0) \end{aligned}$$

if and only if

$$(1/2)x + y = 0$$

$$x + 2y = 0$$

Thus $(x,y) = (2,-1)$ (or any nonzero multiple), is the eigenvector with eigenvalue 0.

Similarly, $T(x,y) = (5/2)(x,y)$ if and only if

$$(1/2)x + y = (5/2)x$$

$$x + 2y = (5/2)y$$

so that $(x,y) = (1,2)$ (or any nonzero multiple) is an eigenvector with eigenvalue $5/2$.

EXERCISES

1.   Find all eigenvalues and, for each eigenvalue, one nonzero eigenvector, for each of the following linear transformations:

(a) $T:R^2 \to R^2$ with matrix $\begin{pmatrix} 11 & 3 \\ 3 & 3 \end{pmatrix}$

(b) $T:R^2 \to R^2$ with matrix $\begin{pmatrix} 3/5 & 4/5 \\ -4/5 & 3/5 \end{pmatrix}$

(c) $T:C^2 \to C^2$ with matrix of (b)

(d) $T:R^2 \to R^2$ with matrix $\begin{pmatrix} 1 & 6 \\ 5 & 2 \end{pmatrix}$

(e) $T:R^2 \to R^2$ with matrix $\begin{pmatrix} 5 & 4 \\ 1 & 2 \end{pmatrix}$

(f) $T:R^2 \to R^2$ with matrix $\begin{pmatrix} 2 & 1 \\ -1 & 4 \end{pmatrix}$

(g) $T:C^2 \to C^2$ with matrix $\begin{pmatrix} 1 & 1 \\ -2 & -1 \end{pmatrix}$

(h) $T:R^3 \to R^3$ with matrix $\begin{pmatrix} 0 & 7 & 5 \\ 0 & 0 & 9 \\ 0 & 0 & 0 \end{pmatrix}$

(i) $T:R^3 \to R^3$ with matrix $\begin{pmatrix} 5 & 4 & 2 \\ 4 & 5 & 2 \\ 2 & 2 & 2 \end{pmatrix}$

(j) $T:R^3 \to R^3$ with matrix $\begin{pmatrix} 1 & 0 & 0 \\ -2 & 1 & 2 \\ -2 & 0 & 3 \end{pmatrix}$

2.   Let $S,T:V \to V$ be linear transformations with $S \circ T = T \circ S$. Suppose
$\vec{v}$ is an eigenvector for S with eigenvalue λ and $T(\vec{v}) \neq \vec{0}$. Then show
that $T(\vec{v})$ is an eigenvector for S with eigenvalue λ.

3.   Prove that if λ is an eigenvalue for a unitary transformation,
then $|\lambda| = 1$.

4.   Prove that if $T:R^3 \to R^3$ is an orthogonal linear transformation
then there is a nonzero $\vec{v}$ in R^3 with $T\vec{v} = \pm \vec{v}$. (Use Exercise 3 above
and Proposition 1.1.)

5. Let $T:V \to V$ be a self adjoint linear transformation and suppose
that λ_1 and λ_2 are distinct real eigenvalues for T with eigenvectors
$\vec{v}_1$ and $\vec{v}_2$ respectively. Prove that $\vec{v}_1$ and $\vec{v}_2$ are orthogonal. (Note:
we prove in the next section that all the eigenvalues of a self adjoint
linear transformation are real.)

6. How do the eigenvalues of -T compare with those of T?

7. How do the eigenvalues of T^2 compare with those of T?

*8. Suppose that $T:V \to V$ is a linear transformation on a complex vec-
tor space and that P(x) is a polynomial with complex coefficients.
How are the eigenvalues of P(T) related to those of T?

9. (a) Show that if $T_1:C^3 \to C^3$ has matrix

$$\begin{pmatrix} -1 & 12 & -4 \\ -1 & 13 & -5 \\ -2 & 30 & -12 \end{pmatrix}$$

then C^3 does not have a basis of eigenvectors of T_1.
    (b) Show that C^3 does have a basis of eigenvectors for T_2, where
$T_2:C^3 \to C^3$ has matrix

$$\begin{pmatrix} 4 & 6 & -3 \\ 6 & 13 & -6 \\ 18 & 36 & -17 \end{pmatrix}$$

(Note that the two matrices have the same characteristic equation.)

10. Suppose that $T:V \to V$ (where dim V = n) is such that V has a basis
of eigenvectors $\{\vec{v}_1,\ldots,\vec{v}_n\}$ of T. Let $\lambda_1,\ldots,\lambda_n$ be the corresponding
eigenvalues. Show that

$$\text{Det } T = \lambda_1\lambda_2 \cdots \lambda_n$$

6.   THE SPECTRAL THEOREM

In this section we prove the Spectral Theorem for self-adjoint linear transformations.   This result states that if T is a self-adjoint linear transformation on the real inner product space V, then V has a basis of eigenvectors of T.   In order to prove this result, we need first to work with complex vector spaces; this is because of the fact that polynomials always have complex roots (by the Fundamental Theorem of Algebra) but need not have real roots.

We begin by relating the eigenvalues of T^* to those of T.

PROPOSITION 6.1.   *Let V be a Hermitian inner product space and* $T:V \to V$ *a linear transformation.   Then the eigenvalues of* T^* *are the complex conjugates of the eigenvalues of* T.

Proof.   We need to prove that if $\bar{\lambda}$ is an eigenvalue of T^*, then λ is an eigenvalue of T, and conversely.   The converse is essentially the same statement (since $T^{**} = T$ and $\bar{\bar{\lambda}} = \lambda$), and therefore we need to prove only one half.   We can rephrase what we have to prove as follows: if λ is *not* an eigenvalue of T, then $\bar{\lambda}$ is not an eigenvalue of T^*.   But if λ is not an eigenvalue of T, then $h(T - \lambda I) = \{0\}$, and $T - \lambda I$ is invertible.   Part (vii) of Proposition 7.5.5 shows that $(T - \lambda I)^* = T^* - \bar{\lambda} I$ is invertible, and therefore $\bar{\lambda}$ is not an eigenvalue of T^*.   This proves the proposition.

If T is self-adjoint, the previous proposition says that the complex eigenvalues of T occur in conjugate pairs.   In fact, even more is true.

PROPOSITION 6.2.   *If T is a self-adjoint linear transformation on a complex vector space, then every eigenvalue of T is real.*

Proof.   Let λ be an eigenvalue of T, and let $\vec{v}$ be a corresponding nonzero eigenvector.   Set $\vec{w} = \vec{v}/\|\vec{v}\|$; then $\|\vec{w}\| = 1$, and $\vec{w}$ is also an eigenvector of T with eigenvalue λ.   Since T is self-adjoint,

$$\lambda = \lambda < \vec{w}, \vec{w} >$$
$$= < \lambda\vec{w}, \vec{w} >$$
$$= < T\vec{w}, \vec{w} >$$
$$= < \vec{w}, T\vec{w} >$$
$$= < \vec{w}, \lambda\vec{w} >$$
$$= \bar{\lambda} < \vec{w}, \vec{w} >$$
$$= \bar{\lambda}$$

Therefore λ is real.

The above proposition is still true if T is a self-adjoint operator on a real inner product vector space. It is also worthless, since the only possible eigenvalues in a real vector space are real numbers. What is not clear for linear transformations on a real vector space is that there are any eigenvalues. Hence the next result.

PROPOSITION 6.3. *If T is a self-adjoint linear transformation on a real inner product space V, then T has at least one eigenvalue.*

Proof. Let $\alpha = \{\vec{v}_1, \ldots, \vec{v}_n\}$ be an orthonormal basis for V, and let $A = A_T(\alpha, \alpha)$ be the matrix for T with respect to this basis. We can think of A as defining a linear transformation T_1 of R^n (relative to the standard basis); the transformation T_1 of R^n thus obtained is self-adjoint, and the eigenvalues are the same as those for T. (In fact, the map taking $\Sigma_{i=1}^n a_i \vec{v}_i$ to the vector $(a_1, \ldots, a_n) \in R^n$ is an isomorphism, and T_1 is the linear transformation on R^n corresponding to T under this isomorphism.) Thus it suffices to prove the proposition in the case where $V = R^n$, and we henceforth assume that $V = R^n$.

Let A be the matrix for T with respect to the standard basis. If $A = (a_{ij})$, then $a_{ij} = \bar{a}_{ji} = a_{ji}$, since T is self-adjoint and all matrix entries are real. Let $T' : C^n \to C^n$ be the linear transformation with matrix A (again with respect to the standard basis). Then T' is self-adjoint. By the Fundamental Theorem of Algebra (Theorem 1.2), the characteristic equation of T' has a root. By Proposition 6.2, this root is real. However, the roots of the characteristic equation of T' are exactly the roots of the characteristic equation of T. This is because $\text{Det}(T - \lambda I) = \text{Det}(T' - \lambda I)$, since the same matrix represents both transformations. The proposition follows, since the roots of the characteristic equation are the eigenvalues of T.

From now on in this section, V may be either a real or a complex inner product space. Notice that if the matrix for a linear transformation $T:V \rightarrow V$ (with respect to some orthonormal basis $\vec{v}_1,\ldots,\vec{v}_n$) is diagonal, then

$$A = \begin{pmatrix} \lambda_1 & & 0 \\ & \ddots & \\ 0 & & \lambda_n \end{pmatrix}$$

Furthermore, if the λ_i are all real, then T is self-adjoint, the $\vec{v}_i$ are all eigenvectors, and the eigenvalue of $\vec{v}_i$ is λ_i. Our goal in the rest of this section is to prove a converse. We need one preparatory lemma.

LEMMA 6.4. *Let* $T:V \rightarrow V$ *be self-adjoint. If W is a subspace of* V *such that* $T(W) \subset W$, *then* $T(W^{\perp}) \subset W^{\perp}$. *Let* $T_1:W \rightarrow W$ *and* $T_2:W^{\perp} \rightarrow W^{\perp}$ *be the restrictions of T to W and* $W^{\perp}$. *Then* T_1 *and* T_2 *are self-adjoint.*

Proof. Suppose that $\vec{v} \in W^{\perp}$ and that $T(W) \subset W$. We need to show that $T(\vec{v}) \in W^{\perp}$, or that $T(\vec{v}) \perp \vec{w}$ for all $\vec{w} \in W$. This is easy:

$$< T(\vec{v}),\vec{w} > = < \vec{v},T(\vec{w}) > = 0$$

because $T(\vec{w}) \in W$ and $\vec{v} \in W^{\perp}$. Thus $T(W^{\perp}) \subset W^{\perp}$, and the first part is proved.

Both halves of the second part are proved the same way; we do only one. To show that T_1 is self-adjoint, we need to show that

$$< T_1(\vec{w}_1),\vec{w}_2 > = < \vec{w}_1,T_1(\vec{w}_2) >$$

for all $\vec{w}_1,\vec{w}_2 \in W$. But this is obvious:

$$< T_1(\vec{w}_1),\vec{w}_2 > = < T(\vec{w}_1),\vec{w}_2 >$$

$$= < \vec{w}_1, T(\vec{w}_2) >$$
$$= < \vec{w}_1, T_1(\vec{w}_2) >$$

as desired.

THEOREM 6.5. (Spectral Theorem). *If* $T:V \rightarrow V$ *is self-adjoint, then* V *has an orthonormal basis of eigenvectors of* T.

Proof. We proceed by induction on n = dim V. If n = 1, the result is easy. Pick a unit vector $\vec{v}_1 \in V$. Now $T\vec{v}_1$ must be a multiple of $\vec{v}_1$, since every vector in V is. Thus $\{\vec{v}_1\}$ is an orthonormal basis of eigenvectors.

Now assume the theorem for all inner product vector spaces of dimension $\leq$ k, and let V have dimension k + 1. We know (from Proposition 6.3 if V is real; from Theorem 5.3 and the Fundamental Theorem of Algebra, Theorem 1.2, if V is complex) that T has an eigenvalue λ. Let $\vec{v}_{k+1}$ be a unit vector which is also an eigenvector for T with eigenvalue λ and let W be the space spanned by $\vec{v}_{k+1}$. Then TW $\subset$ W, since $T\vec{v}_{k+1} = \lambda\vec{v}_{k+1}$. According to Lemma 6.4, $T(W^\perp) \subset W^\perp$, and T_1 (= T restricted to $W^\perp$) is self-adjoint. It now follows from the corollary to Proposition 4.7.2 that dim $W^\perp$ = k, and so, by induction, $W^\perp$ has an orthonormal basis $\{\vec{v}_1, \ldots, \vec{v}_k\}$ of eigenvectors for T_1. It is now easy to see that $\{\vec{v}_1, \ldots, \vec{v}_{k+1}\}$ is an orthonormal basis of eigenvectors for T, and this completes the proof.

Let $\{\vec{v}_1, \ldots, \vec{v}_n\}$ be an orthonormal basis of eigenvectors for T, with eigenvalues $\lambda_1, \ldots, \lambda_n$ respectively. The λ_i are all real, by Proposition 6.2, and the matrix for T with respect to the above basis is

$$\begin{pmatrix} \lambda_1 & & & & 0 \\ & \lambda_2 & & & \\ & & \cdot & & \\ & & & \cdot & \\ & & & & \cdot \\ 0 & & & & \lambda_n \end{pmatrix}$$

This is the desired converse to the statement made before Lemma 6.4.

We note that a vector space can have a basis of eigenvectors for a linear transformation T which is not self-adjoint. For example, the transformation $T:R^2 \to R^2$ with matrix

$$\begin{pmatrix} 2 & 1 \\ 0 & 1 \end{pmatrix}$$

relative to the standard basis is not self-adjoint. (Why?) However, $T\vec{e}_1 = 2\vec{e}_1$ and $T(\vec{e}_1 - \vec{e}_2) = \vec{e}_1 - \vec{e}_2$, so that the vectors $\vec{v}_1 = \vec{e}_1$, $\vec{v}_2 = \vec{e}_1 - \vec{e}_2$ are a basis of R^2 of eigenvectors for T.

EXERCISES

1.   Find an orthonormal basis for C^2 of eigenvectors for the self-adjoint linear transformation $T:C^2 \to C^2$ with matrix

(a) $\begin{pmatrix} 0 & 1 \\ 1 & 0 \end{pmatrix}$ (d) $\begin{pmatrix} a & b \\ b & a \end{pmatrix}$

(b) $\begin{pmatrix} -1 & 1 \\ 1 & -1 \end{pmatrix}$ (e) $\begin{pmatrix} 3 & i \\ -i & 3 \end{pmatrix}$

(c) $\begin{pmatrix} 3 & 1 \\ 1 & 2 \end{pmatrix}$ (f) $\begin{pmatrix} 2 & 1-2i \\ 1+2i & 6 \end{pmatrix}$

2.   Find an orthonormal basis for R^3 of eigenvectors for the self-adjoint linear transformation $T:R^3 \to R^3$ with matrix

(a) $\begin{pmatrix} 0 & 1 & 0 \\ 1 & 0 & 1 \\ 0 & 1 & 0 \end{pmatrix}$ (c) $\begin{pmatrix} \frac{14}{3} & \frac{2}{3} & \frac{14}{3} \\ \frac{2}{3} & -\frac{1}{3} & -\frac{16}{3} \\ \frac{14}{3} & -\frac{16}{3} & \frac{5}{3} \end{pmatrix}$

(b) $\begin{pmatrix} -1 & 0 & 1 \\ 0 & 2 & 0 \\ 1 & 0 & -1 \end{pmatrix}$ (d) $\begin{pmatrix} 5 & 4\sqrt{3} & -\sqrt{3} \\ 4\sqrt{3} & 4 & 4 \\ -\sqrt{3} & 4 & 7 \end{pmatrix}$

3. Let $P:V \to V$ be an orthogonal projection of V onto a subspace $W \subset V$. Prove that P is self-adjoint, determine its eigenvalues, and find a basis for V of eigenvectors of P.

4. Prove that any symmetric $n \times n$ matrix is similar to a diagonal matrix.

*5. Let V be a finite dimensional inner product space, $T:V \to V$ a self-adjoint linear transformation. Let $\lambda_1, \ldots, \lambda_k$ be the distinct eigenvalues for T, V_i the subspace of eigenvectors with eigenvalue λ_i, and $P_i:V \to V$ the orthogonal projection onto V_i, $1 \le i \le k$. Prove the following:

    (i) V_i is orthogonal to V_j for $i \neq j$

    (ii) $\displaystyle\sum_{i=1}^{k} P_i = I$

    (iii) $\displaystyle T = \sum_{i=1}^{k} \lambda_i P_i$

    (iv) $TP = P_i T$ for $1 \le i \le k$

(This result is another form of the Spectral Theorem.)

7.   THE SPECTRAL THEOREM FOR NORMAL LINEAR TRANSFORMATIONS

We now extend the Spectral Theorem of the previous section to a larger class of linear transformations, the normal linear transformations. The linear transformation $T:V \to V$ is called *normal* if T and T^* commute: $TT^* = T^*T$. A typical example is the following: let V be a complex vector space, and let the matrix of T (with respect to some orthonormal basis) be

$$\begin{pmatrix} \lambda_1 & & 0 \\ & \ddots & \\ 0 & & \lambda_n \end{pmatrix}$$

where the λ_i are arbitrary complex numbers. Then the matrix of T^* is

$$\begin{pmatrix} \bar{\lambda}_1 & & 0 \\ & \ddots & \\ 0 & & \bar{\lambda}_n \end{pmatrix}$$

and T^* and T clearly commute.

In what follows, V is always a Hermitian inner product space, and T is a normal linear transformation. Note that if T is normal, so is $T - \lambda I$; this is a simple computation and is left as an exercise.

LEMMA 7.1. *Suppose that* T *is normal and* $\vec{v}$ *is an eigenvector of* T *with eigenvalue* λ. *Let* W *be the subspace of* V *spanned by* $\vec{v}$. *Then:*

(a) $\vec{v}$ *is an eigenvector of* T^*, *with eigenvalue* $\bar{\lambda}$

(b) $T(W) \subset W$ *and* $T^*(W) \subset W$

(c) $T(W^\perp) \subset W^\perp$ *and* $T^*(W^\perp) \subset W^\perp$

(d) *Let* $T_2 = T$ *restricted to* $W^\perp$, $T_2^* = T^*$ *restricted to* $W^\perp$. *Then* $(T_2)^* = T_2^*$, *and* T_2 *is normal*

Proof. (a) Let $S = T - \lambda I$. Then S is normal, and $S\vec{v} = \vec{0}$. Thus

$$0 = \|S\vec{v}\|^2$$

$$= < S\vec{v}, S\vec{v} >$$

$$= < S^* S\vec{v}, \vec{v} >$$

$$= < SS^*\vec{v}, \vec{v} > \qquad \text{(since } S \text{ is normal)}$$

$$= < S^*\vec{v}, S^*\vec{v} > = \|S^*\vec{v}\|^2$$

and $S^*\vec{v} = \vec{0}$. Since $S^* = T^* - \bar{\lambda} I$, we have $T^*\vec{v} = \bar{\lambda}\vec{v}$. This proves the result.

(b) This is obvious.

(c) The two parts are proved the same way; we do only the first. We need to show that if $\vec{w}_1 \in W^\perp$ and $\vec{w} \in W$, then $< T\vec{w}_1, \vec{w} > = 0$. But this is clear:

$$< T\vec{w}_1, \vec{w} > \ = \ < \vec{w}_1, T^*\vec{w} >$$

because $T^*\vec{w} \in W$ and $\vec{w}_1 \in W^\perp$. (Note that $\vec{w}$ is a multiple of $\vec{v}$.)

(d) The proof of this is like that of Lemma 6.4 and is left as an exercise.

THEOREM 7.2. (Spectral Theorem for Normal Operators). *If V is a complex vector space and* T:V $\rightarrow$ V *is normal, then V has an orthonormal basis of eigenvectors of* T.

Proof. The proof now proceeds almost exactly like the proof of Theorem 6.4. We use induction on n = dim V. If n = 1, the result is clear. If n = k + 1, we find a unit vector $\vec{v}_{k+1}$ which is an eigenvector of T, let W = span $\vec{v}_{k+1}$, and use the induction hypothesis and (d) ofthe previous lemma to find an orthonormal basis of eigenvectors, $\{\vec{v}_1, \ldots, \vec{v}_k\}$, for $W^\perp$. Then $\{\vec{v}_1, \ldots, \vec{v}_{k+1}\}$ form an orthonormal basis of eigenvectors for V, and the theorem is proved.

If T:V $\rightarrow$ V is an arbitrary linear transformation on a complex Hermitian inner product space, then Theorem 7.2 need not hold. (In fact, any linear transformation T:V $\rightarrow$ V with an orthonormal basis of eigenvectors is normal. See Exercise 11.) Even if we do not require the eigenvectors to be orthonormal, T need not have enough eigenvectors to form a basis; see Example 2 of Section 5. However, one can say something about such linear transformations T: for instance, the matrix for T is similar to a matrix of a certain prescribed form, the *Jordan Canonical Form*. The interested reader can learn about Jordan Canonical Form in Appendix 3.

EXERCISES

1. Let T:V $\rightarrow$ V be a normal linear transformation and λ a complex number. Prove that T - λI is normal.

2. Let $T:C^2 \rightarrow C^2$ be defined by the matrix

$$\begin{pmatrix} 4 & -1 \\ 1 & 4 \end{pmatrix}$$

Prove that T is normal and find an orthonormal basis of C^2 consisting
of eigenvectors of T.

3.   Prove that $T:C^2 \to C^2$, defined by

$$\begin{pmatrix} 2 & i \\ i & 2 \end{pmatrix}$$

is normal. Find an orthonormal basis of C^2 consisting of eigenvectors
of T.

4.   Prove that $T:C^3 \to C^3$ defined by

$$\begin{pmatrix} 6 & -2 & 3 \\ 3 & 6 & -2 \\ -2 & 3 & 6 \end{pmatrix}$$

is normal and find an orthonormal basis of C^3 consisting of eigenvectors of T.

5.   Let $T:V \to V$ be a normal linear transformation on an inner product
space and let $\vec{v}_1, \vec{v}_2$ be eigenvectors for T corresponding to distinct
eigenvalues. Prove that $\vec{v}_1$ and $\vec{v}_2$ are orthogonal.

6.   Prove Part (d) of Lemma 7.1.

7.   Let $S,T:V \to V$ be normal linear transformations with $S \circ T = T \circ S$.
    *(a)   Prove that $S \circ T^* = T^* \circ S$ and $S^* \circ T = T \circ S^*$.
    (b)   Prove that the composite $S \circ T$ is normal.
    (c)   Prove that $aS + bT$ is normal, where a,b are real (or com-
plex) numbers.

*8.   Prove the results of Exercise 5 of the previous section for nor-
mal linear transformations.

9.   Unitary transformations are normal. What does the Spectral Theorem let one say about them?

10. Let $SU(2,C) = \{T:C^2 \to C^2: \text{ T is unitary and Det T = 1}\}$. (SU
stands for *special unitary*; "special" is sometimes used to denote sets
of matrices of determinant 1.)

(a) Show that if T_1 and T_2 are elements of $SU(2,C)$, then so are
$T_1 \circ T_2$ and T_1^{-1}.

(b) Show that every element in $SU(2,C)$ is conjugate to one of
the form

$$\begin{pmatrix} e^{it} & 0 \\ 0 & e^{-it} \end{pmatrix}$$

$0 \leq t < 2\pi$ (See Exercise 4 of Section 1 for the definition of e^{it}.)

11. Let V be a Hermitian inner product space, and let $T:V \to V$ be a
linear transformation.

(a) Show that if there is an orthonormal basis of eigenvectors
for T, then T is normal.

(b) Show that if in addition all the eigenvalues of T are real,
then T is self-adjoint.

8.   QUADRATIC FORMS

In studying maxima and minima of differentiable functions in Chapter
8, we introduced the notion of a quadratic form. We now show how the
Spectral Theorem can be used to determine the properties of a quadratic form and also give criteria for a quadratic form to be positive
definite, negative definite, and indefinite.

Recall that a quadratic form is a function $Q:R^n \to R$ defined by

$$Q(x_1,\ldots,x_n) = \sum_{i=1}^{n} \sum_{j=1}^{n} a_{ij} x_i x_j$$

where $a_{ij} = a_{ji}$, $1 \leq i, j \leq n$. If $A = (a_{ij})$ and $T = T_A:R^n \to R^n$, we
denote Q by Q_A or by Q_T. (Note that A is a symmetric matrix, so that
T is self-adjoint.) It is immediate that

$$Q_T(\vec{v}) = <T\vec{v},\vec{v}>$$ (8.1)

We call T positive definite, negative definite, or indefinite if Q_T is. The next lemma shows that there is a 1 - 1 correspondence between quadratic forms and linear transformation $T:R^n \to R^n$.

LEMMA 8.1. *If $S,T:R^n \to R^n$ are self-adjoint linear transformation such that $Q_S(x_1,\ldots,x_n) = Q_T(x_1,\ldots,x_n)$ for all $x_1,\ldots,x_n$, then S = T.*

Proof. We know that for all $\vec{v} \in R^n$,

$$< (S - T)\vec{v},\vec{v} > = Q_{S-T}(\vec{v})$$
$$= Q_S(\vec{v}) - Q_T(\vec{v})$$
$$= 0$$

Since S - T is self-adjoint, it has an orthonormal basis of eigenvectors $\vec{v}_1,\ldots,\vec{v}_n$. Let $\lambda_1,\ldots,\lambda_n$ be the corresponding eigenvalues. Then

$$\lambda_i = \lambda_i<\vec{v}_i,\vec{v}_i>$$
$$= < \lambda_i\vec{v}_i,\vec{v}_i >$$
$$= < (S - T)\vec{v}_i,\vec{v}_i >$$
$$= 0$$

It follows that the matrix of S - T with respect to the basis $\{\vec{v}_1,\ldots, \vec{v}_n\}$ is the zero matrix. Thus S - T = 0, or S = T, as claimed.

As can be seen from the proof, the spectral theorem can give useful information about Q_T. One way to summarize this information is in the following theorem.

THEOREM 8.2. (Principal Axis Theorem). *Let Q_T be a quadratic form in n variables. There exists an orthonormal basis $\{\vec{v}_1,\ldots,\vec{v}_n\}$ for R^n and real numbers $\lambda_1,\ldots,\lambda_n$ such that if $\vec{v} = x_1\vec{v}_1 +\cdots+ x_n\vec{v}_n$, then*

$$Q_T(\vec{v}) = \sum_{j=1}^{n} x_i^2 \lambda_i$$

In fact, the $\vec{v}_i$ are eigenvectors of T and the λ_i are the corresponding eigenvalues.

Proof.   The last statement in the theorem gives the proof away. According to Theorem 6.4, T has an orthonormal basis of eigenvectors, $\vec{v}_1,\ldots,\vec{v}_n$, with corresponding eigenvalues $\lambda_1,\ldots,\lambda_n$.   If $\vec{v} = x_1\vec{v}_1 + \cdots + x_n\vec{v}_n$, then

$$< T\vec{v},\vec{v} > = < \sum_{i=1}^{n} T(x_i\vec{v}_i), \sum_{j=1}^{n} x_j\vec{v}_j >$$

$$= \sum_{i=1}^{n}\sum_{j=1}^{n} < x_i\lambda_i\vec{v}_i, x_j\vec{v}_j >$$

$$= \sum_{i=1}^{n} x_i^2 \lambda_i$$

since $\vec{v}_1,\ldots,\vec{v}_n$ are orthonormal.

COROLLARY.   Let $\lambda_1,\ldots,\lambda_n$ be the eigenvalues of T.   Then Q_T is

(a)   positive definite if and only if the λ_i are all > 0

(b)   negative definite if and only if the λ_i are all < 0

(c)   indefinite if and only if some λ_i is > 0 and some λ_j is < 0

(d)   positive semidefinite if and only if the λ_i are all ≥ 0

(e)   negative semidefinite if and only if the λ_i are all ≤ 0

Proof.   We do (d), leaving the rest as an exercise.   If the λ_i are all ≥ 0, then

$$Q_T(\vec{v}) = \sum_{i=1}^{n} x_i^2 \lambda_i \geq 0$$

for any vector $\vec{v} = (x_1,\ldots,x_n) \in R^n$ as required.

If n = 2 and $\{\vec{v}_1,\vec{v}_2\}$ is an orthonormal basis of eigenvectors for T (so that $T(\vec{v}_i) = \lambda_i\vec{v}_i$, i=1,2) and $\vec{v} = x_1\vec{v}_1 + x_2\vec{v}_2$, then

$$T(\vec{v}) = \lambda_1 x_1\vec{v}_1 + \lambda_2 x_2\vec{v}_2$$

It follows that the image of the unit circle in R^2 under T,

$$S = \{\vec{w} = y_1\vec{v}_1 + y_2\vec{v}_2 \in R^2 : \frac{y_1^2}{\lambda_1^2} + \frac{y_2^2}{\lambda_2^2} = 1\}$$

is an ellipse. (See Figure 8.1.)

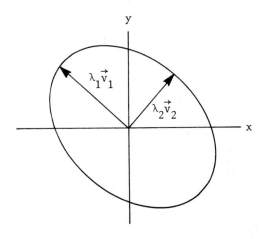

FIGURE 8.1. The ellipse defined by Q_T

The *major axis* of S is the line from $-\lambda_1\vec{v}_1$ to $\lambda_1\vec{v}_1$, and the *minor axis* is the line from $-\lambda_2\vec{v}_2$ to $\lambda_2\vec{v}_2$. They are, respectively, the longest and the shortest diagonal in the ellipse. Thus the principal axis theorem describes the principal axes of the ellipse given by T. In R^3, the corresponding surface turns out to be an ellipsoid, and the principal axes (there are generally three) are the eigenvectors of the linear transformation T.

In studying maxima and minima, it is important for us to have a useful test of when a quadratic form is positive definite. We gave such a test for quadratic forms on R^2 and R^3 in Section 8.2. The next result is an extension of this test to quadratic forms on R^n.

THEOREM 8.3. *Let* Q *be a quadratic form on* R^n *with matrix* A = (a_{ij}). *Let* A_j, $1 \leq j \leq n$, *be the* j × j *matrix formed by the first* j *rows and columns of* A. *Thus* $A_1 = (a_{11})$,

$$A_2 = \begin{pmatrix} a_{11} & a_{12} \\ a_{21} & a_{22} \end{pmatrix}$$

and A_n = A. *Then*

    (a) Q *is positive definite if and only if* Det $A_j > 0$ *for*
$1 \le j \le n$

    (b) Q *is negative definite if and only if* $(-1)^j$Det $A_j > 0$ *for*
$1 \le j \le n$

    (c) *If* Det $A_j \neq 0$ *for* $1 \le j \le n$ *and neither (a) or (b) holds,*
then Q *is indefinite.*

    Proof. (a) Let T_j be the linear transformation on R^n whose ma-
trix is A_j (and let $T_n = T$). We begin with two simple remarks.

    If T is positive definite, then T_j is positive definite
for $1 \le j \le n$.   (8.1)

For if $\vec{v}_j = (x_1,\ldots,x_j) \in R^j$ and $\vec{v}_j \neq 0$, then set $\vec{v} = (x_1,\ldots,x_j,$
$0,\ldots,0) \in R^n$. It is easy to check that

$$< T_j\vec{v}_j,\vec{v}_j > \; = \; < T\vec{v},\vec{v} > \; > 0$$

because T is positive definite, and this proves the result.

    If T_j is positive definite, then Det $T_j > 0$.   (8.2)

This is an immediate consequence of statement (a) of the Corollary to
Theorem 8.2, since Det T_j is the product of its eigenvalues. (See
Exercise 10 of Section 5).

    Remarks (8.1) and (8.2) show that if $Q = Q_T$ is positive definite,
then Det $A_j > 0$ for $1 \le j \le n$. We need to prove the converse.

    The proof proceeds by induction on n. If n = 1, then A = (a),
and A is positive definite if a > 0, (hence if Det A(= Det A_1) > 0).
Thus the theorem holds for n = 1.

    Suppose now that the theorem holds for n = k, and let A be a
symmetric (k + 1) × (k + 1) matrix such that the matrices $A_1,\ldots,A_k,A_{k+1}$

all have positive determinant. We need to show that $A = A_{k+1}$ is positive definite. Let $\{\vec{v}_1, \ldots, \vec{v}_{k+1}\}$ be an orthonormal basis of eigenvectors for A, and let $\lambda_1, \ldots, \lambda_{k+1}$ be the eigenvalues. If the λ_j are
all positive, then A is positive definite (by the Corollary to Theorem
8.2). Assume there is a j with $\lambda_j \leq 0$. We may as well assume (by
permuting the basis elements) that $\lambda_1 \leq 0$. However

$$0 < \text{Det } A = \lambda_1 \cdots \lambda_{k+1}$$

Hence $\lambda_1 \neq 0$, so that $\lambda_1 < 0$. Moreover,

$$\lambda_2 \cdots \lambda_{k+1} < 0$$

and therefore a second λ_j (λ_2, say) is also negative.

Let $\vec{v}_1 = (x_1, \ldots, x_{k+1})$, $\vec{v}_2 = (y_1, \ldots, y_{k+1})$ and let $\vec{w} = c_1\vec{v}_1 + c_1\vec{v}_2$ so that the last component of $\vec{w}$ is 0. Specifically, let

$$c_1 = 1, \ c_2 = 0$$

if $x_{k+1} = 0$ and

$$c_1 = y_{k+1}, \ c_2 = -x_{k+1}$$

if $x_{k+1} \neq 0$. Then $\vec{w} = (z_1, \ldots, z_k, 0)$ is non zero (since $\vec{v}_1$ and $\vec{v}_2$ are
linearly independent) and $z_j = c_1 x_j + c_2 y_j$, $1 \leq j \leq k$. Let $\vec{w}_0 = (z_1, \ldots, z_k) \in R^k$. Then, as we saw in the proof of Remark (8.1),

$$< T\vec{w}, \vec{w} > = < T_k\vec{w}_0, \vec{w}_0 > > 0$$

(this last inequality holds because T_k is positive definite, by the
inductive hypothesis). On the other hand

$$\begin{aligned}
< T\vec{w}, \vec{w} > &= < T(c_1\vec{v}_1 + c_2\vec{v}_2), \ c_1\vec{v}_1 + c_2\vec{v}_2 > \\
&= < c_1\lambda_1\vec{v}_1 + c_2\lambda_2\vec{v}_2, \ c_1\vec{v}_1 + c_2\vec{v}_2 > \\
&= c_1^2\lambda_1 + c_2^2\lambda_2 < 0
\end{aligned}$$

because $\lambda_1 < 0$, $\lambda_2 < 0$, and either c_1 or c_2 is nonzero. This gives a contradiction.   It follows that the λ_j are all positive, and (a) is proved.

(b)   This follows easily from (a) and the fact that T is negative definite if and only if (-T) is positive definite.   We leave the details as an exercise.

(c)   Since Det A = Det $A_n \neq 0$ by hypothesis, 0 is not an eigenvalue of A.   The eigenvalues of A cannot all be positive, since then A would be positive definite and (a) would hold.   Similarly, they cannot all be negative.   Therefore A has both positive and negative eigenvalues, and (c) of the Corollary to Theorem 8.2 says that A is indefinite.

Theorem 8.3 is not very practical to use when n > 3, since the calculations of the determinants soon becomes extremely lengthy.

EXERCISES

1.   Prove statements (a), (b), (c), and (e) of the corollary to Theorem 8.2.

2.   Prove that any positive definite quadratic form on R^n defines an inner product on R^n.

3.   Let $T_1, T_2 : R^n \to R^n$ be self-adjoint linear transformations with $T_1 T_2 = T_2 T_1$.   Prove that there is a basis $\vec{v}_1, \ldots, \vec{v}_n$ such that each $\vec{v}_i$ is an eigenvector for both T_1 and T_2.

4.   Let Q be a quadratic form on R^n.   Prove that there is a basis $\vec{v}_1, \ldots, \vec{v}_n$ for R^n such that, if $\vec{v} = x_1 \vec{v}_1 + \cdots + x_n \vec{v}_n$, then

$$Q(\vec{v}) = x_1^2 + x_2^2 + \cdots + x_k^2 - x_{k+1}^2 - x_{k+2}^2 - \cdots - x_m^2$$

for some k,m with $0 \leq k \leq m \leq n$.

5.   Give the details of the proof of Theorem 8.3 (b).

6.   Use Theorem 8.3 to give a new proof of Proposition 8.2.3.

7.   One might hope that an analogue of Theorem 8.3 holds for semi-definite matrices.

(a)   Show that if Q and A are in Theorem 8.3 and Q is positive semidefinite, then $\text{Det}(A_j) \geq 0$ for $j = 1, \ldots, n$.

(b)   State and prove the corresponding result for negative semi-definite matrices.

*(c)   Show by example that the converses of (a) and (b) are *false*.

9.   ANOTHER PROOF OF THE SPECTRAL THEOREM

In Section 6 we proved the spectral theorem for self-adjoint linear transformations of finite-dimensional inner product spaces: if $T:V \to V$ is a self-adjoint linear transformation, then V has an ortho-normal basis of eigenvectors of T.   In that proof, we needed to use complex vector spaces even if we began by considering a transformation on a real vector space.   We shall now give a proof of the spectral theorem for real vector spaces which obviates the need for any discussion of complex vector spaces.   Instead, we use some of the analysis developed in Chapter 8.

Recall the proof of the spectral theorem.   It proceeded by induction, and may be summarized as follows:

(a)   If $T:V \to V$ is self-adjoint, then T has at least one eigen-vector, $\vec{v}$.   We may assume that $\|\vec{v}\| = 1$.

(b)   Let $W = \{\vec{v}\}^{\perp}$, the space of vectors orthogonal to $\vec{v}$.   Then dim W = dim V - 1, T maps W to itself, and the restriction T_1 of T to W is self-adjoint.

(c)   By the inductive hypothesis, there is an orthonormal basis $\{\vec{v}_1, \ldots, \vec{v}_{n-1}\}$ of eigenvectors for T_1.   Then $\{\vec{v}_1, \ldots, \vec{v}_{n-1}, \vec{v}\}$ is an orthonormal basis of eigenvectors for T.

Steps (b) and (c) involve fairly straightforward reasoning, and are easily proved for either real or complex vector spaces.   The de-tails were given in Section 6.   Step (a) required complex vector spa-ces.   Here is a new proof.

PROPOSITION 9.1. *Let T:V → V be a self-adjoint linear transformation on a finite-dimensional real inner product space. Then T has a nonzero eigenvector.*

Proof. We begin with a lemma.

LEMMA 9.2. *Let V be a finite dimensional inner product space and T:V → V a linear transformation. Define f:V → R by $f(\vec{v}) = <T\vec{v}, \vec{v}>$. Then f is differentiable, with*

$$(D_{\vec{v}}f)(\vec{w}) = <(T + T^*)\vec{v}, \vec{w}>$$

Proof. (See Exercise 10 (c) of Section 6.1.)   Let $\vec{v} \in V$ be a fixed vector and $\vec{w} \in V$ an arbitrary vector.   Then

$$<T(\vec{v} + \vec{w}), \vec{v} + \vec{w}> - <T\vec{v}, \vec{v}> = <T\vec{v}, \vec{w}> + <T\vec{w}, \vec{v}> + <T\vec{w}, \vec{w}>$$

$$= <T\vec{v}, \vec{w}> + <T^*\vec{v}, \vec{w}> + <T\vec{w}, \vec{w}>$$

$$= <(T + T^*)\vec{v}, \vec{w}> + <T\vec{w}, \vec{w}>$$

Let $S = S_{\vec{v}}:V \to R$ be the linear transformation defined by $S(\vec{w}) = <(T + T^*)\vec{v}, \vec{w}>$.   Then, using the computation above, we have

$$|f(\vec{v} + \vec{w}) - f(\vec{v}) - S\vec{w}| = |<T(\vec{v} + \vec{w}), \vec{v} + \vec{w}> - <T\vec{v}, \vec{v}> - S\vec{w}|$$

$$= |<T\vec{w}, \vec{w}>|$$

$$\leq \|T\vec{w}\| \|\vec{w}\|$$

$$\leq \|T\| \|\vec{w}\|^2$$

Hence, if $\vec{w} \neq \vec{0}$,

$$\frac{|f(\vec{v} + \vec{w}) - f(\vec{v}) - S\vec{w}|}{\|\vec{w}\|} \leq \|T\| \|\vec{w}\|$$

and $\|T\| \|\vec{w}\|$ approaches 0 as $\vec{w}$ approaches zero.   This means that S is the derivative of f at $\vec{v}$, and the lemma is proved.

Lemma 9.2 has two useful corollaries.

COROLLARY 1. *Let* $T:V \to V$ *be a self-adjoint linear transformation and define* $f:V \to R$ *by* $f(\vec{v}) = \langle T\vec{v}, \vec{v} \rangle$. *Then* f *is differentiable with*

$$(D_{\vec{v}}f)(\vec{w}) = 2 \langle T\vec{v}, \vec{w} \rangle$$

COROLLARY 2. *Let* $g_0:V \to R$ *be defined by* $g_0(\vec{v}) = \|\vec{v}\|^2$. *Then* g_0 *is differentiable with*

$$(D_{\vec{v}}g_0)(\vec{w}) = 2 \langle \vec{v}, \vec{w} \rangle$$

Corollary 1 is an immediate consequence of Lemma 9.2 and Corollary 2 follows from Corollary 1 by writing

$$g_0(\vec{v}) = \langle I\vec{v}, \vec{v} \rangle$$

where $I:V \to V$ is the identity transformation.

We can now prove Proposition 9.1. As usual, we assume $V = R^n$. Let $T:R^n \to R$ be a self-adjoint linear transformation and define $f, g:R^n \to R$ by

$$f(\vec{v}) = \langle T\vec{v}, \vec{v} \rangle$$
$$g(\vec{v}) = \|\vec{v}\|^2 - 1$$

Then, using Corollaries 1 and 2 above, we have

$$(\text{grad } f)(\vec{w}) = 2T\vec{w}$$
$$(\text{grad } g)(\vec{w}) = 2\vec{w}$$

In particular, $(\text{grad } g)(\vec{w}) \neq \vec{0}$ unless $\vec{w} = \vec{0}$.

We now wish to maximize f on the set S,

$$S = \{\vec{v} \in R^n : g(\vec{v}) = 0\}$$

For this purpose, we use Theorem 8.4.2. We know that f attains its maximum, since S is compact. Let $\vec{v}_0 \in R^n$ be a point at which f is a maximum. Then, according to Theorem 8.4.2, there is a $\lambda \in R$ such that

$$(\text{grad } f)(\vec{v}_0) = \lambda(\text{grad } g)(\vec{v}_0)$$

or, equivalently,

$$2T\vec{v}_0 = 2\lambda\vec{v}_0$$

Thus $T(\vec{v}_0) = \lambda(\vec{v}_0)$, or $\vec{v}_0$ is a nonzero eigenvector of T. This proves the proposition.

It should be clear that $\vec{v}_0$ is an eigenvector corresponding to the largest eigenvalue of T. We could similarly find an eigenvector corresponding to the smallest eigenvalue to T by minimizing f. Note also that $\lambda = f(\vec{v}_0)$.

This last fact leads to an important computational procedure. Suppose that we need to know the maximum eigenvalue λ of T. We can find λ by finding the maximum value of f on S, and we can approximate λ by computing f at lots of points on S and taking the largest value we find. A similar procedure applies to the minimum eigenvalue. There are a number of problems in mathematical physics where it is important to know the size of the largest (or smallest) eigenvalue of a self-adjoint linear transformation, and slightly more sophisticated forms of this procedure are used (under the name of the Rayleigh-Ritz principle). The situation is complicated in these problems by the fact that the inner product spaces which arise have infinite dimension. However, there is a Spectral Theorem for these spaces which generalizes Theorem 6.4. This theorem is one of the important results in the theory of Hilbert Spaces; we shall not be able to discuss it further here.

CHAPTER 11

INTEGRATION

INTRODUCTION

Integration, as covered in a first-year calculus course, divides into two topics: the theory of integration (the definition, general properties, and the like), and techniques of integration (such as integration by parts and substitutions). In this chapter, we begin the task of describing the integral for functions of n variables; in the next chapter, we shall see how to reduce integration problems in R^n to problems in R^1.

The theory of the integral in R^n follows the same lines as for R^1, and we therefore begin with a review of integration of functions of one variable. We then define integrals in R^n, prove some basic properties of integrals, and demonstrate that continuous functions are integrable. The sections defining the integral are important, since the integral often arises in practical applications as a limit of the sorts of sums found in the definition.

The reader may have noticed that we have referred only to integration in R^n, and not to integration in arbitrary inner product spaces. All of our results can be extended to inner product spaces, but R^n seems to be a more natural setting for integration theory. In this and future chapters, therefore, we shall restrict our attention to R^n and its subsets.

463

1. INTEGRATION OF FUNCTIONS OF ONE VARIABLE

In this section, we discuss the theory of integration of functions of one variable. The theory is built on two basic principles. One is that there is an obvious definition for the integral of a step function; the other is that if $f \geq g$, then the integral of f is greater than the integral of g.

We define a *partition* P of a closed interval [a,b] to be a sequence of real numbers $\{t_0, \ldots, t_p\}$ with

$$a = t_0 < t_1 < \ldots < t_p = b$$

For example, $\{0, 1/2, 3/4, 1, 3/2, 2\}$ is a partition of the interval [0,2].

Let $f:[a,b] \to R$ be a bounded function and $P = \{t_0, t_1, \ldots, t_p\}$ a partition of the interval [a,b]. Define numbers M_i and m_i, $1 \leq i \leq p$, by

$$M_i = \text{lub}\{f(x) \,|\, t_{i-1} \leq x \leq t_i\}$$
$$m_i = \text{glb}\{f(x) \,|\, t_{i-1} \leq x \leq t_i\}$$

The *upper sum of f on [a,b] relative to the partition P is defined by*

$$U_a^b(f,P) = \sum_{i=1}^{p} M_i(t_i - t_{i-1})$$

The *lower sum of f on [a,b] relative to the partition P is defined by*

$$L_a^b(f,P) = \sum_{i=1}^{p} m_i(t_i - t_{i-1})$$

Unless there is a need to make explicit the interval of definition, we shall denote $U_a^b(f,P)$ by $U(f,P)$ and $L_a^b(f,P)$ by $L(f,P)$. Notice that $L(f,P) \leq U(f,P)$, since $m_i \leq M_i$, all i.

To understand what these sums mean, assume that f is a non-negative function, and define $h:[a,b] \to R$ by

$$h(x) = \begin{cases} M_i & \text{if } t_{i-1} \leq x < t_i \\ M_p & \text{if } x = b \end{cases}$$

Then h is a *step function* (See Figure 1.1), and its integral, which represents the area under its graph, can be thought of as a sum of

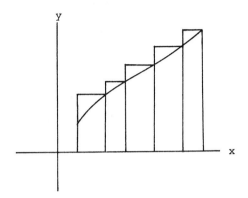

FIGURE 1.1. A step function h $\geq$ f approximating f

areas of rectangles; in fact, the area under the graph of h is precisely U(f,P). Similarly, if we define g:[a,b] → R by

$$g(x) = \begin{cases} = m_i & \text{if } t_{i-1} \leq x < t_i \\ = m_p & \text{if } x = b \end{cases}$$

then g is a step function (See Figure 1.2), and the area under the graph of g is L(f,P). Since g $\leq$ f $\leq$ h, the integral of f from a to b should lie between L(f,P) and U(f,P), the integrals of g and h respectively. The idea behind defining the integral of f is to use a sequence of partitions P_k which make $L(f,P_k)$ - $U(f,P_k)$ arbitrarily small, and then to define the integral as the common limit of $L(f,P_k)$ and $U(f,P_k)$.

Here are some examples of upper and lower sums.

1. Let $f(x) = x^2$ - 2 on the interval [-1,2] and let P be the partition {-1,0,1/2,3/2,2}. (See Figure 1.3.) We have

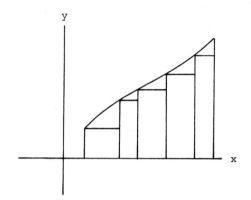

FIGURE 1.2.   A step function g ≤ f approximating f

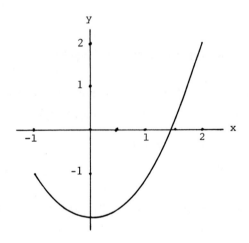

FIGURE 1.3.

$M_1 = -1$, $M_2 = -7/4$, $M_3 = 1/4$, $M_4 = 2$

$m_1 = -2$, $m_2 = -2$, $m_3 = -7/4$, $m_4 = 1/4$

Then

$$U(f,P) = (-1) + (-7/4)1/2 + (1/4)1 + 2(1/2)$$
$$= -3/8$$

$$L(f,P) = (-3)1 + (-2)1/2 + (-7/4)1 + (1/4)1/2$$
$$= - 37/8$$

2. Let $f(x) = x$ on $[0,1]$ and $P_m = \{0,1/m,2/m,\ldots,m-1/m,1\}$. Then $M_i = i/m$ and $m_i = (i-1)/m$, so that

$$U(f,P_m) = \sum_{i=1}^{m} (\frac{i}{m})\frac{1}{m} = \frac{1}{m2} \sum_{i=1}^{m} i = \frac{1}{m^2}(\frac{m(m+1)}{2})$$

$$= \frac{m+1}{2m}$$

(see Proposition 1.3.1), and

$$L(f,P_m) = \sum_{i=1}^{m} (\frac{i-1}{m}) \frac{1}{m} = \frac{1}{m^2} \sum_{i=1}^{m} (i-1)$$

$$= \frac{m-1}{2m}$$

Note that both $U(f,P_m)$ and $L(f,P_m)$ approach 1/2 as m approaches in-
finity.

If P and P' are two partitions of the same interval, we say that
P' is a *refinement* of P whenever P' is obtained from P by introducing
some additional points. For example P' = $\{0,1/2,2/3,1,4/3,3/2,2\}$ is
a refinement of P = $\{0,1/2,3/2,2\}$.

If P_1 and P_2 are any two partitions of the same interval, we say
that P is a *common refinement* of P_1 and P_2 if P is a refinement of
both P_1 and P_2. Given P_1 and P_2, we can obtain a common refinement P
by taking the union of the points of P_1 and P_2. For example, if

$$P_1 = \{0,1/3,1,4/3,2\}, \quad P_2 = \{0,1/4,1,7/4,2\}$$

then

$$P = \{0,1/4,1/3,1,4/3,7/4,2\}$$

is a common refinement of P_1 and P_2.

We now prove two useful results about upper and lower sums.

LEMMA 1.1. *Let* $f:[a,b] \to R$ *be a bounded function and let* P,P' *be two partitions of* $[a,b]$. *Then if* P' *is a refinement of* P, *we have*

$$U(f,P') \leq U(f,P)$$
$$L(f,P') \geq L(f,P)$$

Proof. We shall consider the case in which P' is obtained from P by adding a single point; the general case follows by induction (see Exercise 9).

Let $P = \{t_0,\ldots,t_m\}$ and $P' = \{t_0,\ldots,t_{i-1},s,t_i,\ldots,t_m\}$, where

$$t_{i-1} < s < t_i$$

Most of the terms of $U(f,P)$ and $U(f,P')$ are the same. Subtracting, we see that

$$U(f,P) - U(f,P') = M_i(t_i - t_{i-1}) - M_i'(s - t_{i-1}) - M_i''(t_i - s)$$

where

$$M_i = \text{lub}\{f(x) \mid t_{i-1} \leq x \leq t_i\}$$
$$M_i' = \text{lub}\{f(x) \mid t_{i-1} \leq x \leq s\}$$
$$M_i'' = \text{lub}\{f(x) \mid s \leq x \leq t_i\}$$

(See Figure 1.4.) Then $M_i \geq M_i'$ and $M_i \geq M_i''$ by Exercise 7 of Section 5.4, so that

$$U(f,P) - U(f,P') \geq M_i(t_i - t_{i-1}) - M_i(s - t_{i-1}) - M_i(t_i - s)$$
$$= 0$$

A similar argument holds for the lower sums. (See Exercise 10.)

LEMMA 1.2. *Let* $f:[a,b] \to R$ *be a bounded function and* P_1,P_2 *two partitions of* $[a,b]$. *Then*

$$L(f,P_1) \leq U(f,P_2)$$

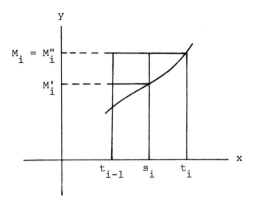

FIGURE 1.4. The upper sum of a refinement

Proof. Let P be a common refinement of P_1 and P_2. Then

$$L(f,P_1) \leq L(f,P)$$

$$U(f,P) \leq U(f,P_2)$$

by Lemma 1.1. Since $L(f,P) \leq U(f,P)$, we have

$$L(f,P_1) \leq L(f,P) \leq U(f,P) \leq U(f,P_2)$$

as claimed.

Define two sets of real numbers $U(f)$ and $L(f)$ by

$$U(f) = \{U(f,P):P \text{ any partition of } [a,b]\}$$

$$L(f) = \{L(f,P):P \text{ any partition of } [a,b]\}$$

If P_0 is any fixed partition of $[a,b]$, then, for any partition P of $[a,b]$,

$$L(f,P) \leq U(f,P_0)$$

$$U(f,P) \geq L(f,P_0)$$

by Lemma 1.2. Thus $L(f)$ is bounded above, and Theorem 5.4.2 implies that the least upper bound of $L(f)$ exists. Similarly, glb $U(f)$ exists.

Definition. A bounded function $f:[a,b] \rightarrow R$ is said to be *integrable* if

$$glb \ U(f) = lub \ L(f)$$

This common value is called the *integral* or *definite integral* of f on the interval $[a,b]$,

$$\int_a^b f(x)dx$$

Here are some examples.

3. Let $f:[a,b] \rightarrow R$ be the constant function, $f(x) = c$. Then $U(f,P) = L(f,P) = c(b-a)$ for any partition P, so that $glb \ U(f) = lub \ L(f) = c(b-a)$. Therefore, f is integrable, with

$$\int_a^b f(x)dx = c(b-a)$$

4. Let $f:[0,1] \rightarrow R$ be defined by $f(x) = x$. We proved in Example 2 earlier in the section that, for any positive integer m, there is a partition P_m with

$$U(f,P_m) = \frac{m+1}{2m} \ , \ L(f,P_m) = \frac{m-1}{2m}$$

We can use these partitions to prove that

$$glb \ U(f) = lub \ L(f) = \frac{1}{2} = \int_0^1 xdx$$

We prove only that $lub \ L(f) = 1/2$; the proof of the fact that $glb \ U(f) = 1/2$ is similar.

First of all, we claim that 1/2 is an upper bound for L(f). If not, then we can find a partition P of $[0,1]$ such that $L(f,P) = r > 1/2$. However, we can now find an integer m with $U(f,P_m) < r = L(f,P)$ which contradicts Lemma 1.2. Explicitly, we choose

$$m > \frac{1}{2r-1}$$

Then

$$U(f,P_m) = \frac{m+1}{2m}$$

$$= \frac{1}{2} + \frac{1}{2m}$$

$$< \frac{1}{2} + \frac{2r-1}{2} = r$$

Next, we claim that $1/2$ is the least upper bound of $L(f)$. If not, then lub $L(f) = r < 1/2$. Choose an integer m such that

$$m > \frac{1}{1-2r}$$

Then

$$L(f,P_m) = \frac{m-1}{2m}$$

$$= \frac{1}{2} - \frac{1}{2m}$$

$$> \frac{1}{2} - \frac{1-2r}{2} = r$$

which contradicts the fact that r is an upper bound for $L(f)$. Thus lub $L(f) = 1/2$.

The argument given in this example suggests the following result.

PROPOSITION 1.3. *Let* $f:[a,b] \to R$ *be a bounded function and* $P_1, P_2, \ldots, P_k, \ldots$ *a sequence of partitions of* $[a,b]$ *such that*

$$\lim_{k\to\infty} U(f,P_k) = \lim_{k\to\infty} L(f,P_k) = I$$

Then f *is integrable and*

$$\int_a^b f(x)dx = I$$

We leave the proof as an exercise.
Here is one more example.

5. Suppose $b > 0$ and $f:[0,b] \to R$ is given by $f(x) = x^2$. For each positive integer k, let P_k be the partition of $[0,b]$ into k equal sub-intervals;

$$P_k = \{0, \frac{b}{k}, \frac{2b}{k}, \ldots, \frac{(k-1)b}{k}, b\}$$

Since f is increasing on $[0,b]$, it follows that

$$M_j = \text{lub}\{f(x): \frac{(j-1)b}{k} \leq x < \frac{jb}{k}\}$$

$$= (\frac{jb}{k})^2$$

$$m_j = \text{glb}\{f(x): \frac{(j-1)b}{j} \leq x < \frac{jb}{k}\}$$

$$= (\frac{(j-1)b}{k})^2$$

Therefore

$$U(f,P_k) = \sum_{j=1}^{k} \frac{j^2 b^2}{k^2} \cdot \frac{b}{k}$$

$$= \frac{b^3}{k^3} \sum_{j=1}^{k} j^2$$

and

$$L(f,P_k) = \sum_{j=1}^{k} \frac{(j-1)^2 b^2}{k^2} \cdot \frac{b}{k}$$

$$= \frac{b^3}{k^3} \sum_{j=1}^{k} (j-1)^2$$

Using Exercise 1 of Section 1.3, we have

$$U(f,P_k) = \frac{b^3}{k^3} \cdot \frac{k(k+1)(2k+1)}{6}$$

$$= \frac{b^3}{6} (1 + \frac{1}{k})(2 + \frac{1}{k})$$

and

$$L(f,P_k) = \frac{b^3}{k^3} \cdot \frac{(k-1)k(2k-1)}{6}$$

$$= \frac{b^3}{6} (1 - \frac{1}{k})(2 - \frac{1}{k})$$

Thus

$$\lim_{k \to \infty} U(f,P_k) = \lim_{k \to \infty} L(f,P_k) = \frac{b^3}{3}$$

so that

$$\int_0^b x^2 dx = \frac{b^3}{3}$$

by Proposition 1.3.

EXERCISES

1. Prove directly from the definition that

$$\int_a^b xdx = \frac{1}{2}(b^2 - a^2)$$

(See Proposition 1.3.1.)

2. Prove directly from the definition that

$$\int_0^1 x^3 dx = \frac{1}{4}$$

(See Exercise 3 of Section 1.3.)

3. Prove directly from the definition that

$$\int_0^b x^3 dx = \frac{b^4}{4}$$

4. (a) Show (using the definition of the derivative) that for $a > 0$, $\lim_{h \to 0}(a^h - 1)/h = \log a$.

(b) Use (a) and the definition of the integral to compute $\int_1^c a^x dx$ (for $a > 0$, $a \neq 1$).

5. Use the result of 4 (a) to compute $\int_1^a 1/x \, dx$, $a > 1$. (Hint: it pays to take the points of the partition in a geometric progression.)

6. Prove that $\lim_{n \to \infty} \sum_{j=1}^n n/(n^2 + j^2) = \pi/4$. (Hint: consider $\int_0^1 dx/(1+x^2)$. Assume the usual calculus formulas for integrals.)

7. (a) Prove that $\lim_{n \to \infty} \sum_{j=1}^n 1/(n+j) = \log 2$.

   *(b) Prove that $\lim_{n \to \infty} \sum_{j=1}^n (-1)^{j-1}/j = \log 2$. (Hint: group all terms of the form $1/2^k \ell$, ℓ odd.)

8. Prove Proposition 1.3.

9. Use induction to complete the proof of the assertion of Lemma 1.1 involving upper sums.

10. Prove the assertion of Lemma 1.1 involving lower sums.

*11. Let $f: [a,b] \to R$ be a function, and define f^+, f^- by

$$f^+(x) = \begin{cases} f(x), & \text{if } f(x) > 0 \\ 0 & \text{otherwise} \end{cases}$$

$$f^-(x) - \begin{cases} -f(x), & \text{if } f(x) < 0 \\ 0 & \text{otherwise} \end{cases}$$

(a) Show that f^+ and f^- are nonnegative and that $f = f^+ - f^-$.

(b) Show that f is integrable if and only if f^+ and f^- are.

*12. (a) Let f,g be nonnegative bounded integrable functions on $[c,d]$. Prove that fg is integrable. (Hint: suppose that on an interval $\lfloor x_j, x_{j+1})$, $\sup(f(x)) - \inf(f(x)) = a$ and $\sup(g(x)) - \inf(g(x)) = b$. Show that

$$\sup(f(x)g(x)) - \inf(f(x)g(x)) \leq K(a+b)$$

where K is a bound for f and g.)

(b)   Use the result of Exercise 11 to show that if F, G are bounded integrable functions, then FG is integrable.

*13.   Let $f:[a,b] \to R$ be integrable and suppose there is a number $m > 0$ such that $|f(x)| \geq m$ for all $x \in [a,b]$. Prove that the function $g(x) = 1/f(x)$ is integrable.

14.   Show that the function $f:[0,1] \to R$ defined by

$$f(x) = \begin{cases} 0, & \text{if } x \text{ is irrational} \\ 1, & \text{if } x \text{ is rational} \end{cases}$$

is not integrable.   (Hint: every interval contains rational and irrational numbers.)

*15.   Show that the function $g:[0,1] \to R$ defined by

$$g(x) = \begin{cases} 0, & \text{if } x \text{ is irrational} \\ \dfrac{1}{q}, & \text{if } x \text{ is the rational number } \dfrac{p}{q} \text{ in lowest terms} \end{cases}$$

is integrable.   (Hint: given $\varepsilon > 0$, $f(x) > \varepsilon/2$ at only finitely many points.   Choose a partition to isolate these points in intervals of total length $< \varepsilon/2$.)

2.   PROPERTIES OF THE INTEGRAL

The definition of the integral which we have given is somewhat cumbersome to use.   The following proposition gives a useful characterization of integrability.

PROPOSITION 2.1.   *A bounded function* $f:[a,b] \to R$ *is integrable if and only if for any* $\varepsilon > 0$, *there is a partition* P *of* $[a,b]$ *such that* $U(f,P) - L(f,P) < \varepsilon$.

Proof. Suppose first of all that f is integrable and set

$$I = \int_a^b f(x)dx = \text{glb } U(f) = \text{lub } L(f)$$

By properties of the least upper bound and greatest lower bound, we can find partitions P_1 and P_2 so that

$$U(f,P_1) - I < \varepsilon/2$$

and

$$I - L(f,P_2) < \varepsilon/2$$

Let P be a common refinement of P_1 and P_2. Then

$$U(f,P) - I < \varepsilon/2$$

and

$$I - L(f,P) < \varepsilon/2$$

by Lemma 1.1. Thus

$$U(f,P) - L(f,P) = (U(f,P) - I) + (I - L(f,P)) < \varepsilon$$

Conversely, if for any $\varepsilon > 0$, we can find a partition P of $[a,b]$ with $U(f,P) - L(f,P) < \varepsilon$, it follows that

$$\text{glb } U(f) - \text{lub } L(f) \leq U(f,P) - L(f,P) < \varepsilon$$

Since ε is arbitrary, it follows that glb $U(f)$ = lub $L(f)$, so that f is integrable.

The next result gives some of the elementary properties of the integral.

PROPOSITION 2.2. *Let* $f,g:[a,b] \to R$ *be integrable functions and* p,q *real numbers.*

(a) $\int_a^b (pf(x) + qg(x))dx = p \int_a^b f(x)dx + q \int_a^b g(x)dx$

(b) *If* $f(x) \geq g(x)$ *for all* $x \in [a,b]$, *then*

$$\int_a^b f(x)dx \geq \int_a^b g(x)dx$$

(c) *If* $m \leq f(x) \leq M$ *for all* $x \in [a,b]$, *then*

$$m(b-a) \leq \int_a^b f(x)dx \leq M(b-a)$$

(d) *If* $c \in [a,b]$, *then*

$$\int_a^c f(x)dx + \int_c^b f(x)dx = \int_a^b f(x)dx$$

Proof. As we shall prove a more general theorem in Section 5, we shall not give a complete proof here. We sketch the proof of (a) in the case $p = q = 1$, leaving the rest as exercises.

Note first of all that, if $[t_{i-1}, t_i]$ is any subinterval of P,

$$\text{lub}\{f(x) + g(x) : t_{i-1} \leq x < t_i\} \leq \text{lub}\{f(x) : t_{i-1} \leq x < t_i\}$$

$$+ \text{lub}\{g(x) : t_{i-1} \leq x < t_i\}$$

and

$$\text{glb}\{f(x) + g(x) : t_{i-1} \leq x < t_i\} \geq \text{glb}\{f(x) : t_{i-1} \leq x < t_i\}$$

$$+ \text{glb}\{g(x) : t_{i-1} \leq x < t_i\}$$

(See Proposition 5.5.4.) Hence if P is any partition of $[a,b]$,

$$L(P,f) + L(P,g) \leq L(P,f+g) \leq U(P,f+g) \leq U(P,f) + U(P,g) \qquad (2.1)$$

Now let $\int_a^b f(x)dx = I$, $\int_a^b g(x)dx = J$. Given any $\varepsilon > 0$, we may choose partitions P_1, P_2 so that

$$I - L(P_1,f) < \frac{\varepsilon}{4}, \quad U(P_1,f) - I < \frac{\varepsilon}{4}$$

and

$$J - L(P_2,g) < \frac{\varepsilon}{4}, \quad U(P_2,g) - J < \frac{\varepsilon}{4}$$

Let P be a common refinement of P_1 and P_2. Then

$$I - L(P,f) < \frac{\varepsilon}{4}, \quad U(P,f) - I < \frac{\varepsilon}{4}$$

and

$$J - L(P,g) < \frac{\varepsilon}{4}, \quad U(P,g) - J < \frac{\varepsilon}{4}$$

We see from (2.1) that

$$(I + J) - L(P,f+g) < \frac{\varepsilon}{2} \qquad\qquad (2.2)$$

and

$$U(P,f+g) - (I + J) < \frac{\varepsilon}{2} \qquad\qquad (2.3)$$

In particular,

$$U(P,f+g) - L(P,f+g) < \varepsilon$$

Thus f + g is integrable, by Proposition 2.1. Moreover, (2.3) implies that

$$I + J - \text{lub}\,\{L(P,f+g) : P \text{ a partition of } [a,b]\} < \frac{\varepsilon}{2}$$

and

$$\text{glb}\,\{U(P,f+g) : P \text{ a partition of } [a,b]\} - (I + J) < \frac{\varepsilon}{2}$$

Hence

$$\left| I + J - \int_a^b (f+g)(x)dx \right| < \frac{\varepsilon}{2}$$

and, since ε is arbitrary,

$$\int_a^b (f+g)(x)dx = I + J = \int_a^b f(x)dx + \int_a^b g(x)dx$$

as claimed.

The basic result for the integral of a function of one variable is the following.

THEOREM 2.3.   (The Fundamental Theorem of Calculus). *Let*
$f:[a,b] \to R$ *be a continuous function and define* $F:[a,b] \to R$ *by*

$$F(t) = \int_a^t f(x)dx$$

Then F is differentiable and $F'(t) = f(t)$.

Note. We have not yet proved that continuous functions are integrable on $[a,b]$. We shall assume this fact here; a proof will be given in Section 6. (Also see Exercise 13.)

Proof. This is an application of parts (c) and (d) of Proposition 2.2. We need to compute

$$\lim_{h \to 0} \frac{F(t+h)-F(t)}{h}$$

Assume for the moment that $h > 0$. Then, from (d) of Proposition 2.2,

$$F(t+h) - F(t) = \int_t^{t+h} f(x)dx$$

Given $\varepsilon > 0$, we may pick $\delta > 0$ such that if $|x - t| < \delta$, then

$$|f(x) - f(t)| < \varepsilon$$

or

$$f(t) - \varepsilon < f(x) < f(t) + \varepsilon$$

(Here, of course, we use the continuity of f.)   Therefore

$$h(f(t) - \varepsilon) < \int_t^{t+h} f(x)dx < h(f(t) + \varepsilon)$$

or

$$f(t) - \varepsilon < \frac{F(t+h)-F(t)}{h} < f(t) + \varepsilon \tag{2.4}$$

when $0 < h < \delta$.

When $h < 0$, we have

$$F(t+h) - F(t) = - \int_{t+h}^t f(x)dx$$

and similar reasoning shows that (2.4) still holds for $-\delta < h < 0$.
Therefore if $0 < |h| < \delta$,

$$\left| \frac{F(t+h)-F(t)}{h} - f(t) \right| < \varepsilon \qquad\qquad (2.5)$$

Since ε is arbitrary, (2.5) implies that

$$\lim_{h \to 0} \frac{F(t+h)-F(t)}{h} = f(t)$$

which proves the theorem.

Theorem 2.3 is useful in evaluating integrals. If $f:[a,b] \to R$
is continuous, a function $G:[a,b] \to R$ is said to be an *antiderivative*
(or a *primitive*) for f if $G'(x) = f(x)$ for $x \in [a,b]$. For example,
$G(x) = x^3/3$ is an antiderivative for $f(x) = x^2$. If $G:[a,b] \to R$ is an
antiderivative for f and $F:[a,b] \to R$ is defined as in Theorem 2.3,
then $(G - F)' = G' - F' = 0$. Therefore F and G differ by a constant,
c:

$$G(t) = F(t) + c$$

(The reason is that if $t \in [a,b]$, then $(G - F)(t) - (G - F)(a) = 0$;
see Exercise 9, Section 3.5.) In particular,

$$G(a) = F(a) + c$$
$$\quad\ = c$$

since $F(a) = 0$. Thus $G(t) - G(a) = F(t)$ for all $t \in [a,b]$. It follows that

$$\int_a^b f(x)dx = F(b) = G(b) - G(a)$$

where G is *any* antiderivative for f. We often use the notation

$$G(x) \Big|_a^b = G(b) - G(a)$$

For example,

$$\int_a^b x^2 dx = x^3/3 \Big|_a^b$$

$$= b^3/3 - a^3/3$$

EXERCISES

(In all exercises except the last, assume that continuous functions are integrable.)

1.   Let $f:[a,b] \to R$ be a continuous function and $g:[c,d] \to [a,b]$ a
differentiable function.   Prove that

$$\int_{g(c)}^{g(d)} f(x)dx = \int_c^d f(g(x))g'(x)dx$$

This formula is the basis for the technique of integration usually
called *substitution*.   (Hint: use the chain rule and the discussion
following Theorem 2.3).   As usual, $\int_a^b = - \int_b^a$ if $a > b$.

2.   Use the formula of Exercise 1 to evaluate the following.

   (a) $\int_1^2 2x(x^2 + 1)^{1/2}dx$ (Let $g(x) = x^2 + 1$ and $f(u) = u^{1/2}$.)

   (b) $\int_0^{\sqrt{\pi/2}} x \sin x^2 dx$

   (c) $\int_0^{\pi/2} \sin x \, e^{\cos x} dx$

   (d) $\int_0^1 e^{x+e^x} dx$

   (e) $\int_0^1 \dfrac{xdx}{x^4+1}$

3.   Prove the "integration by parts" formula:

$$\int_a^b f(x)g'(x)dx = f(x)g(x) \Big|_a^b - \int_a^b f'(x)g(x)dx$$

4.   Evaluate the following:

(a) $\int_0^{\pi/2} x \sin x\, dx$ (d) $\int_0^{\pi/3} x^2 \cos x\, dx$

(b) $\int_0^1 xe^x dx$ (e) $\int_0^{5\pi/6} e^x \sin x\, dx$

(c) $\int_0^1 \text{arc} \tan x\, dx$ (Hint: integrate by parts twice and combine terms.)

5.   Prove Proposition 2.2.

6.   Let $f:[a,b] \to R$ be a continuous function. Prove that there is a number $c \in [a,b]$ such that

$$\int_a^b f(x)\, dx = f(c)(b-a)$$

This is the *First Mean Value Theorem for Integrals*. (Hint: combine the corollary to Theorem 5.5.3 with part (c) of Proposition 2.2 and apply the Intermediate Value Theorem (Theorem 5.8.6).)

7.   Let $f:[a,b] \to R$ be a continuous function with $f(x) \geq 0$ for all $x \in [a,b]$ and $f(x_0) > 0$ for some $x_0 \in [a,b]$. Prove that

$$\int_a^b f(x)\, dx > 0$$

(Hint: Use Exercise 21, Section 3.3.)

8.   Complete the proof of Theorem 2.3 (by dealing with the case $h < 0$).

9.   Show that f is integrable on $[a,b]$ and if $a < c < b$, then f is integrable on $[a,c]$ and on $[c,b]$.

*10.   Let h be continuous and nonnegative, and let f be continuous. Show that there is a number $c \in [a,b]$ such that

$$\int_a^b f(x)h(x)\, dx = f(c) \int_a^b h(x)\, dx$$

(This result is similar to the one in Exercise 6; in fact, Exercise 6 is the special case $h(x) = 1$.) Hint: if m is the minimum value and

M the maximum value of f on $[a,b]$, then $mh(x) \leq f(x)h(x) \leq Mh(x)$.
Now use the hint for Exercise 6.

*11. Let $f:[a,b]$ be monotone and continuously differentiable, and let
$g:[a,b] \to R$ be continuous. Prove that there is a number $c \in [a,b]$
such that

$$\int_a^b f(x)g(x)dx = f(a)\int_a^c g(x)dx + f(b)\int_c^b g(x)dx$$

This is called the *Second Mean Value Theorem for Integrals*. (Hint:
let $G(t) = \int_a^t g(x)dx$. We may assume that f is monotone increasing,
so that $f' \geq 0$. Show that

$$\int_a^b f(x)g(x)dx = f(b)G(b) - \int_a^b f'(x)G(x)dx$$

Apply the result of Exercise 10 and simplify.)

12. What is wrong with the following reasoning? Let f be a function
such that $f(2x) = 1/2\ f(x)$. Then

$$\int_0^2 f(x)dx = \int_0^1 2f(2t)dt \qquad (x = 2t)$$

$$= \int_0^1 f(t)dt$$

from the equation for f. Thus

$$\int_1^2 f(x)dx = \int_0^2 f(x)dx - \int_0^1 f(t)dt$$

$$= 0$$

Now let $f(x) = 1/x$:

$$\int_1^2 \frac{dx}{x} = 0$$

or $\log 2 = 0$. Thus $2 = 1$.

*13. Prove that if f is continuous on $[a,b]$, then f is integrable on
$[a,b]$. (Hint: f is uniformly continuous, by Theorem 5.7.1. Given
$\varepsilon > 0$, partition $[a,b]$ so finely that the lub and glb of f on each
subinterval differ by less than $\varepsilon/b-a$. Now apply Proposition 2.1.)

3. THE INTEGRAL OF A FUNCTION OF TWO VARIABLES

The definition of the integral of a function of several variables is
a straightforward extension of the one variable case. However, as the
number of variables increases, so does the complexity of the notation.
Because of this, we shall give a detailed treatment of the integral of
a function of two variables here. In the next section, we give a
sketch of the definition of the integral for functions of more than
two variables.

Let $\vec{a} = (a_1,a_2)$, $\vec{b} = (b_1,b_2)$ be points in R^2 with $a_1 < b_1$, $a_2 < b_2$
and let $R(\vec{a},\vec{b})$ be the *rectangle*

$$R(\vec{a},\vec{b}) = \{(x,y) \mid a_1 \leq x \leq b_1,\ a_2 \leq y \leq b_2\}$$

For convenience here, we use (x,y) for the coordinates in R^2. If
$P_1 = \{t_0,\ldots,t_p\}$ is a partition of $[a_1,b_1]$ and $P_2 = \{s_0,\ldots,s_k\}$ is a
partition of $[a_2,b_2]$, then the sets R_{ij} given by

$$R_{ij} = \{(x,y) \mid t_{i-1} \leq x \leq t_i \text{ and } s_{j-1} \leq y \leq s_j\}$$

define a decomposition of $R(\vec{a},\vec{b})$ into sub-rectangles. We call this
decomposition a *partition* of $R(\vec{a},\vec{b})$ and denote it by $P_1 \times P_2$ (See
Figure 3.1).

Note that the decomposition pictured in Figure 3.2 is *not* a partition according to our definition. Our partitions of rectangles must
arise from partitions of the edges, as described above.

Suppose now that $f:R(\vec{a},\vec{b}) \to R$ is a bounded function and $P_1 \times P_2$
a partition of $R(\vec{a},\vec{b})$. We define numbers M_{ij}, m_{ij}, $1 \leq i \leq p$, $1 \leq j$
$\leq k$, by

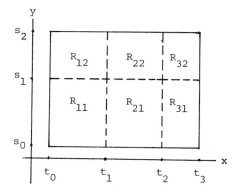

FIGURE 3.1. A partition of a rectangle in R^2

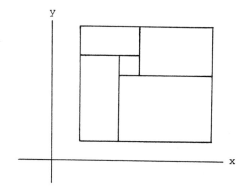

FIGURE 3.2. A decomposition which is not a partition

$$M_{ij} = \text{lub}\{f(\vec{v}) \mid \vec{v} \in R_{ij}\}$$
$$m_{ij} = \text{glb}\{f(\vec{v}) \mid \vec{v} \in R_{ij}\}$$

The *upper sum* of f on $R(\vec{a},\vec{b})$ relative to the partition $P_1 \times P_2$ is defined by

$$U(f,P_1 \times P_2) = \sum_{i=1}^{p} \sum_{j=1}^{k} M_{ij}\mu(R_{ij})$$

where $\mu(R_{ij})$ is the "area" of the rectangle R_{ij},

$$\mu(R_{ij}) = (t_i - t_{i-1})(s_j - s_{j-1})$$

Similarly, the *lower sum* of f on $R(\vec{a},\vec{b})$ relative to the partition $P_1 \times P_2$ is defined by

$$L(f,P_1 \times P_2) = \sum_{i=1}^{p} \sum_{j=1}^{k} m_{ij}\mu(R_{ij})$$

Clearly $L(f,P_1 \times P_2) \leq U(f,P_1 \times P_2)$.

Suppose now that $P_1 \times P_2$ and $P_1' \times P_2'$ are two partitions of $R(\vec{a},\vec{b})$. We say that $P_1' \times P_2'$ is a *refinement* of $P_1 \times P_2$ if P_1' is a refinement of P_1 and P_2' is a refinement of P_2. Equivalently, $P_1' \times P_2'$ is a refinement of $P_1 \times P_2$ if each of the rectangles of $P_1' \times P_2'$ is contained in a rectangle of $P_1 \times P_2$. For example, if

$$P = \{0,1,2\}, \quad P' = \{0,1/2,1,2\}$$

then $P \times P'$ is a refinement of $P \times P$.

If $P_1 \times P_2$ and $P_1' \times P_2'$ are partitions of $R(\vec{a},\vec{b})$, then $P_1'' \times P_2''$ will be a common refinement of $P_1 \times P_2$ and $P_1' \times P_2'$ if P_1'' is a common refinement of P_1 and P_1' and P_2'' is a common refinement of P_2 and P_2'. One way to produce a common refinement of $P_1 \times P_2$ and $P_1' \times P_2'$ is to regard each partition as a transparency with lines on it; we can obtain a common refinement by putting one transparency on the other. (See Figure 3.3.)

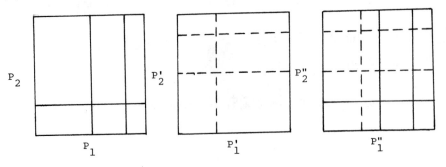

FIGURE 3.3. A common refinement

We can now state the analogue of Lemma 1.1 and 1.2.

LEMMA 3.1. *Let* $f:R(\vec{a},\vec{b}) \to R$ *be a bounded function and let* $P_1 \times P_2$, $P_1' \times P_2'$ *be partitions of* $R(\vec{a},\vec{b})$. *Then*

    (i) $L(f,P_1 \times P_2) \leq U(f,P_1' \times P_2')$

    (ii) *If* $P_1' \times P_2'$ *is a refinement of* $P_1 \times P_2$, *then*

$$U(f,P_1' \times P_2') \leq U(f,P_1 \times P_2)$$
$$L(f,P_1' \times P_2') \geq L(f,P_1 \times P_2)$$

Adapting the proofs of Lemmas 1.1 and 1.2 to this lemma should present no problem. We leave it as an exercise for the reader.

Define two sets $U(f)$ and $L(f)$ by

$$U(f) = \{U(f,P_1 \times P_2):P_1 \times P_2 \text{ a partition of } R(\vec{a},\vec{b})\}$$
$$L(f) = \{L(f,P_1 \times P_2):P_1 \times P_2 \text{ a partition of } R(\vec{a},\vec{b})\}$$

Then, just as in Section 1, $U(f)$ is bounded from below and $L(f)$ is bounded from above, so that glb $U(f)$ and lub $L(f)$ both exist.

Definition. A bounded function $f:R(\vec{a},\vec{b}) \to R$ is said to be *integrable* if

    glb $U(f)$ = lub $L(f)$

This common value is called the *integral* of f on $R(\vec{a},\vec{b})$ and denoted by

$$\int_{R(\vec{a},\vec{b})} f$$

We shall see in Section 6 that, for example, all continuous functions are integrable.

Here are a few examples.

1.   Let $f:R(\vec{a},\vec{b}) \to R$ be a constant function: $f(\vec{v}) = c$. Then $L(f,P)$ = $U(f,P) = c(b_1 - a_1)(b_2 - a_2)$ for all partitions P, and it follows that

$$\int\limits_{R(\vec{a},\vec{b})} f = c(b_1 - a_1)(b_2 - a_2)$$

2. Let $S = R(\vec{a},\vec{b})$, where $\vec{a} = (0,0)$ and $\vec{b} = (1,1)$ and let $f:S \to R$ be defined by $f(x,y) = xy$. Let P_m be the partition of $[0,1]$ defined in Example 2 at the beginning of Section 1: $P_m = \{0,1/m,2/m,\dots,m-1/m,1\}$ and consider the partition $P_m \times P_m$ of S. Then

$$R_{ij} = \{(x,y) \in R^2 : \frac{i-1}{m} \le x \le \frac{i}{m}, \ \frac{j-1}{m} \le y \le \frac{j}{m}\}$$

and $\mu(R_{ij}) = 1/m^2$ for all i,j. Furthermore, it is easy to see that

$$M_{ij} = \frac{i}{m} \cdot \frac{j}{m} = \frac{ij}{m^2}$$

$$m_{ij} = \frac{(i-1)}{m} \cdot \frac{(j-1)}{m} = \frac{(i-1)(j-1)}{m^2}$$

Thus

$$U(f, P_m \times P_m) = \sum_{i=1}^{m} \sum_{j=1}^{m} \frac{ij}{m^2} \cdot \frac{1}{m^2}$$

$$= \frac{1}{m^4} \sum_{i=1}^{m} i \sum_{j=1}^{m} j$$

$$= \frac{1}{m^4} \frac{m(m+1)}{2} \cdot \frac{m(m+1)}{2}$$

$$= \frac{1}{4}(1 + \frac{1}{m})^2$$

and

$$L(f, P_m \times P_m) = \sum_{i=1}^{m} \sum_{j=1}^{m} \frac{(i-1)(j-1)}{m^2} \cdot \frac{1}{m^2}$$

$$= \frac{1}{m^4} \sum_{i=1}^{m} (i-1) \sum_{j=1}^{m} (j-1)$$

$$= \frac{1}{m^4} \frac{(m-1)m}{2} \cdot \frac{(m-1)m}{2}$$

$$= \frac{1}{4}(1 - \frac{1}{m})^2$$

It follows that

$$\lim_{m \to \infty} U(f, P_m \times P_m) = \lim_{m \to \infty} L(f, P_m \times P_m) = \frac{1}{4}$$

and, just as in Example 4 in Section 1, we see that

$$\int_S xy = \frac{1}{4}$$

3.   Suppose $S = R(\vec{a}, \vec{b})$ where $\vec{a} = (0,0)$ and $\vec{b} = (b,c)$ and let $f: S \to R$ be defined by $f(x,y) = x^2 y$. Let P_k be the partition of $[0,b]$ defined in Example 5 of Section 1;

$$P_k = \{0, \frac{b}{k}, \frac{2b}{k}, \ldots, \frac{(k-1)b}{k}, b\}$$

and let P_k' be the analogous partition of $[0,c]$;

$$P_k' = \{0, \frac{c}{k}, \frac{2c}{k}, \ldots, \frac{(k-1)c}{k}, c\}$$

Then, if $P_k'' = P_k \times P_k'$,

$$R_{ij} = \{(x,y) : \frac{(i-1)b}{k} \le x \le \frac{ib}{k}, \frac{(j-1)c}{k} \le y < \frac{jc}{k}\}$$

and $\mu(R_{ij}) = \frac{bc}{k^2}$. Furthermore,

$$M_{ij} = (\frac{ib}{k})^2 \cdot \frac{jc}{k}$$

and

$$m_{ij} = (\frac{(i-1)b}{k})^2 \cdot \frac{(j-1)c}{k}$$

Thus

$$U(f, P_k'') = \sum_{i=1}^{k} \sum_{j=1}^{k} \frac{i^2 b^2}{k^2} \frac{jc}{k} \cdot \frac{bc}{k^2}$$

$$= \frac{b^3 c^2}{k^5} \sum_{i=1}^{k} i^2 \sum_{j=1}^{k} j$$

$$= \frac{b^3 c^2}{k^5} \frac{k(k+1)(2k+1)}{6} \frac{k(k+1)}{2}$$

$$= \frac{b^3 c^2}{12} (1 + \frac{1}{k})^2 (2 + \frac{1}{k})$$

Similarly,

$$L(f, P''_k) = \frac{b^3 c^2}{12} (1 - \frac{1}{k})^2 (2 - \frac{1}{k})$$

and it follows that

$$\lim_{k \to \infty} U(f, P''_k) = \lim_{k \to \infty} L(f, P''_k) = \frac{b^3 c^2}{6}$$

Now, just as in Example 4 of Section 1, we have

$$\int_S x^2 y = \frac{b^3 c^2}{6}$$

One problem with our definition of the integral is that it only lets us integrate functions over rectangles. We may need to integrate over more general sets - for instance, triangles or discs. For integrals over such sets, we proceed as follows. Let S be a closed and bounded subset of R^2, and let $f : S \to R$ be a bounded function. Since S is bounded, we can find a rectangle S_1 containing S. Define $f_1 : S_1 \to R$ by

$$f_1(\vec{v}) = \begin{cases} f(\vec{v}) & \text{if } \vec{v} \in S \\ 0 & \text{if } \vec{v} \notin S \end{cases}$$

We say that $f : S \to R$ is *integrable* if f_1 is and we define the *integral* of f over S by

$$\int_S f = \int_{S_1} f_1$$

We need to show that this definition does not depend on the particular rectangle S_1 which we choose containing S. This is not hard to see. For example, if S_2 is a second rectangle containing S and $f_2:S_2 \to R$ is defined as is f_1, then the intersection $S_0 = S_1 \cap S_2$ is a rectangle containing S. Furthermore, both S_1 and S_2 differ from S_0 by a union of rectangles on which the functions f_1 and f_2 vanish. (See Figure 3.4.) Thus, if $f_0:S_0 \to R$ is defined in the same way as f_1 and f_2, it follows readily that f_i is integrable if and only if f_0 is integrable, $i = 1,2$, and

$$\int_{S_1} f_1 = \int_{S_0} f_0 = \int_{S_2} f_2$$

(See Exercises 7 and 8.)

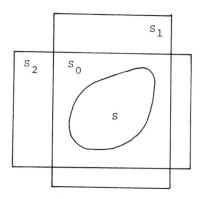

FIGURE 3.4.   Rectangles containing S

We conclude this section by remarking that the result analogous to Proposition 2.1 holds in R^2 as well. We leave its statement and proof as Exercise 2.

EXERCISES

1. Let $R(\vec{a},\vec{b})$ be a rectangle in R^2, and let $f:R(\vec{a},\vec{b}) \to R$ the constant function $f(\vec{v}) = c$, $\vec{v} \in R(\vec{a},\vec{b})$. Prove that f is integrable and that

$$\int_{R(\vec{a},\vec{b})} f = c \ (R(\vec{a},\vec{b}))$$

2. State and prove the analogue of Proposition 2.1 for functions of two variables.

3. Let $f:R^2 \to R$ be defined by $f(x,y) = x$. Use Exercise 2 to prove that

$$\int_S f = \frac{1}{2}$$

where S is the rectangle defined by $(0,0)$ and $(1,1)$. (See Example 2 of Section 1.)

4. Let $S = R(\vec{a},\vec{b})$ be the rectangle in R^2 defined by the points $\vec{a} = (a_1,a_2)$, $\vec{b} = (b_1,b_2)$ and let $f:S \to R$ be the function $f(x,y) = x$. Prove directly from the definition that

$$\int_S f = \frac{1}{2}(b_2 - a_2)(b_1^2 - a_1^2)$$

5. Determine, directly from the definition, the integral of the function f over the rectangle S of Exercise 3, where

(a) $f(x,y) = 2x + y$
(b) $f(x,y) = xy$
(c) $f(x,y) = x^2 + y^2$

6. Prove Lemma 3.1.

*7. Let $S = R(\vec{a},\vec{b})$ be the rectangle in R^2 defined by the points $\vec{a} = (a_1,a_2)$, $\vec{b} = (b_1,b_2)$. The *interior* of S is the set

$$\text{int } S = \{(x,y):a_1 < x < b_1, \ a_2 < y < b_2\}$$

Suppose $S = \bigcup_{j=1}^{m} S_j$ where each S_j is a rectangle and $(\text{int } S_i) \cap$ $(\text{int } S_j) = \phi$ for $i \neq j$. (See Figure 3.5 for an example.) Let $f:S \to R$ be an integrable function. Prove that $f:S_j \to R$ is integrable, $1 \leq j \leq m$, and

$$\int_S f = \sum_{j=1}^{m} \int_{S_j} f$$

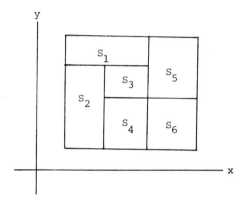

FIGURE 3.5. A decomposition of a rectangle

*8. Suppose $S \subset R^2$ is a bounded set and $f:S \to R$ is a function. Let S_1 and S_2 be rectangles containing S, $S_0 = S_1 \cap S_2$, and define $f_i:S_i \to R$, $i = 0,1,2$, by

$$f_i(\vec{v}) = \begin{cases} f(\vec{v}) & \text{for } \vec{v} \in S \\ 0 & \text{for } \vec{v} \notin S \end{cases}$$

Prove that each of the functions f_1 and f_2 is integrable if and only if f_0 is integrable and that, if the integrals exist,

$$\int_{S_1} f_1 = \int_{S_0} f_0 = \int_{S_2} f_2$$

9.   Let $f(x,y) = x$. Compute $\int_S f$, where S is the triangle bounded by
the x-axis and the lines $x = 1$ and $x = y$. (See Figure 3.6.)

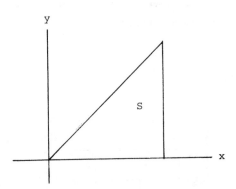

FIGURE 3.6.   A triangular region

4.   THE INTEGRAL OF A FUNCTION OF n VARIABLES

The procedure for defining the integral of a function of n variables
will hardly be a surprise after the previous sections. The main problem is the notation.

Let $\vec{a} = (a_i)$, $\vec{b} = (b_i)$ be points in R^n with $a_i < b_i$, $1 \leq i \leq n$,
and define the rectangle $R(\vec{a},\vec{b})$ by

$$R(\vec{a},\vec{b}) = \{\vec{v} = (x_i) \in R^n | a_i \leq x_i \leq b_i, \ i = 1,2,\ldots,n\}$$

The "volume" $\mu(R(\vec{a},\vec{b}))$ of the rectangle $R(\vec{a},\vec{b})$ is defined to be

$$\mu(R(\vec{a},\vec{b})) = (b_1 - a_1)(b_2 - a_2) \cdots (b_n - a_n)$$

For example, if $r = (0,0,0)$ and $b = (2,1,1)$, the rectangle $R(\vec{a},\vec{b})$
is pictured in Figure 4.1. Its volume is 2.

If P_j is a partition of the interval $[a_j,b_j]$, $j = 1,2,\ldots,n$, we
define a partition $P = P_1 \times P_2 \times \ldots \times P_n$ of $R(\vec{a},\vec{b})$ into sub-rectangles
just as in the case $n = 2$. For example, if $n = 3$ and P_1,P_2,P_3 are
each the partition of the interval $[0,1]$ into two equal pieces,

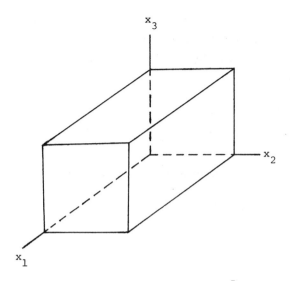

FIGURE 4.1.   A rectangle in R^3

we obtain the partition of the unit cube in R^3 into eight congruent
subcubes.   (See Figure 4.2.)

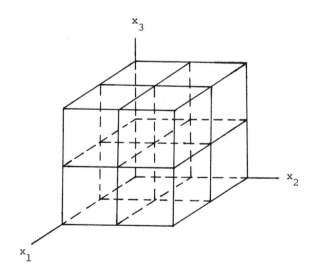

FIGURE 4.2.   A partition of a rectangle in R^3

Suppose that $f:R(\vec{a},\vec{b}) \to R$ is a bounded function and $P = P_1 \times P_2 \times \ldots \times P_n$ a partition of $R(\vec{a},\vec{b})$. Let $R_1,\ldots,R_q$ be all of the subrectangles of $R(\vec{a},\vec{b})$ occurring in the partition P and define m_i, M_i, $1 \leq i \leq q$ by

$$m_i = glb\{f(\vec{v}):\vec{v} \in R_i\}$$
$$M_i = lub\{f(\vec{v}):\vec{v} \in R_i\}$$

Then, just as in the previous section, we define $L(f,P)$ and $U(f,P)$ by

$$L(f,P) = \sum_{i=1}^{q} m_i \mu(R_i)$$
$$U(f,P) = \sum_{i=1}^{q} M_i \mu(R_i)$$

The analogue of Lemma 3.1 is as follows.

LEMMA 4.1. *Let* $f:R(\vec{a},\vec{b}) \to R$ *be a bounded function and let* $P = P_1 \times \cdots \times P_n$, $P' = P_1' \times \cdots \times P_n'$ *be partitions of* $R(\vec{a},\vec{b})$. *Then*

(i) $L(f,P) \leq U(f,P')$

(ii) if P' is a refinement of P, then

$$U(f,P') \leq U(f,P)$$
$$L(f,P') \geq L(f,P)$$

The proof of this result proceeds just as did the proofs of Lemmas 1.1 and 1.2. We leave it as an exercise for the reader.

Finally, we define subsets of R by

$$U(f) = \{U(f,P) \mid P = P_1 \times P_2 \times \cdots \times P_n \text{ is a partition of } R(\vec{a},\vec{b})\}$$
$$L(f) = \{L(f,P) \mid P = P_1 \times P_2 \times \cdots \times P_n \text{ is a partition of } R(\vec{a},\vec{b})\}$$

As before, these sets are bounded. We say that f is *integrable* if

$$glb\ U(f) = lub\ L(f)$$

The *integral* of f is defined to be this common value and denoted by

$$\int_{R(\vec{a},\vec{b})} f$$

The analogue of Proposition 2.1 is true for functions of n variables; we state it here, leaving the proof as an exercise.

PROPOSITION 4.2. *Let S be a rectangle in* R^n *and* $f:S \to R$ *a bounded function. Then f is integrable if and only if for* $\varepsilon > 0$, *there is
a partition P of S such that*

$$U(f,P) - L(f,P) < \varepsilon$$

Here are some examples.

1.   Let $f:R\lceil\vec{a},\vec{b}\rceil \to R$ be the constant function $f(\vec{v}) = c$. Then

$$L(f,P) = U(f,P) = c\mu(R(\vec{a},\vec{b}))$$

for any partition P. Thus (see Exercise 1)

$$\int_{R(\vec{a},\vec{b})} f = c\mu(R(\vec{a},\vec{b}))$$

2.   Suppose S is the rectangle in R^3 defined by the vectors $\vec{a} = (0,0,0)$
and $\vec{b} = (b,c,d)$ and $f:S \to R$ is defined by $f(x,y,z) = xyz$. Let k be
a positive integer and $\widetilde{P}_k$ the partition of S into rectangles $R_{ij\ell}$,
$1 \le i,j,\ell \le k$, where

$$R_{ij\ell} = \{(x,y,z): \frac{(i-1)b}{k} \le x < \frac{ib}{k} , \frac{(j-1)c}{k} \le y < \frac{jc}{k}, \frac{(\ell-1)d}{k}$$
$$\le z < \frac{\ell d}{k}\}$$

One checks easily that $\mu(R_{ij\ell}) = \dfrac{bcd}{k^3}$,

$$M_{ij\ell} = \text{lub}\{f(\vec{v}):\vec{v} \in R_{ij\ell}\}$$

$$= \frac{ib}{k} \cdot \frac{jc}{k} \cdot \frac{\ell d}{k}$$

$$= \frac{bcd}{k^3} ij\ell$$

and

$$m_{ij\ell} = glb\{f(\vec{v}):\vec{v} \in R_{ij\ell}\}$$

$$= \frac{bcd}{k^3} (i-1)(j-1)(\ell-1)$$

Therefore

$$U(f,\widetilde{P}_k) = \sum_{i=1}^{k} \sum_{j=1}^{k} \sum_{\ell=1}^{k} \frac{bcd}{k^3} ij\ell \frac{bcd}{k^3}$$

$$= \frac{b^2c^2d^2}{k^6} \sum_{i=1}^{k} i \sum_{j=1}^{k} j \sum_{\ell=1}^{k} \ell$$

$$= \frac{b^2c^2d^2}{k^6} \frac{k^3(k+1)^3}{8}$$

$$= \frac{b^2c^2d^2}{8} (1 + \frac{1}{k})^3$$

and, similarly,

$$L(f,\widetilde{P}_k) = \frac{b^2c^2d^2}{8} (1 - \frac{1}{k})^3$$

It follows that

$$\lim_{k\to\infty} U(f,\widetilde{P}_k) = \lim_{k\to\infty} L(f,\widetilde{P}_k) = \frac{b^2c^2d^2}{8}$$

and, just as in Example 4 of Section 1,

$$\int_S xyz = \frac{b^2c^2d^2}{8}$$

In R^n, too, one may wish to integrate over non-rectangular regions. The procedure is like that for R^2. If S is a closed, bounded subset of R, we let S_1 be a rectangle containing S. Define $f_1:S_1 \to R$ by

$$f_1(\vec{v}) = \begin{cases} f(\vec{v}) & \text{if } \vec{v} \in S \\ 0 & \text{if } \vec{v} \notin S \end{cases}$$

We then say that $f:S \to R$ is integrable if f_1 is integrable on S_1, and we define

$$\int_S f = \int_{S_1} f$$

It is not hard to show, just as in the case $n = 2$, that this definition does not depend on our choice of S_1. (See Exercise 8.)

EXERCISES

1. Let S be a rectangle in R^n and $f:S \to R$ the constant function $f(\vec{v}) = c$, $\vec{v} \in S$. Prove that f is integrable and that

$$\int_S f = c\mu(S)$$

2. State and prove the analogue of Proposition 2.1 for functions of n variables.

3. Let $f:R^n \to R$ be defined by $f(x_1,\ldots,x_n) = x_1$. Use Exercise 2 to prove that

$$\int_S f = \frac{1}{2}$$

where S is the rectangle defined by $(0,0,\ldots,0)$ and $(1,1,\ldots,1)$. (See Example 2 of Section 1.)

4. Let S be the rectangle in R^n defined by the points $\vec{a} = (a_i)$, $\vec{b} = (b_i)$ and let $f:S \to R$ be the function $f(x_1,\ldots,x_n) = x_1$. Prove directly from the definition that

$$\int_S f = \frac{1}{2}(b_n - a_n)\cdots(b_2 - a_2)(b_1^2 - a_1^2)$$

5. Determine, directly from the definition, the integral of f over the rectangle S in R^3 defined by $(0,0,0)$ and $(1,1,1)$ when
     (a) $f(x,y,z) = 2x + y + z$
     (b) $f(x,y,z) = xyz$
     (c) $f(x,y,z) = x^2 + y^2 + z^2$

6.   Prove Lemma 4.1.

*7.   State and prove the analogue of Exercise 7 of the previous section for R^n.

8.   State and prove the analogue of Exercise 8 of the previous section for R^n.

*9.   Let $f:R^3 \to R$ be given by $f(x,y,z) = x$ and compute $\int_S f$ where S is the region in the first octant bounded by the coordinate planes and the plane $x + y + z = 1$. (See Figure 4.3.)

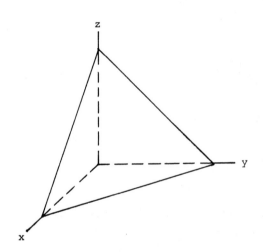

FIGURE 4.3.

10.   Let S be a rectangle in R^n, let $f:S \to R$ be integrable, and define f^+, f^- by

$$f^+(x) = \max(f(x),0)$$

$$f^-(x) = \max(-f(x),0)$$

Show that f^+ and f^- are integrable, that $|f(x)| = f^+(x) + f^-(x)$, and that $f(x) = f^+(x) - f^-(x)$. Use this to prove that if f is integrable, then $|f|$ is integrable.

*11. Show that if f and g are bounded integrable functions on the rectangle S, then fg is also integrable on S. (See Exercises 11 and 12 of Section 1.)

*12. We have defined integrals only for bounded functions on rectangles. In this exercise, we define the integrals of certain unbounded functions.

Let S be a rectangle in R^n, and let $f:S \to R$ be a nonnegative function. For each positive integer n, define $f_n:S \to R$ by

$$f_n(\vec{v}) = \min(f(\vec{v}), n)$$

(a) Show that for every $\vec{v} \in S$, $\lim_{n \to \infty} f_n(\vec{v}) = f(\vec{v})$.

(b) Suppose that for every n, the function f_n is integrable on S, and suppose that the set $\{\int_S f_n : n = 1, 2, \ldots\}$ is bounded. Show that

$$\lim_{n \to \infty} \int_S f_n$$

exists.

We define

$$\int_S f = \lim_{n \to \infty} \int_S f_n$$

when the limit exists. If f is an arbitrary function (not necessarily nonnegative), then we define

$$\int_S f = \int_S f_+ - \int_S f_-$$

(see Exercise 10) if $\int_S f_+$ and $\int_S f_-$ both exist. (Note that both f_+ and f_- are nonnegative.)

(c) Show that this definition coincides with the usual one if f is bounded.

(d) Show that Proposition 4.1 still holds when we extend the definition as above.

Note. The integral defined in this exercise is often called an *improper integral* when f is unbounded.

(e)   Show that $\int_0^1 f(x)dx$ is defined for $f(x) = x^k$ if and only if $k > -1$.   (Define $f(0) = 0$ for $k \leq 0$.)

(f)   Prove the analogue of Exercise 7, Section 3, for these integrals.

(g)   Show that $|f|$ is integrable if f is.

*13.   Another possible definition of integrals for unbounded functions is the following: if f is as in the previous problem, define

$$f^{(n)}(\vec{v}) = \begin{cases} f(\vec{v}) & \text{if } |f(\vec{v})| \leq n \\ n & \text{if } f(\vec{v}) > n \\ -n & \text{if } f(\vec{v}) < -n \end{cases}$$

and let $\int_S' f = \lim_{n \to \infty} \int_S f^{(n)}$ if all the functions $f^{(n)}$ are integrable and the limit exists.   (We write $\int_S' f$ to emphasize that this is *not* a definition we shall use.)

(a)   Show that if $\int_S f$ exists (possibly as an improper integral, in the sense described in the previous problem), then $\int_S' f$ exists and equals $\int_S f$.

(b)   Show that if $S = [-1,1] \subset R$ and $f(x) = 1/x$ $(f(0) = 0)$, then $\int_S' f$ exists, but $\int_S f$ does not.

(c)   Parts (a) and (b) may make it appear that the new definition is better than the old one because it is more extensive. Show, however, that the analogue of Exercise 7, Section 3, fails for $\int'$.   (If $S \subset R^1$, this means that part (a) of Proposition 2.2 fails to hold.)

*14.   Let $f: R^n \to R$ be a function. We can sometimes define a different sort of *improper integral* of f,

$$\int_{R^n} f$$

The difference is that in the previous case (Exercise 11), f became unbounded, while here the rectangle becomes unbounded.

Suppose first that f is nonnegative. Let $\{S_m\}$ be a sequence of rectangles such that $S_1 \subset S_2 \subset S_3 \subset \cdots$ and $\bigcup_{m=1}^{\infty} S_m = R^n$.   (For instance,

S_m might be $R(-\vec{v}_m, \vec{v}_m)$, where $\vec{v}_m = (m, m, \ldots, m)$.) Define

$$\int_{R^n} f = \lim_{m \to \infty} \int_{S_m} f$$

if all the integrals $\int_{S_m} f$ are defined and the limit exists.

(a) Show that $\int_{R^n} f$ is defined if and only if all the integrals $\int_{S_m} f$ are defined and the sequence $\int_{S_m} f$ is bounded.

(b) Show that if $\int_{R^n} f$ is defined, then $\int_S f$ is defined for any rectangle $S \subset R^n$, and the value of $\int_{R^n} f$ does not depend on the particular sequence S_m chosen.

(c) Now let f be arbitrary; define

$$\int_{R^n} f = \int_{R^n} f^+ - \int_{R^n} f^-$$

if both integrals on the right are defined. Prove results like those in parts (d) and (f) of Exercise 11.

(d) Show that $\int_R 1/(1+x^2)dx$ is defined, but that $\int_R x/(1+x^2)dx$ is not.

(e) Show that if f is integrable on R^n, so is $|f|$.

Note. We can define $\int_S f$ similarly when f is an infinite rectangle other than R^n; simply define f to be 0 off S, and then set

$$\int_S f = \int_{R^n} f$$

*15. On $R = R^1$, one sometimes defines improper integrals slightly differently. We shall discuss integrals of the form $\int_a^\infty f(x)dx$; there are similar results for $\int_{-\infty}^a f(x)dx$. What is special about R, incidentally, is that there are only two ways for a rectangle in R to become unbounded.

Let $f:[a,\infty) \to R$ be a function such that $\int_a^t f(x)dx$ exists for all $t \geq a$. We define

$$\int_a^\infty f(x)dx = \lim_{t \to \infty} \int_a^t f(x)dx$$

if the limit exists. (More explicitly: $\int_a^\infty f(x)dx = K$ if and only if for every $\varepsilon > 0$ there is a number T such that

$$\left| K - \int_a^t f(x)dx \right| < \varepsilon$$

if $t > T$.)

(a) Show that Proposition 2.2 holds for this integral.

(b) Find a function f such that $\int_a^\infty f(x)dx$ is defined (for some a), but $\int_a^\infty |f(x)|dx$ is not. (This shows that this new definition differs from the one in the previous problem.)

(c) For what values of n is $\int_1^\infty x^n\,dx$ defined?

(d) If f is defined on all of R, then one defines $\int_R f(x)dx = \int_{-\infty}^\infty f(x)dx$ to be $\int_{-\infty}^a f(x)dx + \int_a^\infty f(x)dx$, if both integrals exist. Show that the choice of the number a does not affect the definition.

(e) Show that

$$\int_{-\infty}^\infty \frac{x\,dx}{1+x^2}$$

is *not* defined, although

$$\lim_{t\to\infty} \int_{-t}^t \frac{x\,dx}{1+x^2}$$

exists. (This shows that it makes a difference in defining $\int_{-\infty}^\infty f(x)dx$ whether or not the end points of the region of integration become infinite independently.)

5. PROPERTIES OF THE INTEGRAL

We now develop some of the elementary properties of the integral of a function of n variables.

PROPOSITION 5.1. *Let S be a bounded subset of* R^n, *c a real number, and* $f,g:S \to R$ *integrable functions. Then the functions* f + g,cf:S → R *are integrable, with*

$$\int_S (f + g) = \int_S f + \int_S g$$

$$\int_S cf = c \int_S f$$

Proof. We prove the part of the proposition relating to the function f + g, leaving the proof of the assertion involving the function cf as Exercise 1. Since the integral of a function over a set that is not a rectangle is defined to be the integral of a related function over a rectangle, it follows that we can assume S is a rectangle.

Let ε be any positive number and let P be a partition of S into subrectangles $S_1,\ldots,S_m$ such that

$$U(f,P) - L(f,P) < \varepsilon/2$$

$$U(g,P) - L(g,P) < \varepsilon/2$$

This is possible because both f and g are integrable. Define numbers M_i, M_i', M_i'', $1 \le i \le m$, by

$$M_i = \text{lub}\{f(\vec{v}) + g(\vec{v}) \,|\, \vec{v} \in S_i\}$$

$$M_i' = \text{lub}\{f(\vec{v}) \,|\, \vec{v} \in S_i\}$$

$$M_i'' = \text{lub}\{g(\vec{v}) \,|\, \vec{v} \in S_i\}$$

so that

$$U(f + g,P) = \sum_{i=i}^{m} M_i \mu(S_i)$$

$$U(f,P) = \sum_{i=1}^{m} M_i' \mu(S_i)$$

$$U(g,P) = \sum_{i=1}^{m} M_i'' \mu(S_i)$$

Since $M_i \le M_i' + M_i''$ (see Proposition 5.5.4),

$$U(f + g,P) \le U(f,P) + U(g,P) \tag{5.1}$$

Similarily, it follows that

$$L(f + g,P) \geq L(f,P) + L(g,P) \tag{5.2}$$

Thus

$$U(f + g,P) - L(f + g,P) \leq (U(f,P) + U(g,P)) - (L(f,P) + L(g,P))$$
$$\leq (U(f,P) - L(f,P)) + (U(g,P) - L(g,P))$$
$$< \varepsilon$$

and f + g is then integrable, by Proposition 4.2.

In addition, the inequalities (5.1) and (5.2) imply that

$$\int_S (f + g) = \text{glb } U(f + g) \leq \text{glb } U(f) + \text{glb } U(g)$$
$$= \int_S f + \int_S g$$

and

$$\int_S (f + g) = \text{lub } L(f + g) \geq \text{lub } L(f) + \text{lub } L(g)$$
$$= \int_S f + \int_S g$$

Therefore

$$\int_S (f + g) = \int_S f + \int_S g$$

as desired.

A straightforward induction argument establishes the following.

COROLLARY. *Let S be a subset of* R^n, $f_j : S \to R$ *integrable functions, and* a_j *real numbers,* $j = 1, \ldots, k$. *Then* $a_1 f_1 + \cdots + a_k f_k$ *is integrable, with*

$$\int_S \sum_{j=1}^{k} a_j f_j = \sum_{j=1}^{k} a_j \int_S f_j$$

We now prove that the integral of a non-negative function is non-negative.

PROPOSITION 5.2. *Let S be a subset of* R^n *and* $f:S \to R$ *a non-negative integrable function on S. Then*

$$\int_S f \geq 0$$

Proof. Again, we can assume S is a rectangle. Let P be the trivial partition of S, the one consisting of exactly one subrectangle, and let

$$m = glb\{f(\vec{v}) \,|\, \vec{v} \in S\}$$

Then $L(f,P) = m\mu(S) \geq 0$ since $m \geq 0$. As a consequence,

$$\int_S f = lub\{L(f) \geq L(f,P) \geq 0$$

and the result is proved.

COROLLARY 1. *Let S be a subset of* R^n *and* $f,g:S \to R$ *integrable functions with* $f(x) \geq g(x)$ *for all* x *in S. Then*

$$\int_S f \geq \int_S g$$

Proof. Apply Proposition 5.2 to the function f - g and use Proposition 5.1.

COROLLARY 2. *Let S be a rectangle in* R^n *and* $f:S \to R$ *an integrable function. Suppose* m *and* M *are numbers with*

$$m \leq f(\vec{v}) \leq M$$

for all $\vec{v} \in S$. *Then*

$$m\mu(S) \leq \int_S f \leq M\mu(S)$$

The proof is immediate.

Let $f:S \to R$ be a non-negative integrable function which is *not* identically zero. One might expect that, in this case, the integral of f over S would be positive. This is not generally true. (See Exercise 4.) However, one can prove the following.

PROPOSITION 5.3. *Let S be a rectangle in* R^n *and* $f: S \to R$ *a non-negative continuous integrable function which is not identically zero. Then*

$$\int_S f > 0$$

Proof. In the interest of simplifying notation, we prove this proposition for the case $n = 2$. The general proof is similar.

Let the rectangle be $R = R(\vec{a}, \vec{b})$, where $\vec{a} = (a_1, a_2)$ and $\vec{b} = (b_1, b_2)$. Suppose that $f(\vec{v}_0) = c > 0$; for simplicity, assume that $\vec{v}_0 = (x_0, y_0)$ is not on the boundary of the rectangle (so that $a_1 < x_0 < b_1$, $a_2 < y_0 < b_2$). Choose $\delta > 0$ so that $|f(\vec{v}) - f(\vec{v}_0)| < c/2$ when $\vec{v} \in B = B_\delta(\vec{v}_0)$. It follows that

$$f(\vec{v}) = |f(\vec{v})| = |f(\vec{v}_0) + f(\vec{v}) - f(\vec{v}_0)|$$
$$\geq |f(\vec{v}_0)| - |f(\vec{v}) - f(\vec{v}_0)|$$
$$\geq b - \frac{b}{2} = \frac{b}{2}$$

for $\vec{v} \in B \cap R$. Furthermore, we may assume (by possibly making δ smaller) that $B \subset R$.

Now let $R_0 = \{\vec{v} = (x,y) : 0 \leq x - x_0 \leq \delta/2, \ 0 \leq y - y_0 \leq \delta/2\}$. Then R_0 is a rectangle, and $R_0 \subset B$. Let $g: S \to R$ be defined by

$$g(\vec{v}) = \begin{cases} b/2 & \text{if } \vec{v} \in R_0 \\ 0 & \text{if } \vec{v} \notin R_0 \end{cases}$$

Then $f(\vec{v}) \geq g(\vec{v})$. Moreover, if we take a partition P of R so that R_0 is one of the rectangles in P, then $U(P,g) = L(P,g) = b\delta^2/8$. By Proposition 3.2, g is integrable. Clearly

$$\int_S g = \frac{b\delta^2}{8} > 0$$

Hence $\int_S f \geq \int_S g > 0$, as claimed.

PROPOSITION 5.4. *Let S be a subset in* R^n *and* $f: S \to R$ *an integrable function. Then the function* $|f|$ *is integrable, with*

$$|\int_S f| \leq \int_S |f|$$

Both assertions of this proposition follow by comparing the upper and lower sums of f and $|f|$. We leave the details to the reader as Exercise 5. (See also Exercise 8.)

EXERCISES

1. Prove the second half of Proposition 5.1.

2. Prove the Corollary to Proposition 5.1.

3. Prove the second Corollary to Proposition 5.2.

4. Let S be the rectangle in R^2 defined by the points $(0,0)$ and $(1,1)$. Define $f:S \rightarrow R$ by

$$f(x,y) = \begin{cases} 1 & \text{if } y = 0 \\ 0 & \text{if } y \neq 0 \end{cases}$$

Prove that f is integrable, with

$$\int_S f = 0$$

5. Prove Proposition 5.4.

6. Let S be a compact pathwise connected subset of R^n, and let $f:S \rightarrow R$ a continuous integrable function. Prove that there is a point $\vec{v}_0 \in S$ such that

$$\int_S f = \mu(S)f(\vec{v}_0)$$

This is the *First Mean Value Theorem for Integrals*. (Hint: combine Corollary 2 to Proposition 5.2 with the Corollary to Theorem 5.5.3 and apply Proposition 5.8.5.)

*7.   Let S be a compact subset of R^n; let $f,h:S \rightarrow R$ be continuous integrable functions with $h \geq 0$. Show that there is a point $\vec{v}_0 \in S$ with

$$\int_S f(\vec{v})h(\vec{v}) = f(\vec{v}_0)\int_S h(\vec{v})$$

(See Exercise 10 of Section 2. Assume that fh is integrable.)

8.   Let S be a rectangle in R^n, and let $f:S \rightarrow R^n$ be a function. Is it true that if $|f|$ is integrable, then f is?

9.   Let S be a rectangle in R^n, and $f:S \rightarrow C$ be a complex-valued function. We say that f is integrable if Re(f) and Im(f) are integrable, and define

$$\int_S f = \int_S Re(f) + i \int_S Im(f)$$

   (a) Prove the analogue of Proposition 5.1 for such functions f. (Note: c is allowed to be complex.)

   *(b) Show that if f is integrable, then so is $|f|$, and $|\int_S f| \leq \int_S |f|$.

6.   INTEGRABLE FUNCTIONS

The basic result of this section is the assertion that a continuous function on a rectangle is integrable. For the integral of a function over a set which is not a rectangle, however, we need to know that certain discontinuous functions are integrable. We shall describe a result of this sort, using the notion of "content zero." (The proof will be given in the next section.)

   THEOREM 6.1.   *Let S be a rectangle in* R^n*, and let* $f:S \rightarrow R$ *be a continuous function. Then* f *is integrable.*

Before proving this result, we introduce two useful notions.

   Let $A \subset R^n$ be a bounded set. The *diameter of* A, diam A, is defined by diam $A = \mathrm{lub}\{\|\vec{v} - \vec{w}\| : \vec{v}, \vec{w} \in A\}$

For example, if $A = B_r(\vec{v}_0)$ is the ball of radius r about $\vec{v}_0$, then diam $A = 2r$. (Think of diametrically opposed points.) Every bounded set A has a diameter, as the set

$$\{\|\vec{v} - \vec{w}\| : \vec{v}, \vec{w} \in A\}$$

is bounded and has a least upper bound by the completeness of the real numbers. (See Theorem 5.4.2.)

Let P be a partition of the rectangle S. We define the *mesh of* P, mesh P, to be the largest of the diameters of the subrectangles making up P. It follows that if $\vec{v}$ and $\vec{w}$ are two points of S in the same subrectangle of P, then

$$\|\vec{v} - \vec{w}\| \leq \text{mesh } P$$

Proof of Theorem 6.1. We prove that, for any $\varepsilon > 0$, there is a partition P of S so that

$$U(f,P) - L(f,P) < \varepsilon$$

Theorem 6.1 will then follow from Proposition 4.2.

Suppose $\varepsilon > 0$. Since S is compact and f is continuous, it follows that f is uniformly continuous. (See Theorem 7.5.1.) Thus there is a $\delta > 0$ such that

$$|f(\vec{v}) - f(\vec{w})| < \frac{\varepsilon}{\mu(S)}$$

whenever $\|\vec{v} - \vec{w}\| < \delta$.

Let P be a partition of S with mesh $P < \delta$. We denote the subrectangles of S by $S_1, \ldots, S_k$. The existence of such a partition follows from the fact that the diameter of a rectangle in R^n is no greater than $\ell\sqrt{n}$, where ℓ is the length of the longest edge of the rectangle. Therefore, we let P be the product of partitions, each of which has mesh $< \delta/\sqrt{n}$.

Define numbers M_i, m_i, and $1 \leq i \leq k$, by

$$M_i = \text{lub}\{f(\vec{v}) : \vec{v} \in S_i\}$$

$$m_i = \text{glb}\{f(\vec{v}) : \vec{v} \in S_i\}$$

Then

$$U(f,P) = \sum_{i=1}^{k} M_i \mu(S_i)$$

and

$$L(f,P) = \sum_{i=1}^{k} m_i \mu(S_i)$$

Now f is continuous and each S_i is compact, so, according to Theorem 5.5.3, we can find points $\vec{v}_i, \vec{w}_i \in S_i$ with $M_i = f(\vec{v}_i)$ and $m_i = f(\vec{w}_i)$, $1 \leq i \leq k$. Furthermore, since mesh $P < \delta$, we have $\|\vec{v}_i - \vec{w}_i\| < \delta$, $1 \leq i \leq k$. It follows that

$$M_i - m_i = f(\vec{v}_i) - f(\vec{w}_i) < \frac{\varepsilon}{\mu(S)}$$

Thus

$$U(f,P) - L(f,P) = \sum_{i=1}^{k} (M_i - m_i)\mu(S_i)$$
$$< \sum_{i=1}^{k} (\varepsilon/\mu(S))\mu(S_i)$$
$$= (\varepsilon/\mu(S)) \sum_{i=1}^{k} \mu(S_i) = \varepsilon$$

This completes the proof of Theorem 6.1.

The requirement in Theorem 6.1 that f be continuous can be weakened; it is only necessary that f be continuous at "most" points of S. To make this precise, we need the notion of content zero.

A set $B \subseteq R^n$ is said to have *content zero (or volume zero)* if, for any $\varepsilon > 0$, we can find a finite collection of rectangles in R^n containing B in their union and having total volume less than ε. For example, any line segment in R^2 has content zero. (See Figure 6.1.)

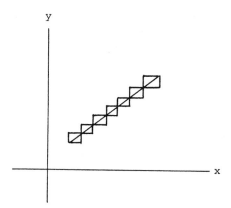

FIGURE 6.1.   A line segment has content zero

In the same way, an arc of a circle in R^2 has content zero.   (See
Figure 6.2).   More generally, "lower-dimensional sets" - curves in R^2,
surfaces in R^3, and the like, have zero content.   (Warning: any theorem of this sort needs to be carefully phrased, since there are many
pathological counterexamples.   See, for instance, Exercise 7 of Section
9.1.)

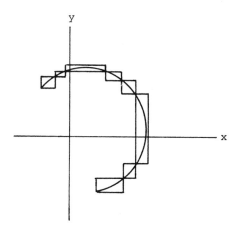

FIGURE 6.2.   A circle has content zero

The strengthened form of Theorem 6.1 is the following.

THEOREM 6.2.   *Let S be a rectangle in R^n and $f:S \to R$ a bounded function.   Suppose that $B \subset S$ is a set of content zero and that f is continuous on S - B.   Then f is integrable.*

We shall prove this theorem in the next section.   The idea behind the proof is, as usual, to use Proposition 4.2.   We enclose B in rectangles of very small volume, so that upper and lower sums for f on B cannot differ by much; on the rest of S, f is continuous and we can reason as in Theorem 6.1.

As we remarked above, Theorem 6.2 can be used to prove the integrability of continuous functions on sets other than rectangles.   For example, let S be the subset in R^2 defined by the curves $y = 1 - x^2$ and $y = x^2$ and let S_1 be a rectangle containing S.   (See Figure 6.3.) Let $f:S \to R$ be any continuous function and define $f_1:S_1 \to R$ by

$$f_1(\vec{v}) = \begin{cases} f(\vec{v}) & \text{if } \vec{v} \in S \\ 0 & \text{if } \vec{v} \notin S \end{cases}$$

Then f_1 is continuous except for the points on the two curves $y = 1 - x^2$ and $y = x^2$.   Since these curves have content zero, Theorem 6.2 applies to say that f_1 is integrable.   Thus, by definition, $f:S \to R$ is integrable.

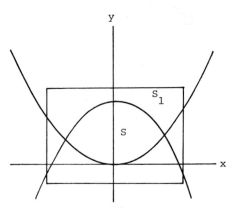

FIGURE 6.3.   A rectangle containing S

EXERCISES

1.   Prove that if $A \subseteq B$, then diam $A \leq$ diam B.

2.   Prove diam $B_r(\vec{v}_0) = 2r$.

3.   Let P, P' be partitions of a rectangle S in R^n.   Prove that if P' is a refinement of P, then mesh $P' \leq$ mesh P.

4.   Prove that a line segment in R^2 has content zero.

5.   Prove that the finite union of sets of content zero has content zero.

*6.   Let $f:[a,b] \to R$ be an integrable function, and let $G(f)$ be the graph of f,

$$G(f) = \{(x,y) \in R^2 : y = f(x)\}$$

Show that $G(f)$ has zero content in R^2.   (Hint: given ε, let P be a partition of $[a,b]$ such that $U(f,P) - L(f,P) < \varepsilon$.   Use the graphs of the step functions for $U(f,P)$ and $L(f,P)$ to enclose $G(f)$ in a union of rectangles of area $< \varepsilon$.)

*7.   Let S be a rectangle in R^n and $f:S \to R$ an integrable function. Prove that the graph of f,

$$G(f) = \{(\vec{x},y) \in R^n \times R^1 : y = f(\vec{x})\}$$

has zero content in $R^{n+1} = R^n \times R^1$.

8.   Let $\{\vec{v}_n\}$ be a convergent sequence in R^n.   Show that the set $\{\vec{v}_1, \vec{v}_2, \ldots, \vec{v}_n, \ldots\}$ has content zero.

*9.   Let $f:[a,b] \to R$ be a bounded increasing function.   (That is, $f(x_1) \geq f(x_2)$ whenever $x_1 \geq x_2$.)   Show that f is integrable.

7.   THE PROOF OF THEOREM 6.2.

The proof uses Proposition 4.2: we wish to show that for every $\varepsilon > 0$ there is a partition P of S such that

$$U(f,P) - L(f,P) < \varepsilon$$

For the moment, let P be any partition, with subrectangles $S_1, \ldots, S_m$, and let $S_1, \ldots, S_k$ be the subrectangles of P whose closures intersect B. Then $B \subset \bigcup_{i=1}^{k} S_k$; and, if

$$M_i = \text{lub}\{f(\vec{v}) : \vec{v} \in S_i\}$$
$$m_i = \text{glb}\{f(\vec{v}) : \vec{v} \in S_i\}$$

we have

$$U(f,P) - L(f,P) = \sum_{i=1}^{k} (M_i - m_i)\mu(S_i) + \sum_{i=k+1}^{m} (M_i - m_i)\mu(S_i) \quad (7.1)$$

We shall show that for an appropriate choice of P, we can make the first sum small because the total volume of $\bigcup_{i=1}^{k} S_i$ can be made small, while the second sum can be made small (as in Theorem 6.1) because f is continuous on the remaining rectangles.

To begin with, let

$$M = \text{lub}\{f(\vec{v}) : \vec{v} \in S\}$$
$$m = \text{glb}\{f(\vec{v}) : v \in S\}$$

and choose the rectangles $S_1, \ldots, S_k$ so that $B \subset S_1 \cup \ldots \cup S_k$ and

$$\sum_{i=1}^{k} \mu(S_i) < \varepsilon/2(M - m) \qquad (7.2)$$

Let $S_{k+1}, \ldots, S_m$ be additional subrectangles so that $S_1, \ldots, S_k, S_{k+1}, \ldots, S_m$ define a partition P of S. We can, by slightly enlarging the rectangles $S_1, \ldots, S_k$ if necessary, assume that $B \subset \text{Int}(S_1 \cup \ldots \cup S_k)$ so that f is continuous on the compact set $S' = S_{k+1} \cup \ldots \cup S_m$. (See Figure 7.1.)

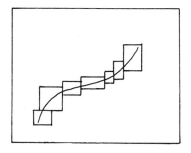

FIGURE 7.1.

Now, according to Theorem 5.7.1, f is uniformly continuous on S', so we can choose $\delta > 0$ such that

$$|f(\vec{v}) - f(\vec{w})| < \varepsilon/2\mu(S)$$

for $\vec{v}, \vec{w} \in S'$, $\|\vec{v} - \vec{w}\| < \delta$. We can assume, subdividing if necessary, that $\mathrm{diam}(S_i) < \delta$ for $i = k+1, \ldots, m$. (This subdivision may subdivide the rectangles $S_1, \ldots, S_k$, too, but for notational convenience we assume that $S_1, \ldots, S_k$ remain unchanged.) Then

$$(M_i - m_i) < \varepsilon/2\mu(S) \tag{7.3}$$

for $i = k+1, \ldots, m$. Using (7.2) and (7.3), we have

$$
\begin{aligned}
U(f,P) - L(f,P) &= \sum_{i=1}^{k} (M_i - m_i)\mu(S_i) + \sum_{i=k+1}^{m} (M_i - m_i)\mu(S_i) \\
&< \sum_{i=1}^{k} (M - m)\mu(S_i) + \sum_{i=k+1}^{m} (\varepsilon/2\mu(S))\mu(S_i) \\
&= (M - m) \sum_{i=1}^{k} \mu(S_i) + (\varepsilon/2\mu(S)) \sum_{i=k+1}^{m} \mu(S_i) \\
&< (M - m)(\varepsilon/2(M - m)) + (\varepsilon/2\mu(S))\mu(S) \\
&= \varepsilon
\end{aligned}
$$

This completes the proof of Theorem 6.2.

Theorem 6.2 can also be extended. Define a set $B \subseteq S$ to be of *measure zero* if for every $\varepsilon > 0$ we can find a set of rectangles

$R_1, R_2, \ldots$, possibly overlapping, such that

$$B \subset \bigcup_{i=1}^{\infty} R_i \quad \text{and} \quad \sum_{i=1}^{\infty} \mu(R_i) \leq \varepsilon$$

(The second statement means that $\sum_{i=1}^{n} \mu(R_i) \leq \varepsilon$ for every n.)

THEOREM 7.1. *The bounded function* $f:S \to R$ *is integrable if and only if the set of points on which* f *is discontinuous is of measure* 0.

The interested reader may wish to prove Theorem 7.1 by working through Exercises 3-7 below.

EXERCISES

1.   Prove that a compact set with measure zero has content zero.

2.   Show that the set $S = \{(x,y) \in R^2 : x = 0\}$ has measure zero but does not have content zero.

*3.   Let S be a compact rectangle in R^n, and let $f:S \to R$ be bounded. We define the *oscillation* of f at $\vec{v}_0$, $\omega(f,\vec{v}_0)$, by

$$\omega(f,\vec{v}_0) = \lim_{\delta > 0} (\text{lub}\{f(\vec{v}) : \|\vec{v} - \vec{v}_0\| < \delta\} - \text{glb}\{f(\vec{v}) : \|\vec{v} - \vec{v}_0\| < \delta\})$$

Prove that the above limit always exists and that f is continuous at $\vec{v}_0$ if and only if $\omega(f,\vec{v}_0) = 0$. (See Exercises 9 and 10 of Section 5.4.)

*4.   Let f and S be as in Exercise 3, and suppose that f is integrable. For $\varepsilon > 0$, set

$$K_\varepsilon = \{\vec{v}_0 \in S : \omega(f,\vec{v}_0) \geq \varepsilon\}$$

Show that K_ε has content 0. (Hint: if not, then $\exists \delta > 0$: any finite union of rectangles containing K_ε has volume $\geq \delta$. Show that any upper and lower sums for f differ by at least $\delta\varepsilon$.)

*5.   Let f and S be as in Exercise 3, and suppose that $K \subset S$ is a compact set such that

$$\omega(f, \vec{v}_0) < \varepsilon$$

for all $\vec{v}_0 \in K$. Show that $\exists \delta > 0$: if $\vec{v}_0, \vec{v}_1 \in K$ and $\|\vec{v}_0 - \vec{v}_1\| < \delta$, then $|f(\vec{v}_0) - f(\vec{v}_1)| < \varepsilon$. (Note the similarity of this result to uniform continuity. That should give a hint of how to do this problem.)

*6.   Let f and S be as in Exercise 3, and let K_ε be defined as in Exercise 4. Show that if K_ε has content 0 for all $\varepsilon > 0$, then f is integrable on S. (Hint: the proof is like that for Theorem 7.2, but the result of Exercise 5 replaces Theorem 5.7.1.) Note that this is the converse of Exercise 4.

*7.   Prove Theorem 7.1. (Hint: the set of points on which f is discontinuous is $\bigcup_{n=1}^{\infty} K_{1/n}$. It is also useful to remember that $\sum_{n=1}^{\infty} \varepsilon/2^n = \varepsilon$.)

*8.   (a) Show that $\{x \in R : 0 \le x \le 1,$ and x is rational$\}$ has measure 0 but not content 0.

      (b) Let g be the function of Exercise 15, Section 1. Use (a) and Theorem 7.1 to show that g is integrable.

      (c) Let f be the function of Exercise 14, Section 1. Use Theorem 7.1 to show that f is not integrable.

CHAPTER 12

ITERATED INTEGRALS AND THE FUBINI THEOREM

INTRODUCTION

In the previous chapter we introduced the notion of the integral for a function of n variables. We now take up the problem of computing this integral. The basic result is the Fubini Theorem, which effectively reduces the problem of integrating a function of n variables to the evaluation of n ordinary integrals of a function of one variable. We give two forms of the Fubini theorem and show how to apply it to some problems.

We also discuss two other topics in this chapter. One is that of differentiating a function defined by an integral; we show that, under certain circumstances, differentiating an integral is the same as integrating a derivative. The other, the formula for changing variables in an integral, is an n-dimensional analogue of integration by substitution.

1. THE FUBINI THEOREM

In Section 11.6 we proved that a large class of functions, the continuous functions, are integrable. However, we still have no convenient way of evaluating the integral. We rectify this situation now by reducing the problem of integrating a function of n variables to that of performing n integrations of a function of a single variable. We begin with functions of two variables.

521

Let $S = R(\vec{a},\vec{b})$ be the rectangle with $\vec{a} = (a_1,a_2)$ and $\vec{b} = (b_1,b_2)$, as in Figure 1.1, and let $f:S \to R$ be a continuous function. If we fix a point y in $[a_2,b_2]$, we can think of $f(x,y)$ as a function from $[a_1,b_1]$ into R. In fact, this function is continuous so we can integrate it. The integral depends on y; thus it gives a function $g:[a_2,b_2] \to R$,

$$g(y) = \int_{a_1}^{b_1} f(x,y)dx$$

For example, if $f(x,y) = 2xy - y^2$, then

$$g(y) = \int_{a_1}^{b_1}(2xy - y^2)dx$$

$$= (x^2y - xy^2)\Big|_{x=a_1}^{x=b_1}$$

$$= b_1^2 y - b_1 y^2 - a_1^2 y + a_1 y^2$$

$$= (b_1^2 - a_1^2)y - (b_1 - a_1)y^2$$

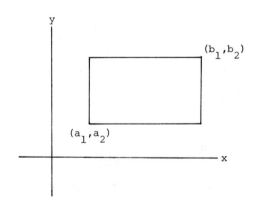

FIGURE 1.1. A rectangle in R_2

LEMMA 1.1. *Let f be a continuous function or a rectangle S =* $R(\vec{a},\vec{b})$, $\vec{a} = (a_1,a_2)$, $\vec{b} = (b_1,b_2)$. *Define* $g:[a_2,b_2] \to R$ *by*

$$g(y) = \int_{a_1}^{b_1} f(x,y)dx$$

Then g is continuous.

Proof. Let ε be an arbitrary positive number. Since f is uniformly continuous on S (Theorem 5.7.1), we can find a $\delta > 0$ so that $|f(x,y) - f(x',y')| < \varepsilon/(b_1 - a_1)$ whenever $\|(x,y) - (x',y')\| < \delta$. Then, if $|y - y'| < \delta$, we have

$$|g(y) - g(y')| = |\int_{a_1}^{b_1} (f(x,y) - f(x,y'))dx|$$

$$\leq \int_{a_1}^{b_1} |f(x,y) - f(x,y')|dx$$

(by Proposition 11.5.4)

$$< \int_{a_1}^{b_1} \frac{\varepsilon}{b_1 - a_1} dx = \varepsilon$$

since $\|(x,y) - (x,y)\| = |y - y'| < \delta$. It follows that g is continuous.

Now, since $g:[a_2,b_2] \to R$ is continuous, it is integrable. The integral of g,

$$\int_{a_2}^{b_2} g(y)dy = \int_{a_2}^{b_2}(\int_{a_1}^{b_1} f(x,y)dx)dy$$

is called an *iterated integral*.

In the example above,

$$\int_{a_2}^{b_2} (\int_{a_1}^{b_1} (2xy - y^2)dx)dy = \int_{a_2}^{b_2} [(b_1^2 - a_1^2)y - (b_1 - a_1)y^2]dy$$

$$= (b_1^2 - a_1^2) \frac{y^2}{2} - (b_1 - a_1) \frac{y^3}{3} \Big|_{a_2}^{b_2}$$

$$= \frac{1}{2}(b_1^2 - a_1^2)(b_2^2 - a_2^2)$$

$$- \frac{1}{3}(b_1 - a_1)(b_2^3 - a_2^3)$$

Of course, we could also have integrated f with respect to y first and then with respect to x.   In that case, the calculation goes as follows:

$$\int_{a_2}^{b_2}(2xy - y^2)dy = (xy^2 - \frac{y^3}{3})\,|_{y=a_2}^{y=b_2}$$

$$= b_2^2x - \frac{b_2^3}{3} - a_2^2x + \frac{a_2^3}{3}$$

$$= (b_2^2 - a_2^2)x - (\frac{b_2^3}{3} - \frac{a_2^3}{3})$$

$$\int_{a_1}^{b_1}(\int_{a_2}^{b_2}(2xy - y^2)dy)dx = \int_{a_1}^{b_1}[(b_2^2 - a_2^2)x - (\frac{b_2^3}{3} - \frac{a_2^3}{3})]dx$$

$$= ((b_2^2 - a_2^2)\frac{x^2}{2} - (\frac{b_2^3}{3} - \frac{a_2^3}{3})x)\,|_{a_1}^{b_1}$$

$$= \frac{1}{2}(b_2^2 - a_2^2)(b_1^2 - a_1^2) - \frac{1}{3}(b_2^3 - a_2^3)(b_1 - a_1)$$

the same answer as above.   The next result shows that this is not a coincidence.

THEOREM 1.2.   (The Fubini Theorem).   *Let f be a continuous function on the rectangle* $S = R(\vec{a},\vec{b})$ *in* R^n*, where* $\vec{a} = (a_1,\ldots,a_n)$ *and* $\vec{b} = (b_1,\ldots,b_n)$*.   Let* $S' = R(\vec{a}',\vec{b}')$ *be the rectangle in* R^{n-1} *where* $\vec{a}' = (a_1,\ldots,a_{n-1})$ *and* $\vec{b}' = (b_1,\ldots,b_{n-1})$ *and let* $g:S' \to R$ *be defined by*

$$g(x_1,\ldots,x_{n-1}) = \int_{a_n}^{b_n} f(x_1,\ldots,x_n)dx_n$$

Then g is continuous and

$$\int_S f = \int_{S'} g = \int_{S'}(\int_{a_n}^{b_n} f(x_1,\ldots,x_n)dx_n) \qquad (1.1)$$

We shall prove this theorem in Section 4.   First, we examine what it means in some special cases.

1. Assume that n = 2. Then the rectangle S = R($\vec{a},\vec{b}$) is the one in Figure 1.1, and S' is the interval $[a_1,b_1]$. The theorem says that

$$\int_S f = \int_{a_1}^{b_1}(\int_{a_2}^{b_2} f(x_1,x_2)dx_2)dx_1$$

or, if we use x and y as the variables,

$$\int_S f = \int_{a_1}^{b_1}(\int_{a_2}^{b_2} f(x,y)dy)dx \qquad (1.2)$$

Usually we write (1.2) with fewer parentheses:

$$\int_S f = \int_{a_1}^{b_1} \int_{a_2}^{b_2} f(x,y)dydx$$

Actually, Theorem 1.2 gives more information. As the proof makes clear, we can treat x and y symmetrically; that is, we can get the same result by integrating first with respect to x:

$$\int_S f = \int_{a_2}^{b_2} \int_{a_1}^{b_1} f(x,y)dxdy$$

In particular, we see that the two iterated integrals of f are equal (assuming that f is continuous). We saw one example of this earlier. It is useful to state this as a separate result:

COROLLARY TO THEOREM 1.2. *Let f be a continuous function on the rectangle* S = R($\vec{a},\vec{b}$) *in* R^2, *where* $\vec{a}$ = (a_1,a_2) *and* $\vec{b}$ = (b_1,b_2) . *Then*

$$\int_S f = \int_{a_1}^{b_1} \int_{a_2}^{b_2} f(x,y)dydx = \int_{a_2}^{b_2} \int_{a_1}^{b_1} f(x,y)dxdy$$

2. Assume now that n = 3. Then S = R($\vec{a},\vec{b}$) is a rectangular solid, as in Figure 1.2, and S' is the rectangle given by (a_1,a_2) and (b_1,b_2). Using x, y, and z as the variables, we get

$$\int_S f = \int_{S'}(\int_{a_3}^{b_3} f(x,y,z)dz)$$

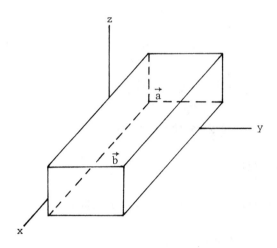

FIGURE 1.2. A rectangle in R^3

However, using the discussion above, we can write the integral over S' as a double integral:

$$\int_S f = \int_{a_1}^{b_1} (\int_{a_2}^{b_2} (\int_{a_3}^{b_3} f(x,y,z)dz)dy)dx$$

or, without parentheses,

$$\int_S f = \int_{a_1}^{b_1} \int_{a_2}^{b_2} \int_{a_3}^{b_3} f(x,y,z)dzdydx$$

The order dzdydx means that one integrates first with respect to the z variable (holding x and y fixed), then with respect to y (with x fixed), and finally with respect to x. For instance, if $\vec{a} = (0,0,0)$, $\vec{b} = (1,2,3)$, and $f(x,y,z) = xy^2z^3$, then

$$\int_S f = \int_0^1 \int_0^2 \int_0^3 xy^2z^3 dzdydx$$

$$= \int_0^1 \int_0^2 \frac{xy^2z^4}{4} \Big|_{z=0}^{z=3} dydx$$

$$= \int_0^1 \int_0^2 \frac{81}{4} xy^2 dydx$$

$$= \int_0^1 \left(\frac{27}{4} xy^3\right) \Big|_{y=0}^{y=2} dx$$

$$= \int_0^1 54x \, dx = 27x^2 \Big|_0^1 = 27$$

Again, we can take the iterated integrals in other orders: for instance,

$$\int_S f = \int_{a_2}^{b_2} \int_{a_1}^{b_1} \int_{a_3}^{b_3} f(x,y,z) dz dx dy$$

$$= \int_{a_3}^{b_3} \int_{a_1}^{b_1} \int_{a_2}^{b_2} f(x,y,z) dy dx dz$$

That is, a result like the Corollary to Theorem 1.2 holds for n = 3.

We could continue in the same way, giving examples for n = 4, n = 5, and so on. However, the idea is the same and the cases n = 2 and n = 3 are the most important in practice. The same principle holds in higher dimensions, too: as long as f is continuous, all the iterated integrals are equal. That is, *the order in which we integrate the variables does not affect the result.*

Our statement of Theorem 1.2 is by no means the strongest possible; for instance, it is not necessary to assume that f is continuous. (We shall give a stronger result in the next section.) However, it is not sufficient merely to assume that f is integrable. For instance, let S be the rectangle in R^2 defined by (0,0) and (1,1), and define f:S → R by

$$f(x,y) = \begin{cases} 1 & \text{if either x or y is irrational} \\ 1 - \frac{1}{q} & \text{if x,y are rational and } x = \frac{p}{q} \text{ in lowest terms} \end{cases}$$

Then f is integrable (see Exercise 8), but $\int_0^1 f(x,y) dy$ does not exist when x is rational. Hence we cannot even define the iterated integral $\int_0^1 \int_0^1 f(x,y) dy dx$.

EXERCISES

1. Evaluate the following.

(a) $\int_3^5 \int_1^2 (y - 2x)dxdy$

(e) $\int_0^2 \int_{-1}^1 e^{2x+y}dxdy$

(b) $\int_1^2 \int_3^5 (y - 2x)dydx$

(f) $\int_0^\pi \int_0^{\pi/2}\cos(x + y)dydx$

(c) $\int_0^2 \int_0^3 xy^2dxdy$

(g) $\int_0^1 \int_0^1 \frac{x}{1+xy} dxdy$

(d) $\int_0^3 \int_0^2 xy^2dydx$

(h) $\int_0^1 \int_0^1 \frac{x}{1+xy} dydx$

2. Evaluate the following.

(a) $\int_{-2}^4 \int_4^8 \int_0^2 xyz\ dxdydz$

(b) $\int_{-2}^6 \int_{-4}^4 \int_0^1 (xy + z^2)dxdzdy$

(c) $\int_0^2 \int_0^{\pi/3} \int_0^{\pi/2} \cos(x + z)dzdxdy$

(d) $\int_1^3 \int_0^2 \int_{-1}^1 yze^{xz}dxdzdy$

(e) $\int_0^{\pi/3} \int_1^2 \int_{-\pi/2}^0 y\cos x\cos z\ dzdydx$

(f) $\int_0^1 \int_{-3}^0 \int_0^2 \frac{yz^2}{1+x^2} dydzdx$

3. Use Theorem 1.2 to find $\int_{R(a,b)}f$ where

(a) $f(x,y) = y - 2x$, $\vec{a} = (1,3)$, $\vec{b} = (2,5)$

(b) $f(x,y) = y - 2x$, $\vec{a} = (-2,4)$, $\vec{b} = (1,5)$

(c) $f(x,y) = xy^2$, $\vec{a} = (0,0)$, $\vec{b} = (2,2)$

(d) $f(x,y) = xy^2$, $\vec{a} = (1,1)$, $\vec{b} = (2,2)$

(e) $f(x,y) = x(1 + xy)^{-1}$, $\vec{a} = (0,0)$, $\vec{b} = (1,1)$

(f) $f(x,y) = y \sin xy$, $\vec{a} = (0,0)$, $\vec{b} = (\pi/4,1)$

(g) $f(x,y,z) = xyz$, $\vec{a} = (0,4,-2)$, $\vec{b} = (2,8,4)$

(h) $f(x,y,z) = xy + z^2$, $\vec{a} = (0,-2,-4)$, $\vec{b} = (1,6,4)$

(i) $f(x,y,z) = x \sin(xy + z)$, $\vec{a} = (0,1,0)$, $\vec{b} = (\pi/6,2,\pi/2)$

(j) $f(x,y,z) = yz^2(1 + x^2)^{-1}$, $\vec{a} = (0,-3,0)$, $\vec{b} = (1,0,2)$

(k) $f(x_1,x_2,x_3,x_4) = x_1 x_3 - x_2 x_4$, $\vec{a} = (0,0,0,0)$, $\vec{b} = (1,2,2,1)$

(ℓ) $f(x_1,x_2,x_3,x_4) = 2x_1 x_3^2 - x_3 x_4 \sin x_2$, $\vec{a} = (0,0,0,0)$,

$$\vec{b} = (1,\pi/3,2,4)$$

*4.   Prove the analogue of Lemma 1.1 for functions of n variables.

5.   Let $f(x,y) = f_1(x)f_2(y)$, where f_1 and f_2 are continuous. Verify directly that

$$\int_{a_2}^{b_2}(\int_{a_1}^{b_1} f(x,y)dx)dy = \int_{a_1}^{b_1}(\int_{a_2}^{b_2} f(x,y)dy)dx$$

*6.   Prove that in the situation of Exercise 5,

$$\int_S f = \int_{a_2}^{b_2}(\int_{a_1}^{b_1} f(x,y)dx)dy$$

where S is the rectangle $R((a_1,a_2),(b_1,b_2))$. (That is, prove Theorem 1.2 in this special case.)

7.   Let $f:[-1,1] \to R$ be continuous, and define g by

$$g(x) = \int_0^1 f(xy)dy, \qquad -1 \le x \le 1$$

Show that g is continuous.

*8.   (a) Show that the function f, defined on the unit square in R^2 by

$$f(x,y) = \begin{cases} 1 \text{ if } x \text{ or } y \text{ is irrational} \\ 1 - \frac{1}{q} \text{ if } x \text{ and } y \text{ are both rational and } x = \frac{p}{q} \text{ in} \\ \text{lowest terms} \end{cases}$$

is integrable. Hint: use Exercise 6 or 7 of Section 11.7. Show that

$$\omega(f;(x_0,y_0)) = \begin{cases} 0 & \text{if } x_0 \text{ is irrational} \\ \dfrac{1}{q} & \text{if } x_0 = \dfrac{p}{q} \text{ in lowest terms} \end{cases}$$

    (b) Show that $\int_0^1 \int_0^1 f(x,y)dxdy$ exists and equals $\int_S f$.

    (c) Show that $\int_0^1 f(x,y)dy$ is not defined if x is rational.

2.   INTEGRALS OVER NON-RECTANGULAR REGIONS

We saw in the previous section how to integrate continuous functions
over rectangles. There are many cases, however, in which we wish to
integrate over regions which are not rectangular - over circular discs,
for instance. In Chapter 11, we defined such integrals in terms of
integrals over rectangles: if we want to find $\int_S f$, where S is a bounded set, we simply put S in a rectangle S_1, extend f by defining

$$f_1(\vec{v}) = \begin{cases} f(\vec{v}) & \text{if } \vec{v} \in S \\ 0 & \text{if } \vec{v} \notin S \end{cases}$$

and find $\int_{S_1} f_1$. Actually computing $\int_{S_1} f_1$, though, involves us in
two difficulties, one theoretical and one more practical. We take care
of the theoretical one with a theorem; the practical one will occupy
us for this section and the next.

    Consider the following problem: find the integral of the func-
tion f(x,y) = xy over the triangular region S bounded by the x-axis,
y-axis and the line x + y = 1. (See Figure 2.1.) We may take S_1 to
be the rectangle defined by (0,0) and (1,1); if we define f_1 as above,
then, by definition,

$$\int_S f = \int_{S_1} f_1$$

and we now need to integrate f over a rectangle. However, Theorem 1.2
does not apply, because the function f_1 is not continuous.

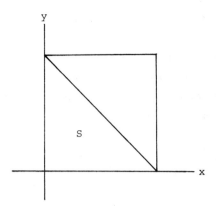

FIGURE 2.1.

This is the theoretical difficulty we spoke of earlier. To get
around it, we give an extension of Theorem 1.2.

THEOREM 2.1. *Let* f *be an integrable function on the rectangle*
$S = R(\vec{a},\vec{b})$ *in* R^n, *where* $\vec{a} = (a_1,\ldots,a_n)$ *and* $\vec{b} = (b_1,\ldots,b_n)$. *Let*
$S' = R(\vec{a}',\vec{b}')$ *be the rectangle in* R^{n-1} *with* $\vec{a}' = (a_1,\ldots,a_{n-1})$ *and*
$\vec{b}' = (b_1,\ldots,b_{n-1})$. *Suppose that*

$$g(x_1,\ldots,x_{n-1}) = \int_{a_n}^{b_n} f(x_1,\ldots,x_n)dx_n$$

exists for all $(x_1,\ldots,x_{n-1}) \in S'$; *that is, suppose that* f *can always*
be integrated in the nth *coordinate. Suppose further that* g *is inte-*
grable. Then

$$\int_S f = \int_{S'} g = \int_{S'} (\int_{a_n}^{b_n} f(x_1,\ldots,x_n)dx_n)$$

Remarks: 1. As with Theorem 1.2, we have stated the theorem in a
way which singles out the last variable. The same result holds if we
integrate out any other variable; that is, any iterated integral is
equal to $\int_S f$ (assuming that f is integrable and the iterated integral
makes sense.) We have not stated the theorem in this more general form
simply because the notation becomes unwieldy.

2.   Theorem 2.1 can be improved somewhat; for instance, the hypothesis that g is integrable is unnecessary, since it can be proved from the other hypotheses.   (An even better result is given in Exercise 3 of Section 4.)   However, Theorem 2.1 is good enough for most cases arising in practice.   We shall prove Theorem 2.1 in Section 4.

We now return to the example above, namely

$$f_1(x,y) = \begin{cases} xy & \text{if } (x,y) \in S \\ 0 & \text{if } (x,y) \notin S \end{cases}$$

where S is the triangle of Figure 2.1.   To apply Theorem 2.1, we first need to know that, for fixed x, the integral

$$\int_0^1 f_1(x,y)\,dy$$

exists.   However, for fixed x, $f_1(x,y)$ is a continuous function of y except at y = 1 - x, so this integral exists by Theorem 10.6.2.   In fact,

$$\int_0^1 f_1(x,y)\,dy = \int_0^{1-x} xy\,dy$$

(since $f_1(x,y) = 0$ for y > 1 - x)

$$= \frac{xy^2}{2}\Big|_{y=0}^{y=1-x}$$

$$= \frac{1}{2}(x - 2x^2 + x^3)$$

This function is clearly integrable, so

$$\int_S f = \int_{S_1} f_1 = \int_0^1 \int_0^1 f(x,y)\,dy\,dx$$

$$= \int_0^1 \frac{1}{2}(x - 2x^2 + x^3)\,dx$$

$$= (\frac{x^2}{4} - \frac{x^3}{3} + \frac{x^4}{8})\Big|_0^1 = \frac{1}{24}$$

and the problem is solved.

In practice, one rarely bothers to extend f to a rectangle as we did above; instead, one writes down the iterated integral and computes directly. It is here that the practical problem arises. The limits of integration in the iterated integral may depend on some of the variables which have not been integrated out, and it may require some care to set up the integral correctly.

Here is how the above problem would generally be done: we want to write the integral in the form

$$\int\int f(x,y)dydx$$

where we have not yet determined the limits of integration. Since the first integral is with respect to y, we fix x. Now we notice that for fixed $x \in [0,1]$, the piece of the line through $(x,0)$ and parallel to the y-axis which lies in S corresponds to the interval $0 \le y \le 1 - x$. (See Figure 2.2.) Thus y varies between 0 and 1 - x in the first

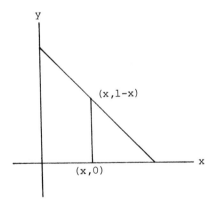

FIGURE 2.2.   Integrating first with respect to y

integral. The x-coordinates of points in S vary from 0 to 1, and therefore the second integral (over x) should be from 0 to 1. In short,

$$\int_S f = \int_0^1 \int_0^{1-x} xy \; dydx$$

Now the computation is much as before:

$$\int_0^1 (\int_0^{1-x} xy \; dy) dx = \int_0^1 (\tfrac{1}{2} xy^2 \big|_{y=0}^{y=1-x}) dx$$

$$= \int_0^1 \tfrac{1}{2}(x - 2x^2 + x^3) dx = \frac{1}{24}$$

as above. We could also do this integral by integrating first with respect to x. In this case, we first fix y. We now need to integrate $f(x,y)$ as a function of x. The integral is over the values of x which makes (x,y) lie in S - that is, for $0 \le x \le 1 - y$. (See Figure 2.3).

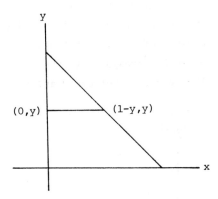

FIGURE 2.3.   Integrating first with respect to x

The possible y-coordinates in S vary from 0 to 1. Therefore

$$\int_S f = \int_0^1 \int_0^{1-y} xy \; dxdy$$

and the computation proceeds just as above.

The major problem, then, in these computations is getting the limits of integration correct. (Of course, there is also the problem of actually performing the integrations, but that problem is treated in elementary calculus.) We shall see a number of further examples, but the method is always the same. Given an integral

$$\int_S\!\int f(x,y) dxdy$$

for example, where x is integrated first, we fix y and determine the limits on the integral in x by seeing which points (x,y) are in S. Then we see which values of y appear in S. For integrals in R^3, the method is similar, but we shall postpone this discussion until the next section.

We close this section with another example. More examples are given in the next section.

Suppose S is the subset of R^2 defined by the two curves $y = 1 - x^2$ and $y = x^2$ (see Figure 2.4), and $f(x,y) = x + y$. The curves intersect

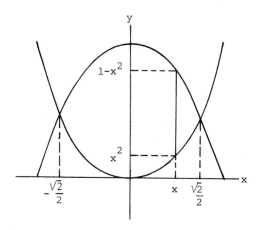

FIGURE 2.4. A region defined by two parabolas

where $1 - x^2 = x^2$, or $x = \pm \sqrt{2}/2$. (The y-coordinate is 1/2, but we shall not need this fact.) For fixed x in $[-\sqrt{2}/2, \sqrt{2}/2]$, y ranges from x^2 to $1 - x^2$, so

$$h(x) = \int_{x^2}^{1-x^2} (x + y)\,dy$$

$$= xy + y^2/2 \Big|_{y=x^2}^{y=1-x^2}$$

$$= x(1 - x^2) + \frac{1}{2}(1 - x^2)^2 - x^3 - \frac{1}{2}x^4$$

$$= 1/2 + x - x^2 - 2x^3$$

It follows that

$$\int_{-\sqrt{2}/2}^{\sqrt{2}/2} (\int_{x^2}^{1-x^2} (x + y)dy)dx = \int_{-\sqrt{2}/2}^{\sqrt{2}/2} (\frac{1}{2} + x - x^2 - 2x^3)dx$$

$$= \frac{\sqrt{2}}{2} - \frac{\sqrt{2}}{6} = \frac{\sqrt{2}}{3}$$

We can do this integral as an iterated integral in the other order, but it is considerably more complicated. The reason is that for
fixed y, x varies between $-\sqrt{y}$ and $\sqrt{y}$ if $0 \leq y \leq 1/2$ and between $-\sqrt{1-y}$
and $\sqrt{1-y}$ if $1/2 \leq y \leq 1$. (See Figure 2.4 again; if $y = 1 - x^2$, then
$x^2 = 1 - y$, or $x = \pm \sqrt{1-y}$.) Thus the iterated integral needs to be
broken into two pieces:

$$\int_S f = \int_0^{1/2} \int_{-\sqrt{y}}^{\sqrt{y}} (x + y)dxdy + \int_{1/2}^1 \int_{-\sqrt{1-y}}^{\sqrt{1-y}} (x + y)dxdy$$

Of course, evaluating the right-hand side gives the same answer as
above. (See Exercise 4.) It often pays in problems like this to
experiment with setting up both iterated integrals, in order to see
which will be less troublesome to compute.

EXERCISES

1.   Evaluate the following.

(a) $\int_1^2 \int_1^{x^2} (x^2 + y^2)dydx$ (f) $\int_{-1}^1 \int_{x^2}^1 \sqrt{1-y}\ dydx$

(b) $\int_{-1}^1 \int_{-x^2}^{x^2} (x^2 - y)dydx$ (g) $\int_0^1 \int_0^{1-y} \int_0^{1-x-y} y\ dzdxdy$

(c) $\int_{-1}^1 \int_{x^2}^{2-x^2} x^2 y^2 dydx$ (h) $\int_0^1 \int_0^{1-x} \int_0^{1-y^2-x} xzdzdydx$

(d) $\int_0^1 \int_0^{2-2y} (x^2 + y^2)dxdy$ (i) $\int_{-1}^1 \int_0^{1-z^2} \int_{-\sqrt{y}}^{\sqrt{y}} xyz^2 dxdydz$

(e) $\int_{-1}^1 \int_0^{\sqrt{1-y^2}} xy^2 dxdy$ (j) $\int_0^1 \int_0^x \int_{3-y}^3 (x + y)^2 zdzdydx$

2. Let S_1 be the region in the plane bounded by the x-axis, the
y-axis, and the line $x + y = 1$, and let S_2 be the region bounded by
the curves $y = 1 - x^2$ and $y = x^2$. Compute:

(a) $\int_{S_1} (x^2 - xy)$

(d) $\int_{S_2} x \sin y$

(b) $\int_{S_2} (x^2 + y^3 + 2xy)$

(e) $\int_{S_2} (1 - x^2)(1 + y^2)$

(c) $\int_{S_1} (x - 2x^2) e^{xy}$

(f) $\int_{S_1} \frac{y(1-x)}{1+y^2}$

3. Compute $\int_S (xy - y^2)$, where

(a) S is the region bounded by the x-axis, the y-axis, and the
line $x + 2y = 4$

(b) S is the region bounded by the x-axis, the line $y = x$, and
the line $x = 2$

(c) S is the region bounded by the y-axis, the line $y = x$, and
the line $y = 2$

(d) S is the region bounded by the lines $y = x$, $y = -2x$, and
$y = 3$

4. Complete the calculation of $\int_S f$ in the last example of the section, integrating first with respect to x.

3. MORE EXAMPLES

We give here some further examples of iterated integrals of functions
of two and three variables.

1. Let S be the region in R^2 defined by the curves $y = x^2$ and
$x = y^2$ (see Figure 3.1), and let $f:S \to R$ be the function defined by
$f(x,y) = 3x + 2y$. The curves intersect at $(0,0)$ and $(1,1)$. For fixed
x (with $0 \le x \le 1$), the points (x,y) which are in S are those with

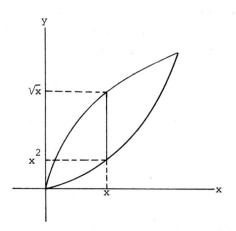

FIGURE 3.1.   A region defined by two parabolas

$x^2 \leq y \leq \sqrt{x}$, as the figure shows.   Thus

$$\int_S f = \int_0^1 \int_{x^2}^{\sqrt{x}} (3x + 2y)\,dy\,dx$$

We compute the inner integral:

$$\int_{x^2}^{\sqrt{x}} (3x + 2y)\,dy = (3xy + y^2)\Big|_{y=x^2}^{y=\sqrt{x}}$$

$$= 3x^{3/2} + x - 3x^3 - x^4$$

It follows that

$$\int_S f = \int_0^1 (3x^{3/2} + x - 3x^3 - x^4)\,dx$$

$$= \frac{6}{5} x^{5/2} + \frac{x^2}{2} - \frac{3}{4} x^4 - \frac{1}{5} x^5 \Big|_0^1 = 3/4$$

2.   According to the remark at the end of Section 1, the order of integration in the iterated integral of a continuous function is immaterial: one gets the same answer either way.   However, it can happen that one order is preferable over the other for computation.

We saw an example of this at the end of the previous section. Here is another.

Let S be the region in the first quadrant in R^2 defined by $2y = x^2$ and $x = 2$, and $f:S \to R$ be the function defined by

$$f(x,y) = \frac{2x}{(1+x^2 + y^2)^2}$$

(See Figure 3.2.) In this case (as in Example 1),

$$\int_S f = \int_0^2 \int_0^{x^2/2} \frac{2x}{(1+x^2 + y^2)^2} \, dy\,dx$$

or

$$\int_S f = \int_0^2 \int_{\sqrt{2y}}^2 \frac{2x}{(1+x^2 + y^2)^2} \, dx\,dy$$

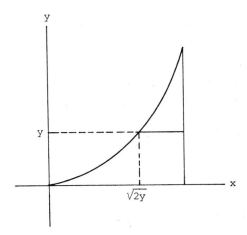

FIGURE 3.2.

The first of these iterated integrals involves integrating the function

$$\frac{2x}{(1+x^2 + y^2)^2}$$

with respect to y. This is extremely complicated. The second iterated integral is much easier:

$$\int_0^2 \int_{\sqrt{2y}}^2 \frac{2x}{(1+x^2 + y^2)^2}\, dxdy = \int_0^2 \left(\frac{-1}{1+x^2 + y^2} \Big|_{x=\sqrt{2y}}^{x=2} \right) dy$$

$$= \int_0^2 \left(\frac{-1}{5+y^2} + \frac{1}{1 + 2y + y^2} \right) dy$$

$$= \left(-\frac{1}{\sqrt5} \arc\tan\left(\frac{y}{\sqrt5}\right) - \frac{1}{1+y} \right) \Big|_0^2$$

$$= -\frac{1}{\sqrt5} \arc\tan \frac{2}{\sqrt5} - 1/3 + 1$$

3. If S is a region in R^2 and $f:S \to R$ is a non-negative integrable function, we can think of the upper and lower sums of f as
giving approximations for the volume of the region under the graph of
f and over S. It follows that the integral of f over S can be interpreted as giving the exact value for this volume.

For example, let $f(x,y) = 2 - x - y$ and S be the region in the
first quadrant bounded by the curve $x^2 + y = 1$. (See Figure 3.3.)
The volume of the region under the graph of f over S is

$$\int_S f = \int_0^1 \int_0^{1-x^2} (2 - x - y)\, dydx$$

$$= \int_0^1 \left((2 - x)y - y^2/2 \Big|_{y=0}^{y=1-x^2} \right) dx$$

$$= \int_0^1 \left(\frac{3}{2} - x - x^2 + x^3 - \frac{1}{2} x^4 \right) dx$$

$$= \frac{3}{2} - \frac{1}{2} - \frac{1}{3} + \frac{1}{4} - \frac{1}{10} = \frac{49}{60}$$

4. The integration of functions of three variables is a bit more
complicated. Let S be the region in R^3 defined by

$$S = \{(x,y,z)\,|\,x + y + z \le 1,\ x \ge 0,\ y \ge 0,\ z \ge 0\}$$

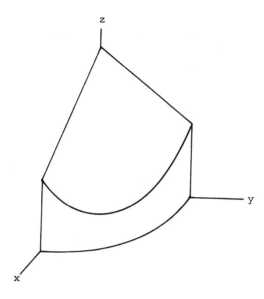

FIGURE 3.3.

and $f:S \to R$ the function defined by $f(x,y,z) = y - z$. Then, for fixed y and z, x ranges along the line parallel to the x axis from the point $(0,y,z)$ to the point $(1 - y - z, y, z)$ which is on the plane $x + y + z = 1$. (See Figure 3.4.) Once the integration with respect to x has been performed, if we fix z, y will range from 0 to $1 - z$. (One way of looking at it is this: after the first integral in x, we have a function of y and z defined for those pairs (y,z) in the projection S' of S on the y - z plane. For fixed z, the points $y \in S'$ range from 0 to $1 - z$.) Finally, z ranges from 0 to 1. Thus

$$\int_S f = \int_0^1 \int_0^{1-z} \int_0^{1-y-z} (y - z) dx dy dz$$

$$= \int_0^1 \int_0^{1-z} ((y - z)x \Big|_{x=0}^{x=1-y-z}) dy dz$$

$$= \int_0^1 \int_0^{1-z} (y - z - y^2 + z^2) dy dz$$

$$= \int_0^1 ((\frac{1}{2} y^2 - zy - \frac{1}{3} y^3 + z^2 y) \Big|_{y=0}^{y=1-z}) dz$$

$$\int_0^1 (\frac{1}{6} - z + \frac{3}{2} z^2 - \frac{2}{3} z^3) dz$$

$$= \frac{1}{6} - \frac{1}{2} + \frac{1}{2} - \frac{1}{6} = 0$$

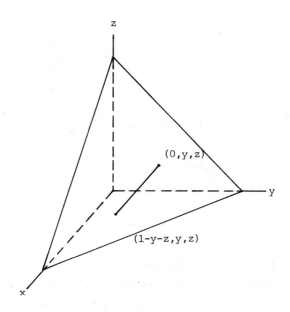

FIGURE 3.4.

In setting up these limits, it is useful to remember that the limits of integration can depend on the variables which have not yet been integrated out, but not on the variables which have been integrated out or the variable being integrated. For instance,

$$\int_0^1 \int_0^{1-z} \int_0^{1-y^2-z^2} f(x,y,z) dxdydz$$

is possible, but

$$\int_0^1 \int_0^{1-x} \int_0^{1-y^2-z^2} f(x,y,z) dxdydz$$

is not (because the x variable has been integrated out by the time one integrates over y), and

$$\int_0^z \int_0^{1-z} \int_0^{1-y^2z^2} f(x,y,z)dxdydz$$

is not (because z may not be a limit of an integral over z).

5. In some cases, it is necessary to divide a region into pieces in order to compute an integral. Here is an example.

Let S be the region bounded by the x and y-axes and the lines x + y = 1, x + y = 2. (See Figure 3.5.) We compute the integral of f over S where $f(x,y) = x + y$. We shall write it as an iterated integral,

$$\iint f(x,y)dydx$$

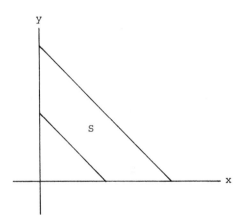

FIGURE 3.5.

When $0 \le x \le 1$, y varies from 1 - x to 2 - x. When $1 \le x \le 2$, however, y varies from 0 to 2 - x. Thus we need to break S into two pieces S_1 and S_2 (see Figure 3.6):

$$\int_S f = \int_0^1 \int_{1-x}^{2-x}(x + y)dydx + \int_1^2 \int_0^{2-x}(x + y)dydx$$

$$= \int_0^1 (xy + \frac{y^2}{2})\Big|_{y=1-x}^{y=2-x}dx + \int_1^2 (xy + \frac{y^2}{2})\Big|_{y=0}^{y=2-x}dx$$

$$= \int_0^1 \frac{3}{2} dx + \int_1^2 \frac{1}{2}(4 - x^2)dx$$

$$= \frac{3}{2} + (2x - \frac{x^3}{6})\big|_1^2 = \frac{7}{3}$$

We could, of course, try this as an iterated integral in the other order; however, it would still be necessary to divide S into two pieces. (See Exercise 3.)

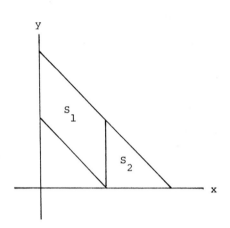

FIGURE 3.6.

EXERCISES

1.   Sketch the region S and compute the integral of f over S, where

(a) $f(x,y) = x^2 - y^2 + xy - 3$, $S = \{(x,y)\,|\,1 \leq x \leq 2,\ 4 \leq y \leq 5\}$

(b) $f(x,y) = x^2 + xy$, S is the region in R^2 defined by the curves $y = x^2$ and $y = x^3$

(c) $f(x,y) = x \sin y$, S is the region in R^2 defined by the curves $y = x^2$, $y = 2x^2$, and $x = 2$

(d) $f(x,y) = x$, $S = \{(x,y)\,|\,x^2 + y^2 \leq 1\}$

(e) $f(x,y) = x \cos y$, S is the region in the first quadrant of R^2 defined by the curves $x = \sqrt{\pi/2}$ and $y = x^2$

(f) $f(x,y) = \log y$, S the region in the first quadrant of R^2 defined by the curves $x = 1/2$, $y = 1$, and $y = x$

(g) $f(x,y) = x(x^2 + y^2)^{-1/2}$, S is the region in the first quadrant of R^2 defined by the curves $y = x$, $x = 1$, and $x = 2$

2. Compute the integral of f over S where

(a) $f(x,y,z) = x + y + z$, S is the region in the first octant with $x + y + z \leq 1$

(b) $f(x,y,z) = x^2 + xyz$, S is the region defined in 2 (a)

(c) $f(x,y,z) = x + y^2 - xz$, S is the region bounded by the x,y plane, the plane $z = 2$, and the cylinder $x^2 + y^2 = 1$

(d) $f(x,y,z) = z$, S is the region in the first octant bounded by $x^2 + y^2 + z^2 = 4$

(e) $f(x,y,z) = 2$, S is the region in R^3 defined by the surfaces $z = x^2 + y^2$ and $z = 27 - 2x^2 - 2y^2$

(f) $f(x,y,z) = 1$, S is the region in R^3 defined by the surfaces $z^2 = 8x$, $x^2 + y^2 = 4x$

(g) $f(x,y,z) = x^2 + y^2$, S is the region in R^3 defined by the surfaces $x^2 + z^2 = a^2$ and $y^2 + z^2 = a^2$

3. Compute the integral in Example 5 of the text by integrating in the other order.

4. Find the volume under the graph of $f(x,y) = 2x + 3y + 6$ over the region defined by the curves $x = 4 - y^2$ and $x = 1$.

5. Find the volume under the graph of $f(x,y) = 2x^2 + xy$ over the triangular region with vertices $(1,1)$, $(2,3)$, $(-1,2)$.

6. Let B be the region obtained by cutting a hole of radius $(r^2 - a^2)^{1/2}$ through a ball of radius r;

$$B = \{(x,y,z) \in R^3 : x^2 + y^2 + z^2 \leq r^2, x^2 + y^2 + a^2 \geq r^2\}$$

Compute the volume of B.

7. Find the volume of the region defined by the two cylinders $x^2 + y^2 = 9$ and $y^2 + z^2 = 9$.

8.   Find the volume of the region defined by the parabolic cylinders
$y = x^2$ and $y = z^2$ and the plane $y = 1$.

9.   Find the volume of the region defined by the three cylinders in
R^3 defined by $x^2 + y^2 = 1$, $x^2 + z^2 = 1$, and $y^2 + z^2 = 1$.

10.   Suppose that the density of a body in the region $S \subset R^3$ is given
by a function $\rho(\vec{v})$.   The *mass* of the body is

$$m = \int_S \rho$$

and the *center of mass* is the point $(\bar{x}, \bar{y}, \bar{z})$, where

$$\bar{x} = \frac{1}{m} \int_S x\rho, \quad \bar{y} = \frac{1}{m} \int_S y\rho, \quad \bar{z} = \frac{1}{m} \int_S z\rho$$

Let S be the region bounded by the three coordinate planes and the
plane $x + y + 2z = 4$.   Find the center of mass of a body in the region
S whose density is given by

   (a) $\rho(\vec{v}) = 1$
   (b) $\rho(x,y,z) = x + z$

11.   Find the center of mass of the region defined by $x^2 + y^2 + z^2 \leq 1$
$z \geq 0$ if the density is given by $\rho(x,y,z) = 3z$.

12.   Find the center of mass of the region bounded by the cylinder
$y^2 + z^2 = 16$ and the planes $x = 0$ and $x = 5$ if the density is given
by $\rho(x,y,z) = 2x$.

*13.   Let A be an elementary n × n matrix which corresponds to one of
the elementary row (or column) operations (see Section 7.6) and let
$\vec{v}_1, \ldots, \vec{v}_n$ be the columns of A.   Let $c_1, \ldots, c_n$ be positive real numbers and define $f: R^n \to R$ by

$$f(\vec{v}) = \begin{cases} 1 & \text{if } \vec{v} = \sum_{j=1}^{n} a_j \vec{v}_j \text{ with } 0 \leq a_j \leq c_j, \text{ all } j \\ 0 & \text{otherwise} \end{cases}$$

Define $S \subset R^n$ by

$$S = \{\vec{v} \in R^n : f(\vec{v}) > 0\}$$

(For instance, if

$$A = \begin{pmatrix} 1 & 2 \\ 0 & 1 \end{pmatrix}$$

and $c_1 = c_2 = 1$, then S is the parallelogram defined by the vectors $(1,0)$ and $(2,1)$.) Prove that

$$\int_{R^n} f = c_1 c_2 \cdots c_n |\text{Det } A|$$

Note that one can regard this integral as the volume of S. (Hint: this exercise can easily be reduced to one in two variables. See what it means there.)

*14. Let B be any invertible $n \times n$ matrix, $c_1, \ldots, c_n$ positive real numbers, and define $f: R^n \to R$ and $S \subset R^n$ as in the previous exercise. Prove that

$$\int_{R^n} f = c_1 c_2 \cdots c_n |\text{Det } B|$$

(Hint: write $B = A_1 A_2 \cdots A_k$, where each A_j is an elementary matrix, and proceed by induction on k.)

4.   THE PROOF OF FUBINI'S THEOREM

This section is devoted to the proof of Theorems 1.2 and 2.1. For simplicity of notation, we shall deal only with the case $n = 2$; the proof in the general case is quite similar.

We begin with Theorem 1.2. We have seen (Lemma 1.1) that if $g: [a_1, b_1] \to R$ is defined by

$$g(x) = \int_{a_2}^{b_2} f(x,y) dy$$

then g is continuous. Since we are assuming that n = 2, we need to prove that

$$\int_S f = \int_{a_1}^{b_1} g(x)dx = \int_{a_1}^{b_1} (\int_{a_2}^{b_2} f(x,y)dy)dx \qquad (4.1)$$

We shall show that for any partition P of S = $R(\vec{a},\vec{b})$, we have

$$L(f,P) \le \int_{a_1}^{b_1} g(x)dx \le U(f,P) \qquad (4.2)$$

As a result,

$$\int_S f = \text{lub } L(f,P) \le \int_{a_1}^{b_1} g(x)dx \le \text{glb } U(f,P) = \int_S f$$

and equation (4.1) will be proved.

Let $P = P_1 \times P_2$ be any partition of S, where $P_1 = \{t_0, t_1, \ldots, t_p\}$ and $P_2 = \{s_0, s_1, \ldots, s_q\}$. Define numbers M_{ij} and m_{ij}, $1 \le i \le p$, $1 \le j \le q$, by

$$M_{ij} = \text{lub}\{f(x,y): x \in [t_{i-1}, t_i], y \in [s_{j-1}, s_j]\}$$
$$m_{ij} = \text{glb}\{f(x,y): x \in [t_{i-1}, t_i], y \in [s_{j-1}, s_j]\}$$

Then

$$U(f,P) = \sum_{i=1}^{p} \sum_{j=1}^{q} M_{ij}(t_i - t_{j-1})(s_j - s_{j-1})$$
$$L(f,P) = \sum_{i=1}^{p} \sum_{j=1}^{q} m_{ij}(t_i - t_{i-1})(s_j - s_{j-1})$$

For each $x \in [a_1, b_1]$ define the function $f_x: [a_2, b_2] \to R$ by $f_x(y) = f(x,y)$. Then, by the definition of the integral,

$$g(x) = \int_{a_2}^{b_2} f_x(y)dy \le U(f_x, P_2)$$

The next step is to prove that $U(g,P_1) \le U(f,P)$ so that

$$\int_{a_1}^{b_1} g(x)dx \le U(g,P_1) \le U(f,P)$$

which is one of the inequalities (4.2). We do this by looking at the definition of the upper sums. If $x \in [t_{i-1},t_i]$, we have

$$f_x(y) = f(x,y) \le M_{ij}$$

for all $y \in [s_{j-1},s_j]$. Thus, for all $x \in [t_{i-1},t_i]$,

$$\text{lub}\{f_x(y):y \in [s_{j-1},s_j]\} \le M_{ij}$$

Multiply this inequality by $(s_j - s_{j-1})$ and sum over j:

$$g(x) \le U(f_x,P_2) \le \sum_{j=1}^{q} M_{ij}(s_j - s_{j-1})$$

for all $x \in [t_{i-1},t_i]$. Let $N_i = \text{lub}\{g(x):x \in [t_{i-1},t_i]\}$; then

$$N_i \le \sum_{j=1}^{q} M_{ij}(s_j - s_{j-1})$$

and

$$U(g,P_1) = \sum_{i=1}^{p} N_i(t_i - t_{i-1}) \le \sum_{i=1}^{p} (\sum_{j=1}^{q} M_{ij}(s_j - s_{j-1}))(t_i - t_{i-1})$$

$$= U(f,P) \qquad\qquad (4.3)$$

We know, of course, that

$$\int_{a_1}^{b_1} g(x)dx \le U(g,P_1) \qquad\qquad (4.4)$$

Combining (4.3) and (4.4), we get

$$\int_{a_1}^{b_1} g(x)dx \le U(f,P)$$

which is half of (4.2). The same sort of reasoning proves that $L(g,P_1) \ge L(f,P)$, so that

$$\int_{a_1}^{b_1} g(x)dx \geq L(g,P_1) \geq L(f,P)$$

Therefore (4.2) is proved and the theorem follows.

The proof of Theorem 2.1 is identical with the proof of Theorem 1.2, except that the first sentence (concerning the continuity of g) should be omitted. From there on, the argument requires no change.

EXERCISES

1.   Give the Proof of Theorem 1.2 for arbitrary n.

2.   Justify the Remark after Theorem 1.2 by proving that theorem with g defined by integrating with respect to x instead of y.

*3.   Let $S = R((a_1,a_2), (b_1,b_2))$, let $f:S \to R$ be an integrable function, and define $h_x:[a_2,b_2] \to R$ by

$$h_x(y) = f(x,y)$$

Suppose h_x is integrable for all but a finite number of values $x_1,\ldots,x_r$ and define $\tilde{h}_x:[a_2,b_2] \to R$ by

$$\tilde{h}_x(y) = \begin{cases} 0 & \text{if } x = x_j, \ 1 \leq j \leq r \\ h_x(y) & \text{otherwise} \end{cases}$$

Prove that $\tilde{h}_x$ is integrable and that

$$\int_S f = \int_{a_1}^{b_1}(\int_{a_2}^{b_2} h_x(y)dy)dx$$

(The proof is similar to the proof of Theorem 2.1.)

4.   State and prove the analogue of Exercise 3 for functions of n variables.

5. DIFFERENTIATING UNDER THE INTEGRAL SIGN

Let $S = R(\vec{a},\vec{b})$ be a rectangle in R^2 (as before, $\vec{a} = (a_1,a_2)$ and $\vec{b} = (b_1,b_2)$) and let U be an open set in R^2 containing S. If $f:U \rightarrow R$ is a continuously differentiable function, we can get a new function g, defined on the interval $a_1 \leq x \leq b_1$, by integrating f against the y variable:

$$g(x) = \int_{a_2}^{b_2} f(x,y)dy$$

For example, if $f(x,y) = x^2y^3$ and $S = R((0,0),(1,1))$, then

$$g(x) = \int_0^1 x^2y^3dy = \frac{x^2}{4}$$

In this case, g is differentiable, and $g'(x) = x/2$. Notice also that

$$\int_0^1 \frac{\partial f}{\partial x}(x,y)dy = \int_0^1 2xy^3dy = \frac{x}{2}$$

$$= g'(x)$$

The main result of this section is the statement that this last equality holds quite generally.

THEOREM 5.1. *Let f and g be as above.* Then g *is continuously differentiable on the interval* (a_1,b_1) *and*

$$g'(x) = \int_{a_2}^{b_2} \frac{\partial f}{\partial x}(x,y)dy$$

Proof. The quickest proof is indirect; we do not differentiate g(x), but rather integrate g'(x). Define $h:[a_1,b_1] \rightarrow R$ by

$$h(t) = \int_{a_1}^{t} (\int_{a_2}^{b_2} \frac{\partial f}{\partial x}(x,y)dy)dx \tag{5.1}$$

The Fundamental Theorem of Calculus (Theorem 11.2.3) states that

$$h'(x) = \int_{a_2}^{b_2} \frac{\partial f}{\partial x}(x,y)dy \tag{5.2}$$

Note that h' is continuous, by Lemma 1.1.

We can evaluate h(t) another way. Let S_t be the rectangle given by (a_1, a_2) and (t, b_2). (See Figure 5.1.) Then

$$h(t) = \int_{S_t} \frac{\partial f}{\partial x} (x, y) dx$$

by Theorem 1.2, since (5.1) is just an iterated integral over S_t.

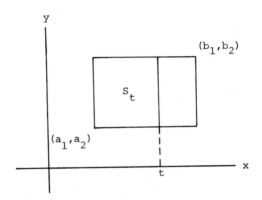

FIGURE 5.1.   The rectangle S_t

Iterate in the other order:

$$h(t) = \int_{a_2}^{b_2} (\int_{a_1}^{t} \frac{\partial f}{\partial x} (x, y) dx) dy$$

$$= \int_{a_2}^{b_2} (f(x, y) \Big|_{x=a_1}^{x=t}) dy$$

(again by the Fundamental Theorem of Calculus)

$$= \int_{a_2}^{b_2} (f(t, y) - f(a_1, y)) dy$$

$$= g(t) - g(a_1)$$

Therefore g and h differ by a constant, and so $g'(x) = h'(x)$. Now formula (5.2) completes the proof.

We have proved this theorem for functions in R^2 because this case includes all of the usual situations which arise. Suppose, for instance, that $f:R^n \to R^1$ is continuously differentiable, and that $g:R^{n-1} \to R^1$ is defined by

$$g(x_1,\ldots,x_{n-1}) = \int_{a_n}^{b_n} f(x_1,\ldots,x_n)dx_n$$

Then g is continuously differentiable, and

$$\frac{\partial g}{\partial x_j} = \int_{a_n}^{b_n} \frac{\partial f}{\partial x_j}(x_1,\ldots,x_n)dx_n$$

The easiest way to verify this claim is to regard f as a function of x_j and x_n alone, fixing all the other coordinates. Then Theorem 5.1 applies directly. Since the partial derivatives of g are continuous, Proposition 6.7.1 implies that g is continuously differentiable.

We give some other variations on this theme in the Exercises.

EXERCISES

1. Compute $g'(x)$ in two ways where

    (a) $g(x) = \int_0^1 \sin xy\,dy$

    (b) $g(x) = \int_0^1 \frac{e^{xy}}{x}\,dy$

    (c) $g(x) = \int_2^3 \frac{dy}{y+x}$

    (d) $g(x) = \int_0^1 x\cos xy\,dy$

2. Let $f:R^2 \to R^1$ be continuously differentiable, and let u, $v:R^1 \to R^1$ be continuously differentiable functions with $u(x) < v(x)$, for all x. Define $h:R^1 \to R^1$ by

$$h(x) = \int_{u(x)}^{v(x)} f(x,y)dy$$

Show that $h(x)$ is differentiable, and that

$$h'(x) = f(x,v(x))v'(x) - f(x,u(x))u'(x) + \int_{u(x)}^{v(x)} \frac{\partial f}{\partial x}(x,y)dy$$

(Hint: what if $u(x)$ or $v(x)$ were constant?   Use the chain rule.)

3.   Let $f: R^3 \to R^1$ be continuously differentiable, and let

$$h(x) = \int_{a_2}^{b_2} (\int_{a_3}^{b_3} f(x,y,z)dy)dz$$

Show that h is differentiable and that

$$h(x) = \int_{a_2}^{b_2} (\int_{a_3}^{b_3} \frac{\partial f}{\partial x} (x,y,z)dy)dz$$

*4.   Given that $\int_0^1 e^{xy^2} t^2 dt = 1$, compute $\frac{dy}{dx}$.

*5.   Given that $\int_{-y}^{x} e^{t^3} dt = 2$, compute $\frac{dy}{dx}$.

*6.   Given that $\int_{-y^2}^{x^2} e^{xyt} dt = 3$, compute $\frac{dy}{dx}$.

*7.   Given that $\int_1^2 \log(x+y+z+t)dt = 2$, compute $\frac{\partial z}{\partial x}$ and $\frac{\partial z}{\partial y}$.

6.   THE CHANGE OF VARIABLE FORMULA

We now discuss the so-called change of variable formula. This formula describes the behavior of the integral of a function when we change coordinates and will be useful to us later in the text. We state the result and give two examples here; the proof is given in the next section.

      One of the standard techniques of integration, so-called substitution, states that if $f:[a,b] \to R$ and $h:[c,d] \to [a,b]$ are differentiable functions with h surjective and $h'(x) \neq 0$ for x in $[c,d]$, then

$$\int_{h(c)}^{h(d)} f(x)dx = \int_{c}^{d} f(h(x))h'(x)dx \tag{6.1}$$

The proof of this formula involves the Fundamental Theorem of Calculus and the Chain Rule. Explicitly, if $F(x)$ is an antiderivative for $f(x)$, then $F(h(x))$ is an antiderivative for $f(h(x))h'(x)$. Thus, using the Fundamental Theorem of Calculus, we have

$$\int_{h(c)}^{h(d)} f(x)dx = F(h(d)) - F(h(c))$$

$$= \int_{c}^{d} f(h(x))h'(x)dx$$

We wish to generalize formula (6.1) to the integral of a function of n-variables. To do this, we first reinterpret it.

First of all, we note that $h:[c,d] \to [a,b]$ is a diffeomorphism; that is, h is a differentiable function, h is invertible, and h^{-1} is differentiable. (See Theorem 9.1.3.) If $h'(x) > 0$ for x in $[c,d]$, then $h(c) = a$ and $h(d) = b$, and (6.1) becomes

$$\int_{a}^{b} f(x)dx = \int_{c}^{d} f(h(x))h'(x)dx \tag{6.2}$$

If $h'(x) < 0$, then $h(c) = b$ and $h(d) = a$ and (6.1) becomes

$$\int_{b}^{a} f(x)dx = \int_{c}^{d} f(h(x))h'(x)dx$$

or

$$\int_{a}^{b} f(x)dx = \int_{c}^{d} f(h(x))(-h'(x))dx \tag{6.3}$$

Combining (6.2) and (6.3), we have the desired restatement of (6.1).

PROPOSITION 6.1. *Let* $f:[a,b] \to R$ *be a differentiable function and let* $h:[c,d] \to [a,b]$ *be a diffeomorphism. Then*

$$\int_{a}^{b} f(x)dx = \int_{c}^{d} f(h(x)) |h'(x)|dx \tag{6.4}$$

Proposition 6.1 does not quite make sense as it stands for functions of several variables; we need an appropriate meaning for $|h'(x)|$. It turns out that the right generalization is the absolute value of the *determinant* of the derivative, $|\mathrm{Det}(Dh)|$.

THEOREM 6.2. (The change of variable formula). *Suppose that S and S' are compact subsets of* R^n, $f:S \to R$ *is a differentiable function, and* $h:S' \to S$ *a* C^1-*diffeomorphism.*[*] *Then*

$$\int_S f = \int_{S'} (f \circ h) \, |\mathrm{Det}(Dh)|$$

The integrand on the right side of the equation is the function taking a vector $\vec{v} \in S'$ into the number

$$f(h(\vec{v})) \, |\mathrm{Det}(D_{\vec{v}}h)|$$

As we mentioned above, the proof of this result is given in the next section. Here are two examples of Theorem 6.2.

(1) Let $h:R^2 \to R^2$ be defined by

$$h(x,y) = (u,v)$$
$$= (x + y, x - y)$$

and let S' be the unit disc

$$S' = \{(x,y):x^2 + y^2 \le 1\}$$

The map h is certainly a diffeomorphism, since

$$h^{-1}(u,v) = (\frac{u+v}{2}, \frac{u-v}{2})$$

Also,

$$S = h(S') = \{(x,y):x^2 + y^2 \le 2\}$$

---

[*] We say that $h:S' \to S$ is a C^r-*diffeomorphism* if both h and h^{-1} are of class C^r; that is, if each of the component functions of both h and h^{-1} have continuous partial derivatives of orders $\le r$.

since

$$(x + y)^2 + (x - y)^2 = 2(x^2 + y^2)$$

Thus $\|h(x,y)\|^2 = 2\|(x,y)\|^2$, and one effect of h is to multiply the norm of every vector by $\sqrt{2}$.

We need one other calculation:

$$D_{(x,y)}h = \begin{pmatrix} 1 & 1 \\ 1 & -1 \end{pmatrix}$$

so

$$|Det_{(x,y)}h| = 2$$

Now let f:S → R be continuous. Then Theorem 6.2 states that

$$\int_S f = \int_{S'} 2(f \circ h)$$

or (as iterated integrals)

$$\int_{-\sqrt{2}}^{\sqrt{2}} \int_{(2-u^2)^{1/2}}^{(2-u^2)^{1/2}} f(u,v)dvdu = \int_{-1}^{1} \int_{(1-x^2)^{1/2}}^{(1-x^2)^{1/2}} 2f(x+y,x-y)dxdy$$

For instance, let f(u,v) = uv. Then

$$f(x + y, x - y) = (x + y)(x - y) = x^2 - y^2$$

and we get

$$\int_{-\sqrt{2}}^{\sqrt{2}} \int_{-(2-u^2)^{1/2}}^{(2-u^2)^{1/2}} uv\, dvdu = \int_{-1}^{1} \int_{-(1-x^2)^{1/2}}^{(1-x^2)^{1/2}} (x^2 - y^2)dxdy$$

The reader should verify that the two sides are indeed equal (both, in fact, are 0).

(2) Polar coordinates are defined in R^2 by

$$\begin{aligned}(x,y) &= h(r,\theta) \\ &= (r \cos \theta,\ r \sin \theta)\end{aligned}$$

Since

$$h'(r,\theta) = \begin{pmatrix} \cos\theta & \sin\theta \\ -r\sin\theta & r\cos\theta \end{pmatrix}$$

we have

$$|\text{Det } h'(r,\theta)| = |r|$$

The map h is *not* a diffeomorphism of R^2 to R^2; for one thing, h' is not invertible whenever $r = 0$, and for another, $h(r,0) = h(r,2\pi)$. However, the theorem does hold on the set

$$U = \{(r,\theta):r > 0, \ 0 < \theta < 2\pi\}$$

since h is a diffeomorphism from U onto

$$U' = \{(x,y):y \neq 0 \ \text{ or } \ y = 0, \ x < 0\}$$

The set U' is the space R^2 with the positive x-axis removed; because the set removed has content zero, we can generally ignore the fact that some points are missing. (We return to this matter below.)

Now let S' be a set in U and let $S = h(S')$. We can think of S as a set in the plane and S' as the same set in polar coordinates. For instance, if S' is the piece of the annulus shown in Figure 6.2,

$$S' = \{(x,y):1 \leq x^2 + y^2 \leq 2, \ x \geq 0, \ y \geq 0\}$$

then the points in S' can be expressed in polar coordinates as

$$S = \{(r,\theta):1 \leq r^2 \leq 2, \ 0 \leq \theta \leq \frac{\pi}{2}\}$$

Theorem 6.2 now says that if f is continuous, then

$$\int_S f = \int_{S'} f \circ h \, |\text{Det } h|$$

or

$$\int_S f(x,y)\,dydx = \int_{S'} (f \circ h)(r,\theta)r \ drd\theta$$

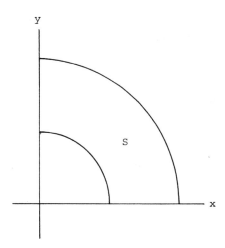

FIGURE 6.2.   An annulus

For instance, if $f(x,y) = xy$, then

$$(f \circ h)(r, \theta) = (r \cos \theta)(r \sin \theta)$$
$$= r^2 \sin \theta \cos \theta$$

and the right-hand integral (for S', S as above) is

$$\int_0^{\pi/2} \int_1^{\sqrt{2}} r^3 \sin \theta \cos \theta \, dr d\theta = \int_0^{\pi/2} \frac{r^4}{4} \sin \theta \cos \theta \, d\theta \Big|_{r=1}^{r=\sqrt{2}}$$

$$= \frac{3}{4} \int_0^{\pi/2} \sin \theta \cos \theta \, d\theta$$

$$= \frac{3 \sin^2 \pi}{8} \Big|_0^{\pi/2} = \frac{3}{8}$$

Computing the left-hand integral is harder, since it is not easy to write the limits of integration. (See Exercise 3.) Polar coordinates are often useful when symmetry about the origin is present in the problem.

The fact that $U' \neq R^2$ may seem to cause problems when one uses polar coordinates. In fact, it causes none at all. The missing part of R^2 has content 0, and whether it is included or not does not affect the integral. For instance, if S is the unit disc,

$$S = \{(x,y):x^2 + y^2 \le 1\}$$

then $S = h(S')$, where

$$S' = \{(r,\theta):0 \le r \le 1, \ 0 \le \theta \le 2\pi\}$$

and the change-of-variables formula still applies:

$$\iint_S f(x,y)\,dxdy = \iint_{S'} f(r\cos\theta, \ r\sin\theta)r\,drd\theta$$

EXERCISES

1. Compute the integral of $f(x,y) = (x^2 + y^2)^{-3}$ over the region S where

    (a) $S = \{(x,y):1 \le x^2 + y^2 \le 4\}$

    (b) $S = \{(x,y):1 \le x^2 + y^2 \le 4, \ x \ge 0, \ y \ge 0\}$

    (c) $S = \{(x,y):1 \le x^2 + y^2 \le 4, \ y \ge 0 \ \text{ and } \ y \le x\}$

2. Compute the integral of $f(x,y) = x^2 y$ over the regions of Exercise 1.

3. Compute the integral of $f(x,y) = xy$ over the set

$$S = \{(x,y):1 \le x^2 + y^2 \le 2, \ x \ge 0, \ y \ge 0\}$$

without transforming coordinates.

4. *Cylindrical coordinates* in R^3 are defined by

$$(x,y,z) = h(r,\theta,z)$$
$$= (r\cos\theta, \ r\sin\theta,z)$$

(See Figure 6.3.) Show that $|\text{Det Dh}| = |r|$.

5. Use cylindrical coordinates to compute the integral of $f(x,y,z) = z(1 + x^2 + y^2)^{-1}$ over the region S where

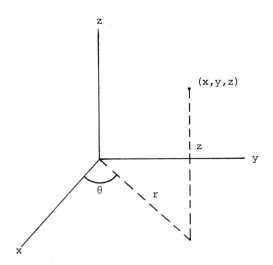

FIGURE 6.3. Cylindrical coordinates

(a) $S = \{(x,y,z):1 \leq x^2 + y^2 \leq 3,\ 1 \leq z \leq 5\}$

(b) $S = \{(x,y,z):1 \leq x^2 + y^2 \leq 3,\ x \geq 0,\ y \geq 0,\ 1 \leq z \leq 5\}$

(c) $S = \{(x,y,z):1 \leq x^2 + y^2 \leq 3,\ x \geq 0,\ y \geq x,\ 1 \leq z \leq 5\}$

6 .   Compute the integral of $f(x,y,z) = xz(1 + x^2 + y^2)^{-1}$ over the region S where

(a) $S = \{(x,y,z):1 \leq x^2 + y^2 \leq 3,\ 0 \leq z \leq 3\}$

(b) $S = \{(x,y,z):1 \leq x^2 + y^2 \leq 3,\ x \geq 0,\ 0 \leq z \leq 3\}$

(c) $S = \{(x,y,z):1 \leq x^2 + y^2 \leq 3,\ x\sqrt{3}/3 \leq y \leq x\sqrt{3},\ 0 \leq z \leq 3\}$

7 .   *Spherical coordinates* in R^3 are defined by

$(x,y,z) = h(\rho,\theta,\phi)$

$\qquad = (\rho \sin \phi \cos \theta, \rho \sin \phi \sin \theta, \rho \cos \phi)$

(See Figure 6.4.) Show that $|\mathrm{Det}\ h| = \rho^2 |\sin \phi|$.

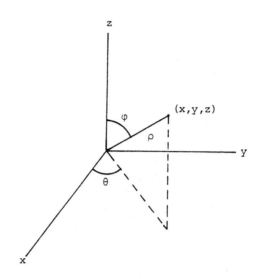

FIGURE 6.4. Spherical coordinates

8.   Use spherical coordinates to compute the integral of $f(x,y,z)$ = $x^2 + y^2 + z^2$ over the region S where

   (a) $S = \{(x,y,z):1 \le x^2 + y^2 + z^2 \le 5, \ z \ge 0\}$

   (b) $S = \{(x,y,z):1 \le x^2 + y^2 + z^2 \le 5, \ z \ge 0, \ x\sqrt{3}/3 \le y \le x\sqrt{3}\}$

   (c) $S = \{(x,y,z):1 \le x^2 + y^2 + z^2 \le 5, \ z \ge 0, \ z^2 \ge x^2 + y^2\}$

9.   Compute the integral of $f(x,y,z) = xy(x^2 + y^2 + z^2)^{-1}$ over the regions of Exercise 8.

7.   THE PROOF OF THEOREM 6.2

We now prove the change of variable formula[*]. We begin with a useful observation.

---

[*]This proof is an expanded version of the proof given by J. Schwartz, "The formula for change of variable in a multiple integral," *American Mathematical Monthly,* 61 (1954), pp. 81-85.

LEMMA 7.1. *Let* S_1, S_2 *and* S *be compact subsets of* R^n *and* $h_2 : S_2 \to S_1$, $h_1 : S_1 \to S$ *diffeomorphisms. Suppose*

$$\int_{S_1} f_1 = \int_{S_2} (f_1 \circ h_2) \, |\text{Det}(Dh_2)| \tag{7.1}$$

for any continuous function $f_1 : S_1 \to R$ *and*

$$\int_S f = \int_{S_1} (f \circ h_1) \, |\text{Det}(Dh_1)| \tag{7.2}$$

for any continuous function $f : S \to R$. *Then*

$$\int_S f = \int_{S_2} (f \circ (h_1 \circ h_2)) \, |\text{Det}(D(h_1 \circ h_2))|$$

for any continuous function $f : S \to R$.

Proof. First of all, we note that, for any $\vec{v} \in S_2$,

$$\text{Det}(D_{\vec{v}}(h_1 \circ h_2)) = (\text{Det}((D_{h_2(\vec{v})} h_1) \circ (D_{\vec{v}} h_2)))$$

$$= (\text{Det}(D_{h_2(\vec{v})} h_1))(\text{Det}(D_{\vec{v}} h_2))$$

by the Chain Rule and the multiplicative property of determinants (Proposition 10.4.1). Thus, if we define $g_1 : S_1 \to R$ by

$$g_1(\vec{w}) = (f \circ h_1)(\vec{w}) \, |\text{Det}(D_{\vec{w}} h_1)|$$

we have

$$\int_{S_2} (f \circ (h_1 \circ h_2)) \, |\text{Det}(D(h_1 \circ h_2))| = \int_{S_2} (g_1 \circ h_2) \, |\text{Det}(Dh_2)|$$

$$= \int_{S_1} g_1 = \int_S f$$

by (7.1), (7.2), and the definition of g_1. This proves Lemma 7.1.

Remark. The analogue of Lemma 7.1 for the composite of more than two diffeomorphisms follows by induction. We leave the statement and proof as Exercise 1.

We next verify the change of variable formula when the diffeomorphism h is a linear transformation.

PROPOSITION 7.2.   *Let* $T:R^n \to R^n$ *be an invertible linear trans-*
formation, S' *a compact subset of* R^n *and* $f:S = T(S') \to R$ *a continuous*
function.   Then

$$\int_S f = \int_{S'} (f \circ T) |\mathrm{Det}(T)| \tag{7.3}$$

Proof.   Let A be the matrix of T (relative to the standard bases).   We know from Proposition 7.6.2 (and the discussion following
the statement of that proposition) that there are invertible matrices
$A_1, \ldots, A_k$, each of which is either a permutation matrix, a diagonal
matrix, or an elementary matrix, such that

$$A = A_1 A_2 \ldots A_k$$

Let $T_j:R^n \to R^n$ be the linear transformation with matrix A_j, $1 \le j \le k$.
Then T is the composite $T_1 \circ T_2 \circ \cdots \circ T_k$ (by Proposition 7.1.2).   It
now follows from Lemma 7.1 that equation (7.3) will hold for arbitrary
T if it can be proved in case the matrix of T is either a permutation
matrix, a diagonal matrix, or an elementary matris.   These special
cases were Exercise 13 of Section 3.   We sketch the proof in the case
where T has an elementary matrix; the other cases are easier.

Suppose the matrix of T is the elementary matrix $\begin{pmatrix} 1 & \lambda \\ 0 & 1 \end{pmatrix}$ (the general case in R^n is similar).   Then Det T = 1.   We may assume that S'
is a rectangle; if S' is $R((a_1,a_2),(b_1,b_2))$, then

$$S = T(S') = \{(x_1 + \lambda x_2, x_2) : a_1 \le x_1 \le b_1, \ a_2 \le x_2 \le b_2\}$$

and, by the Fubini Theorem (Theorem 1.2)

$$\int_{S'} (f \circ T) \det T = \int_{S'} f \circ T$$
$$= \int_{a_2}^{b_2} \int_{a_1}^{b_1} f(x_1 + \lambda x_2, x_2) dx_1 dx_2$$
$$= \int_{a_2}^{b_2} \int_{a_1 + \lambda x_2}^{b_1 + \lambda x_2} f(y_1, x_2) dy_1 dx_2$$

(by substitution)

$$= \int_S f$$

Thus (7.3) holds in this case; the proof in the case of an arbitrary elementary matrix is essentially the same.

If S is the subset of R^n, we define the *characteristic function* of S to be the function $\chi_S : R^n \to R$ given by

$$\chi_S(\vec{v}) = \begin{cases} 1 & \text{if } \vec{v} \in S \\ 0 & \text{if } \vec{v} \in S \end{cases}$$

The *volume* of the set S is defined to be the number $\mu(S)$,

$$\mu(S) = \int_{R^n} \chi_S$$

when this integral exists. (See Exercise 14 of Section 11.4.)

The following is an immediate consequence of Proposition 7.2 and is left as Exercise 2.

COROLLARY. *Let S be a compact subset of R^n and $T : R^n \to R^n$ an invertible linear transformation. Then*

$$\mu(T(S)) = |\text{Det } T| \mu(S)$$

In proving Theorem 6.2, we shall need a different method of obtaining integrals, one useful in other applications as well. Let S be a rectangle in R^n, $f : S \to R$ an integrable function, and P a partition of S into rectangles $S_1, \ldots, S_m$. For each j, $1 \leq j \leq m$, we choose a point $\vec{v}_j \in S_j$. We then form the sum

$$\sum_{j=1}^{m} f(\vec{v}_j) \mu(S_j)$$

Such an expression is called a *Riemann sum* for the function f relative to the partition P. Of course, (7.4) depends on the choice of the points $\vec{v}_1, \ldots, \vec{v}_m$.

If we fix the partition P, then for any choice of points $\vec{v}_1, \ldots, \vec{v}_m$, the Riemann sum (7.4) lies between the upper sum and lower sum for f relative to P. Thus the Riemann sums can be thought of as approximations for the integral of f over S. In fact, we have the following.

PROPOSITION 7.3.   *Let S and f be as above, and let* $P_1, \ldots, P_j, \ldots$
be a sequence of partitions for S.   Let R_j *be a Riemann sum for f re-*
lative to P_j.

(a)   *If f is continuous and* $\lim_j \text{mesh } P_j = 0$, *then*

$$\lim_{j \to \infty} R_j = \int_S f$$

(b)   *Suppose f is continuous off a set* S_1 *of content* 0, *and let*
q_j *be the total volume of the rectangles in the partition* P_j *which*
intersect S_1.   *If*

$$\lim_{j \to \infty} \text{mesh } P_j = 0, \quad \lim_{j \to \infty} q_j = 0$$

then

$$\lim_{j \to \infty} R_j = \int_S f$$

The proofs are similar.   In each case, we show that

$$L(f, P_j) \leq R_j \leq U(f, P_j)$$

and then show that

$$\lim_{j \to \infty} (U(f, P_j) - L(f, P_j)) = 0 \tag{7.5}$$

Thus both $U(f, P_j)$ and $L(f, P_j)$ converge to $\int_S f$, and so R_j must converge
to the same limit.   The details of proving (7.5) are given in the
proofs of Theorem 6.1 and 6.2.   The task of assembling these ingre-
dients into a proof is left as Exercise 7.

The proof of Theorem 6.2 is based on the following result.

PROPOSITION 7.4.   *Suppose that S and S' are compact subsets of*
R^n, *f:S → R is a nonnegative function continuous off a set of content*
0, *and h:S' → S is a diffeomorphism.   Then*

$$\int_S f \leq \int (f \circ h) \text{Det}(Dh) \tag{7.6}$$

Before proving this inequality, we use it to prove the change of variable formula. Define $g:S' \to R$ by

$$g(\vec{w}) = f(h(\vec{w}))\,|\mathrm{Det}\,(D_{\vec{w}}h)|$$

and let $h^{-1}:S \to S'$ be the inverse of the diffeomorphism h. Then $g \geq 0$ if $f \geq 0$. Applying Proposition 7.4 with g replacing f and h^{-1} replacing h, we have

$$\int_{S'} g \leq \int_S (g \circ h^{-1})\,|\mathrm{Det}\,(Dh^{-1})| \tag{7.7}$$

Now

$$
\begin{aligned}
(g \circ h^{-1})(\vec{v})\,|\mathrm{Det}\,(D_{\vec{v}}h^{-1})| &= f(\vec{v})\,|\mathrm{Det}\,(D_{h^{-1}(\vec{v})}h)| \cdot |\mathrm{Det}\,(D_{\vec{v}}h^{-1})| \\
&= f(\vec{v})\,|\mathrm{Det}\,((D_{h^{-1}(\vec{v})}h)\,(D_{\vec{v}}h^{-1}))| \\
&= f(\vec{v})\,|\mathrm{Det}\,(D_{\vec{v}}(hh^{-1}))| = f(\vec{v})
\end{aligned}
$$

using the Chain Rule. Therefore, the inequality (7.7) becomes

$$\int_S (f \circ h)\,|\mathrm{Det}\,(Dh)| \leq \int_S f$$

from the definition of g. This inequality, together with (7.6), gives Theorem 6.2 when $f \geq 0$. From here, it is not hard (using Exercise 10 of Section 11.4) to get Theorem 6.2 in full generality. (See Exercise 5.)

We now prove Proposition 7.4. For convenience, we use slightly different norms on R^n and $\mathcal{L}(R^n, R^n)$ in the remainder of this section:

$$
\begin{aligned}
\|(x_1, \ldots, x_n)\| &= \max_i |x_i| \\
\|T\| &= \max_i \sum_{j=1}^{n} |a_{ij}|
\end{aligned}
$$

It follows from Exercise 4 of Section 5.6 that these norms are equivalent to the usual norms. Note also that $\|T\vec{v}\| \leq \|T\|\,\|\vec{v}\|$.

We shall use rectangles with equal sides, and shall call them cubes: if $\vec{a} = (a_1, \ldots, a_n)$ and $s > 0$,

$$C = C(\vec{a};s) = \{\vec{v} \in R^n : \|\vec{v} - \vec{a}\| \leq s\}$$

is a cube centered at $\vec{a}$ with side length $2s$. Thus $\mu(C) = (2s)^n$.

If $\vec{v} \in \vec{C}$, Proposition 9.3.1 says that there exist $\vec{c}_1, \ldots, \vec{c}_n$ on the line segment from $\vec{a}$ to $\vec{v}$ with

$$h_i(\vec{v}) - h_i(\vec{a}) = \sum_{j=1}^{n} \frac{\partial h_i}{\partial x_j}(c_i)(x_j - \vec{a}_j)$$

Thus

$$\|h(\vec{v}) - h(\vec{a})\| \leq s \max_{\vec{v} \in C} \|D_{\vec{v}} h\|$$

and

$$h(C) \subset C' = C(h(\vec{a}); s \max_{\vec{v} \in C} \|D_{\vec{v}} h\|)$$

In particular,

$$\mu(h(C)) \leq \mu(C) (\max_{\vec{v} \in C} \|D_{\vec{v}} h\|)^n \tag{7.8}$$

Now let $T: R^n \to R^n$ be a linear isomorphism. Then $D_{\vec{v}}(T^{-1}h) = T^{-1} D_{\vec{v}} h$ (by the Chain Rule), and, by the Corollary to Proposition 7.2,

$$\mu(T^{-1}h(C)) = |\text{Det } T|^{-1} \mu(h(C))$$

(recall that $\text{Det}(T^{-1}) = (\text{Det } T)^{-1}$). Replacing h in (7.8) by $T^{-1}h$, we get

$$\mu(h(C)) \leq |\text{Det } T| (\max_{\vec{v} \in C} \|T^{-1} D_{\vec{v}} h\|)^n \tag{7.9}$$

Next we subdivide C into non-overlapping cubes $C_1, \ldots, C_k$, with centers $\vec{v}_1, \ldots, \vec{v}_k$, apply (7.9) to each C_j (with $T = D_{\vec{v}_j} h$), and sum:

$$\mu(h(C)) \leq \sum_{j=1}^{k} |\text{Det}(D_{\vec{v}_j} h)| (\max_{\vec{v} \in C_j} \|(D_{\vec{v}_j} h)^{-1} (D_{\vec{v}} h)\|)^n \mu(C_j) \tag{7.10}$$

Choose $\varepsilon > 0$. Since $D_{\vec{v}} h$ is continuous, we can choose the cubes small enough that

$$(\max_{\vec{v} \in C_j} \|(D_{\vec{v}_j} h)^{-1} D_{\vec{v}} h\|)^n \leq 1 + \varepsilon$$

Thus (7.10) becomes

$$\mu(h(C)) \leq (1 + \varepsilon) \sum_{j=1}^{k} |\text{Det}(D_{\vec{v}_j} h)| \mu(C_j)$$

However, the sum on the right is a Riemann sum for the integral of the function $J:C \to R$,

$$J(\vec{v}) = |\text{Det}(D_{\vec{v}} h)|$$

Since ε can be taken to approach zero as the mesh of the partition $C_1, \ldots, C_k$ approaches zero, we have (using Proposition 7.3)

$$\mu(h(C)) \leq \int_C |\text{Det}(Dh)| \tag{7.11}$$

for any cube $C \subset S'$.

Suppose that f is non-negative and continuous and that S' is the union of cubes $C_1, \ldots, C_m$ with centers $\vec{v}_1, \ldots, \vec{v}_m$. Then, multiplying the inequality (7.11) by $f(h(\vec{v}_j))$ (with C_j replacing C) and adding, we obtain the inequality

$$\sum_{j=1}^{m} f(h(\vec{v}_j)) \mu(h(C_j)) \leq \sum_{j=1}^{m} \int_{C_j} f(h(\vec{v}_j)) |\text{Det}(Dh)| \tag{7.12}$$

It is not difficult (using Proposition 7.3) to show that, given any $\varepsilon > 0$, the cubes $C_1, \ldots, C_m$ can be chosen small enough so that

$$\left| \int_S f - \sum_{j=1}^{m} f(h(\vec{v}_j)) \mu(h(C_j)) \right| < \varepsilon$$

and

$$\left| \int_{S'} (f \circ h) |\text{Det}(Dh)| - \sum_{j=1}^{m} \int_{C_j} f(h(\vec{v}_j)) |\text{Det}(Dh)| \right| < \varepsilon$$

(See Exercise 6.) Since ε is arbitrary, the inequality (7.6) now follows from (7.12). If f is only continuous off a set of content 0, the proof is the same except that we must insure that the rectangles on which f is discontinuous have total volume less than $\varepsilon/\sup\{f(x):x \in S'\}$. (See Exercise 6.)

To complete the proof of Proposition 7.4, we must show that we can assume S' to be the union of cubes.

LEMMA 7.5. *Let* S, S' $\subset R^n$ *be compact and* h:S' $\to$ S *a* C^1-*diffeomorphism. Then there exists a compact set* S_1' $\subset R$ *and a* C^1-*diffeomorphism* h_1:S_1' $\to S_1$ *such that*

(a) $S' \subset S_1'$

(b) S_1' *is the union of a finite number of cubes*

(c) $S \subset S_1$

(d) $h_1(\vec{v}) = h(\vec{v})$ *for* $\vec{v} \in S'$

The proof of this lemma is outlined in Exercise 7.

If we now define f_1:$S_1 \to R$ by

$$f_1(\vec{v}) = \begin{cases} f(\vec{v}) & \text{for} \quad \vec{v} \in S \\ 0 & \text{for} \quad \vec{v} \notin S \end{cases}$$

(this is why we need to work with discontinuous functions), then

$$\int_S f = \int_{S_1} f_1$$

$$\leq \int_{S_1'} (f_1 \circ h_1) |Det(Dh_1)|$$

(by the argument above)

$$= \int_{S_1'} (f \circ h) |Det(Dh)|$$

and the proposition follows.

EXERCISES

1.   State and prove the analogue of Lemma 7.1 for the composite of more than two diffeomorphisms.

2.   Prove the corollary to Proposition 7.2.

3.   (a) Prove that $\mu(S_1) \leq \mu(S_2)$ if $S_1 \subset S_2$.

      (b) Is it true that $S_1 \subset S_2$ and $\mu(S_1) = \mu(S_2)$ implies $S_1 = S_2$?

      (c) Suppose that $S \subset R^n$ has nonempty interior and that $\mu(S)$ is
defined. Prove $\mu(S) > 0$.

*4.   Prove Proposition 7.3.

5.   Complete the derivation of Theorem 6.2 from Proposition 7.4.

*6.   Give a detailed proof that the inequality (7.6) follows from
the inequality (7.12).

*7.   Supply the details for the following outline of a proof of
Lemma 7.5.

      Since $h:S' \to S$ is a C^1-function on a closed set, there is, by
definition, an open set U containing S' and a C^1-function $g:U \to R^n$
such that $g(\vec{v}) = h(\vec{v})$ for $\vec{v} \in S'$.

      (a) Prove that there is an open set U' in R^n with $S' \subset U' \subset U$
and such that $D_{\vec{v}}g$ is an isomorphism for all $\vec{v} \in U'$. (Hint: consider
the function $\vec{v} \to \mathrm{Det}(D_{\vec{v}}g)$ on U.)

      (b) Define open sets U(r) containing S' by

$$U(r) = \{\vec{v} \in R^n : \exists \vec{w} \in S' \text{ with } \|\vec{v} - \vec{w}\| < r\}$$

Prove that there is an $r_1 > 0$ such that $U(r_1) \subset U'$. (Hint: Show that
$\inf\{\|\vec{v} - \vec{w}\| : \vec{v} \in S', \vec{w} \notin U'\} > 0$.)

      (c) Prove that there is an r_0, $0 < r_0 < r_1$, such that the func-
tion g is injective on $U(r_0)$. (Hint: Suppose not and use the fact
that S' is compact and the Inverse Function Theorem (Theorem 9.1.4).)

      (d) Let C be defined by

$$C = \{\vec{v} \in R^n : \exists \vec{w} \in S' \text{ with } \|\vec{v} - \vec{w}\| \leq r_0/2\}$$

Prove that C is compact. (Hint: Use Theorem 5.6.1.)

(e)   Prove that, for each $\vec{v} \in C$, there is an open cube centered at $\vec{v}$ contained in $U(r_0)$.

(f)   For each $\vec{v} \in C$, let $U(\vec{v})$ be an open cube centered at $\vec{v}$ contained in $U(r_0)$.   This defines an open covering of the compact set C, so it has a finite subcovering.   (See Exercise 9-11 of Section 5.4.) That is, there are points $\vec{v}_1, \ldots, \vec{v}_m \in C$ such that

$$C \subset \bigcup_{i=1}^{m} U(\vec{v}_j)$$

Let C_j be the closure of $U(\vec{v}_j)$; C_j is a closed cube centered at $\vec{v}_j$. Define S_1' by

$$S_1' = \bigcup_{j=1}^{m} C_j$$

and prove that S_1' is compact.   (Use Theorem 5.6.1 here.)

(g)   Let h_1 be the restriction of g to S_1'.   Prove that h_1 is a diffeomorphism of S_1' onto $S_1 = h_1(S_1')$.

CHAPTER 13

LINE INTEGRALS

INTRODUCTION

In this chapter, we discuss line integrals of functions and of vector
fields along curves. As we shall see in this chapter and the next,
these ideas are important in the physical sciences. The key result
is Green's Theorem, which relates the line integral of a vector field
around a closed curve in R^2 to the ordinary integral of a related
function over the region enclosed by the curve.

1. CURVES

In order to discuss line integrals, we need the notion of a curve in
R^n. We begin with some definitions.

A *parametrized curve* in R^n is a continuously differentiable function $\alpha:[a,b] \to R^n$. (Recall that this means that there is a continuously differentiable function $\tilde{\alpha}$, defined on an open interval containing $[a,b]$, whose restriction to $[a,b]$ is α. As a result, we can differentiate α even at the ends of the interval.) For example, if $\vec{u}$
and $\vec{v}$ are distinct points in R^n, then the parametrized curve $\alpha:[0,1]$
$\to R^n$ given by

$$\alpha(t) = (1 - t)\vec{u} + t\vec{v}$$

defines the *line segment from* $\vec{u}$ *to* $\vec{v}$. (See Figure 1.1.) In fact,
if we let t vary over all of R, this same equation defines the *line
determined by* $\vec{u}$ *and* $\vec{v}$.

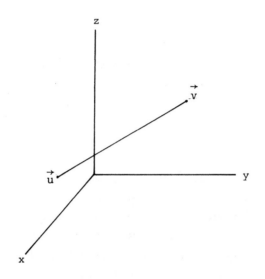

FIGURE 1.1. The line segment from $\vec{u}$ to $\vec{v}$

Another example of a parametrized curve is

$$\alpha(t) = (\cos t, \sin t)$$

$t \in [0,2\pi]$. The image set in this case is the *unit circle* in R^2 with
center the origin. (See Figure 1.2.)

Note that two different parametrized curves can have the same
image set. For instance, the function $\beta:[1,2] \to R^2$ defined by

$$\beta(s) = (\cos 2\pi s, \sin 2\pi s)$$

is also a parametrization for the unit circle with center at the origin. For many purposes, it is useful to regard these curves as the
same. We therefore make the following definition.

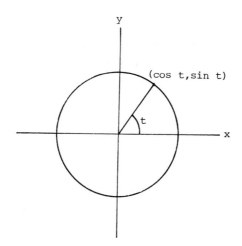

FIGURE 1.2. The unit circle

Let $\alpha:[a,b] \to R^n$, $\beta:[c,d] \to R^n$ be two parametrized curves. We
say that α is *equivalent* to β if there is a diffeomorphism $h:[c,d]$
$\to [a,b]$ such that

$$\alpha(h(s)) = \beta(s)$$

for all s in $[c,d]$.

Since we have used the word "equivalent" in this definition, we
should verify that this relation is an equivalence relation.

PROPOSITION 1.1. *Let* $\alpha:[a,b] \to R^n$, $\beta:[c,d] \to R^n$, *and* $\gamma:[e,f]$
$\to R^n$ *be the parametrized curves. Then:*

(a) α *is equivalent to* α

(b) *if* α *is equivalent to* β, *then* β *is equivalent to* α

(c) *if* α *is equivalent to* β *and* β *is equivalent to* γ, *then* α
is equivalent to γ

Thus equivalence is an equivalence relation.

We leave the proof as Exercise 5.

We now wish to define a *curve* as a parametrized curve or any other parametrized curve equivalent to it. As a matter of mathematical convenience and custom, we phrase the definition differently: *a curve is an equivalence class of parametrized curves with respect to the relation of equivalence given above.* (See Section 1.5.) Since any two equivalent parametrized curves clearly have the same image set (Exercise 6), it makes sense to speak of the image set of a curve. We denote the image set of the curve Γ by Im Γ.

Remarks. 1. The two parametrizations of the unit circle given above are equivalent with, $h:[1,2] \to [0,2\pi]$ given by $h(s) = 2\pi(s - 1)$.

2. Two different curves can have the same image set. For example, the parametrized curve $\gamma:[0,2\pi] \to R^2$ given by $\gamma(t) = (\sin t, \cos t)$ also has the unit circle as image set. However, γ is not equivalent to α (or β), as given above. One way to see this is to notice that the "end points" of α, $\alpha(0)$ and $\alpha(2\pi)$, are at $(1,0)$, while those of γ are at $(0,1)$. It is not hard to show that two equivalent parametrized curves have the same end points.

3. It is possible for the image set of a parametrized curve to have "corners." For example, if $\alpha:[-1,1] \to R^2$ is defined by

$$\alpha(t) = (t^2, t^3)$$

then the image set has a corner at $(0,0)$. (This point is called a *cusp*; see Figure 1.3.)

A parametrized curve $\alpha:[a,b] \to R^n$ is said to be *regular* if $\alpha'(t) \neq 0$ for all t in $[a,b]$. Since a regular curve has a continuous nowhere vanishing tangent vector, it cannot have corners. In what follows, we shall deal only with regular curves.

EXERCISES

1. Find a parametrized curve in R^2 whose image set is:
   (a) the circle of radius 2 with center at the origin
   (b) the line segment from $(2,3)$ to $(5,8)$

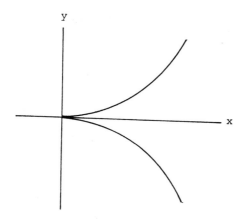

FIGURE 1.3. A cusp

(c) the piece of the parabola $y = \pm\sqrt{x}$, $0 \le x \le 1$

(d) the unit circle with center $(1,0)$

(e) the ellipse $x^2/4 + y^2/9 = 1$

(f) the graph of $y = x^3 - 7x^2 + 3x - 2$, $0 \le x \le 3$

2. Prove that the following pairs of parametrized curves are equivalent:

(a) $\alpha(s) = (s, s + 1)$, $1 \le s \le 8$; $\beta(t) = (t^3, t^3 + 1)$, $1 \le t \le 2$

(b) $\alpha(s) = (s, s^2)$, $2 \le s \le 3$; $\beta(t) = (\sqrt{t}, t)$, $4 \le t \le 9$

(c) $\alpha(s) = (\sin s, \cos s)$, $0 \le s \le 2\pi$; $\beta(t) = (\cos t, \sin t)$, $\pi/2 \le t \le 5\pi/2$

(d) $\alpha(s) = (2 \cos s, 2 \sin s)$, $0 \le s \le \pi/2$; $\beta(t) = ((2 - 2t^2)/(1 + t^2), 4t/(1 + t^2))$, $0 \le t \le 1$

3. Find a parametrized curve in R^3 whose image set is:

(a) the line segment from $(1,1,4)$ to $(2,3,2)$

(b) the curve $y = x^2$ and $z = x^3$, $0 \le x \le 1$

(c) the intersection of the planes $x + y + z = 1$ and $z = x - 2y$

(d)   the circle with radius 1 and center at the origin lying in the plane $z = 2x - 2y$

4.   Prove the following pairs of parametrized curves equivalent:

(a) $\alpha(s) = (s+2, s+3, 2s-4)$, $-1 \leq s \leq 2$; $\beta(t) = (t^2, t^2 + 1, 2t^2 - 8)$, $1 \leq t \leq 2$

(b) $\alpha(s) = (e^s, e^{2s}, e^{3s})$, $0 \leq s \leq 1$; $\beta(t) = (t+1, (t+1)^2, (t+1)^3)$, $0 \leq t \leq e - 1$

(c) $\alpha(s) = (\sin s, \cos s, s)$, $0 \leq s \leq 7\pi/2$; $\beta(t) = (-\cos t, \sin t, t-\pi/2)$, $\pi/2 \leq t \leq 4\pi$

5.   Prove Proposition 1.1.

6.   Prove that equivalent curves have the same image set.

2.   LINE INTEGRALS OF A FUNCTION

In this section, we show how to integrate a function along a curve. Besides being important in their own right, the notions we develop in this section will help when we deal with more general sorts of line integrals in the next section.

Suppose that U is an open subset of R^n and Γ is the curve defined by the parametrized curve $\alpha:[a,b] \to U$. If $f:U \to R$ is a continuous function, we would like to integrate f along Γ. One way of doing this would be to pick a rectangle S containing Im α and define $F:S \to R$ by

$$F(\vec{v}) = \begin{cases} f(\vec{v}) & \text{if } \vec{v} \in \text{Im } \alpha \\ 0 & \text{if } \vec{v} \notin \text{Im } \alpha \end{cases}$$

and integrate F over S as described in Chapter 11. However, if $n > 1$, Γ has content zero in R^n. (See Exercise 7.) It follows easily that this integral is zero in these cases.

In order to motivate the proper definition of the integral of f
along the curve Γ, we consider the problem of determining the mass of
an inhomogeneous piece of wire. We imagine Γ to be a piece of wire
and f as giving the density (mass per unit length) of the wire. We
can get an approximation for the total mass of wire by choosing suc-
cessive points $\vec{v}_0, \vec{v}_1, \ldots, \vec{v}_m$ on Γ and constructing the polygonal curve
Γ' with these vertices. (See Figure 2.1.) If we assign the density

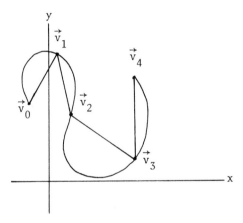

FIGURE 2.1. A polygonal curve approximating

$f(\vec{v}_i)$ to the segment of Γ' from $\vec{v}_{i-1}$ to $\vec{v}_i$, then the total mass of
the wire represented by Γ' is

$$\sum_{i=1}^{m} f(\vec{v}_i) \|\vec{v}_i - \vec{v}_{i-1}\| \qquad (2.1)$$

If the points $\vec{v}_0, \ldots, \vec{v}_m$ are close together, then the expression (2.1)
ought to be close to the mass of the wire represented by the curve Γ.

We can rewrite (2.1) in terms of the parametrization for Γ as
follows. Let $a = t_0 < \cdots < t_m = b$ be a partition of $[a,b]$ with the
property that $\alpha(t_i) = \vec{v}_i$. Assume that the wire is in R^3, so that we
can write

$$\alpha(t) = (\alpha_1(t), \ \alpha_2(t), \ \alpha_3(t))$$

Then we can rewrite (2.1) as

$$\sum_{i=1}^{m} f(\alpha(t_i)) \left(\sum_{j=1}^{3} (\alpha_j(t_i) - \alpha_j(t_{i-1}))^2 \right)^{1/2} \tag{2.2}$$

According to the Mean Value Theorem, we can find points c_i, c_i', c_i'' in $[t_{i-1}, t_i]$ such that

$$\alpha_1(t_i) - \alpha_1(t_{i-1}) = \alpha_1'(c_i)(t_i - t_{i-1})$$

$$\alpha_2(t_i) - \alpha_2(t_{i-1}) = \alpha_2'(c_i')(t_i - t_{i-1})$$

$$\alpha_3(t_i) - \alpha_3(t_{i-1}) = \alpha_3'(c_i'')(t_i - t_{i-1})$$

Thus, (2.2) becomes

$$\sum_{i=1}^{m} f(\alpha(t_i)) \left((\alpha_1'(c_i))^2 + (\alpha_2'(c_i'))^2 + (\alpha_3'(c_i''))^2 \right)^{1/2} (t_i - t_{i-1})$$

This is approximately equal to

$$\sum_{i=1}^{m} f(\alpha(t_i)) \|\alpha'(t_i)\| (t_i - t_{i-1}) \tag{2.3}$$

since t_i, c_i, c_i', and c_i'' all lie in $[t_{i-1}, t_i]$

As in the partition of $[a,b]$ gets finer, we expect that the expression (2.3) should "converge" to the total mass of the wire. On the other hand, the expression (2.3) is a Riemann Sum for the function $f(\alpha(t))\|\alpha'(t)\|$, and so converges to the integral

$$\int_a^b f(\alpha(t)) \|\alpha'(t)\| dt \tag{2.4}$$

by Proposition 12.7.3. (Here, $\alpha'(t) = (\alpha_1'(t), \ \alpha_2'(t), \ \alpha_3'(t))$.) Thus the total mass of the wire is given by the integral (2.4).

This discussion may explain the definition which follows. If Γ is a curve in the open subset U of R^n, if $\alpha:[a,b] \to U$ is a parametrization of Γ, and if $f:U \to R$ is a continuous function, we define the *line integral* of f along Γ by

$$\int_\Gamma f = \int_a^b f(\alpha(t)) \|\alpha'(t)\| dt$$

(In this case, $\alpha'(t) = (\alpha_1'(t), \ldots, \alpha_n'(t))$.)

For example, let Γ be the piece of the *circular helix* in R^3 (see Figure 2.2) defined by

$$\alpha(t) = (\cos t, \sin t, t) \quad 0 \le t \le \pi/2$$

and $f: R^3 \to R$ the function

$$f(x,y,z) = x^2 + y^2 + z^2$$

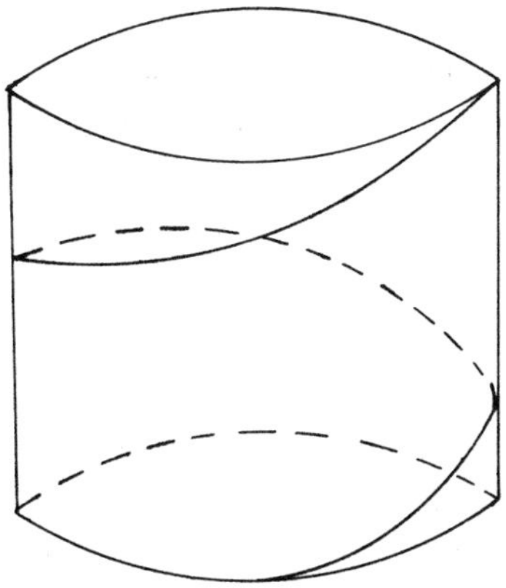

FIGURE 2.2.   A circular helix

Then

$$\int_\Gamma f = \int_0^{\pi/2} f(\alpha(t)) \|\alpha'(t)\| dt$$

$$= \int_0^{\pi/2} (\cos^2 t + \sin^2 t + t^2)(\sin^2 t + \cos^2 t + 1)^{1/2} dt$$

$$= \int_0^{\pi/2} (1 + t^2)\sqrt{2}\, dt = \sqrt{2}(\pi/2 + \pi^3/24)$$

As we have defined it, the line integral seems to depend upon the parametrization chosen. The next result shows that this is not the case.

PROPOSITION 2.1. *Let* $\alpha:[a,b] \to U$, $\beta:[c,d] \to U$ *be equivalent parametrized curves, and let* $f:U \to R$ *a continuous function.* *Then*

$$\int_a^b f(\alpha(t))\|\alpha'(t)\|dt = \int_c^d f(\beta(t))\|\beta'(t)\|dt$$

Proof. The idea behind the proof is the change-of-variables formula, or the method of integration by substitution. Let $h:[c,d] \to [a,b]$ be the diffeomorphism with $\alpha(h(s)) = \beta(s)$ for all s in $[c,d]$. Then

$$\int_c^d f(\beta(s))\|\beta'(s)\|ds = \int_c^d f(\alpha(h(s))\|\alpha'(h(s))h'(s)\|ds \qquad (2.5)$$

since $\beta'(s) = \alpha'(h(s))h'(s)$ (by the Chain Rule). We know that $h'(s)$ is never 0 (since h is a diffeomorphism); therefore, $h'(s)$ is either always > 0 or always < 0. If $h'(s) > 0$, then $h(c) = a$ and $h(d) = b$, and the second integral in (2.5) is

$$\int_c^d f(\alpha(h(s)))\|\alpha'(h(s))\|h'(s)ds \qquad (2.6)$$

If $F(t)$ is an antiderivative for the function $f(\alpha(t))\|\alpha'(t)\|$, then $F(h(s))$ is an antiderivative for $f(\alpha(h(s)))\|\alpha'(h(s))\|h'(s)$. (To see this, use the Chain Rule.) It follows from the Fundamental Theorem of Calculus (or see Section 12.6) that the integral in (2.6) is equal to

$$\int_{h(c)}^{h(d)} f(\alpha(t))\|\alpha'(t)\|dt = \int_a^b f(\alpha(t))\|\alpha'(t)\|dt$$

If $h(s) < 0$, then $h(c) = b$ and $h(d) = a$ and a similar argument shows that

$$\int_c^d f(\beta(s))\|\beta'(s)\|ds = - \int_{h(c)}^{h(d)} f(\alpha(t))\|\alpha'(t)\|dt$$

$$= \int_a^b f(\alpha(t))\|\alpha'(t)\|dt$$

Thus Proposition 2.1 is proved.

Remarks. 1. If f is the constant function 1, the line integral of f along Γ is simply the arc length of Γ.

2. We note that, as a consequence of Proposition 2.1, the line
integral of a function f along a curve Γ does not depend on which
direction we traverse Γ; we can start at either end. This is *not* true
of the line integral discussed in the next section.

3. Finally we observe that the requirement that the function
α be continuously differentiable can be relaxed. Suppose $\alpha:[a,b] \to R^n$
is a function and that there are points $a = a_0 < a_1 < \cdots < a_m = b$ such
that the restriction of α to each of the intervals $[a_{i-1},a_i]$ is continuously differentiable, $i = 1,\ldots,m$. In this case, α is called a
piecewise continuously differentiable parametrized curve. We can define the line integral of a function F along the "curve" Γ represented by α to be the sum of the line integrals of F along the curves
Γ_i defined by the restriction of α to the intervals $[a_{i-1},a_i]$, $i = 1$
,$\ldots$,m.

We close this section with a list of some of the properties of
the line integral.

PROPOSITION 2.2. *Let U be an open subset of* R^n, $\alpha:[a,b] \to U$ *a*
parametrization for the curve Γ, *f,g:U* $\to$ R *continuous functions and*
c,d real numbers. Then

(a) $\int_\Gamma (cf + dg) = c\int_\Gamma f + d\int_\Gamma g$

(b) *If* $f \geq g$ *on U, then*

$\int_\Gamma f \geq \int_\Gamma g$

(c) *Let c be a real number,* $a < c < b$. *Let* $\alpha_1:[a,c] \to U$,
$\alpha_2:[c,b] \to U$ *be the restrictions of* α *and* Γ_1,Γ_2 *the corresponding*
curves. Then

$\int_\Gamma f = \int_{\Gamma_1} f + \int_{\Gamma_2} f$

These results follow immediately from the corresponding properties of the standard integral. For example,

$$\int_\Gamma (cf + dg) = \int_a^b (cf + dg)(\alpha(t)) \|\alpha'(t)\| dt$$

$$= \int_a^b (cf(\alpha(t)) + dg(\alpha(t))) \|\alpha'(t)\| dt$$

$$= \int_a^b f(\alpha(t)) \|\alpha'(t)\| dt + d\int_a^b g(\alpha(t)) \|\alpha'(t)\| dt$$

$$= c\int_\Gamma f + d\int_\Gamma g$$

The remainder of the proof is left as Exercise 6.

EXERCISES

1.   Compute the line integral of $f: R^2 \to R$ over the curve Γ with parametrization $\alpha: [a,b] \to R^2$ where

(a) $f(x,y) = 2xy$ and $\alpha(t) = (t, t + 1)$, $a = 1$, $b = 8$

(b) $f(x,y) = 2xy$ and $\alpha(t) = (-t^3, 1-t^3)$, $a = -2$, $b = -1$

(c) $f(x,y) = x(1 + 4y)$ and $\alpha(t) = (t, t^2)$, $a = 2$, $b = 3$

(d) $f(x,y) = x(1+4y)$ and $\alpha(t) = (\sqrt{t}, t)$, $a = 4$, $b = 9$

(e) $f(x,y) = 3x + y$ and $\alpha(t) = (t - (1/3)t^3, t^2 - 1)$, $a = 0$, $b = 3$

(f) $f(x,y) = 2x + y$ and $\alpha(t) = (\cos t, \sin t)$, $a = 0$, $b = \pi$

2.   Compute the line integral of $f: R^3 \to R$ over the curve Γ with parametrization $\alpha: [a,b] \to R^3$ where

(a) $f(x,y,z) = xyz$ and $\alpha(t) = (t + 2, t + 3, 2t - 4)$, $a = -1$, $b = 2$

(b) $f(x,y,z) = xyz$ and $\alpha(t) = (t^2, t^2 + 1, 2t^2 - 8)$, $a = 1$, $b = 2$

(c) $f(x,y,z) = x^2 y + y^2 z$ and $\alpha(t) = (e^t, e^{2t}, \frac{2e^{3t}}{3})$, $a = 0$, $b = 1$

(d) $f(x,y,z) = x^2 y + y^2 z$ and $\alpha(t) = (t + 1, (t + 1)^2, \frac{2}{3}(t + 1)^3)$, $a = 0$, $b = e - 1$

(e) $f(x,y,z) = x$ and $\alpha(t) = (t^2 - t, 2 - t^2, (4/3)t^3 - t^2)$, $a = 0$, $b = 1$

(f) $f(x,y,z) = 2x - y + z - 1$ and $\alpha(t) = (\cos t, \sin t, t)$, $a = 0$, $b = \pi$

3.   Find a parametrization for the curve Γ and evaluate the line integral of the function $f:R^2 \to R$ over Γ, where $f(x,y) = y$ and

(a)   Im Γ is the line segment from $(0,1)$ to $(2,3)$

(b)   Im Γ is the piece of the parabola $y = x^2$ between $(0,0)$ and $(2,4)$

(c)   Im Γ is the piece of the circle $x^2 + y^2 = 4$ between $(2,0)$ and $(- \sqrt{2}, \sqrt{2})$

(d)   Im Γ is the polygon whose successive vertices are $(1,0)$, $(1,1)$, and $(2,2)$

4.   Find the total weight of a piece of wire 7 inches long if

(a)   the density is proportional to the distance from the center of the wire

(b)   the density is proportional to the square of the distance from one end of the wire

5.   Find a parametrization for the curve Γ and evaluate the line integral of the function $f:R^3 \to R$ over Γ, where $f(x,y,z) = xyz$ and

(a)   Im Γ is the line segment from $(0,1,1)$ to $(1,1,0)$

(b)   Im Γ is the polygon whose successive vertices are $(0,0,1)$, $(0,1,1)$, and $(1,2,3)$

(c)   Im Γ is the unit circle in the plane $z = 1$ with center $(0,0,1)$

(d)   Im Γ is the piece of the unit circle in the plane $z = 0$ between the points $(1,0,0)$ and $(0,1,0)$ together with the line segment from $(0,1,0)$ to $(1,1,1)$

6.   Prove statements (b) and (c) of Proposition 2.2.

*7.   Let $\alpha:[a,b] \to R^n$ be a regular parametrized curve, $n > 1$.   Prove that $Im(\alpha)$ has content zero.

*8.   Let $\alpha:[a,b] \to R^n$ be a parametrized curve.   We say that α is *parametrized by arc length* if, for each $t_0 \in [a,b]$, the length of the piece of α between $t = a$ and $t = t_0$ is exactly $t_0 - a$.

(a) Prove that α is parametrized by arc length if and only if $\|\alpha'(t)\| = 1$.

(b) Prove that if α is parametrized by arc length, then $< \alpha'(t), \alpha''(t) > = 0$ for all $t \in [a,b]$. (Here, $\alpha''(t) = (\alpha_1''(t), \ldots, \alpha_n''(t))$.)

(c) Let α be a regular curve. Prove that there is an equivalent curve β which is parametrized by arc length.

3. LINE INTEGRALS OF VECTOR FIELDS

We now define another line integral, namely the integral of a vector field along a curve.

Let U be an open subset of R^n. A *vector field* [*] on U is a continuously differentiable function $\vec{F}: U \to R^n$. We think of a vector field as assigning to each point of U a vector based at the point. (See Figure 3.1). In physics, for instance, a vector field could represent the force experienced by a particle in U. (In this case, a vector field is sometimes called a *force field*.)

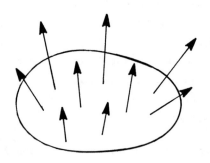

FIGURE 3.1. A vector field on U

For example, if we place a positive charge at the origin in R^2, a negative charge at the point $\vec{v} \neq 0$ in R^2 will feel a force directed

---

[*] In this and future chapters, we shall write vector fields with an arrow.

towards the origin with magnitude

$$\frac{q}{\|\vec{v}\|^2}$$

where q is a positive constant. The corresponding force field is given by

$$\vec{F}:U \rightarrow R^2$$

where U is the complement of the origin in R^2 and

$$\vec{F}(x,y) = (\frac{-qx}{(x^2+y^2)^{3/2}}, \frac{-qy}{(x^2+y^2)^{3/2}})$$

(See Figure 3.2). (We exclude the origin because $q/\|\vec{v}\|^2$ is not defined when $\vec{v} = \vec{0}$.)

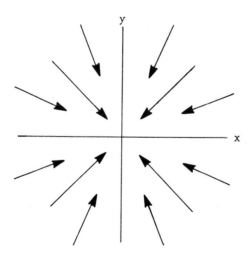

FIGURE 3.2. The force field from a charged particle

Suppose now that $\vec{F}$ is a vector field on the open set $U \subset R^2$ and $\alpha:[a,b] \rightarrow U$ is a parametrized curve in U. We think of F as a force field and ask for the work done by the field in moving a particle

from $\alpha(a)$ to $\alpha(b)$ along Γ. If the force is constant and the curve
is a straight line, then the work is the product of the distance tra-
veled and the force along the line:

$$W = <\vec{F},\alpha(b) - \alpha(a)>$$

In general, the curve is not a line and the force varies. In this
case, we follow a procedure like that used in the last section for
finding the total weight of a piece of wire. We first choose successive points along Γ,

$$\alpha(a) = \vec{v}_0,\vec{v}_1,\ldots,\vec{v}_m = \alpha(b)$$

and construct the polygon Γ' approximating Γ. The work done in moving
a particle along Γ' is approximately

$$\sum_{i=1}^{m} <\vec{F}(\vec{v}_i),\vec{v}_i - \vec{v}_{i-1}> \tag{3.2}$$

(Here we are approximating the force along the segment of Γ' from $\vec{v}_{i-1}$
to $\vec{v}_i$ by $\vec{F}(\vec{v}_i)$.)

We can rewrite the expression (3.2) by choosing a partition for
the interval $[a,b]$,

$$a = t_0 < t_1 <\cdots< t_m = b$$

with the property that $\alpha(t_i) = \vec{v}_i$, $0 \le i \le m$. Then (3.2) becomes

$$\sum_{i=1}^{m} <\vec{F}(\alpha(t_i)),\alpha(t_i) - \alpha(t_{i-1})> \tag{3.3}$$

We can now apply the Mean Value Theorem just as in Section 2 to show
that the expression (3.3) is approximately equal to

$$\sum_{i=1}^{m} <\vec{F}(\alpha(t_i)),\alpha'(t_i)> (t_i - t_{i-1}) \tag{3.4}$$

As the partition gets finer, we expect (3.4) to converge to the work
done in moving the particle from $\alpha(a)$ to $\alpha(b)$ along Γ. On the other
hand, (3.4) is a Riemann sum for the function $<\vec{F}(\alpha(t)),\alpha'(t)>$, so,
as the partition gets finer, (3.4) converges to the integral

$$\int_a^b < \vec{F}(\alpha(t)), \alpha'(t) > dt \qquad\qquad (3.5)$$

by Proposition 12.7.3. Therefore this integral gives the work done
in moving a particle along Γ in the force field F.[*]

For example, let $\vec{F}:U \to R^2$ be the force field of (3.1) arising
from the positive charge at the origin and $\alpha:[0,\pi/2] \to R^2$ the piece
of the unit circle given by

$$\alpha(t) = (\cos t, \sin t)$$

The work done in moving a negative charge along this curve from $(1,0)$
to $(0,1)$ is zero, since

$$< \vec{F}(\alpha(t)), \alpha'(t) > = < (-q \cos t, -q \sin t), (-\sin t, \cos t) >$$

$$= 0$$

and therefore

$$\int_a^b < \vec{F}(\alpha(t)), \alpha'(t)) > dt = 0$$

Suppose we try another path from $(1,0)$ to $(0,1)$. Let $\beta:[0,1] \to U$
describe the straight line segment from $(1,0)$ to $(0,1)$;

$$\beta(t) = (1 - t, t)$$

Then $\beta'(t) = (-1,1)$ and the work done in moving the particle along
this curve is given by

$$\int_0^1 < \vec{F}(\beta(t)), \beta'(t) > dt$$

$$= q\int_0^1 < (\frac{-(1-t)}{((1-t)^2 + t^2)^{3/2}}, \frac{-t}{((1-t)^2 + t^2)}, (-1,1) > dt$$

_____

[*] Often one wants to compute the work required to move the particle
along Γ in the force field $\vec{F}$. This is $\int_a^b < -\vec{F}(\alpha(t)), \alpha'(t) > dt$, since
the force that must be applied to the particle must counteract the
force field (and must therefore be opposite in direction to $\vec{F}$).

$$= q \int_0^1 \frac{1-2t}{(1-2t+2t^2)^{3/2}} \, dt$$

$$= q(1 - 2t + 2t^2)^{-1/2} \Big|_0^1$$

$$= 0$$

Therefore the work done in moving a particle from $(1,0)$ to $(0,1)$ in this force field is the same for these two paths. We shall see in the next section that this is no accident.

We use (3.5) to define the integral of a vector field $\vec{F}$ along a parametrized curve. Let $\alpha:[a,b] \to R^n$ be the curve, let $\vec{F}$ be defined on the open set $U \subset R^n$ with $\alpha([a,b]) \subset U$. The *line integral of* $\vec{F}$ along Γ is defined to be

$$\int_a^b < \vec{F}(\alpha(t)), \alpha'(t) > dt \tag{3.6}$$

As the above examples show, the actual computation of (3.6) boils down to computing an ordinary integral.

We shall give various properties of line integrals in the next section. For the moment, we give only one result concerning the line integrals for different parametrizations of a curve. It is *not* necessarily true that all equivalent parametrizations give the same value for the line integral. In fact, it is not hard to see that the work required to move a particle along a path depends on the direction in which we traverse the path. By an argument like that used to prove Proposition 2.1, we can show that the integral (3.5) changes sign if we move along Γ in the opposite direction. This fact may help to explain the following definition.

The parametrized curve $\alpha:[a,b] \to R^n$ is 0-*equivalent* (or *oriented equivalent*) to the parametrized curve $\alpha:[c,d] \to R^n$ if there is a diffeomorphism $h:[c,d] \to [a,b]$ with $h'(s) > 0$ and $\alpha(h(s)) = \beta(s)$ for all s in $[c,d]$. Roughly speaking, α is 0-equivalent to β if α is equivalent to β *and both* α *and* β *are traversed in the same direction* (as t goes from a to b and s from c to d). The relation of being 0-equi-

valent is easily seen to be an equivalence relation (see Exercise 8);
as in Section 1, an *oriented curve* is defined to be an equivalence
class of parametrized curves under this relation.

For instance, the parametrized curves $\alpha, \beta : [0,2] \rightarrow R^2$ given by

$$\alpha(t) = (\cos t, \sin t)$$
$$\beta(t) = (\cos t, -\sin t)$$

both define the unit circle in R^2 with center the origin and are equivalent with $\alpha(h(t)) = \beta(t)$, $h(t) = 2\pi - t$, $h'(t) = -1$. However, α
is *not* 0-equivalent to β; in fact, α corresponds to traversing the
unit circle in a counter clockwise direction and β corresponds to traversing the unit circle in a clockwise direction. Thus α and β define
distinct oriented curves.

We say that $\alpha : [a,b] \rightarrow R^n$ is a *parametrization for the oriented*
curve Γ *or a parametrization for* Γ *as an oriented curve* if α is in
the equivalence class of parametrized curves defining the oriented
curve Γ. For example, if Γ is the oriented curve defined by $\alpha(t) =$
$(\cos t, \sin t)$, $t \in [0,2\pi]$, then $\beta(t) = (\cos t, -\sin t)$, $t \in [0,2\pi]$,
is *not* a parametrization of Γ as an oriented curve.

Suppose now that Γ is an oriented curve in the open subset $U \subset R^n$
and $\vec{F} : U \rightarrow R^n$ is a vector field on U. The *line integral of* $\vec{F}$ *along* Γ
is defined by

$$\int_{\Gamma} F = \int_a^b < \vec{F}(\alpha(t)), \alpha'(t) > dt$$

where $\alpha : [a,b] \rightarrow U$ is a parametrization for Γ as an oriented curve.
The next result shows that this integral does not depend on the parametrization chosen for Γ.

PROPOSITION 3.1. *Let* $\alpha : [a,b] \rightarrow U$, $\beta : [c,d] \rightarrow U$ *be two parametri-*
zations for the oriented curve Γ *and* $\vec{F} : U \rightarrow R^n$ *a vector field on* U.
Then

$$\int_a^b < \vec{F}(\alpha(t)), \alpha'(t) > dt = \int_c^d < \vec{F}(\beta(s)), \beta'(s) > ds$$

Here $\alpha'(t) = (\alpha_1'(t), \ldots, \alpha_n'(t))$ where $\alpha(t) = (\alpha_1(t), \ldots, \alpha_n(t))$.

Proof. This proof is analogous to that of Proposition 2.1; we give only an outline here.

Suppose $\alpha(t) = \beta(h(t))$ where $h:[a,b] \to [c,d]$ is a diffeomorphism with $h'(t) > 0$, $t \in [a,b]$. Then

$$\int_a^b < \vec{F}(\alpha(t)), \alpha'(t) > dt = \int_a^b < \vec{F}(\beta(h(t))), \beta'(h(t))h'(t) > dt$$

$$= \int_a^b < \vec{F}(\beta h(t))), \beta'(h(t)) > h'(t)dt$$

$$= \int_c^d < \vec{F}(\beta(s)), \beta'(s) > ds$$

just as in the proof of Proposition 2.1. (In the last step, we let $s = h(t)$. Note that $h(a) = c$ and $h(b) = d$ because $h' > 0$.)

There is another notation for the line integral of a vector field. Let $\alpha(t) = (\alpha_1(t), \ldots, \alpha_n(t))$ be a parametrization for the oriented curve Γ in U and $\vec{F}:U \to R^n$ the vector field given by $\vec{F}(v) = (f_1(\vec{v}), \ldots, f_n(\vec{v}))$. Then the line integral of $\vec{F}$ along Γ is given by

$$\int_a^b \sum_{i=1}^n f_i(\alpha(t))\alpha_i'(t)dt = \sum_{i=1}^n \int_a^b f_i(\alpha(t))\alpha_i'(t)dt$$

If we think of x_i as $\alpha_i(t)$, then we can set $dx_i = \alpha_i'(t)dt$ and write

$$\int_\Gamma \vec{F} = \int_\Gamma f_1 dx_1 + \cdots + f_n dx_n$$

This notation will make more sense when we deal with the integration of differential forms in Chapter 16.

EXERCISES

1. Evaluate the line integral of the vector field $\vec{F}:R^2 \to R^2$, where $\vec{F}(x,y) = (-y,x)$, along each of the curves of Exercise 1 of the previous section.

2. Find a parametrization for the curve Γ and determine the work done on a particle moving along Γ in R^3 through the force field $\vec{F}:R^3 \to R^3$, where $\vec{F}(x,y,z) = (1,-x,z)$ and

(a) Im Γ is the line segment from $(0,0,0)$ to $(1,2,1)$

(b) Im Γ is the polygonal curve with successive vertices $(1,0,0)$, $(0,1,1)$, and $(2,2,2)$

(c) Im Γ is the unit circle in the plane $z = 1$ with center $(0,0,1)$ beginning and ending at $(1,0,1)$, and starting towards $(0,1,1)$

(d) Im Γ is the piece of the unit circle in the plane $z = 0$ between $(1,0,0)$ and $(0,1,0)$ together with the line segment from $(0,1,0)$ to $(1,1,1)$

3. Evaluate the line integral of the vector field $\vec{F}:R^3 \to R^3$ along the curve Γ where

(a) $\vec{F}(x,y,z) = (x^2,xy,1)$ and Γ is defined by $\alpha:[0,1] \to R^3$, $\alpha(t) = (t,t^2,1)$

(b) $\vec{F}(x,y,z) = (\cos z,e^x,e^y)$ and Γ is defined by $\alpha:[0,2] \to R^3$, $\alpha(t) = (1,t,e^t)$

(c) $\vec{F}(x,y,z) = (2x,z,y)$ and Γ is the *twisted cubic* defined by $\alpha:[0,1] \to R^3$, $\alpha(t) = (t,t^2,t^3)$

(d) $\vec{F}(x,y,z) = (\sin z, \cos z, -(xy)^{1/3})$ and Γ is defined by $\alpha:[0,\pi/2] \to R^3$, $\alpha(t) = (\cos^3 t, \sin^3 t, t)$

4. Compute $\int_\Gamma x dx + xy dy$, where Γ is given by

(a) $\alpha(t) = (\sin t, \cos t)$, $0 \le t \le 2\pi$

(b) $\alpha(t) = (t^2,t^3)$, $1 \le t \le 2$

(c) $\alpha(t) = (e^{2t},e^{3t})$, $0 \le t \le \log 2$

(d) $\alpha(t) = (t^2 - 1, t^2 + 1)$, $1 \le t \le 5$

5. Compute $\int_\Gamma y dx - z dy + x dz$, where Γ is given by

(a) $\alpha(t) = (t,t^2,t^3)$, $0 \le t \le 3$

(b) $\alpha(t) = (\sin t, \cos t, 2 \cos t), \ 0 \le t \le \pi$

(c) $\alpha(t) = (2t, 1 - t^2, 1 + t^2), \ -1 \le t \le 1$

6.   Let Γ be a parametrization of the unit circle centered at the origin, starting and ending at $(1,0)$ and traversed counterclockwise. Compute $\int_\Gamma x\,dx + y\,dy$, $\int_\Gamma y\,dx + x\,dy$, and $\int_\Gamma (y^3 - 2xy)\,dx + (3xy^2 - x^2)\,dy$.

*7.   Let Γ be a curve in R^n with parametrization $\alpha : [a,b] \to R^n$ and let $\vec{F} : U \to R^n$ be a vector field defined on an open set containing Γ. Suppose that Γ is 1 - 1 and that, for every $\vec{u} \in \Gamma$, $\vec{u} = \alpha(t)$, we have $\vec{F}(\vec{u}) = \alpha'(t)/\|\alpha'(t)\|$. Show that $\int_\Gamma \vec{F}$ is the arc length of Γ.

8.   Prove that the relation "α is 0-equivalent to β" is an equivalence relation.

9.   Give a detailed proof of Proposition 3.1.

*10. Let Γ be the boundary of the unit square in R^2, and let S be the unit square. Show that if $\vec{F}$ is a continuously differentiable vector field in R^2,

$$\vec{F}(\vec{v}) = (f_1(\vec{v}), f_2(\vec{v}))$$

then

$$\int_\Gamma \vec{F} = \int_S \left(\frac{\partial f_2}{\partial x} - \frac{\partial f_1}{\partial y} \right)$$

(the second integral is an ordinary area integral). Here, Γ goes around the square counterclockwise. Hint: compute the line integral along each line segment; show that the two horizontal ones give

$$\int_S -\left(\frac{\partial f_1}{\partial y} \right)$$

and that the two vertical ones give

$$\int_S \left(\frac{\partial f_2}{\partial x} \right)$$

4.   CONSERVATIVE VECTOR FIELDS

In this section, we study an important class of vector fields, the
conservative vector fields. These are the vector fields whose line
integrals do not depend on the particular curve chosen, only its end
points.

We begin with some of the properties of line integrals.

PROPOSITION 4.1. *Let* U *be an open subset in* R^n *and* $\alpha:[a,b] \to U$
a parametrized curve defining the oriented curve Γ. *Let* $\vec{F}, \vec{G}:U \to R^n$
be vector fields and c,d *real numbers. Then*

$$\int_\Gamma (c\vec{F} + d\vec{G}) = c\int_\Gamma \vec{F} + d\int_\Gamma \vec{G}$$

Furthermore, let e *lie between* a *and* b *and let* $\alpha_1:[a,e] \to U$,
$\alpha_2:[e,b] \to U$ *be the restrictions of* α *(see Figure 4.1). Then, if* Γ_1
and Γ_2 *are the oriented curves defined by* α_1 *and* α_2 *we have*

$$\int_\Gamma \vec{F} = \int_{\Gamma_1} \vec{F} + \int_{\Gamma_2} \vec{F}$$

Finally, if $-\Gamma$ *represents the curve* Γ *with its opposite orientation,*
then

$$\int_{-\Gamma} \vec{F} = - \int_\Gamma \vec{F}$$

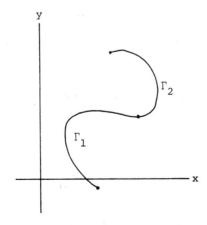

FIGURE 4.1.

Proof. We prove only the last part; see Exercise 3. One parametrization for -Γ is obtained by "running backwards along Γ"; if
$\alpha:[a,b] \to R^n$ parametrizes Γ, then $\beta:[-b,-a] \to R^n$, with

$$\beta(t) = \alpha(-t)$$

parametrizes -Γ. Thus

$$\int_{-\Gamma} \vec{F} = \int_{-b}^{-a} < \vec{F}(\beta(t)), \beta'(t) > dt$$

$$= \int_{-b}^{-a} < \vec{F}(\alpha(-t)), -\alpha'(-t) > dt$$

(by the chain rule and the definition of β)

$$= \int_{b}^{a} - < \vec{F}(\alpha(s)), -\alpha'(s) > ds$$

(letting s = - t)

$$= \int_{a}^{b} < \vec{F}(\alpha(s)), -\alpha'(s) > ds = - \int_{\Gamma} \vec{F}$$

as claimed.

Recall that in the previous section we computed line integrals
of a function $\vec{F}$ along two different curves connecting the same points
and found that the answers agreed. This is not always the case. For
instance, suppose that $\vec{F}:R^2 \to R^2$ is defined by

$$\vec{F}(x,y) = (y,-x)$$

and let Γ be the first quadrant of the unit circle from $(1,0)$ to $(0,1)$.
Then Γ is parametrized by $\alpha_1:[0,\pi/2] \to R^2$, where

$$\alpha_1(t) = (\cos t, \sin t)$$

and

$$\int_{\Gamma_1} \vec{F} = \int_0^{\pi/2} < (\sin t, -\cos t), (-\sin t, \cos t) > dt$$

$$= \int_0^{\pi/2} -(\sin^2 t + \cos^2 t)dt = \int_0^{\pi/2} -1 \, dt = -\frac{\pi}{2}$$

If, however, Γ_2 is the straight line from $(1,0)$ to $(0,1)$, so that Γ_2 is parametrized by $\alpha_2:[0,1] \to R^2$ where

$$\alpha_2(t) = (1 - t,t)$$

then

$$\int_{\Gamma_2} \vec{F} = \int_0^1 < (t,t - 1),(-1,1) > dt$$

$$= \int_0^1 -1dt = - 1$$

We now consider the question of when the line integral of F along Γ depends only on the end points of Γ. A vector field $\vec{F}$ with this property is called *conservative*. Such fields are of importance in physics because of the law of conservation of energy. For example, we have seen that the work done in moving a particle along a curve Γ in a force field $\vec{F}$ is

$$\int_\Gamma \vec{F}$$

If Γ_1 and Γ_2 are two curves connecting the same points $\vec{v}_1$ and $\vec{v}_2$ (as in Figure 4.2), then the work done in going from $\vec{v}_1$ to $\vec{v}_2$ along Γ_1 and back to $\vec{v}_1$ along Γ_2 is

$$\int_{\Gamma_1} \vec{F} + \int_{-\Gamma_2} \vec{F}$$

(we traverse Γ_2 in the reverse direction)

$$= \int_{\Gamma_1} \vec{F} - \int_{\Gamma_2} \vec{F}$$

by Proposition 4.1. The law of conservation of energy says that when we return to the original state, the total work done should be 0; that is,

$$\int_{\Gamma_1} \vec{F} - \int_{\Gamma_2} \vec{F} = 0$$

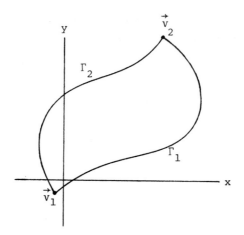

FIGURE 4.2. Two curves from $\vec{v}_1$ to $\vec{v}_2$

or

$$\int_{\Gamma_1} \vec{F} = \int_{\Gamma_2} \vec{F}$$

That is, the law of conservation of energy requires force fields to
be conservative fields.[*]

Let $\vec{F}$ be a vector field on the open set $U \subset R^n$. A function
$V:U \to R$ of class C^2 is called a *potential function* for the vector field
$\vec{F}$ if $\vec{F}$ is the gradient of V,

$$\vec{F} = \text{grad } V$$

For example, let U be the complement of the origin in R^2 and
$\vec{F}:U \to R^2$ the force field arising from a positive change at the origin,
given by (3.1). Then $V:U \to R$ defined by

$$V(x,y) = \frac{q}{(x^2+y^2)^{1/2}}$$

---

[*]The law of conservation of energy may fail for some physical systems
because of friction, or because energy is being added from outside.
In such cases, force fields need not be conservative.

is a potential function for $\vec{F}$. On the other hand, if $\vec{F}:R^2 \to R^2$ is given by

$$\vec{F}(x,y) = (y,-x)$$

then $\vec{F}$ has no potential function V. For if $\vec{F} = \text{grad } V$, then

$$\frac{\partial V}{\partial x} = y, \ \frac{\partial V}{\partial y} = -x$$

Thus

$$\frac{\partial^2 V}{\partial y \partial x} = \frac{\partial}{\partial y} y = 1$$

and

$$\frac{\partial^2 V}{\partial x \partial y} = \frac{\partial}{\partial x} (-x) = -1$$

which contradicts Theorem 6.3.3.

PROPOSITION 4.2. *Suppose that* $\vec{F}$ *is a vector field on the open set* $U \subset R^n$ *with a potential function* V. *Then* $\vec{F}$ *is conservative.*

Proof. Let $\vec{v}_0$ and $\vec{v}_1$ be points in U, and let Γ be an oriented curve with parametrization $\alpha:[a,b] \to U$, $\alpha(a) = \vec{v}_0$, $\alpha(b) = \vec{v}_1$. Write

$$\alpha(t) = (\alpha_1(t),\ldots,\alpha_n(t))$$

We compute:

$$\int_\Gamma \vec{F} = \int_a^b < \vec{F}(\alpha(t)),\alpha'(t) > dt$$

$$= \int_a^b < (\text{grad } V)(\alpha(t)),\alpha'(t) >$$

$$= \int_a^b (\sum_{j=1}^n \frac{\partial V}{\partial x_j} (\alpha(t)) \frac{d\alpha_j}{dt}) dt$$

$$= \int_a^b (\frac{d}{dt} V(\alpha(t))) dt$$

(by the Chain Rule)

$$= V(\alpha(a)) - V(\alpha(b))$$

(by the Fundamental Theorem of Calculus

$$= V(\vec{v}_1) - V(\vec{v}_0)$$

Thus if the vector field $\vec{F}$ has a potential function, the line integral of $\vec{F}$ along Γ depends only on the end points of Γ, not on the curve itself. Hence $\vec{F}$ is conservative, as we claimed.

Proposition 4.2 has an almost complete converse, as we now see.

THEOREM 4.3. *Let* U *be an open subset of* R^n *with the property that any two points in* U *can be connected by a curve. (For instance,* U *could be an open ball.) A vector field* $\vec{F}$ *on* U *is conservative if and only if it has a potential function.*

A subset U of R^n satisfying the condition of Theorem 4.3 (any two points can be connected by a curve) is called *pathwise connected.**

Proof. We have already seen that a vector field is conservative if it has a potential function. We prove the converse.

Let $\vec{v}_0$ be a fixed point of U. Define $V: U \to R$ by

$$V(\vec{v}) = \int_\Gamma \vec{F}$$

where Γ is a curve on U from $\vec{v}_0$ to $\vec{v}$. The curve Γ exists by hypothesis and V is well defined (that is, independent of the curve Γ) since $\vec{F}$ is conservative (using Proposition 4.1). We now compute $\frac{\partial V}{\partial x_i}$.

Let $\vec{v} = (x_1, \ldots, x_n)$ and Γ_h the curve with parametrization $\alpha_h : [0,1] \to U$,

$$\alpha_h(t) = (x_1, \ldots, x_i + th, \ldots, x_n)$$

If Γ is a curve from $\vec{v}_0$ to $\vec{v}$, we let Γ_h' be the curve from $\vec{v}_0$ to the

---

* The usual definition of pathwise connected only requires the existence of a continuous curve between any two points of the set. (See Section 5.8.) Here, we require the curve to be piecewise continuously differentiable. However, it can be shown that these two definitions are equivalent for open sets. (See Exercise 4 of Section 6.)

point $\vec{v}_h = \alpha_h(1) = (x_1,\ldots,x_i + h,\ldots,x_n)$ obtained by first traversing Γ and then traversing Γ_h; Γ_h' has the parametrization $\beta_h:[a,b+1] \to U$ given by

$$\beta_h(t) = \begin{cases} \alpha(t), & a \le t \le b \\ \alpha_h(t-b), & b \le t \le b + 1 \end{cases}$$

(See Figure 4.3.) Then

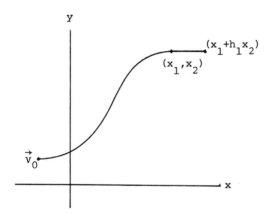

FIGURE 4.3.

$$\frac{\partial V}{\partial x_i}(\vec{v}) = \lim_{h\to 0} \frac{1}{h} (V(\vec{v}_h) - V(\vec{v}))$$

$$= \lim_{h\to 0} \frac{1}{h} (\textstyle\int_{\Gamma_h'} \vec{F} - \int_\Gamma \vec{F})$$

$$= \lim_{h\to 0} \frac{1}{h} \int_{\Gamma_h} \vec{F}$$

by Proposition 4.1. Thus

$$\frac{\partial V}{\partial x_i}(\vec{v}) = \lim_{h\to 0} \frac{1}{h} \int_0^1 < \vec{F}(\alpha_h(t)), \alpha_h'(t) > dt$$

$$= \lim_{h \to 0} \frac{1}{h} \int_0^1 \; < \vec{F}(\alpha_h(t)), h\vec{e}_i > dt$$

$$= \lim_{h \to 0} \int_0^1 \vec{F}_i(x_1, \ldots, x_i + th, \ldots, x_n) dt$$

where $\vec{e}_i$ is the ith basis vector in R^n and F_i is the ith component of $\vec{F}$. Now

$$\int_0^1 F_i(x_1, \ldots, x_i + th, \ldots, x_n) dt$$

is a continuous function of h (see Exercise 7 of Section 12.1). Thus

$$\frac{\partial V}{\partial x_i} (\vec{v}) = \lim_{h \to 0} \int_0^1 F_i(x_1, \ldots, x_i + th, \ldots, x_n) dt$$

$$= \int_0^1 F_i(x_1, \ldots, x_n) dt$$

$$= F_i(x_1, \ldots, x_n)$$

Therefore grad $V = \vec{F}$, and Theorem 4.3 is proved.

We still need a criterion for determining when a force field is conservative. For the first results along those lines, however, we need Green's theorem, the subject of the next section.

EXERCISES

1. Find a potential function for each of the following force fields and determine the work required to move a particle from the point $\vec{v}_0$ to the point $\vec{v}$, where

   (a) $\vec{F}(x,y) = (2xy^3, 3x^2y^2)$, $\vec{v}_0 = (1,0)$, $\vec{v}_1 = (2,1)$

   (b) $\vec{F}(x,y) = (y \cos xy, x \cos xy)$, $\vec{v}_0 = (0,1)$, $\vec{v}_1 = (\pi/2, 1)$

   (c) $\vec{F}(x,y,z) = (2x + y, x, 1)$, $\vec{v}_0 = (0,0,0)$, $\vec{v}_1 = (1,2,-1)$

   (d) $\vec{F}(x_1, \ldots, x_n) = (2x_1, \ldots, 2x_n)$, $\vec{v}_0 = (0, \ldots, 0)$, $\vec{v}_1 = (1, \ldots, 1)$

   (e) $\vec{F}(x,y,z) = (yz \cos(xyz), xz \cos(xyz), xy \cos(xyz))$, $\vec{v}_0 = (1,1,0)$, $\vec{v}_1 = (1/3, 1/2, 2\pi)$

(f) $\vec{F}(x,y) = (2xye^{x^2y}, x^2e^{x^2y})$, $\vec{v}_0 = (0,0)$, $\vec{v}_1 = (2,1)$

(g) $\vec{F}(x,y) = (x^2 + 4xy + 4y^2, 2x^2 + 8xy + 8y^2)$, $\vec{v}_0 = (2,-1)$, $\vec{v}_1 = (-4,2)$

(h) $\vec{F}(x,y,z) = (2xy, x^2 + z, y)$, $\vec{v}_0 = (1,3,2)$, $\vec{v}_1 = (2,3,1)$

(The work done is $-\int_\Gamma F$, since one pushes *against* the force.)

2.   Show that if $\vec{F} = (f_1, \ldots, f_n)$ is a conservative force field on R^n, then, for all i,j with $1 \le i \le j \le n$,

$$\frac{\partial f_i}{\partial x_j} = \frac{\partial f_j}{\partial x_i}$$

3.   Prove the first two assertions of Proposition 4.1.

4.   Let U be an open subset of R^n and $\vec{F}:U \to R^n$ a vector field.   Prove that $\vec{F}$ is conservative if and only if the line integral around any closed curve is zero.

5.   GREEN'S THEOREM

Two chapters ago, we defined the integral of a function of n variables over a region in R^n; two sections ago, we defined the line integral of a vector field over a curve in R^n.   In this section, we show that, when n = 2, these two integrals are closely related.   The precise result is called Green's Theorem.   Its proof will be given in the next section; extending it to R^n will occupy us for the next three chapters.

Let $\alpha:[a,b] \to R^n$ be a parametrization for a curve Γ.   We say that Γ is *simple* if it does not intersect itself; that is, if α is injective.   For instance, the curve Γ_1 in Figure 5.1 is simple, but Γ_2 and Γ_3 are not.   The curve Γ is said to be *closed* if $\alpha(a) = \alpha(b)$, so that Γ does indeed close up on itself.   In Figure 5.1, Γ_2 is the only closed curve.

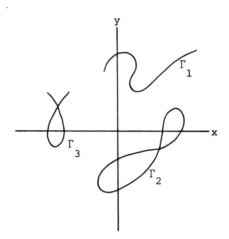

FIGURE 5.1. Three curves

It is clearly impossible for α to be injective and for α(a) to equal α(b) - that is, for Γ to be simple and closed. Nevertheless, we define Γ to be a *simple closed curve* if Γ is closed and if α is injective on the interval [a,b), as in Figure 5.2.

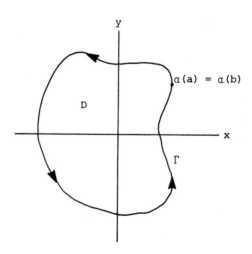

FIGURE 5.2. A simple closed curve bounding D

We can now state our main theorem for this section.

THEOREM 5.1.   (Green's Theorem) *Let* Γ *be an oriented piecewise continuously differentiable simple closed curve in* R^2 *bounding a region* D. [*] *Suppose that* Γ *is oriented so that the region* D *is on the left as we traverse* Γ. *(See Figure 5.2.) Then, if* $\vec{F}:D \to R^2$ *is a vector field on* D, $\vec{F}(\vec{v}) = (f_1(\vec{v}), f_2(\vec{v}))$, *we have*

$$\int_\Gamma \vec{F} = \int_D (\frac{\partial f_2}{\partial x} - \frac{\partial f_1}{\partial y}) \tag{5.1}$$

As mentioned above, the proof will be given in the next section.

Example.   Let $\vec{F}(x,y) = (y,x)$, and let Γ be the unit circle in R^2, traversed counterclockwise. We can compute $\int_\Gamma \vec{F}$ quite easily with the help of Green's Theorem. Let D be the unit disc. Then, from Green's Theorem,

$$\int_\Gamma \vec{F} = \int_D (\frac{\partial}{\partial x} x - \frac{\partial}{\partial y} y)$$

$$= 0$$

We use the idea of this example to derive a criterion for a vector field to be conservative. Suppose that U is a piecewise connected open subset of R^2 and that $\vec{F}$ is a conservative vector field on U, $\vec{F}(\vec{v}) = (f_1(\vec{v}), f_2(\vec{v}))$. We know from Proposition 4.3 that F has a potential function V. If $\vec{F}$ has a continuous derivative, then

$$\frac{\partial f_1}{\partial y} = \frac{\partial^2 V}{\partial y \partial x}$$

$$= \frac{\partial^2 V}{\partial x \partial y}$$

$$= \frac{\partial f_2}{\partial x}$$

---

[*] A theorem (known as the *Jordan Curve Theorem*) says that a simple closed curve divides the plane into two regions, only one of which is bounded. This result seems fairly obvious. Proving it, however, turns out to be surprisingly difficult.

by Theorem 6.1.1. Thus if $\vec{F}$ is conservative and continuously differentiable, then $\dfrac{\partial f_1}{\partial y} = \dfrac{\partial f_2}{\partial x}$.

The converse, unfortunately, is not true in general. For example, let U be the complement of the origin in R^2 and define $\vec{F}:U \to R^2$ by

$$\vec{F}(x,y) = (\frac{-y}{x^2+y^2}, \frac{x}{x^2+y^2})$$

Then

$$f_1(x,y) = \frac{-y}{x^2+y^2} , \quad f_2(x,y) = \frac{x}{x^2+y^2}$$

and

$$\frac{\partial f_1}{\partial y} = \frac{y^2-x^2}{x^2+y^2}$$

$$= \frac{\partial f_2}{\partial x}$$

However, if Γ is the closed curve defined by the parametric curve $\alpha:[0,2\pi] \to U$,

$$\alpha(t) = (\cos t, \sin t)$$

then

$$\int_\Gamma \vec{F} = \int_0^{2\pi} < \vec{F}(\alpha(t)),\alpha'(t) > dt$$

$$= \int_0^{2\pi} < (-\sin t, \cos t),(-\sin t, \cos t) > dt$$

$$= \int_0^{2\pi} dt = 2\pi$$

Thus $\vec{F}$ is not conservative.

In spite of this example, we can prove that a vector field in R^2 is conservative whenever $\dfrac{\partial f_1}{\partial y} = \dfrac{\partial f_2}{\partial x}$ if we suitably restrict the open set U. We say that a pathwise connected set $X \subseteq R^2$ is *simply connected* if the region enclosed by any simple closed curve in X is entirely contained in X. For example, any (open or closed) ball in R^2 is simply connected. However, the set

$$U = \{(x,y) \in R^2 : (x,y) \neq (0,0)\}$$

is not simply connected; the region enclosed by the unit circle is the unit disc and is not contained in U.

We can now state the result, promised last section, which gives a criterion for a vector field on R^2 to be conservative.

PROPOSITION 5.2. *Let* $\vec{F} = (f_1, f_2)$ *be a vector field on the simply connected open set* $U \subseteq R^2$. *Then* $\vec{F}$ *is conservative if and only if*

$$\frac{\partial f_1}{\partial y} = \frac{\partial f_2}{\partial x} \tag{5.2}$$

Proof. We have proved above that equation (5.2) holds if $\vec{F}$ is conservative. For the converse, we use Green's Theorem.

Suppose $\vec{F}$ is a vector field on the simply connected open set U satisfying (5.2), and let Γ' and Γ'' be two curves joining $\vec{v}_0$ to $\vec{v}_1$. We must show that

$$\int_{\Gamma'} \vec{F} = \int_{\Gamma''} \vec{F}$$

Let Γ be the closed curve obtained by first traversing Γ' from $\vec{v}_0$ to $\vec{v}_1$ and then traversing $-\Gamma''$ from $\vec{v}_1$ to $\vec{v}_0$. Then

$$\int_{\Gamma} \vec{F} = \int_{\Gamma'} \vec{F} - \int_{\Gamma''} \vec{F}$$

Thus we need to show that the line integral of $\vec{F}$ over Γ vanishes whenever Γ is a closed curve.

Suppose, first of all, that Γ is a simple closed curve enclosing the region D on its left as in Figure 5.2. Then $D \subseteq U$, since U is simply connected, and we can apply Green's Theorem to obtain the equation

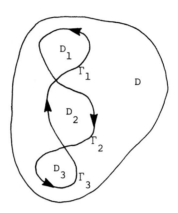

FIGURE 5.3.

$$\int_\Gamma \vec{F} = \int_D (\frac{\partial f_2}{\partial x} - \frac{\partial f_1}{\partial y})$$

However, the integrand is 0, by equation (5.2). Therefore $\int_\Gamma \vec{F} = 0$, as desired.

If Γ is not a simple closed curve we break Γ into a number of simple closed curves and apply the argument above to each of these. For example, if Γ is the closed curve of Figure 5.3 then

$$\int_\Gamma \vec{F} = \int_{\Gamma_1} \vec{F} + \int_{\Gamma_2} \vec{F} + \int_{\Gamma_3} \vec{F} \tag{5.3}$$

Now each of the regions D_1, D_2, D_3 is contained in U since U is simply connected and Green's Theorem gives us the equations

$$\int_{\Gamma_1} \vec{F} = \int_{D_1} (\frac{\partial f_2}{\partial x} - \frac{\partial f_1}{\partial y})$$

$$\int_{\Gamma_2} \vec{F} = - \int_{D_2} (\frac{\partial f_2}{\partial x} - \frac{\partial f_1}{\partial y})$$

$$\int_{\Gamma_3} \vec{F} = - \int_{D_2} (\frac{\partial f_2}{\partial x} - \frac{\partial f_1}{\partial y}$$

However, each of these integrals in zero, since equation (5.2) holds. Thus

$$\int_\Gamma \vec{F} = 0$$

by equation (5.3).

This reasoning clearly proves the proposition in the case where Γ intersects itself finitely many times. It is possible for Γ to intersect itself infinitely often. The proposition remains true, but the proof requires some extra work. We shall omit this case here.

EXERCISES

1.   Let Γ be the boundary curve of the rectangle $R(\vec{a},\vec{b})$ (traversed so that $R(\vec{a},\vec{b})$ is on the left) where $\vec{a} = (0,0)$, $\vec{b} = (2,3)$. Use Green's Theorem to compute the integral of the vector field $\vec{F}$ over Γ where

   (a) $\vec{F}(x,y) = (0,x)$

   (b) $\vec{F}(x,y) = (x + y, y^2)$

   (c) $\vec{F}(x,y) = (xy^2, 2x - y)$

   (d) $\vec{F}(x,y) = (\sin \pi xy/2, 2x)$

   (e) $\vec{F}(x,y) = (xy, x + y)$

   (f) $\vec{F}(x,y) = (x \sin y, \frac{1}{2} x^2 \cos y)$

2.   Determine which of the following vector fields $\vec{F}$ on R^2 are conservative. If $\vec{F}$ is conservative, find a potential function for $\vec{F}$.

   (a) $\vec{F}(x,y) = (x^2 - y^2, 2xy)$

   (b) $\vec{F}(x,y) = (x^2, 2y^3)$

   (c) $\vec{F}(x,y) = (x \sin y, y \cos x)$

   (d) $\vec{F}(x,y) = (x^2 + y^2, 2xy)$

   (e) $\vec{F}(x,y) = (xe^{xy}, ye^{xy})$

   (f) $\vec{F}(x,y) = (y^2 - 2x, 2xy + 6y^2)$

3.   A subset X in R^n is said to be *convex* if the line segment converting any two points of X is entirely contained in X; that is, if $\vec{v}_1, \vec{v}_2 \in X$, then $t\vec{v}_1 + (1-t)\vec{v}_2$ is in X for all $t \in [0,1]$.

(a) Prove that any convex subset of R^n is connected.

(b) Prove that any convex subset of R^2 is simply connected.

4. Let U be a convex open subset of R^2 and $\vec{F}:U \to R^2$ a vector field, $\vec{F}(\vec{v}) = (f_1(\vec{v}), f_2(\vec{v}))$, with

$$\frac{\partial f_2}{\partial x} = \frac{\partial f_1}{\partial y}$$

Prove that the function $V:U \to R$ defined by

$$V(\vec{u}) = (x-a)\int_0^1 f_1(t\vec{u} + (1-t)\vec{u}_0)dt + (y-b) \int f_2(t\vec{u} + (1-t)\vec{u}_0)dt$$

(where $\vec{u}_0 = (a,b)$ is a fixed point of U and $\vec{u} = (x,y)$) is a potential function for $\vec{F}$. Where is the fact that U is convex needed?

5. (a) Let Γ be a simple closed curve in R^2 bounding a region U. Show that the area of U is

$$-\int_\Gamma y\, dx$$

where Γ is oriented so that U is to the left of Γ.

(b) Show that the area of U is also $\int_\Gamma x\, dy$.

6. THE PROOF OF GREEN'S THEOREM

We now give the proof of Green's Theorem. The proof involves several cases. The first two deal with the result when the region D has a simple form. We give fairly complete details in these cases. Two others are similar to the first two, and are left as Exercises 2 and 3. The last is the general case and we give only an indication of the proof here.

   Case 1. Suppose D is the rectangle defined by the points (a,b), (c,d) (see Figure 6.1);

$$D = \{(x,y) \,|\, a \leq x \leq c, b \leq y \leq d\}$$

We prove that

$$\int_\Gamma f_1 dx = - \int_D \left(\frac{\partial f_1}{\partial y}\right) \tag{6.1}$$

and leave the proof of equality

$$\int_\Gamma f_2 dy = \int_D \left(\frac{\partial f_2}{\partial x}\right)$$

to the reader.

Recall that if Γ_1 is parametrized by $\alpha:[a,b] \to R^2$, where $\alpha(t) = (\alpha_1(t), \alpha_2(t))$, then

$$\int_{\Gamma_1} f_1 dx = \int_a^b f_1(\alpha(t))\alpha_1'(t)dt$$

If Γ_1 is a vertical line segment, then $\alpha_1(t)$ is constant and

$$\int_{\Gamma_1} f_1 dx = 0$$

Thus

$$\int_\Gamma f_1 dx$$

is equal to the line integrals over the horizontal line segments:

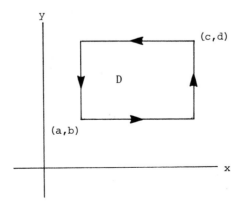

FIGURE 6.1. An oriented curve bounding a rectangle in R^2

$$\int_\Gamma f_1 dx = \int_a^c f_1(x,b)dx + \int_c^a f_1(x,d)dx$$

$$= \int_a^c (f_1(x,b) - f_1(x,d))dx$$

$$= \int_a^c (- \int_b^d \frac{\partial f_1}{\partial y} dy)dx$$

$$= - \int_D \frac{\partial f_1}{\partial y}$$

by Theorem 12.1.2. (Here, we have parametrized the line from (a,b) to (c,b) by $\alpha:[a,c] \to R^2:\alpha(t) = (t,b)$. The parametrization for the other line is similar.)

Case 2. In this case we assume that D is defined by

$$D = \{(x,y) | a \le x \le b, 0 \le y \le h(x)\}$$

where $h:[a,b] \to R$ is a positive differentiable function. (See Figure 6.2.) Let S be the rectangle in R^2 defined by

$$S = \{(x,y) | a \le x \le b, 0 \le y \le 1\}$$

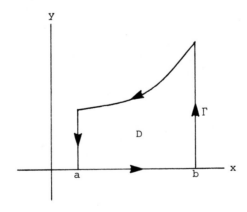

FIGURE 6.2.

and define $\varphi:S \to D$ by

$$\varphi(x,y) = (x,yh(x))$$

Then φ is bijective and differentiable. Furthermore, if $\alpha:[c,d] \to R^2$ is a parametrization of the boundary curve Δ to S such that S is on the left of Δ, then $\beta = \varphi \circ \alpha$ is a parametrization of Γ with D on the left. Thus

$$\int_\Gamma f_1 dx + f_2 dy = \int_c^d < f(\beta(t)), \beta'(t) > dt$$

$$= \int_c^d < f(\varphi(\alpha(t))), (D_{\alpha(t)}\varphi)\alpha'(t) > dt \qquad (6.2)$$

Since

$$\varphi(x,y) = (x,yh(x)$$

the linear transformation $D_{\vec{v}}\varphi$ has matrix

$$\begin{pmatrix} 1 & 0 \\ yh'(x) & h(x) \end{pmatrix}$$

In particular, $D_{\vec{v}}\varphi$ is invertible, so φ^{-1} is differentiable by Theorem 9.1.4.

Now define $g: \Delta \to R^2$, $g = (g_1(\vec{v}), g_2(\vec{v}))$, by

$$g(\vec{v}) = (D_{\vec{v}}\varphi)^*(f \circ \varphi(\vec{v}))$$

where the matrix for $(D_{\vec{v}}\varphi)^*$ is

$$\begin{pmatrix} 1 & yh'(x) \\ 0 & h(x) \end{pmatrix}$$

Thus $(D_{\vec{v}}\varphi)^*$ is the adjoint of $D_{\vec{v}}\varphi$ (see Section 7.5) so that

$$< (D_{\vec{v}}\varphi)^* \vec{w}_1, \vec{w}_2 > = < \vec{w}_1, (D_{\vec{v}}\varphi)\vec{w}_2 >$$

for any two vectors $\vec{w}_1, \vec{w}_2 \in R^2$. It follows from (6.2) that

$$\int_\Gamma f_1 dx + f_2 dy = \int_c^d < (D_{\alpha(t)}\varphi)^* f \circ \varphi(\alpha(t)), \alpha'(t) > dt$$

$$= \int_\Delta g_1 dx + g_2 dy$$

From the matrix expression for $(D_{\vec{v}}\varphi)^*$, we can compute g_1 and g_2:

$$g_1(x,y) = f_1(x,yh(x)) + yh'(x)f_2(x,yh(x))$$
$$g_2(x,y) = h(x)f_2(x,yh(x))$$

Therefore

$$\frac{\partial g_1}{\partial y} = h(x)\frac{\partial f_1}{\partial x}(x,yh(x)) + h'(x)f_2(x,yh(x))$$
$$+ yh'(x)\frac{\partial f_2}{\partial y}(x,yh(x))h(x)$$

$$\frac{\partial g_2}{\partial x} = h'(x)f_2(x,yh(x)) + h(x)\frac{\partial f_2}{\partial x}(x,yh(x))$$
$$+ h(x)\frac{\partial f_2}{\partial y}(x,yh(x))yh'(x)$$

We can now apply Case 1:

$$\int_\Delta g_1 dx + g_2 dy = \int_S \left(\frac{\partial g_2}{\partial x} - \frac{\partial g_1}{\partial y}\right)$$
$$= \int_S \left(\frac{\partial f_2}{\partial x}(x,yh(x)) - \frac{\partial f_1}{\partial y}(x,yh(x))\right)h(x)$$
$$= \int_S \left(\frac{\partial f_2}{\partial x}\circ\varphi - \frac{\partial f_1}{\partial y}\circ\varphi\right)\det(D\varphi)$$
$$= \int_D \left(\frac{\partial f_2}{\partial x} - \frac{\partial f_1}{\partial y}\right)$$

(The last equality is by the change of variable formula, Theorem 12.6.2; recall that $h > 0$.) This finishes the proof of Case 2.

General Case. Suppose now that Γ is an arbitrary oriented closed piecewise continuously differentiable curve parametrized by the function $\alpha(t) = (\alpha_1(t),\alpha_2(t))$, $a \leq t \leq b$. We partition the interval $[a,b]$ into a finite number of subintervals,

$$a = t_0 < t_1 < \ldots < t_m = b$$

and let Γ_j be the curve obtained by restricting α to the interval $[t_{j-1},t_j]$, $1 \leq j \leq m$. We construct this partition so that the following are true.

(i)   The function α is differentiable on $[t_{j-1}, t_j]$ for $1 \le j \le m$.

(ii)   For each j, $1 \le j \le m$, we have either $\alpha_1'(t) \ne 0$ for all $t \in (t_{j-1}, t_j)$ or $\alpha_2'(t) \ne 0$ for all $t \in (t_{j-1}, t_j)$.   (See Figure 6.3.)

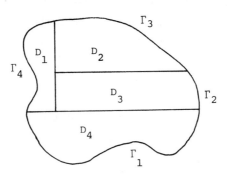

FIGURE 6.3.

We can then divide the region D into subregions whose boundary curves are either pieces of Γ or horizontal or vertical line segments. Each of these subregions is, up to a rotation of ± 90 or 180, of the type considered in Case 2 (see also Exercise 3), so

$$\int_{\Delta_j} f_1 dx + f_2 dy = \int_{D_j} (\frac{\partial f_2}{\partial x} - \frac{\partial f_1}{\partial y})$$

where Δ_j is the bounding curve of D_j. Since each of the interior horizontal or vertical line segments occurs as the boundary of exactly two regions, one sees easily that their contributions cancel out and we have

$$\int_\Gamma f_1 dx + f_2 dy = \sum_{j=1}^{m} \int_{\Gamma_j} f_1 dx + f_2 dy$$

$$= \sum_{j=1}^{m} \int_{D_j} (\frac{\partial f_2}{\partial x} - \frac{\partial f_1}{\partial y})$$

$$= \int_D (\frac{\partial f_2}{\partial x} - \frac{\partial f_1}{\partial y})$$

This completes the proof of Green's Theorem.

Note. The reason that this discussion does not constitute a proof is that we have not explicitly described how to sub-divide the region D. It is easy to convince oneself that D can be split into regions of the desired shape, but harder to describe a procedure for doing the splitting. Such a description leads us to considerations rather far removed from the main topics of our text.

EXERCISES

1.   Complete the other half of Case 1 of the proof of Theorem 4.1, namely that

$$\int_\Gamma f_2 dy = \int_D \frac{\partial f_2}{\partial x}$$

2.   Give a direct proof of Theorem 4.1 for regions defined by a right triangle with two sides parallel to the coordinate axes.

*3.   Some of the regions in Figure 6.3, like D_2 are distorted triangles; that is, they are (up to a rotation) regions like those in Figure 6.2, except that $h(a) = 0$ (and $h(t) > 0$ for $a < t \leq b$). Prove Green's Theorem for such regions. (Hint: modify the proof of Case 2 and use Exercise 2.)

*4.   Show that if $U \subset R^n$ is a connected open set, then any two points $\vec{v}, \vec{w} \in U$ can be connected by a continuous, piecewise differentiable path. (Hint: use a polygonal path.)

CHAPTER 14

SURFACE INTEGRALS

INTRODUCTION

In this chapter, we begin the task of extending the results of the
previous section to higher dimensions. The first step is defining
surface integrals for surfaces in R^3. As with line integrals, there
are two sorts of surface integrals: integrals of functions and integrals of vector fields. We next relate the surface integral of a
vector field to the line integral of a related vector field along the
boundary of the surface. This result, Stokes' Theorem, is an analogue of Green's theorem. We state it here and give an application to
hydrodynamics. The proof is given in Chapter 16.

As will be seen, surface integrals are considerably more difficult to work with than line integrals. For further generalizations,
we need to introduce differential forms. This, however, must wait
until the next chapter.

1. SURFACES

In this section, we define a surface and give some examples.

Let S be a closed subset of R^2. A *parametrized surface* is a
continuously differentiable function $\alpha : S \to R^n$.

For example, let $S = R^2$ and let $\vec{u}, \vec{v}, \vec{w}$ be vectors in R^3, with
$\vec{v}$ and $\vec{w}$ independent. Then

617

$$\alpha(x,y) = \vec{u} + x\vec{v} + y\vec{w} \qquad\qquad (1.1)$$

defines a parametrized surface whose image is a *plane*. (See Figure
1.1). If $\vec{v}'$ and $\vec{w}'$ are any two linearly independent vectors in the
span of $\vec{v},\vec{w}$, then

$$\beta(x,y) = \vec{u} + x\vec{v}' + y\vec{w}'$$

defines the same plane.

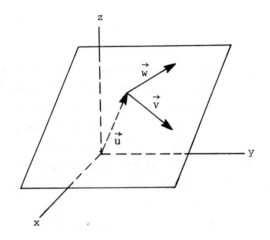

FIGURE 1.1. A plane in R^3

Geometrically, the plane defined by α is obtained by translating
the subspace W of R^3 spanned by $\vec{v}$ and $\vec{w}$ to the end point of the vector
$\vec{u}$. We can find a nonzero vector $\vec{n}$ orthogonal to this plane; in fact,
$\vec{n}$ is unique except for multiplication by a scalar. (See Exercise 8.)
The vector $\vec{n}$ is called a *normal* vector to the plane. One choice for
$\vec{n}$ is the vector $\vec{v} \times \vec{w}$.

We can also describe the plane given by α in another way: it is
the only plane containing (the end point of) $\vec{u}$ and normal to $\vec{v} \times \vec{w}$.
This statement should be clear geometrically. We can verify it as
follows: if $(x_1,x_2,x_3) = \vec{v}_1$ is a point in the plane P containing

$\vec{u} = (c_1, c_2, c_3)$ and normal to $\vec{v} \times \vec{w}$, then

$$< \vec{v}_1 - \vec{u}, \vec{v} \times \vec{w} > = 0 \qquad (1.2)$$

If we let $\vec{v} = (a_1, a_2, a_3)$ and $\vec{w} = (b_1, b_2, b_3)$ and compute, we can verify that the points satisfying (1.2) are exactly the points in the image of α. (See Exercise 7.)

If we restrict the functions α of equation (1.1) to a rectangle $S = R(\vec{a}, \vec{b})$, we get a *parallelogram*. (See Figure 1.2.)

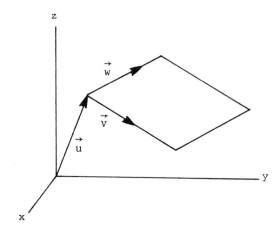

FIGURE 1.2. A Parallelogram in R^3

In general, if $\alpha : S \to R^3$ is a parametrized surface and $\vec{z}_0 \in S$ is a point at which $(D_{\vec{z}_0} \alpha)\vec{e}_1$ and $(D_{\vec{z}_0} \alpha)\vec{e}_2$ are independent, we define the *tangent plane* at $\vec{u} = \alpha(z_0)$ to be the plane

$$\beta(x,y) = \vec{u} + x(D_{\vec{z}_0} \alpha)\vec{e}_1 + y(D_{\vec{z}_0} \alpha)\vec{e}_2 \qquad (1.3)$$

(See Figure 1.3.) One sees easily that this notion of tangent plane corresponds to the one defined in Exercise 8 of Section 3.7. (See Exercise 6.)

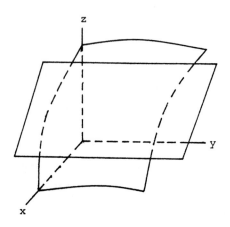

FIGURE 1.3. A tangent plane

For a more interesting example of a surface, let $S = B_1(\vec{0})$ be the closed ball of radius 1 about the origin in R^2 and let $\alpha: S \to R^3$ be defined by

$$\alpha(x,y) = (\frac{2x}{1+x^2+y^2}, \frac{2y}{1+x^2+y^2}, \frac{1-x^2-y^2}{1+x^2+y^2}) \qquad (1.4)$$

(Here (x,y) are coordinates in R^2.) The image of α is a unit upper hemisphere in R^3. (See Figure 1.4.) This mapping is obtained by

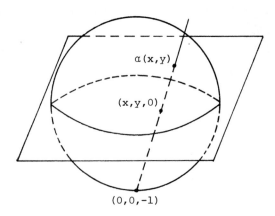

FIGURE 1.4. Stereographic projection of the upper hemisphere

stereographic projection; $\alpha(x,y)$ is the point of intersection of the upper hemisphere with the line through the points $(0,0,-1)$ and $(x,y,0)$. Similarly,

$$\beta(x,y) = (\frac{2x}{1+x^2+y^2}, \frac{2y}{1+x^2+y^2}, \frac{x^2+y^2-1}{1+x^2+y^2})$$

defines the lower hemisphere.

In fact, we can consider both α and β above as defined on all of R^2. If we do, then the image of α is the complement of the "south pole" $(0,0,-1)$ and the image of β is the complement of the "north pole" $(0,0,1)$. (See Figure 1.5). Computing the tangent plane is not hard. If $\vec{v} = (x,y)$, we see from (1.4) that $D_{\vec{v}}\alpha$ has matrix

$$\begin{pmatrix} \dfrac{2(1-x^2+y^2)}{d^2} & \dfrac{-4xy}{d^2} \\[2em] \dfrac{-4xy}{d^2} & \dfrac{2(1+x^2-y^2)}{d^2} \\[2em] \dfrac{-4x}{d^2} & \dfrac{-4y}{d^2} \end{pmatrix}$$

where $d = (1 + x^2 + y^2)$.

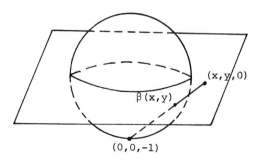

FIGURE 1.5. Stereographic projection of the lower hemisphere

So, for example, if $x = y = 1/2$, so that $\alpha(x,y) = (2/3, 2/3, 1/3)$, then $D_{\vec{v}}\alpha$ has matrix

$$\begin{pmatrix} \frac{8}{9} & \frac{-4}{9} \\ \frac{-4}{9} & \frac{8}{9} \\ \frac{-8}{9} & \frac{-8}{9} \end{pmatrix}$$

and

$$(D_{\vec{v}}\alpha)\vec{e}_1 = (\frac{8}{9}, \frac{-4}{9}, \frac{-8}{9}) \qquad (D_{\vec{v}}\alpha)\vec{e}_2 = (\frac{-4}{9}, \frac{8}{9}, \frac{-8}{9})$$

Therefore, the tangent plane is given by

$$\gamma(x,y) = (\frac{2}{3}, \frac{2}{3}, \frac{1}{3}) + (\frac{8}{9}, \frac{-4}{9}, \frac{-8}{9})x + (\frac{-4}{9}, \frac{8}{9}, \frac{-8}{9})y$$

$$= \frac{1}{9}(6 + 8x - 4y, \; 6 - 4x + 8y, \; 3 - 8x - 8y)$$

An easier point at which to visualize the tangent plane is $\vec{v} = (0,0,1)$ $= \alpha(0,0)$. There, $D_{\vec{v}}\alpha$ has matrix

$$\begin{pmatrix} 2 & 0 \\ 0 & 2 \\ 0 & 0 \end{pmatrix}$$

and

$$(D_{\vec{v}}\alpha)\vec{e}_1 = (2,0,0) \qquad (D_{\vec{v}}\alpha)\vec{e}_2 = (0,2,0)$$

Thus the tangent plane is

$$\delta(x,y) = (0,0,1) + (2,0,0)x + (0,2,0)y = (2x, 2y, 1)$$

This is the plane through $(0,0,1) = \alpha(0,0)$ parallel to the x,y-plane, as it should be. (See Figure 1.6.)

Just as in the case of curves, we say that two parametrized surfaces $\alpha: S \to R^n$, $\beta: S' \to R^n$ are *equivalent* if there is a diffeomorphism

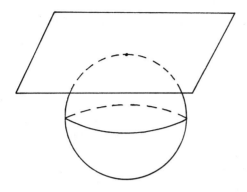

FIGURE 1.6. The tangent plane to the sphere at the north pole

$h:S' \to S$ such that $\alpha(h(\vec{v})) = \beta(\vec{v})$ for all $\vec{v}$ in S'. This is easily seen to be an equivalence relation. A *surface*[*] is an equivalence class Σ of parametrized surfaces under this relation. We say that $\alpha:S \to R^n$ is a *parametrization* for Σ if α is in the equivalence class defining Σ.

To illustrate, let $\vec{u}, \vec{v}, \vec{w}$ be three vectors in R^n with $\vec{v}$ and $\vec{w}$ independent, and let $\beta:R^2 \to R^n$,

$$\beta(x,y) = \vec{u} + x\vec{v} + y\vec{w}$$

be the corresponding plane. Define $\alpha:R^2 \to R^n$ by

$$\alpha(s,t) = \vec{u} + s\vec{v} + t(\vec{w} + \vec{v})$$

Then α defines the same plane. In fact, α is equivalent to β with $h:R^2 \to R^2$ defined by $h(x,y) = (x - y, y)$, since

$$\alpha \circ h(x,y) = \alpha(x - y, y) = \vec{u} + (x - y)\vec{v} + y(\vec{w} + \vec{v})$$
$$= \vec{u} + x\vec{v} + y\vec{w} = \beta(x,y)$$

and h has a differentiable inverse $(h^{-1}(x,y) = (x + y, y))$.

A surface is said to be *simple* if it has a parametrization which is injective. All of the examples given above are simple surfaces.

---

[*] We shall need a more general notion of surface in the next chapter. It will involve piecing together surfaces of the kind considered here.

An example of a surface which is not simple is the one with parametrization

$$\alpha:S \rightarrow R^3$$

where S is the rectangle

$$S = \{(x,y) \mid - \pi/2 \le x \le \pi/2, \ 0 \le y \le 1\}$$

and $\alpha(x,y) = (\sin x \cos 2x, \cos x \cos 2x, y)$. (See Figure 1.7.) Note that $\alpha(- \pi/4,y) = \alpha(\pi/4,y)$ for all y.

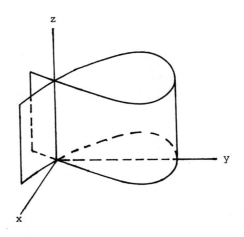

FIGURE 1.7. A nonsimple surface

As with curves, a surface can have corners. For example, the surface with parametrization

$$\beta:S' \rightarrow R^3$$

where S' is the rectangle

$$S' = \{(x,y) \mid -1 \le x \le 1, \ 0 \le y \le 1\}$$

and $\alpha(x,y) = (x^3, x^2, y)$, has a corner along the z-axis. (See Figure 1.8.)

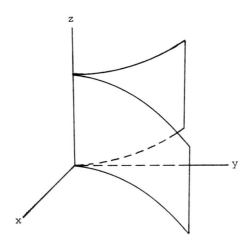

FIGURE 1.8. A nonregular surface

A surface is said to be *regular* if it has a parametrization $\alpha : S \to R^n$
with Jacobian of rank 2 at each point of S. A regular surface cannot
have corners, since it has a well defined tangent plane at each of
its points. In the future, we shall generally restrict attention to
regular surfaces.

EXERCISES

1. (a) Find a parametrization for the plane through $(1,0,0)$ and
orthogonal to $\vec{v}_0 = (1,-2,2)$.

(b) Find a parametrization for the plane through $(1,3,4)$ and
orthogonal to $\vec{v}_0 = (2,1,-1)$.

(c) Find a parametrization for the plane containing the points
$(1,3,2)$, $(2,-5,4)$, and $(1,0,1)$.

(d) Find a parametrization for the plane containing the points
$(1,-1,1)$, $(3,-2,3)$, and $(4,0,7)$.

(e) Find a parametrization for the parallelogram two of whose
sides are the lines connecting $(1,-2,1)$ with $(1,4,3)$ and $(1,-2,1)$ with
$(2,3,-1)$.

(f) Find a parametrization for the parallelogram two of whose sides are the lines connecting (3,2,1) with (4,7,5) and (3,2,1) with (5,-2,-3).

(g)   Find a parametrization for the points on the sphere of radius 2 centered about the origin.

(h)   Find a parametrization for the points on the ellipsoid

$$\frac{x^2}{4} + \frac{y^2}{9} + \frac{z^2}{16} = 1$$

(i)   Find a parametrization for the triangle with vertices at (1,0,0), (0,1,0), and (0,0,1).

(j)   Find a parametrization for the points on the cylinder $x^2 + y^2 = 1/4$ lying inside the unit sphere.

2.   Show that the following parametrizations are equivalent:

(a) $\alpha(s,t) = (s^2, t^2, s^2 t^2)$, $8 \le s$, $t \le 27$; $\beta(u,v) = (u^3, v^3, u^3 v^3)$, $4 \le u, v \le 9$.

(b) $\alpha(s,t) = (s, t, (1 - s^2 - t^2)^{1/2})$, $s^2 + t^2 \le 1/4$; $\beta(u,v) = (v, u, (1 - u^2 - v^2)^{1/2})$, $u^2 + v^2 \le 1/4$.

(c) $\alpha(s,t) = (s^2 - t^2, 2s, s + t)$, $s^2 + t^2 \le 1$; $\beta(u,v) = (uv, u + v, v)$, $u^2 + v^2 \le 2$.

(d) $\alpha(s,t) = (e^s, e^{s+2t}, e^{t+2s})$, $0 \le s \le 1$, $-1 \le t \le 0$; $\beta(u,v) = (v, u^2 v, v^2 u)$, $1/e \le u \le 1$, $1 \le v \le e$.

3.   Write the equations of the parametrization $\beta:U \to R^3$ for the upper unit hemisphere in R^3 in which U is the unit disc and $\beta(\vec{v})$ is the point where the line from $(0,0,-2)$ through $\vec{v}$ meets the hemisphere.

4.   Let $U = \{\vec{v} = (x,y): 1 \le \|\vec{v}\| \le 2, y \ge 0\}$, and let a parametrization for the surface Σ be given by $\alpha:U \to R^3$; $\alpha(x,y) = (x, y, x^2 + y^2)$. Find an equivalent parametrization $\beta:V \to R^3$, where $V = \{(s,t): 1 \le s \le 2, 0 \le t \le \pi\}$.

5.   Let $\alpha:R^2 \to R^3$ be given by $\alpha(x,y) = (x, y, x^2 + y^2)$. Compute the tangent plane to the surface Σ parametrized by α at $(1,2)$.

6. Let $f:R^2 \to R$ be a differentiable function, and define $\alpha(x,y) = (x,y,f(x,y))$. Compute the tangent plane to the surface Σ parametrized by α at (x_0,y_0).

7. Verify that the points satisfying Equation 1.2 are exactly the points in the image of the function α defined in Equation 1.1.

8. Show that if $\vec{w}'$ is non zero and orthogonal to the linearly independent vectors $\vec{v}$ and $\vec{w}$ in R^3, then every vector orthogonal to $\vec{v}$ and $\vec{w}$ is a multiple of $\vec{w}'$.

2. SURFACE AREA

A natural approach in attempting to define the area of a simple surface is to approximate it by surfaces whose areas we can compute and then pass to some kind of limit. For example, suppose the simple surface Σ is defined by the injective mapping $\alpha:S \to R^3$, where S is the rectangle

$$S = \{(x,y) \,|\, a_1 \le x \le a_2, \ b_1 \le y \le b_2\}$$

The points $\alpha(a_1,b_1)$, $\alpha(a_2,b_1)$, and $\alpha(a_1,b_2)$ span a planar triangle in R^2 as do the points $\alpha(a_2,b_1)$, $\alpha(a_1,b_2)$, and $\alpha(a_2,b_2)$. (See Figure 2.1). These two triangles fit together to form a polygonal surface which can be considered to be an approximation to the surface Σ, admittedly a rather poor one.

We can improve this approximation as follows. Let P be a partition of S and S_{ij} one of the subrectangles of P. Just as we did above, we form a pair of planar triangles from the vertices of S_{ij}. Repeating this construction for each of the subrectangles of P, we obtain a polygonal surface which approximates Σ. We expect that the approximation gets better as the partition gets finer and that the surface area of Σ should be the least upper bound over all partitions of S of the areas of these polygonal surfaces. Unfortunately, this does not work.

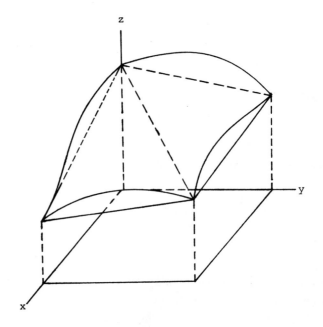

FIGURE 2.1. An approximating polygonal surface

Examples of smooth surfaces can be given for which the least upper bound of the areas of the polygonal approximations does not exist.[*]
    To give the proper definition of surface area for surfaces in R^3, we use the fact that the derivative of α at $\vec{v} \in S$ approximates the function α near $\vec{v}$.
    Let Σ be the surface defined by the function $\alpha : S \to R^3$, where $S = R(\vec{a}, \vec{b})$ is a rectangle, and let P be a partition of S defined by

$$a_1 = t_0 < t_1 < \cdots < t_m = a_2$$
$$b_1 = s_0 < s_1 < \cdots < s_n = b_2$$

Let S_{ij} be the subrectangle given by

---

[*] Such an example can be found in the article "What is Surface Area?" by T. Rado, *American Mathematical Monthly* 50 (1943), pp. 139-141.

$$S_{ij} = \{(x,y) \,|\, t_{i-1} \leq x \leq t_i, \, s_{j-1} \leq y \leq s_j\}$$

and let $\vec{c}_{ij}$ be any point in S_{ij}. If $\beta : S_{ij} \to R^3$ is defined by

$$\beta(\vec{v}) = \alpha(\vec{c}_{ij}) + (D_{\vec{c}_{ij}} \alpha)(\vec{v} - \vec{c}_{ij}) \tag{2.1}$$

then β is a good approximation for α near $\vec{c}_{ij}$. Consequently, if the partition P is fine enough, the area of the image of S_{ij} under β should be approximately the area of the image of S_{ij} under α. We can compute the area of the image of S_{ij} under β by using Proposition 4.8.3; it turns out that the area is

$$(t_i - t_{i-1})(s_j - s_{j-1}) \| (D_{\vec{c}_{ij}} \alpha)(\vec{e}_1) \times (D_{\vec{c}_{ij}} \alpha)(\vec{e}_2) \|$$

(See Exercise 10.) Thus, the expression

$$\sum_{i=1}^{m} \sum_{j=1}^{n} (t_i - t_{i-1})(s_j - s_{j-1}) \| (D_{\vec{c}_{ij}} \alpha)(\vec{e}_1) \times (D_{\vec{c}_{ij}} \alpha)(\vec{e}_2) \|$$

should converge to the area of the surface Σ as the partition P gets finer. This expression is a Riemann sum for the function taking $\vec{v}$ into $\| (D_{\vec{v}} \alpha)(\vec{e}_1) \times (D_{\vec{v}} \alpha)(\vec{e}_2) \|$, and so converges to the integral

$$\int_S \| (D\alpha)(\vec{e}_1) \times (D\alpha)(\vec{e}_2) \| \tag{2.2}$$

by Proposition 12.7.3, as the partition P gets finer (provided that the function $\| (D\alpha)(\vec{e}_1) \times (D\alpha)(\vec{e}_2) \| : S \to R$ is integrable). We therefore define the *area* of the simple surface Σ to be the integral (2.2) when it exists. (Note that this definition applies only to surfaces in R^3; indeed, (2.2) makes sense only in R^3, since the cross product is defined only in R^3.)

For example, let Σ be the piece of the cylinder with parametrization $\alpha : S \to R^3$ where

$$S = \{(x,y) \,|\, 0 \leq x \leq \pi/2, \, 0 \leq y \leq 1\}$$

and

$$\alpha(x,y) = (\cos x,\ \sin x, y)$$

(See Figure 2.2.)

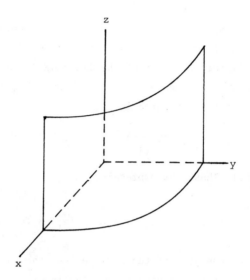

FIGURE 2.2.   A cylinder

The Jacobian matrix of α is

$$J_\alpha(x,y) = \begin{pmatrix} -\sin x & 0 \\ \cos x & 0 \\ 0 & 1 \end{pmatrix}$$

Thus, if $\vec{v} = (x,y)$,

$$(D_{\vec{v}}\alpha)\vec{e}_1 = (-\sin x,\ \cos x,\ 0)$$
$$(D_{\vec{v}}\alpha)\vec{e}_2 = (0,0,1)$$

and

$$(D_{\vec{v}}\alpha)\vec{e}_1 \times (D_{\vec{v}}\alpha)\vec{e}_2 = (\cos x,\ \sin x,\ 0) \tag{2.3}$$

The surface area of Σ is then

$$\int_S \|(\cos x, \sin x, 0)\| = \int_S 1$$

$$= \frac{\pi}{2}$$

As a more complicated example, we compute the surface area of a hemisphere. As we saw in Section 1, we can parametrize the hemisphere by $\alpha: S \to R^3$ where S is the closed ball of radius 1 about the origin and

$$\alpha(x,y) = (\frac{2x}{1+x^2+y^2}, \frac{2y}{1+x^2+y^2}, \frac{1-x^2-y^2}{1+x^2+y^2})$$

We see from (1.5) that, if $\vec{v} = (x,y)$, then

$$(D_{\vec{v}}\alpha)(e_1) = \vec{u}, \quad (D_{\vec{v}}\alpha)(e_2) = \vec{w}$$

where

$$\vec{u} = \frac{2}{(1+x^2+y^2)^2} (1 - x^2 + y^2, -2xy, -2x)$$

$$\vec{w} = \frac{2}{(1+x^2+y^2)^2} (-2xy, 1 + x^2 - y^2, -2y)$$

Thus

$$\vec{u} \times \vec{v} = \frac{4}{(1+x^2+y^2)^4}(2x + 2xy^2 + 2x^3, 2y + 2yx^2 + 2y^3,$$

$$1 - x^4 - 2x^2y^2 - y^4) \qquad (2.4)$$

$$= \frac{4}{(1+x^2+y^2)^3}(2x, 2y, 1 - x^2 - y^2),$$

and

$$\|\vec{u} \times \vec{v}\| = \frac{4}{(1+x^2+y^2)^3}((2x)^2 + (2y)^2 + (1 - x^2 - y^2)^2)^{1/2}$$

$$\qquad (2.5)$$

$$= \frac{4}{(1+x^2+y^2)^2}$$

Hence the area of the surface is

$$\int_S \frac{4}{(1+x^2+y^2)^2}$$

This integral is hard to evaluate in this form, but in polar coordinates, it becomes easy. On the surface S, r varies from 0 to 1 and θ from 0 to 2π, so the area of the hemisphere is given by

$$\int_0^1 \int_0^{2\pi} \frac{4}{(1+r^2)^2} \, r \, dr \, d\theta = \int_0^1 \frac{8\pi r}{(1+r^2)^2} \, dr$$

$$= \frac{-4\pi}{(1+r^2)} \Big|_0^1$$

$$= 2\pi$$

(See the example at the end of Section 12.6.) Thus the area of the sphere of radius one is 4π.

We may note one odd fact about this computation, which serves as a partial check. Notice that $\vec{u} \times \vec{v}$ (given in (2.4)) is a multiple of $\alpha(x,y)$. We know that $\vec{u}$ and $\vec{v}$ are vectors generating the tangent plane to the surface at $\alpha(x,y)$; since $\vec{u} \times \vec{v}$ is orthogonal to $\vec{u}$ and to $\vec{v}$, $\vec{u} \times \vec{v}$ is orthogonal to the tangent plane. That is, $\vec{u} \times \vec{v}$ points in the direction of the normal to the hemisphere at (x,y). But the radius line to a point on a sphere is normal to the sphere - that is, $\alpha(x,y)$ is normal to the surface. Therefore $\vec{u} \times \vec{v}$ and $\alpha(x,y)$ should be proportional, and they are.

The next result shows that the area of a simple surface Σ does not depend on how we parametrize it.

PROPOSITION 2.1.   *Let* $\alpha:S \to R^3$, $\beta:S' \to R^3$ *be two parametrizations for the simple surface* Σ.   *Then*

$$\int_S \| (D\alpha)(\vec{e}_1) \times (D\alpha)(\vec{e}_2) \| = \int_{S'} \| (D\beta)(\vec{e}_1) \times (D\beta)(\vec{e}_2) \|$$

Proof.   Let h:S → S' be a diffeomorphism with $\beta(h(\vec{v})) = \alpha(\vec{v})$ for all $\vec{v}$ in S. Let f:S → R, g:S' → R be defined by

$$f(\vec{v}) = \|(D_{\vec{v}}\alpha)\vec{e}_1 \times (D_{\vec{v}}\alpha)\vec{e}_2\|$$

$$g(\vec{w}) = \|(D_{\vec{w}}\beta)\vec{e}_1 \times (D_{\vec{w}}\beta)\vec{e}_2\|$$

We must show

$$\int_S f = \int_{S'} g$$

Now

$$f(\vec{v}) = \|(D_{\vec{v}}\alpha)\vec{e}_1 \times (D_{\vec{v}}\alpha)\vec{e}_2\|$$

$$= \|(D_{\vec{v}}(\beta \circ h))\vec{e}_1 \times (D_{\vec{v}}(\beta \circ h))\vec{e}_2\|$$

$$= \|(D_{h(\vec{v})}\beta) \circ (D_{\vec{v}}h)(\vec{e}_1) \times (D_{h(\vec{v})}\beta) \circ (D_{\vec{v}}h)(\vec{e}_2)\|$$

by the Chain Rule. (See Section 6.1.) The matrix of $D_{\vec{v}}h$ is the Jacobian of h,

$$J_h = \begin{pmatrix} a_{11} & a_{12} \\ a_{21} & a_{22} \end{pmatrix}$$

where the a_{ij} are the partial derivatives of the components of h. Thus, setting $T = D_{h(\vec{v})}\beta$, we have

$$f(\vec{v}) = \|T(a_{11}\vec{e}_1 + a_{21}\vec{e}_2) \times T(a_{12}\vec{e}_1 + a_{22}\vec{e}_2)\|$$

$$= \|a_{11}a_{12}(T\vec{e}_1 \times T\vec{e}_1) + a_{11}a_{22}(T\vec{e}_1 \times T\vec{e}_2)$$

$$+ a_{21}a_{12}(T\vec{e}_2 \times T\vec{e}_1) + a_{21}a_{22}(T\vec{e}_2 \times T\vec{e}_2)\|$$

$$= \|(a_{11}a_{22} - a_{21}a_{12})(T\vec{e}_1 \times T\vec{e}_2)\|$$

$$= |\text{Det } J_h| \cdot \|(D_{h(\vec{v})}\beta)\vec{e}_1 \times (D_{h(\vec{v})}\beta)\vec{e}_2\|$$

$$= |\text{Det } J_h| \cdot g(h(\vec{v}))$$

It follows that

$$\int_S f = \int g \circ h |\text{Det } J_h|$$

$$= \int_{S'} g$$

by the change of variable formula, Theorem 12.6.2. This proves the
proposition.

Let $\alpha : S \rightarrow R^3$ be a parametrization for the regular surface Σ so
that $D_{\vec{v}}\alpha : R^2 \rightarrow R^3$ has rank 2 for each $\vec{v} \in S$. Then $(D_{\vec{v}}\alpha)(\vec{e}_1)$ and
$(D_{\vec{v}}\alpha)(\vec{e}_2)$ are independent and span the tangent plane to the surface
Σ at $\alpha(\vec{v})$. Thus $\vec{n}(\vec{v}) = (D_{\vec{v}}\alpha)(\vec{e}_1) \times (D_{\vec{v}}\alpha)(\vec{e}_2)$ is non-zero and orthogonal to the tangent plane to Σ at $\alpha(\vec{v})$. (We saw an example of this
earlier, when considering the hemisphere.) For any surface Σ, regular
or not, we call the function $\vec{n} : S \rightarrow R^3$ given by

$$\vec{n}(\vec{v}) = (D_{\vec{v}}\alpha)(\vec{e}_1) \times (D_{\vec{v}}\alpha)(\vec{e}_2)$$

the normal field to Σ defined by α. (See Figure 2.3.) Thus   the
surface area of Σ is given by

$$\int_S \|\vec{n}\|$$

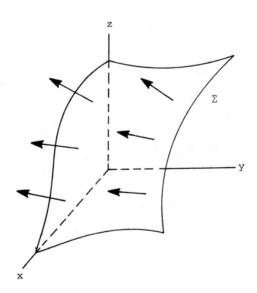

FIGURE 2.3.   A normal field to a surface

We sometimes write $\vec{n}_\alpha(\vec{v})$ for $\vec{n}(\vec{v})$ to emphasize the dependence of $\vec{n}$ on α.

EXERCISES

1.   (a) Let S be a closed set in R^2 and $f:S \to R$ a continuously differentiable function. Show that the graph of f,

$$G(f) = \{(x,y,z) \in R^3 | z = f(x,y)\}$$

is the image of a parametrized surface.

(b) Show that the area of the surface $G(f)$ is given by

$$\int_S (1 + (\frac{\partial f}{\partial x})^2 + (\frac{\partial f}{\partial y})^2)^{1/2}$$

2.   Find the area of the surface $x^2 + y^2 + 2z - 4 = 0$ over the region S in the xy plane where

$$S = \{(x,y):1 \leq x^2 + y^2 \leq 2\}$$

3.   Find the area of the surface $z = xy$ over the unit disc in the xy plane with center the origin.

4.   A *surface of revolution* is obtained by rotating a curve $z = f(x)$, $a \leq x \leq b$, in the xz plane about the z axis.

(a) Show that this surface has the equation $z = f(r)$ in cylindrical coordinates. (See Exercise 4 of Section 12.6.)

(b) Prove that the area of the surface is given by

$$2\pi \int_a^b (1 + (f'(r))^2)^{1/2} \, rdr$$

(c) Find the area of the surface obtained by rotating the curve $z = 2x + 3$, $0 \leq x \leq 4$, about the z-axis.

5.   Let $S = \{(x,y) | 0 \leq x \leq 2\pi, 0 \leq y \leq \pi\}$ and define the parametrized surface $\alpha:S \to R^3$ by

$$\alpha(x,y) = (\cos x \sin y, \sin x \sin y, \cos y)$$

Show that the image of α is the unit 2-sphere,

$$\{(x,y,z) \in R^3 | x^2 + y^2 + z^2 = 1\}$$

and use this parametrization to find the surface area of the unit 2-sphere.

6.   Find the area of the surface of the sphere of radius r.

7.   Find the area of the piece of the paraboloid $z = x^2 + y^2$ which is cut out by the region between the cylinder $x^2 + y^2 = 2$ and the cylinder $x^2 + y^2 = 6$.

8.   Set up the iterated integral for the surface area of the ellipsoid

$$\frac{x^2}{a^2} + \frac{y^2}{b^2} + \frac{z^2}{c^2} = 1$$

9.   Find the area of the surface obtained by rotating the circle $z^2 + (x - a)^2 = r^2$, $a > r$, about the z-axis. (This surface is called a *torus*.)

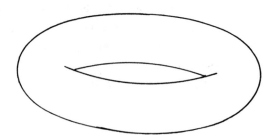

FIGURE 2.4. A torus

10.   Modify the result in Proposition 4.8.3 to show that if $S = R(\vec{a}_i, \vec{b}_i)$ is a rectangle in R^2 with $\vec{a}_i = (s_{i-1}, t_{i-1})$, $\vec{b}_i = (s_i, t_i)$,

and if $\beta:S \to R^3$ is given by (2.1), then the area of $\beta(s)$ is given by

$$(t_i - t_{i-1})(s_i - s_{i-1})\|D_{\vec{c}_{ij}}(\vec{e}_1) \times D_{\vec{c}_{ij}}(\vec{e}_2)\|$$

3. SURFACE INTEGRALS

Suppose that Σ is a surface in the open subset U of R^3 and $f:U \to R$ is a continuous function. In analogy with the situation for line integrals, we define the *surface integral of f over Σ* to be

$$\int_\Sigma f = \int_S (f\circ\alpha)\|(D\alpha)\vec{e}_1 \times (D\alpha)\vec{e}_2\|$$

$$= \int_S (f\circ\alpha)\|\vec{n}_\alpha\| \tag{3.1}$$

where $\alpha:S \to R^3$ is a parametrization for the surface Σ and the integrals on the right are the ordinary integrals of the function

$$v \to f(\alpha(\vec{v}))\|\vec{n}_\alpha(\vec{v})\| = f(\alpha(\vec{v}))\|(D_{\vec{v}}\alpha)\vec{e}_1 \times (D_{\vec{v}}\alpha)\vec{e}_2\|$$

over the region $S \subset R^2$. If we think of the surface as representing a thin sheet of material and f (restricted to the surface) as giving the density (weight per unit area), then (3.1) gives the total weight of the sheet.

As an example, let Σ be the unit hemisphere, and let $f(x,y,z) = z$. Then we may use the parametrization of (1.1). From (2.5), we get

$$\|\vec{n}(x,y)\| = \frac{4}{(1+x^2+y^2)^2}$$

while

$$(f\circ\alpha)(x,y) = \frac{1-x^2-y^2}{(1+x^2+y^2)}$$

Thus

$$\int_{\Sigma} f = \int_S \frac{4(1-x^2-y^2)}{(1+x^2+y^2)^3}$$

where S is the unit disc. Again, this integral is easier in polar coordinates:

$$\int_{\Sigma} f = \int_0^1 \int_0^{2\pi} \frac{(1-r^2)}{(1+r^2)^3} \, rd\theta dr$$

$$= 8\pi \int_0^1 \frac{(1-r^2)}{(1+r^2)^3} \, rdr$$

Let $u = r^2$; the integral becomes

$$4\pi \int_0^1 \frac{1-u}{(1+u)^3} \, du = 4\pi \int_0^1 \left(\frac{2}{(1+u)^3} - \frac{1}{(1+u)^2}\right) du$$

$$= 4\pi \left(\frac{1}{1+u} - \frac{1}{(1+u)^2}\right) \Big|_0^1$$

$$= \pi$$

As in the case of surface area, the integral (3.1) does not depend upon the parametrization of the surface. Explicitly:

PROPOSITION 3.1: *Let* $\alpha:S \to R^3$ *and* $\beta:S' \to R^3$ *be two parametrizations for the surface* Σ. *Then*

$$\int_S (f\circ\alpha)\|(D\alpha)\vec{e}_1 \times (D\alpha)\vec{e}_2\| = \int_{S'} (f\circ\beta)\|(D\beta)\vec{e}_1 \times (D\beta)\vec{e}_2\|$$

The proof is almost exactly the same as the proof of Proposition 2.1. We leave the details to the reader.

The surface integral of a function has the usual properties of integrals.

PROPOSITION 3.2. *Let* Σ *be a surface in the open subset* U *in* R^3, f,g:U $\to$ R *continuous functions, and* a,b *real numbers. Then*

(a) $\int_\Sigma (af + bg) = a\int_\Sigma f + b\int_\Sigma g$

(b) *If $f(\vec{v}) \geq g(\vec{v})$ for all $\vec{v}$ in U, then*

$\int_\Sigma f \geq \int_\Sigma g$

(c) *Let $\alpha:S \to R^3$ be a parametrization of Σ where*

$S = \{(x,y) \,|\, a \leq x \leq b, \ c \leq y \leq d\}$

and let

$S_1 = \{(x,y) \,|\, a \leq x \leq e, \ c \leq y \leq d\}$

$S_2 = \{(x,y) \,|\, e \leq x \leq b, \ c \leq y \leq d\}$

where $a < e < b$. Let $\alpha_1:S_1 \to R^3$, $\alpha_2:S_2 \to R^3$ be the restrictions of α, and Σ_1 and Σ_2 the corresponding surfaces. Then

$\int_\Sigma f = \int_{\Sigma_1} f + \int_{\Sigma_2} f$

This proposition follows from the properties of the standard integral. For example, to prove (a), we have

$$\int_\Sigma (af + bg) = \int_S ((af + bg)\circ\alpha)\|\vec{n}\|$$

$$= \int_S (a(f\circ\alpha) + b(g\circ\alpha))\|\vec{n}\|$$

$$= a\int_S (f\circ\alpha)\|\vec{n}\| + b\int_S (g\circ\alpha)\|\vec{n}\|$$

$$= a\int_\Sigma f + b\int_\Sigma g$$

The remainder of the proof is left as an exercise.

Suppose now that U is an open subset of R^3 and $\alpha:S \to U$ is a parametrization for the surface Σ. Let $\vec{F}:U \to R^3$ be a vector field on U. *The surface integral of the vector field $\vec{F}$ along Σ* (or simply the integral of $\vec{F}$ along Σ) is defined by

$$\int_\Sigma \vec{F} = \int_S <\vec{F}\circ\alpha, \vec{n}_\alpha> = \int_S <\vec{F}\circ\alpha, (D\alpha)\vec{e}_1 \times (D\alpha)\vec{e}_2> \qquad (3.2)$$

Here are two examples. In both cases, we let $\vec{F}(x,y,z) = (x,y,z)$.
(1) Let Σ be the piece of the cylinder with parametrization

$$\alpha(x,y) = (\cos x, \sin x, y)$$

$0 \le x \le \pi/2$, $0 \le y < 1$. (See Figure 3.1.) We computed $\vec{n}_\alpha$ in Section
2 (see Equation 2.2):

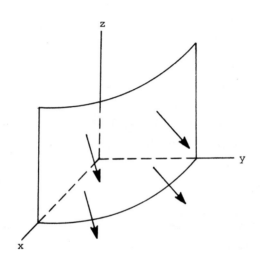

FIGURE 3.1. A normal field to a cylinder

$$\vec{n}_\alpha(x,y) = (\cos x, \sin x, 0)$$

Furthermore

$$(\vec{F}\circ\alpha)(x,y) = (\cos x, \sin x, y)$$

Thus

$$\int_\Sigma \vec{F} = \int_S (\cos^2 x + \sin^2 x + 0y)$$

$$= \int_0^1\int_0^{\pi/2} 1 \; dxdy = \frac{\pi}{2}$$

(2) Let Σ be the unit hemisphere, parametrized as in (1.4).
Then

$$(\vec{F} \circ \alpha)(x,y) = (\frac{2x}{1+x^2+y^2}, \frac{2y}{1+x^2+y^2}, \frac{1-x^2-y^2}{1+x^2+y^2})$$

and (from (2.4))

$$\vec{n}_\alpha = \frac{4}{(1+x^2+y^2)^2} (F \circ \alpha)(x,y)$$

Thus

$$< \vec{F} \circ \alpha(x,y), \vec{n}_\alpha(x,y) > = \frac{4}{(1+x^2+y^2)^2} < \vec{F} \circ \alpha(x,y), \vec{F} \circ \alpha(x,y) >$$

$$= \frac{4}{(1+x^2+y^2)^2}$$

since $\| (\vec{F} \circ \alpha)(x,y) \| = 1$. Hence

$$\int_\Sigma \vec{F} = \int_S \frac{4}{(1+x^2+y^2)^2}$$

This is the same integral we evaluated in Section 2; thus

$$\int_\Sigma \vec{F} = 2\pi$$

In order to determine the dependence of this integral on the
parametrization, we need the notion of an *oriented surface*.

Let D_1, D_2 be two subsets of R^n. A diffeomorphism $h:D_1 \to D_2$ is
said to be *orientation preserving* if $\text{Det}(D_{\vec{v}}h) > 0$ for all $\vec{v}$ in D_1 and
orientation reversing if $\text{Det}(D_{\vec{v}}h) < 0$ for all $\vec{v}$ in D_1. Two simple
parametrized surfaces $\alpha:S \to R^n$, $\beta:S' \to R^n$ are *0-equivalent* if there
is an orientation preserving diffeomorphism $h:S' \to S$ such that $\alpha(h(\vec{v}))$
$= \beta(\vec{v})$ for all $\vec{v}$ in S'. This is easily seen to be an equivalence relation. We define an *oriented surface* Σ to be an equivalence class
of simple parametrized surfaces under the relation of 0-equivalence.
(This is like the definition for curves given in Section 13.3.)

We say that α is a parametrization for the oriented surface Σ
if α is one of the parametrized surfaces in the equivalence class determining Σ as an oriented surface.

Suppose that $\alpha:S \to R^n$ is a parametrization for the oriented surface Σ where

$$S = \{(x,y) \mid a \le x \le b,\ c \le y \le d\}$$

Define $\tilde{\alpha}:S \to R^n$ by

$$\tilde{\alpha}(x,y) = \alpha(a + b - x,y)$$

Then, if $h:S \to S$ is defined by

$$h(x,y) = (a + b - x,y)$$

we have $\tilde{\alpha}(x,y) = \alpha(h(x,y))$, so that α and $\tilde{\alpha}$ are equivalent. However,
the matrix of $D_{\vec{v}}h$ is

$$\begin{pmatrix} -1 & 0 \\ 0 & 1 \end{pmatrix}$$

so $\mathrm{Det}(D_{\vec{v}}h) = -1$. It follows that α and $\tilde{\alpha}$ are not 0-equivalent and
represent different oriented surfaces. We let $-\Sigma$ denote the surface
determined by $\tilde{\alpha}$.

We can now prove that the integral of a vector field along an
oriented surface is independent of which parametrization we choose.
(This is the analogue of Proposition 3.1.)

PROPOSITION 3.3. *Let U be an open subset of* R^3 *and let* $\alpha:S \to U$,
$\beta:S' \to U$ *be two parametrizations for the oriented surface* Σ. *Then, if*
$\vec{F}:U \to R^3$ *is a vector field on U, we have*

$$\int_S \langle \vec{F} \circ \alpha, (D\alpha)\vec{e}_1 \times (D\alpha)\vec{e}_2 \rangle = \int_{S'} \langle \vec{F} \circ \beta, (D\beta)\vec{e}_1 \times (D\beta)\vec{e}_2 \rangle$$

Proof. Since this proof is similar to the proof of Proposition
3.1, we give only a sketch here. Define $f:S \to R$, $g:S' \to R$ by

$$f(\vec{v}) = \langle \vec{F} \circ \alpha(\vec{v}), (D_{\vec{v}}\alpha)\vec{e}_1 \times (D_{\vec{v}}\alpha)\vec{e}_2 \rangle$$
$$g(\vec{w}) = \langle \vec{F} \circ \beta(\vec{w}), (D_{\vec{w}}\beta)\vec{e}_1 \times (D_{\vec{w}}\beta)\vec{e}_2 \rangle$$

We must show that

$$\int_S f = \int_{S'} g \tag{3.3}$$

Suppose now that $h:S \to S'$ is an orientation preserving diffeomorphism with $\text{Det}(D_{\vec{v}}h) > 0$ and $\beta \circ h(\vec{v}) = \alpha(\vec{v})$. Then

$$f(\vec{v}) = \langle \vec{F}(\beta \circ h)\vec{v}, \; D_{\vec{v}}(\beta \circ h)\vec{e}_1 \times D_{\vec{v}}(\beta \circ h)\vec{e}_2 \rangle$$

But the calculation in Proposition 3.1 shows that

$$D_{\vec{v}}(\beta \circ h)\vec{e}_1 \times D_{\vec{v}}(\beta \circ h)\vec{e}_2 = \text{Det}(D_{\vec{v}}h)\left((D_{h(\vec{v})}\beta)\vec{e}_1 \times (D_{h(\vec{v})}\beta)\vec{e}_2\right) \tag{3.4}$$

and therefore that

$$f(\vec{v}) = \text{Det}(D_{\vec{v}}h) \langle \vec{F} \circ \beta(h(\vec{v})), \; (D_{h(\vec{v})}\beta)\vec{e}_1 \times (D_{h(\vec{v})}\beta)\vec{e}_2 \rangle$$

$$= \text{Det}(D_{\vec{v}}h)g(h(\vec{v}))$$

(Notice that because we are not taking norms, we end up with $\text{Det}(D_{\vec{v}}h)$ rather than $|\text{Det}(D_{\vec{v}}h)|$. This explains the need for oriented surfaces.) Equation (3.3) now follows from the change of variable formula, Theorem 12.6.2.

It is not difficult to show that properties analogous to (a) and (c) of Proposition 3.2 hold for the line integral of a vector field along a curve. We leave the explicit statement of these properties and their proofs to the reader.

Let $\alpha:S \to R^3$ and $\beta:S' \to R^3$ be two parametrizations for the oriented surface Σ and let $h:S \to S'$ be an orientation preserving diffeomorphism with $\beta \circ h(\vec{v}) = \alpha(\vec{v})$. We know from (3.4) that

$$(D_{\vec{v}}\alpha)(\vec{e}_1) \times (D_{\vec{v}}\alpha)(\vec{e}_2) = (\text{Det}(D_{\vec{v}}h))(D_{h(\vec{v})}\beta)(\vec{e}_1) \times (D_{h(\vec{v})}\beta)(\vec{e}_2)$$

Equivalently,

$$\vec{n}_\alpha(\vec{v}) = (\text{Det}(D_{\vec{v}}h))\vec{n}_\beta(h(\vec{v}))$$

Since $\text{Det}(D_{\vec{v}}h) > 0$, we see that *any two parametrizations for the same regular oriented surface define normal fields that point in the same direction.* Thus we can speak of the two "sides" of an oriented surface.

We note that not all surfaces have two sides.   The Mobius Band, pictured in Figure 3.4, is an example of a surface with only one side. Thus, the Mobius Band cannot be considered as an oriented surface. (Of course, it is not a simple surface in our sense.)

FIGURE 3.4.   A Mobius Band

EXERCISES

1.   Compute the surface integral of the function $f(x_1,x_2,x_3) =$ $x_1 x_3 + x_2^2$ over the surface Σ, where

    (a) Σ is the cylinder with parametrization $\alpha(x,y) =$ $(\cos x, \sin x, y)$, $0 \le x \le \pi/2$, $0 \le y \le 2$

    (b) Σ has the parametrization $\alpha(x,y) = (x^2, y^2/2, xy)$, $0 \le x \le 1$, $0 \le y \le 2$

    (c) Σ has the parametrization $\alpha:S \to R^3$, $S = \{(x,y) \,|\, 0 \le x,y \le 1\}$ and $\alpha(x,y) = (x, x^2, y)$

    (d) Σ is the parallelogram with vertices $(0,0,0)$, $(1,0,2)$, $(0,1,1)$, and $(1,1,3)$.   (Find a parametrization for Σ)

2.   Compute the surface integral of the vector field $\vec{F}(x_1,x_2,x_3) =$ $(x_1 - x_2, x_3, 1)$ over the surfaces Σ of Exercise 1.

3.   Let $h_1, h_2 : S' \to S$ be orientation preserving diffeomorphisms. Prove that the composite $h_1 \circ h_2$ is also orientation preserving.   What

can be said about $h_1 \circ h_2$ if h_1 and h_2 are orientation reversing? if h_1 is orientation preserving and h_2 orientation reversing?

4.   Prove that the relation between surfaces "α is 0-equivalent to β" is an equivalence relation.

5.   Prove Proposition 3.1.

6.   Prove Proposition 3.2.

7.   Give a detailed proof of Proposition 3.3.

8.   Let Σ be an oriented surface in the open set U of R^3 and $f:U \to R^3$ a vector field.   Prove that

$$\int_{-\Sigma} \vec{F} = - \int_{\Sigma} \vec{F}$$

9.   State and prove the analogue of Proposition 3.2 for the surface integral of a vector field.

4.   STOKES' THEOREM

We now state an analogue of Green's Theorem known as Stokes' Theorem and give an application to hydrodynamics.   The proof of Stokes' Theorem is given in Chapter 16.

In order to state Stokes' Theorem, we need to introduce the *curl* of a vector field.

Let $\vec{F}$ be a vector field on the open set $U \subset R^3$, $\vec{F}(\vec{v}) = (f_1(\vec{v}), f_2(\vec{v}), f_3(\vec{v}))$.   The *curl of* $\vec{F}$ is the vector field $\text{curl}\,\vec{F}:U \to R^3$ defined by

$$\text{curl } \vec{F} = (\frac{\partial f_3}{\partial x_2} - \frac{\partial f_2}{\partial x_3}, \frac{\partial f_1}{\partial x_3} - \frac{\partial f_3}{\partial x_1}, \frac{\partial f_2}{\partial x_1} - \frac{\partial f_1}{\partial x_2})$$

For example, if $\vec{F}(x,y) = (xy^2, x + 3, yz)$, then

$$(\text{curl } \vec{F})(x,y,z) = (z - 1, 0, 1 - 2xy)$$

Note the similarity between the formula for the curl and the formula
for the cross product. In fact, if we use ∇, the symbol for the gradient, as a "symbolic vector",

$$\nabla = (\frac{\partial}{\partial x}, \frac{\partial}{\partial y}, \frac{\partial}{\partial z})$$

then

$$\text{curl } \vec{F} = \nabla \times \vec{F}$$

as one can verify by expanding the formula for the cross product. (Of
course, this formula is purely formal; ∇ is not really a vector.)

We can now state Stokes' Theorem. As we see, it relates line
integrals to surface integrals.

THEOREM 4.1. (Stokes' Theorem) *Let* U *be an open subset of* R^3,
S *a rectangle in* R^2*, and* $\alpha : S \to U$ *a parametrization for the oriented
surface* Σ*. The restriction of* α *to the boundary of* S *defines a piecewise differentiable oriented curve which we denote by* Γ*. If* $\vec{F} : U \to R^3$
is a vector field on U*, we have*

$$\int_\Gamma \vec{F} = \int_\Sigma \text{curl } \vec{F}$$

As we indicated above, the proof of this result is given in
Chapter 16; we give an interpretation of it here.

Suppose now that U is an open subset of R^3 containing a fluid
which is in motion and $\vec{F}$ is the vector field on U which assigns to
each point $\vec{v} \in U$ the velocity of the fluid at $\vec{v}$; $\vec{F}$ is called a *velocity field*. In this case, curl $\vec{F}$ measures the angular velocity or
rotation of the fluid. For example, if $U = R^3$ and

$$\vec{F}(x,y,z) = (0,1,0)$$

the fluid is flowing uniformly in the positive y-direction and has
no rotation. (See Figure 4.1.) This is reflected in the fact that
curl $\vec{F} = (0,0,0)$.

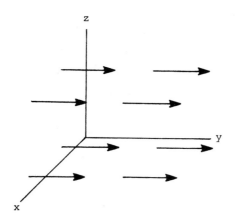

FIGURE 4.1.   Uniform fluid flow

On the other hand, if $U = R^3$ and

$$\vec{F}(x,y,z) = (-y,x,0)$$

we have a "whirlpool" along the z-axis.   (See Figure 4.2.)   This is
reflected in the fact that

$$\text{curl } \vec{F} = (0,0,-2)$$

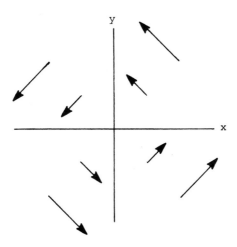

FIGURE 4.2.   The Whirlpool - looking down along the z-axis

The fact that the nonzero entry in curl $\vec{F}$ is in the z coordinate indicates that the rotation is about the z-axis. The size of the entry indicates something about the rate of rotation. In general, the curl varies from point to point, because the field rotates at different rates near different points.

A velocity field $\vec{F}$ is said to be *irrotational* if curl $\vec{F} = \vec{0}$

Let Γ be a closed curve in the open set U. We define the *circulation of the velocity field* $\vec{F}$ *around* Γ to be the line integral of $\vec{F}$ around Γ. For instance, if $\vec{F}$ is the velocity field for the whirlpool described above,

$$\vec{F}(x,y,z) = (-y,x,0)$$

and if Γ is defined by the parametrized curve $\alpha: 0,2\pi \to U$,

$$\alpha(t) = (\cos t, \sin t, 0)$$

the circulation of $\vec{F}$ around Γ is given by

$$\int_\Gamma \vec{F} = \int_0^{2\pi} < \vec{F}(\alpha(t)),\alpha'(t) > dt$$

$$= 2\pi$$

We can use this notion to interpret the curl.

Let $\vec{n}$ be a unit vector (a "direction") at the point $\vec{v}_0 \in U$. It is not difficult to show that, for any sufficiently small $r > 0$, a parametrized surface $\alpha_r : B_r \to U$ can be found whose image is the flat disk of radius r, center $\vec{v}_0$, and normal

$$(D_{\vec{v}}\alpha_r)(\vec{e}_1) \times (D_{\vec{v}}\alpha_r)(\vec{e}_2) = \vec{n}$$

for all $\vec{v} \in B_r$. (See Figure 4.3 and Exercise 5; here $B_r = \bar{B}_r(\vec{0})$ is the closed disk of radius r about $\vec{0}$ in R^2.) Let Σ_r denote the image surface and Γ_r its boundary curve. Then, by Stokes' Theorem, the circulation of $\vec{F}$ around Γ_r is given by

$$\int_{\Gamma_r} \vec{F} = \int_{\Sigma_r} \text{curl } \vec{F}$$

$$= \int_{B_r} < \text{curl } \vec{F}, \vec{n} >$$

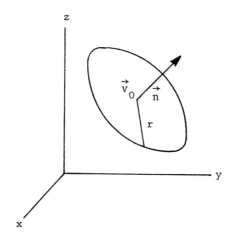

FIGURE 4.3. A disk with normal $\vec{n}$

According to the Mean Value Theorem for integrals (see Exercise 6, Section 11.5), we can find a point $\vec{v}_r \in B_r$ such that

$$\int_{B_r} < \text{curl } \vec{F}, \vec{n} > \; = \; \mu(B_r) < (\text{curl } \vec{F})(\vec{v}_r), \vec{n} >$$

where $\mu(B_r) = \pi r^2$ is the area of B_r. Thus,

$$< (\text{curl } \vec{F})(\vec{v}_r), \vec{n} > \; = \; \frac{1}{\mu(B_r)} \int_{\Gamma_r} \vec{F}$$

Now, as r approaches zero, $\vec{v}_r$ approaches $\vec{v}_0$, and we have

$$< (\text{curl } \vec{F})(\vec{v}_0), \vec{n} > \; = \; \lim_{r \to 0} \frac{1}{\mu(B_r)} \int_{\Gamma_r} \vec{F}$$

Therefore, the quantity $< (\text{curl } \vec{F})(\vec{v}_0), \vec{n} >$ is the limiting value of the circulation of $\vec{F}$ per unit area on a surface orthogonal to $\vec{n}$ at $\vec{v}_0$. Because this quantity is clearly maximized when

$$\vec{n} = \frac{(\text{curl } \vec{F})(\vec{v}_0)}{\|(\text{curl } \vec{F})(\vec{v}_0)\|}$$

(that is, when the circulation is measured in the direction of curl $\vec{F}$), curl $\vec{F}$ is called the *vorticity (vector) field of the velocity field* $\vec{F}$. (Note that "vortex" means "whirlpool.")

EXERCISES

1.   Let U be an open subset of R^3, $\vec{F},\vec{G}:U \to R^3$ vector fields, and
a,b real numbers. Prove that

$$curl(a\vec{F} + b\vec{G}) = a \ curl \ \vec{F} + b \ curl \ \vec{G}$$

2.   Let U be an open subset of R^3 and $f:U \to R$ a C^2 function. Prove
that curl(grad f) = $\vec{0}$.

3.   Compute the vorticity field of the following velocity fields.
      (a) $\vec{F}(x,y,z) = (xy,yz,xyz)$
      (b) $\vec{F}(x,y,z) = (x^2 + y^2,xz - y,2xyz^2)$
      (c) $\vec{F}(x,y,z) = (z \sin x, \sin x, x + y + z)$

*4.   A subset $X \subseteq R^3$ is said to be *simply connected* if, whenever Γ
is a closed curve in X, there is a surface Σ in X whose boundary curve
is Γ. Let U be an open subset of R^3 which is both pathwise connected
and simply connected and let $\vec{F}$ be a vector field on U. Prove that U
has a potential function if and only if curl $\vec{F}$ = $\vec{0}$.

*5.   Let $\vec{n}$ be a unit vector at $\vec{v}_0 \in R^3$ and $B_r = \bar{B}_r(\vec{0})$ the closed ball
of radius r about $\vec{0}$ in R^2. Prove that there is a parametrized sur-
face $\alpha_r:B_r \to R^3$ whose image is the planar disk of radius r with cen-
ter $\vec{v}_0$ and whose normal $(D_{\vec{v}_0} \alpha_r)(\vec{e}_1) \times (D_{\vec{v}_0} \alpha_r)(\vec{e}_2)$ is $\vec{n}$.

CHAPTER 15

DIFFERENTIAL FORMS

INTRODUCTION

In this chapter and the next we show how to generalize line and surface integrals to more general sorts of integrals on R^n. At the end,
we shall give a far-reaching generalization of Green's Theorem and
Stokes' Theorem. Before then, however, we need to develop a good deal
of material, much of it formal and not obviously connected with integration. This chapter is concerned with the more formal aspects of
the theory; in the next chapter we show how to give it substance. As
a result, this chapter may seem rather abstract and artificial.

The subject matter of this chapter is algebraic operations on
differential forms. We define differential forms, show how to add,
multiply, and differentiate them, and discuss the relationships among
these operations.

1.   THE ALGEBRA OF DIFFERENTIAL FORMS

In this section, we introduce the notion of a differential form on an
open subset of Euclidean space. Eventually it will be clear what the
relevance of this section and those that follow is to Green's and Stokes'
theorems, but for some time the connection will be mysterious. The
best procedure for the moment is simply to regard differential forms
as completely new mathematical objects and to learn how to compute

651

with them, using the rules for algebraic manipulations presented here.
From time to time, we shall give illustrations of the relation between
differential forms and the notions of the past few chapters.

Let U be an open subset of R^n. A *differential 0-form* is a differentiable function $f:U \to R$. A *differential 1-form* on U is an expression of the kind

$$f_1 dx_1 + f_2 dx_2 + \cdots + f_n dx_n$$

where each f_i is a differentiable function, $f_i:U \to R$, $1 \le i \le n$. For
example

$$xy dx + 2 dy$$

is a differentiable 1-form on R^n,

$$(1/z)dy + (2 \sin x)dz$$

is a differentiable 1-form on the open set $U = \{(x,y,z) \mid z \ne 0\}$ in R^3,
and

$$x_1 dx_2 + x_4^2 dx_7$$

is a differentiable 1-form on R^7.

For convenience, we drop the word "differential" from "differential form" and speak simply of *forms*. For example, a differential
0-form will be called an 0-form and a differential 1-form will be
called a 1-form.

Notice that the expressions we integrated when working with line
integrals (see the end of Section 13.3) were 1-forms. The reason for
introducing differential forms is, in fact, to allow us to generalize
line integrals to higher-dimensional surfaces.

We can add two 1-forms on U and multiply a 1-form on U by an
0-form on U in the obvious way:

$$\left(\sum_{i=1}^{n} f_i dx_i \right) + \left(\sum_{i=1}^{n} g_i dx_i \right) = \sum_{i=1}^{n} (f_i + g_i) dx_i$$

$$g \left(\sum_{i=1}^{n} f_i dx_i \right) = \sum_{i=1}^{n} (gf_i) dx_i$$

These operations satisfy the usual formal properties (for example, the commutative and distributive laws).

A 2-*form* ω *on* U is an expression of the kind

$$\omega = \sum_{i,j=1}^{n} f_{ij} dx_i dx_j \tag{1.1}$$

where each f_{ij} is a differentiable function on U. Here we think of the term $dx_i dx_j$ as the "product" of the 1-form dx_i with the 1-form dx_j. This multiplication is defined to have the property

$$dx_i dx_j = - dx_j dx_i \tag{1.2}$$

for all $i,j=1,\ldots,n$. In particular, $dx_i dx_i = - dx_i dx_i$, so that $dx_i dx_i = 0$ for $i=1,\ldots,n$.

We note that, using (1.2), we can write the 2-form ω of (1.1) as

$$\omega = \sum_{1 \le i < j \le n} g_{ij} dx_i dx_j$$

where $g_{ij} = f_{ij} - f_{ji}$. For instance,

$$\sum_{i,j=1}^{3} x_i^2 x_j^3 dx_i dx_j = (x_1^2 x_2^3 - x_1^3 x_2^2) dx_1 dx_2 + (x_1^2 x_3^3 - x_1^3 x_3^2) dx_1 dx_3$$

$$+ (x_2^2 x_3^3 - x_2^3 x_3^2) dx_2 dx_3$$

since the $dx_1 dx_1, dx_2 dx_2$, and $dx_3 dx_3$ terms are all 0.

We define the product of two 1-forms by

$$\left(\sum_{i=1}^{n} f_i dx_i \right) \left(\sum_{j=1}^{n} g_j dx_j \right) = \sum_{i,j=1}^{n} f_i g_j dx_i dx_j$$

$$= \sum_{1 \le i < j \le n} (f_i g_j - f_j g_i) dx_i dx_j$$

Here are two examples:

1. If ω = xdy + xydz and η = 2dx - 4 sin zdy + x^2dx, then

$\omega\eta$ = 2xdydx - 4x sin zdydy + x^3dydz

   + 2xydzdx - 4xy sin zdzdy + x^3ydzdz

= 2xdydx + x^3dydz + 2xydzdx - 4xy sin zdzdy

(since dydy = dzdz = 0)

= - 2xdxdy - 2xydxdz + (x^3 + 4xy sin z)dydz

(since dydx = - dxdy, dzdx = - dxdz, and dzdy = - dydz).

2. If

$$\omega = \sum_{i=1}^{3} a_i dx_i, \quad \eta = \sum_{i=1}^{3} b_j dx_j$$

are 1-forms on R^3, then

$$\omega\eta = (a_1 b_2 - a_2 b_1)dx_1 dx_2 + (a_1 b_3 - a_3 b_1)dx_1 dx_3$$

$$+ (a_2 b_3 - a_3 b_2)dx_2 dx_3$$

Note the similarity here with the cross product of the vectors (a_1, a_2, a_3) and (b_1, b_2, b_3). (See Section 3.8.)

It is easy to see that the multiplication of 1-forms defined above satisfies all the usual properties of multiplication, except that

$$\omega\eta = - \eta\omega$$

for any two 1-forms ω and η. (See Exercise 4; to make sense of this exercise, we need to be able to add 2-forms and to multiply them by 0-forms. The definitions should be obvious; in any case we shall give them soon.)

More generally, if p is any integer, we define a p-form on U to be an expression of the kind

$$\omega = \Sigma f_{i_1 i_2 \ldots i_p} dx_{i_1} dx_{i_2} \ldots dx_{i_p} \tag{1.3}$$

where the sum ranges over all sets of p integers $i_1, \ldots, i_p$ with $1 \le i_1, i_2, \ldots, i_p \le n$, and where each $f_{i_1 i_2 \ldots i_p}$ is a differentiable real valued function on U. For example,

$$x_1 x_7 dx_1 dx_3 dx_4 - \cos x_3 dx_3 dx_5 dx_8$$

is a 3-form on R^8 and

$$(\log x_4) dx_1 dx_3 dx_4 dx_5 + (x_2 - x_4^2) dx_1 dx_2 dx_3 dx_4$$

is a 4-form on the set $U = \{(x_i) \in R^5 | x_4 > 0\}$.

As above, we think of the terms $dx_{i_1} \ldots dx_{i_p}$ as the product of the 1-forms $dx_{i_1}, \ldots, dx_{i_p}$, again with the condition that $dx_i dx_j = - dx_j dx_i$. We can also add two p-forms, multiply a p-form by an 0-form, and multiply a p-form and a q-form to obtain a (p+q)-form. For addition, we simply sum all the terms of the two forms: if

$$\omega_1 = \Sigma f_{i_1 \ldots i_p} dx_{i_1} \ldots dx_{i_p}$$

and

$$\omega_1 = \Sigma g_{i_1 \ldots i_p} dx_{i_1} \ldots dx_{i_p}$$

then

$$\omega_1 + \omega_2 = \Sigma (f_{i_1 \ldots i_p} + g_{i_1 \ldots i_p}) dx_{i_1} \ldots dx_{i_p}$$

If we multiply ω_1 by the 0-form h, we obtain

$$\Sigma h f_{i_1 \ldots i_p} dx_{i_1} \ldots dx_{i_p}$$

To multiply a p-form by a q-form, we multiply each term of the p-form by each term of the q-form, remembering the rule that $dx_i dx_j = - dx_j dx_i$, and add the products together. Here is an example. If

$$\omega = x_1 dx_1 dx_3 - \sin x_2 dx_2 dx_5$$

and

$$\eta = 2x_3^2 dx_1 dx_3 + 2x_4 dx_2 dx_3 - (x_1 + x_2) dx_4 dx_5$$

then

$$\omega\eta = 2x_1 x_3^2 dx_1 dx_3 dx_1 dx_3 + 2x_1 x_4 dx_1 dx_3 dx_2 dx_3$$

$$- x_1(x_1 + x_2) dx_1 dx_3 dx_4 dx_5$$

$$- 2x_3^2 \sin x_2 dx_2 dx_5 dx_1 dx_3$$

$$- 2x_4 \sin x_2 dx_2 dx_5 dx_2 dx_3$$

$$+ (x_1 + x_2)\sin x_2 dx_2 dx_5 dx_4 dx_5$$

Now

$$2x_1 x_3^2 dx_1 dx_3 dx_1 dx_3 = - 2x_1 x_3^2 dx_1 dx_1 dx_3 dx_3$$

since $dx_1 dx_3 = - dx_3 dx_1$. However $dx_1 dx_1 = 0$, so this term vanishes.
Similarly,

$$2x_1 x_4 dx_1 dx_3 dx_2 dx_3 = 0$$

$$2x_4 \sin x_2 dx_2 dx_5 dx_2 dx_3 = 0$$

and

$$(x_1 + x_2)\sin x_2 dx_2 dx_5 dx_4 dx_5 = 0$$

(More generally, any term which repeats a dx_j factor is 0.   See Exercise 3.)   Thus

$$\omega\eta = - x_1(x_1 + x_2) dx_1 dx_3 dx_4 dx_5 - 2x_3^2 \sin x_2 dx_2 dx_5 dx_1 dx_3$$

$$= - x_1(x_1 + x_2) dx_1 dx_3 dx_4 dx_5 + 2x_3^2 \sin x_2 dx_1 dx_2 dx_3 dx_5$$

since $dx_2 dx_5 dx_1 dx_3 = - dx_1 dx_2 dx_3 dx_5$.   (Three interchanges are needed to get $dx_1 dx_2 dx_3 dx_5$ from $dx_2 dx_5 dx_1 dx_3$:

$$dx_2dx_5dx_1dx_3 = - dx_2dx_1dx_5dx_3$$
$$= dx_1dx_2dx_5dx_3$$
$$= - dx_1dx_2dx_3dx_5)$$

EXERCISES

1. Let ω_1, ω_2, ω_3 be the 1-forms on R^3 given by

$$\omega_1 = x_1dx_2 - dx_3$$

$$\omega_2 = 2x_3^2dx_1$$

$$\omega_3 = dx_1 - x_2x_3dx_2$$

Compute the following.

(a) $x_2\omega_1 + \omega_2$

(b) $x_3\omega_2 - x_1\omega_3$

(c) $\omega_1\omega_3$

(d) $(2\omega_1 - x_2\omega_3)\omega_2$

(e) $\omega_1\omega_2\omega_3$

(f) $(\omega_1 - x_1\omega_2)(2\omega_3 - \omega_1)(x_2^2\omega_2 - \omega_3)$

2. Let ω_1, ω_2, ω_3 and ω_4 be the 1-forms on R^4 given by

$$\omega_1 = x_2dx_1 + \sin x_1dx_3 + 2dx_4$$

$$\omega_2 = x_3^2dx_2 + x_2x_4dx_3$$

$$\omega_3 = (x_1 - x_4)dx_1 + x_2dx_4$$

$$\omega_4 = x_2dx_2 - x_1x_4dx_3$$

Compute the following.

(a) $x_2\omega_1 + 3\omega_2$

(b) $2x_4\omega_2 - x_3^2\omega_4$

(c) $\omega_1\omega_3$

(d) $(2\omega_1 + x_4\omega_2)\omega_4$

(e) $\omega_1\omega_2\omega_3$

(f) $(x_4\omega_1 + \omega_3)(2\omega_1 - \omega_2)\omega_4$

(g) $3\omega_2(\omega_2 + \omega_3)(\omega_1 - x_2\omega_4)$

(h) $\omega_1\omega_2\omega_3\omega_4$

(i) $(\omega_1 - x_2\omega_3)(\omega_2 + \omega_3)\omega_4(\omega_2 - \omega_1)$

3.   Let $\omega = fdx_{i_1}\ldots dx_{i_p}$ and $\eta = gdx_{j_1}\ldots dx_{j_q}$. Show that if $i_k = j_\ell$ for any h,ℓ, then $\omega\eta = 0$. (For example,

$$(fdx_1 dx_4 dx_5)(g\,dx_2 dx_3 dx_4) = 0$$

because $i_2 = 4 = j_3$.)

4.   Prove that $\omega\eta = -\eta\omega$ for any 1-forms ω and η.

5.   (a) Prove that $\omega^2 = \omega\omega = 0$ for any 1-form ω.
     (b) Show that there is a 2-form η on R^4 with $\eta^2 \neq 0$.

2.   BASIC PROPERTIES OF THE SUM AND PRODUCT OF FORMS

We now derive some of the basic properties of the addition and multi-plication of differential forms. Our first result states that the associative law and distributive laws hold for these operations.

PROPOSITION 2.1.   Let ω_1,ω_2 be p-forms, η_1,η_2 q-forms, and Υ an r-form.   Then

(a) $(\omega_1 + \omega_2)\eta_1 = \omega_1\eta_1 + \omega_2\eta_1$

(b) $\omega_1(\eta_1 + \eta_2) = \omega_1\eta_1 + \omega_1\eta_2$

(c) $\omega_1(\eta_1\gamma) = (\omega_1\eta_1)\gamma$

Proof. This result is an immediate consequence of the definitions; we give the proof of (a), leaving the proofs of (b) and (c) as
exercises.

Suppose

$$\omega_1 = \Sigma_I \, f_{i_1\ldots i_p} dx_{i_1}\ldots dx_{i_p}$$

$$\omega_2 = \Sigma_I \, g_{i_1\ldots i_p} dx_{i_1}\ldots dx_{i_p}$$

$$\eta_1 = \Sigma_J \, h_{j_1\ldots j_q} dx_{j_1}\ldots dx_{j_q}$$

where Σ_I denotes the sum over all $i_1\ldots i_p$ with $1 \leq i_1,\ldots,i_p \leq n$ and
Σ_J denotes the sum over all $j_1,\ldots,j_q$ with $1 \leq j_1,\ldots,j_q \leq n$. Then

$$(\omega_1 + \omega_2)\eta_1 = (\Sigma_I \, f_{i_1\ldots i_p} dx_{i_1}\ldots dx_{i_p} + \Sigma_I \, g_{i_1\ldots i_p} dx_{j_1}\ldots dx_{i_p})$$

$$\Sigma_J \, h_{j_1\ldots j_q} dx_{j_1}\ldots dx_q$$

$$= (\Sigma_I (f_{i_1\ldots i_p} + g_{i_1\ldots i_p}) dx_{i_1}\ldots dx_{i_p})$$

$$\Sigma_J \, h_{j_1\ldots j_q} dx_{j_1}\ldots dx_{j_q}$$

(by the definition of the sum of forms)

$$= \Sigma_I \Sigma_J (f_{i_1\ldots i_p} + g_{i_1\ldots i_p}) h_{j_1\ldots j_q} dx_{i_1}\ldots dx_{i_p} dx_{j_1}\ldots dx_{j_q}$$

(by the definition of the product of forms)

$$= \Sigma_I \Sigma_J (f_{i_1\ldots i_p} h_{j_1\ldots j_q} + g_{i_1\ldots i_p} h_{j_1\ldots j_q}) dx_{i_1}\ldots dx_{i_p} dx_{j_1}\ldots dx_{j_q}$$

$$= \Sigma_I \Sigma_J \; f_{i_1 \ldots i_p} h_{j_1 \ldots j_q} \, dx_{i_1} \ldots dx_{i_p} \, dx_{j_1} \ldots dx_{j_q}$$

$$+ \; \Sigma_I \Sigma_J \; g_{i_1 \ldots i_p} h_{j_1 \ldots j_q} \, dx_{i_1} \ldots dx_{i_p} \, dx_{j_1} \ldots dx_{j_q}$$

(by the definition of the sum of forms)

$$= \omega_1 \eta_1 + \omega_2 \eta_1$$

(by the definition of the product of forms). Thus, statement (a) of Proposition 2.1 is proved.

The next result is a version of the commutative law for the multiplication of forms (sometimes called the *anti-commutative law*).

PROPOSITION 2.2. *Let ω be a p-form on* U *and η a q-form on* U. *Then*

$$\omega\eta = (-1)^{pq} \eta\omega \qquad\qquad (2.1)$$

Proof. Suppose $\omega = f dx_{i_1} \ldots dx_{i_p}$ and $\eta = g dx_{j_1} \ldots dx_{j_q}$. Then

$$\omega\eta = fg dx_{i_1} \ldots dx_{i_p} \, dx_{j_1} \ldots dx_{j_q}$$

and

$$\eta\omega = fg dx_{j_1} \ldots dx_{j_q} \, dx_{i_1} \ldots dx_{i_p}$$

To get from $\omega\eta$ to $\eta\omega$ clearly involves interchanging pq of the dx_{i_ℓ} and dx_{j_k}, since each of the q dx_j's must pass p dx_i's. Each of these interchanges introduces a factor of -1. Therefore equation (2.1) holds in this case.

If ω and η are arbitrary forms, then each is a sum of terms of the kind considered above, and equation (2.1) follows from (a) and (b) of Proposition 2.1.

We now show that p-forms on R^n for $p \geq n$ have a particularly simple expression.

PROPOSITION 2.3. *Let ω be an n-form on the open set U of R^n.* *Then*

$$\omega = f dx_1 dx_2 \ldots dx_n$$

If ω is a p-form on U with $p > n$, then $\omega = 0$.

Proof. As is not hard to see, any p-form ω can be written

$$\omega = \Sigma_I \, f_{i_1 \ldots i_p} \, dx_{i_1} \ldots dx_{i_p}$$

where $1 \leq i_1 < i_2 < \cdots < i_p < n$. If $p = n$, it follows that $i_1 = 1$, $i_2 = 2, \ldots, i_n = n$, and the sum consists of exactly one term, $f_{1 \ldots n} dx_1 \ldots dx_n$.

If ω is a p-form with $p > n$, then, in each of the summands of ω, at least one of the dx_i must occur twice. Thus $\omega = 0$, since $dx_i dx_i = 0$.

Our final result expresses the determinant of an $n \times n$ matrix in terms of the product of 1-forms.

PROPOSITION 2.4. *Let $A = (a_{ij})$ be an $n \times n$ matrix and define* *1-forms $\omega_1, \ldots, \omega_n$ by*

$$\omega_j = \sum_{j=1}^{n} a_{ij} dx_i$$

Then

$$\omega_1 \omega_2 \ldots \omega_n = (\text{Det } A) dx_1 dx_2 \ldots dx_n$$

Proof. We give a sketch of the proof, leaving the details as Exercise 4.

For any $n \times n$ matrix $A = (a_{ij})$, we define a number $\Delta(A)$ by the equation

$$\omega_1 \omega_2 \ldots \omega_n = \Delta(A) dx_1 dx_2 \ldots dx_n$$

where $\omega_1, \ldots, \omega_n$ are defined as in Proposition 2.4. Using the properties of addition and multiplication of forms, we see that the function
Δ satisfies the properties (3.1), (3.2), and (3.3) of Section 10.3.
Proposition 2.4 now follows from Theorem 3.2.

EXERCISES

1.   Prove statement (b) of Proposition 2.3.

2.   Prove statement (c) of Proposition 2.3.

3.   Verify Proposition 2.4 directly for

      (a) 2×2 matrices
      (b) 3×3 matrices

4.   Give the details of the proof of Proposition 2.4.   (In particular, verify that the function Δ satisfies (3.1), (3.2) and (3.3) of
Section 10.3.)

3.   THE EXTERIOR DIFFERENTIAL

In this section, we describe differentiation on differential forms:
given a p-form ω, we produce a (p+1)-form $d\omega$, called the exterior differential of ω.

    To begin with, recall that a real valued function f defined on
an open subset U in R^n is said to be a C^r-*function* (or of *class* C^r) if
all partial derivatives of f of order r exist and are continuous.
(Here, $1 \leq r < \infty$.) We say f is a C^∞-function if f is C^r for all r.
Similarly, the p-form ω of (1.3) is called a C^r-*form* if each of the
functions $f_{i_1 \ldots i_p}$ is C^r, $1 \leq r \leq \infty$. We assume all forms in this section to be C^r for some $r \geq 2$.

Let f be a C^2-function on the open set U in R^n. The *exterior differential* of f, df, is the 1-form defined by

$$df = \sum_{j=1}^{n} \frac{\partial f}{\partial x_j} dx_j$$

For example, if $f: R^3 \to R$ is given by $f(x_1, x_2, x_3) = x_1^2 + x_2 x_3$, then

$$df = 2x_1 dx_1 + x_3 dx_2 + x_2 dx_3$$

In defining 1-forms in the previous section, we introduced the formal expressions dx_i, $1 \leq i \leq n$. In fact, dx_i can be identified with the exterior differential of the ith coordinate function. More explicitly, let $x_i: R^n \to R$ be the function that assigns to any point in R^n its ith coordinate; $x_i(y_1, \ldots, y_n) = y_i$. Then

$$d(x_i) = \sum_{j=1}^{n} \frac{\partial x_i}{\partial x_j} dx_j$$

$$= dx_i$$

since $\frac{\partial x_i}{\partial x_j} = 1$ if $i = j$ and 0 otherwise.

If ω is a p-form,

$$\omega = \Sigma_I f_{i_1 \ldots i_p} dx_{i_1} \ldots dx_{i_p}$$

with $1 \leq i_1, \ldots, i_p \leq n$, then $d\omega$ is the (p+1)-form defined by

$$d\omega = \Sigma_I d(f_{i_1 \ldots i_p}) dx_{i_1} \ldots dx_{i_p}$$

$$= \Sigma_I \Sigma_j \frac{\partial f_{i_1 \ldots i_p}}{\partial x_j} dx_j dx_{i_1} \ldots dx_{i_p}$$

This last sum is over all $j, i_1, \ldots, i_p$ with $1 \leq j \leq n$ and $1 \leq i_1, \ldots, i_p \leq n$.

To illustrate, let ω be the 1-form on R^4 given by

$$= x_2 x_4 dx_1 + x_1^2 dx_3$$

Then

$$dw = d(x_2 x_4) dx_1 + d(x_1^2) dx_3$$

$$= (x_4 dx_2 + x_2 dx_4) dx_1 + (2x_1 dx_1) dx_3$$

$$= - x_4 dx_1 dx_2 + 2x_1 dx_1 dx_3 - x_2 dx_1 dx_4$$

Another example; let ω be the 2-form on R^4 given by

$$\omega = x_4 dx_1 dx_2 + x_2 x_3 dx_1 dx_3$$

Then

$$dw = dx_4 dx_1 dx_2 + (x_3 dx_2 + x_2 dx_3) dx_1 dx_3$$

$$= dx_1 dx_2 dx_4 - x_3 dx_1 dx_2 dx_3$$

since $dx_3 dx_1 dx_3 = dx_1 dx_3 dx_3 = 0$.

We shall derive some of the properties of the exterior differential in the next section. We conclude this section with a discussion of the relationship between the exterior differential and certain important operations on vector fields and functions.

Let U be an open subset of R^n, $f:U \to R$ a differentiable function, and $\vec{F}:U \to R^n$ a vector field, $\vec{F}(\vec{v}) = (f_1(\vec{v}),\ldots,f_n(\vec{v}))$. The gradient of f and the curl of $\vec{F}$ (when n = 3) are vector fields on U defined earlier:

$$\text{grad } f = (\frac{\partial f}{\partial x_1},\ldots,\frac{\partial f}{\partial x_n})$$

$$\text{curl } \vec{F} = (\frac{\partial f_3}{\partial x_2} - \frac{\partial f_2}{\partial x_3}, \frac{\partial f_1}{\partial x_3} - \frac{\partial f_3}{\partial x_1}, \frac{\partial f_2}{\partial x_1} - \frac{\partial f_1}{\partial x_2})$$

We define the *divergence* of $\vec{F}$ to be the function div $\vec{F}:U \to R$ given by

$$\text{div } \vec{F} = \frac{\partial f_1}{\partial x_1} + \frac{\partial f_2}{\partial x_2} + \cdots + \frac{\partial f_n}{\partial x_n}$$

We have already seen how the gradient and curl arise in applications; we shall give an interpretation of the divergence in the next chapter. Notice that (formally) the divergence of $\vec{F}$ behaves like an inner product of $\nabla = (\frac{\partial}{\partial x_1}, \ldots, \frac{\partial}{\partial x_n})$ with $\vec{F} = (f_1, \ldots, f_n)$, and the curl of $\vec{F}$ like the cross product $\nabla \times \vec{F}$.

Let $\Omega^p(U)$ be the set of all p-forms on U (where, as above, U is an open subset of R^n). It is easily verified that $\Omega^p(U)$ is a vector space with addition and scalar multiplication defined as in Section 1. It follows immediately from Proposition 2.3 that $\Omega^0(U)$ and $\Omega^n(U)$ are isomorphic vector spaces with $f \in \Omega^0(U)$ corresponding to the n-form

$$f dx_1 \ldots dx_n$$

Similarly, we define a mapping from $\Omega^1(U)$ into $\Omega^{n-1}(U)$ by taking the 1-form

$$\omega = \sum_{i=1}^{n} f_i dx_i$$

into the (n-1)-form

$$\tilde{\omega} = \sum_{i=1}^{n} (-1)^{i+1} f_i dx_1 \ldots dx_{i-1} dx_{i+1} \ldots dx_n$$

(The factor $(-1)^{i+1}$ is included for convenience; this will be explained below.) This mapping is easily seen to be an isomorphism between the vector spaces $\Omega^1(U)$ and $\Omega^{n-1}(U)$. (See Exercise 7.)

For example, the 1-form

$$x_2^3 dx_1 + 3x_1 x_2 dx_4$$

in R^4 corresponds to the 3-form

$$x_2^3 dx_2 dx_3 dx_4 - 3x_1 x_2 dx_1 dx_2 dx_3$$

under this isomorphism.

In the same way, it can be shown that $\Omega^p(U)$ and $\Omega^{n-p}(U)$ are isomorphic vector spaces for $1 \leq p \leq n$. (See Exercise 8.)

Now, let $\Phi(U)$ be the vector space of all vector fields on U (with the obvious addition and scalar multiplication). Define a mapping from $\Phi(U)$ into $\Omega^1(U)$ by assigning the 1-form

$$\sum_{i=1}^{n} f_i dx_i$$

to the vector field $\vec{F} = (f_1, \ldots, f_n)$. This mapping is an isomorphism (Exercise 9) and, as a result, we can identify the vector spaces $\Omega^1(U)$, $\Omega^{n-1}(U)$, and $\Phi(U)$. Under this identification, the vector field

$$\vec{F} = (f_1, \ldots, f_n) \tag{3.1}$$

the 1-form

$$\omega = \sum_{i=1}^{n} f_i dx_i \tag{3.2}$$

and the $(n-1)$-form

$$\tilde{\omega} = \sum_{i=1}^{n} (-1)^{i+1} f_i dx_1 \ldots dx_{i-1} dx_{i+1} \ldots dx_n \tag{3.3}$$

each correspond to one another.

We can now reinterpret the divergence, gradient, and curl in terms of operations on differential forms.

Suppose, first of all, that f is a differentiable function on U. Then grad f is the vector field

$$\text{grad } f = \left(\frac{\partial f}{\partial x_1}, \ldots, \frac{\partial f}{\partial x_n}\right)$$

which corresponds to the 1-form

$$\sum_{i=1}^{n} \frac{\partial f}{\partial x_i} dx_i$$

under the identification described above. However, this 1-form is exactly df, the exterior differential of f. Thus we see that grad f *is the vector field corresponding to the 1-form* df.

Next, let $\vec{F} = (f_1, \ldots, f_n)$ and let $\tilde{\omega}$ be the $(n-1)$-form corresponding to $\vec{F}$ (given in (3.3)). Then $d\tilde{\omega}$ is an n-form;

$$d\tilde{\omega} = \sum_{i,j=1}^{n} (-1)^{i+1} \frac{\partial f_i}{\partial x_j} dx_j dx_1 \ldots dx_{i-1} dx_{i+1} \ldots dx_n$$

$$= \sum_{i=1}^{n} (-1)^{i+1} \frac{\partial f_i}{\partial x_i} dx_i dx_1 \ldots dx_{i-1} dx_{i+1} \ldots dx_n$$

(see Exercise 5)

$$= (\sum_{i=1}^{n} \frac{\partial f_i}{\partial x_i}) dx_1 \ldots dx_n$$

(since $dx_i dx_j = -dx_j dx_i$)

$$= (\text{div } \vec{F}) dx_1 \ldots dx_n$$

Thus div $\vec{F}$ *is the function corresponding to the exterior differential of the (n-1)-form* $\tilde{\omega}$ *corresponding to* $\vec{F}$.

Finally, let $\vec{F}$ be a vector field on the open subset U of R^3 and let ω be the corresponding 1-form (given in (3.2)). Then, a straight-forward computation shows that

$$d\omega = (\frac{\partial f_2}{\partial x_1} - \frac{\partial f_1}{\partial x_2}) dx_1 dx_2 + (\frac{\partial f_3}{\partial x_1} - \frac{\partial f_1}{\partial x_3}) dx_1 dx_3$$

$$+ (\frac{\partial f_3}{\partial x_2} - \frac{\partial f_2}{\partial x_3}) dx_2 dx_3$$

which is exactly the 2-form corresponding to the vector field curl $\vec{F}$. (Note that n - 1 = 2 here since n = 3.) Therefore, curl $\vec{F}$ *is the vector field corresponding the exterior differential of the 1-form* ω *corresponding to* $\vec{F}$.

We note that the factor $(-1)^{i+1}$ which occurred in defining the correspondence between 1-forms and (n-1)-forms was needed in reinterpreting both div $\vec{F}$ and curl $\vec{F}$.

EXERCISES

1. Compute $d\omega$ where ω is the form on R^3 given by
    (a) $\omega = xy + z^2$

(b) $\omega = xdy - ydz$

(c) $\omega = xydx + (x - 2z)ydy + xyz^2dz$

(d) $\omega = 7xzdxdy + \sin xydxdz - 2zdydz$

(e) $\omega = \cos xydxdy - yz^2dxdz + (x^2y - yz^2)dydz$

2.   Compute $d\omega$ where ω is the form on R^4 given by

(a) $\omega = x_1x_3 - x_2^2x_4$

(b) $\omega = x_2dx_1 + x_1x_3dx_2 - x_2x_4^2dx_4$

(c) $\omega = \sin x_1x_2dx_1 + x_3 \cos x_4dx_3 - (x_1^2 + x_2^2)dx_4$

(d) $\omega = x_2^2x_4dx_1dx_4 + \sin x_1x_2dx_2dx_3$

(e) $\omega = (x_3^2 - x_1x_4)dx_1dx_3 + 7 \cos(x_2^2 + x_3^2)dx_2dx_3$

(f) $\omega = x_1x_3^2dx_1dx_2dx_4 - \cos x_2x_3dx_1dx_3dx_4$

3.   Verify that $d(d\omega) = 0$ for the forms in Exercises 2 and 3.

4.   Let U be an open subset of R^n, $\vec{F}$, $\vec{G}:U \rightarrow R^n$ vector fields on U, and a,b real numbers. Prove that

$$\text{div}(a\vec{F} + b\vec{G}) = a \text{ div } \vec{F} + b \text{ div } \vec{G}$$

5.   Verify that if $\omega = (-1)^{j+1}dx_1...dx_{j-1}dx_{j+1}...dx_n$, then

$$(dx_k)\omega = \begin{cases} dx_1...dx_n & \text{if } j = k \\ 0 & \text{otherwise} \end{cases}$$

6.   Verify that for n = 3, if the vector field $\vec{F}$ corresponds to the 1-form ω and $\vec{G}$ corresponds to η, then $\vec{F} \times \vec{G}$ corresponds to $\omega\eta$ and curl $\vec{F}$ corresponds to $d\omega$.

7.   Verify that the mapping taking the 1-form

$$\omega = \sum_{i=1}^{n} f_i dx_i$$

into the (n-1)-form

$$\widetilde{\omega} = \Sigma(-1)^{i+1}f_i dx_1 \ldots dx_{i-1} dx_{i+1} \ldots dx_n$$

is an isomorphism between $\Omega^1(U)$ and $\Omega^{n-1}(U)$ for U open in R^n.

8.   Prove that $\Omega^p(U)$ and $\Omega^{n-p}(U)$ are isomorphic vector spaces.

9.   Verify that the mapping taking the vector field $\vec{F} = (f_1, \ldots, f_n)$ into the 1-form

$$\omega = \sum_{i=1}^{n} f_i dx_i$$

is an isomorphism of $\Phi(U)$ onto $\Omega^1(U)$.

4.   BASIC PROPERTIES OF THE EXTERIOR DIFFERENTIAL

We now derive some of the important properties of the exterior differential. We begin by showing that it preserves the sum of forms.

PROPOSITION 4.1.   *If ω_1 and ω_2 are p-forms on the open set U of* R^n, *then*

$$d(\omega_1 + \omega_2) = d\omega_1 + d\omega_2$$

Proof.   Let ω_1 and ω_2 be given by

$$\omega_1 = \Sigma_I f_{i_1 \ldots i_p} dx_{i_1} \ldots dx_{i_p}$$
$$\omega_2 = \Sigma_I g_{i_1 \ldots i_p} dx_{i_1} \ldots dx_{i_p}$$

Then

$$
\begin{aligned}
d(\omega_1 + \omega_2) &= d(\Sigma_I f_{i_1 \ldots i_p} dx_{i_1} \ldots dx_{i_p} + g_{i_1 \ldots i_p} dx_{i_1} \ldots dx_{i_p}) \\
&= d(\Sigma_I (f_{i_1 \ldots i_p} + g_{i_1 \ldots i_p}) dx_{i_1} \ldots dx_{i_p}) \\
&= \Sigma_I d(f_{i_1 \ldots i_p} + g_{i_1 \ldots i_p}) dx_{i_1} \ldots dx_{i_p}) \\
&= \Sigma_I (df_{i_1 \ldots i_p} + dg_{i_1 \ldots i_p}) dx_{i_1} \ldots dx_{i_p})
\end{aligned}
$$

since the partial derivative of the sum of two functions is the sum
of the partial derivatives. This last expression is just $d\omega_1 + d\omega_2$.

An easy induction proves the following extension of Proposition
4.1.

COROLLARY. *Let* $\omega_1, \ldots, \omega_k$ *be* p-*forms on the open set* U *of* R^n.
Then

$$d\left(\sum_{i=1}^{k} \omega_i\right) = \sum_{i=1}^{k} d\omega_i$$

We now determine the behavior of the exterior differential on
the product of two forms.

PROPOSITION 4.2. *Let* ω *be a* p-*form on* U *and* η *a* q-*form on* U.
Then

$$d(\omega\eta) = (d\omega)\eta + (-1)^p \omega d\eta$$

Proof. Note first of all that, for 0-forms f and g,

$$d(fg) = gdf + fdg$$

by the product rule for differentation. Thus, if

$$\omega = fdx_{i_1} \ldots dx_{i_p}$$

and

$$\eta = gdx_{j_1} \ldots dx_{j_q}$$

then

$$d(\omega\eta) = d(fgdx_{i_1} \ldots dx_{i_p} dx_{j_1} \ldots dx_{j_q})$$

$$= (gdf + fdg)(dx_{i_1} \ldots dx_{i_p} dx_{j_1} \ldots dx_{j_q})$$

$$= (df)dx_{i_1} \ldots dx_{i_p} gdx_{j_1} \vdots \ldots dx_{j_q}$$

$$+ fdg(dx_{i_1} \ldots dx_{i_p} dx_{j_1} \ldots dx_{j_q})$$

Now dg is a 1-form and $dx_{i_1} \ldots dx_{i_p}$ a p-form, so

$$dg(dx_{i_1} \ldots dx_{i_p}) = (-1)^p (dx_{i_1} \ldots dx_{i_p}) dg$$

by Proposition 2.2. Thus

$$d(\omega\eta) = ((df) dx_{i_1} \ldots dx_{i_p}) g dx_{j_1} \ldots dx_{j_q}$$
$$+ (-1)^p (f dx_{i_1} \ldots dx_{i_p}) dg dx_{j_1} \ldots dx_{j_q}$$
$$= (d\omega)\eta + (-1)^p \omega d\eta$$

Since any p-form and any q-form are the sum of terms of the type treated above, the proof of Proposition 4.2 for general ω and η follows from Proposition 4.1 and the special case above.

The final result of this section describes the result of applying the exterior differential twice.

PROPOSITION 4.3. *For any* p-*form* ω, $d(d\omega) = 0$.

Sketch of proof. It is easily seen that Proposition 2.3 for an arbitrary p-form will follow if we prove $d(df) = 0$ where f is a zero form. In this case, direct computation shows that

$$d(df) = \sum_{1 \leq i < j \leq n} \left(\frac{\partial^2 f}{\partial x_i \partial x_j} - \frac{\partial^2 f}{\partial x_j \partial x_i}\right) dx_i dx_j$$

which, vanishes by Theorem 6.3.3 (since f is assumed to be of class c^2).

EXERCISES

1.   Prove the Corollary to Proposition 4.1.

2.   Give the details of the proof of Proposition 4.3 when $p > 0$.
(The problem is to keep track of the subscripts.)

3.   Let U be an open subset of R^3, $f:U \to R$ a function, and $\vec{F}:U \to R^3$ a vector field. Use Proposition 4.3 to prove

$$\text{curl grad } f = 0$$
$$\text{div curl } \vec{F} = 0$$

5. THE ACTION OF DIFFERENTIABLE FUNCTIONS ON FORMS

Suppose that U is an open subset of R^n, V an open subset of R^m, and $\varphi : U \to V$ a differentiable function of class C^2. (For the definition of class C^2, see Section 3.) In coordinates,

$$\varphi(\vec{v}) = (\varphi_1(\vec{v}), \ldots, \varphi_m(\vec{v}))$$

We now associate to each p-form ω on V a p-form $\varphi^*\omega$ on U.

If f is an 0-form on V, $\varphi^* f$ is defined to be the composite $f \circ \varphi$. If $y_1, \ldots, y_m$ are coordinates in R^m, we define $\varphi^* dy_i$ by

$$\varphi^* dy_i = \sum_{j=1}^{n} \frac{\partial \varphi_i}{\partial x_j} dx_j \tag{5.1}$$

For an arbitrary 1-form ω on U,

$$\omega = f_1 dy_1 + \cdots + f_m dy_m$$

we define $\varphi^* \omega$ by

$$\varphi^* \omega = \sum_{i=1}^{m} \varphi^*(f_i) \varphi^* dy_i$$

$$= \sum_{i=1}^{m} \sum_{j=1}^{n} (f_i \circ \varphi) \frac{\partial \varphi_i}{\partial x_j} dx_j \tag{5.2}$$

Finally, if

$$\omega = \Sigma f_{i_1 \ldots i_p} dy_{i_1} \ldots dy_{i_p}$$

is a p-form, then

$$\varphi^* \omega = \Sigma (\varphi^*(f_{i_1 \ldots i_p}) \varphi^*(dy_{i_1}) \ldots \varphi^*(dy_{i_p}) \tag{5.3}$$

We call φ^* the *mapping induced by* φ *on forms* or simply the *mapping induced by* φ.

The definition of φ^* given above may seem somewhat artificial. The following remarks may help motivate it.

If ω is a 0-form, say $\omega = f:V \to R$, then defining $\varphi^*\omega$ to be the composite $f \varphi$ is quite natural. The definition of $\varphi^*\omega$ when ω is a p-form, $p > 0$, is dictated by the desire to have φ^* satisfy the following:

$$d(\varphi^*\omega) = \varphi^* d\omega \tag{5.4}$$

$$\varphi^*(\omega + \omega') = \varphi^*\omega + \varphi^*\omega' \tag{5.5}$$

$$\varphi^*(\omega\eta) = (\varphi^*\omega)(\varphi^*\eta) \tag{5.6}$$

Here ω and ω' are p-forms and η is a q-form. (We prove that these properties do actually hold in the next section.)

For example, if $x_j : R^n \to R$, $1 \le j \le n$, are the coordinate functions in R^n and $y_i : R^m \to R$, $1 \le i \le m$, the coordinate function in R^m, then, if (5.4) is to hold, we must have

$$\varphi^*(dy_i) = \varphi^*(d(y_i))$$
$$= d(\varphi^*(y_i))$$
$$= d(y_i \circ \varphi)$$
$$= d\varphi_i$$
$$= \sum_{j=1}^{n} \frac{\partial \varphi_i}{\partial x_j} dx_j$$

This is exactly equation (5.1). To define φ^* on an arbitrary 1-form, we use properties (5.5) and (5.6) to obtain (5.2).

The same reasoning shows that equation (5.3), which defines φ^* on an arbitrary p-form, is also a consequence of properties (5.4), (5.5), and (5.6). (See Exercise 9.)

Here are some examples.

1.   Let $\varphi : R \to R$ and let $\omega = f\,dy$ be a 1-form on R.   Then

$$\varphi^*\omega = \varphi^*(f)\varphi^*(dy)$$

$$= (f\circ\varphi)\,\frac{d\varphi}{dx}\,dx$$

2.   Let $\varphi:R^2 \to R^2$ be given by

$$\varphi(x_1,x_2) = (ax_1 + bx_2, cx_1 + ex_2)$$

and let $\omega = dx_1 dx_2$.   Then

$$\varphi^*dx_1 = \frac{\partial\varphi_1}{\partial x_1}\,dx_1 + \frac{\partial\varphi_1}{\partial x_2}\,dx_2$$

$$= adx_1 + bdx_2$$

$$\varphi^*dx_2 = \frac{\partial\varphi_2}{\partial x_1}\,dx_1 + \frac{\partial\varphi_2}{\partial x_2}\,dx_2$$

$$= cdx_1 + edx_2$$

Thus

$$\varphi^*\omega = (\varphi^*dx_1)(\varphi^*dx_2)$$

$$= (adx_1 + bdx_2)(cdx_1 + edx_2)$$

$$= (ac - be)dx_1 dx_2$$

We generalize this example in Exercise 7.

3.   Let $\varphi:R^2 \to R^2$ be defined by

$$\varphi(x_1,x_2) = (x_1 x_2, x_1^2 - x_2^2, 2x_1 + 3x_2)$$

and set

$$\omega = y_1 dy_1 dy_2 - y_2^2 dy_2 dy_3$$

Then

$$\varphi^*(y_1) = y_1\circ\varphi = x_1 x_2$$

$$\varphi^*(y_2) = x_1^2 - x_2^2$$

$$\varphi^*(y_3) = 2x_1 + 3x_2$$

$$\varphi^*(dy_1) = x_2 dx_1 + x_1 dx_2$$
$$\varphi^*(dy_2) = 2x_1 dx_1 - 2x_2 dx_2$$

and

$$\varphi^*(dy_3) = 2dx_1 + 3dx_2$$

Therefore

$$\varphi^*\omega = x_1 x_2 (x_2 dx_1 + x_1 dx_2)(2x_1 dx_1 - 2x_2 dx_2)$$
$$- (x_1^2 - x_2^2)^2 (2x_1 dx_1 - 2x_2 dx_2)(2dx_1 + 3dx_2)$$
$$= - 2(x_1 x_2^3 + x_1^3 x_2 + 3x_1(x_1^2 - x_2^2) + 2x_2(x_1^2 - x_2^2)^2)dx_1 dx_2$$

We give the basic properties of the mapping induced by φ in the next section.

EXERCISES

1.   Let $\varphi : R^2 \rightarrow R^2$ be given by $\varphi(x,y) = (x^2 + y^2, 2xy)$ and determine $\varphi^*\omega$ where

    (a) $\omega = st^2$ (Here, s and t are the coordinates in the second R^2.)

    (b) $\omega = tds$

    (c) $\omega = t^2 ds - sdt$

    (d) $\omega = dsdt$

    (e) $\omega = \sin(st)dsdt$

2.   (a)-(e). Let $\varphi : R^3 \rightarrow R^2$ be given by $\varphi(x,y,z) = (xy, yz)$ and determine $\varphi^*\omega$ for the forms ω of Exercise 1.

3.   Let $\varphi : R^3 \rightarrow R^3$ be given by $\varphi(x_1, x_2, x_3) = (2x_1 + x_2, x_1 x_3, x_2 - x_3)$ and determine $\varphi^*\omega$ where

    (a) $\omega = y_1 y_2 y_3$ (Here, y_1, y_2, and y_3 are the coordinates in the second R^3.)

    (b) $\omega = dy_2$

(c) $\omega = y_2 dy_1 - y_1 dy_3$

(d) $\omega = dy_1 dy_3$

(e) $\omega = y_1 y_3^2 dy_1 dy_2 - y_1^3 dy_1 dy_3 + \sin y_1 dy_2 dy_3$

(f) $\omega = dy_1 dy_2 dy_3$

(g) $\omega = (y_1 y_2 - y_3^2) dy_1 dy_2 dy_3$

4.   (a)-(g). Let $\varphi : R^3 \to R^3$ be given by $\varphi(x_1, x_2, x_3) = (x_1^2 x_3, x_2 x_3, x_2^3)$ and determine $\varphi^* \omega$ for the forms in Exercise 3.

5.   (a)-(e). Show that $\varphi^* d\omega$ and $d\varphi^* \omega$ are equal for φ and ω as in Exercise 1.

6.   (a)-(g). Show that $\varphi^* d\omega$ and $d\varphi^* \omega$ are equal for φ and ω as in Exercise 3.

7.   Let U and V be open subsets of R^n, $\varphi : U \to V$ a C^2 function, and $\omega = g dx_1 \ldots dx_n$ an n-form on V. Prove that

$$\varphi^* \omega = (g \circ \varphi)(\text{Det}(D\varphi)) dx_1 \ldots dx_n$$

(Hint: recall Proposition 2.4.)

8.   Let $\beta : I^2 \to R^3$ be a C^2-function and $\vec{F}(\vec{v}) = (f_1(\vec{v}), f_2(\vec{v}), f_3(\vec{v}))$ a vector field. Let ω be the 1-form defined by

$$\omega = f_1 dx_1 + f_2 dx_2 + f_3 dx_3$$

and prove that

$$\beta^* d\omega = \; < (\text{curl } \vec{F}) \circ \beta, (D\beta)\vec{e}_1 \times (D\beta)\vec{e}_2 > dx dy$$

where x and y are coordinates in I^2.

9.   Show that, given the definition of φ^* on 0-forms and 1-forms, the definition (5.3) of φ^* on p-forms, $p > 1$, follows from (5.4), (5.5), and (5.6).

6.   BASIC PROPERTIES OF THE INDUCED MAPPING

We give here some of the important properties of the induced mapping on forms.

PROPOSITION 6.1. *Let U be an open subset of* R^n, *V an open subset of* R^m, *and* $\varphi:U \to V$ *a differentiable function. Then*

$$\varphi^*(\omega + \omega') = \varphi^*\omega + \varphi^*\omega' \tag{6.1}$$

and

$$\varphi^*(\omega\eta) = (\varphi^*\omega)(\varphi^*\eta) \tag{6.2}$$

for any p-forms ω,ω' *and q-form* η *on V. Furthermore, if W is an open subset of* R^k *and* $\psi:V \to W$ *a differentiable function, then*

$$(\psi\circ\varphi)^* = \varphi^* \circ \psi^* \tag{6.3}$$

Proof. The first two assertions of the proposition are immediate consequences of the definition of φ^* and are left as Exercises 1 and 2. We prove the third.

If μ is the zero form $f:W \to R$, then

$$(\psi\circ\varphi)^* f = f \circ \psi \circ \varphi$$

$$= (f\circ\psi) \circ \varphi$$

$$= \varphi^*(f\circ\psi)$$

$$= \varphi^*(\psi^* f)$$

Let $z_1,\ldots,z_k$ be coordinates in R^k and suppose that $\varphi(\vec{v}) = (\varphi_1(\vec{v}),\ldots,\varphi_m(\vec{v}))$ and $\psi(\vec{w}) = (\psi_1(\vec{w}),\ldots,\psi_k(\vec{w}))$. Then

$$(\psi\circ\varphi)^* dz_i = \sum_{j=1}^{n} \frac{\partial(\psi_i\circ\varphi)}{\partial x_j} \, dx_j$$

$$= \sum_{j=1}^{n} \sum_{\ell=1}^{m} (\frac{\partial\psi_i}{\partial y_\ell}\circ\varphi)\frac{\partial\varphi_\ell}{\partial x_j} \, dx_j$$

by the Chain Rule. Therefore,

$$(\psi \circ \varphi)^* dz_i = \sum_{\ell=1}^{m} (\frac{\partial \psi_i}{\partial y} \circ \varphi) \sum_{j=1}^{n} \frac{\partial \varphi_\ell}{\partial x_j} dx_j$$

$$= \sum_{\ell=1}^{m} \varphi^* \frac{\partial \psi_i}{\partial y} \varphi^* (dy_\ell)$$

$$= \varphi^* \sum_{\ell=1}^{m} \frac{\partial \psi_i}{\partial y} dy_\ell$$

$$= \varphi^* (\psi^* dz_i) = (\varphi^* \circ \psi^*) dz_i$$

Thus, if $\mu = f_1 dz_1 + \cdots + f_k dz_k$, we have

$$(\psi \circ \varphi)^* \mu = \sum_{i=1}^{k} ((\psi \circ \varphi)^* f_i)(\psi \circ \varphi)^* dz_i$$

$$= \sum_{i=1}^{k} (\varphi^* \circ \psi^* f_i)(\varphi^* \circ \psi^* dz_i)$$

$$= \varphi^* \sum_{i=1}^{k} (\psi^* f_i)(\psi^* dz_i)$$

$$= (\varphi^* \circ \psi^*) \mu$$

The verification of (6.3) when μ is a p-form now follows from the definition of φ^*, formulas (6.1) and (6.2), and the cases p = 0,1 above.
The details are left as Exercise 3.

The next proposition shows that the action of φ^* and the exterior
differential commute.

PROPOSITION 6.2. *Let U be an open subset of* R^n, *V an open subset of* R^m, *and* $\varphi : U \to V$ *a differentiable function. Then*

$$d(\varphi^* \omega) = \varphi^* (d\omega)$$

for any p-form ω *on V.*

Proof. Let $x_1, \ldots, x_n$ be coordinates in R^n, $y_1, \ldots, y_m$ coordinates
in R^m, and $\varphi(\vec{v}) = (\varphi_1(\vec{v}), \ldots, \varphi_m(\vec{v}))$. The proof breaks up into four

cases. We proceed directly by computing both $\varphi^*(d\omega)$ and $d(\varphi^*\omega)$.

Case 1. $\omega = f$ is a 0-form.

$$\varphi^* d\omega = \varphi^* \sum_{i=1}^{m} \frac{\partial f}{\partial y_i} dy_i$$

$$= \sum_{i=1}^{m} \varphi^*(\frac{\partial f}{\partial y_i})\varphi^*(dy_i)$$

$$= (\sum_{i=1}^{m} \frac{\partial f}{\partial y_i}\circ\varphi) \sum_{j=1}^{m} \frac{\partial \varphi_i}{\partial x_j} dx_j$$

and

$$d\varphi^* \omega = d(f\circ\varphi)$$

$$= \sum_{j=1}^{n} \frac{\partial(f\circ\varphi)}{\partial x_j} dx_j$$

$$= \sum_{j=1}^{n} \sum_{i=1}^{m} (\frac{\partial f}{\partial y_i}\circ\varphi)\frac{\partial \varphi_j}{\partial x_j} dx_j$$

by the Chain Rule. Thus the formula holds in this case.

Case 2. $\omega = dy_k$.

First of all, $\varphi^* d\omega = 0$, since $d\omega = d(dy_k) = 0$ (by Proposition 4.3). On the other hand,

$$d\varphi^* \omega = d \sum_{j=1}^{n} \frac{\partial \varphi_k}{\partial x_j} dx_j$$

$$= \sum_{i,j=1}^{n} \frac{\partial^2 \varphi_k}{\partial x_i \partial x_j} dx_i dx_j$$

$$= \sum_{1\leq i<j\leq n} (\frac{\partial^2 \varphi_k}{\partial x_i \partial x_j} - \frac{\partial^2 \varphi_k}{\partial x_j \partial x_i}) dx_i dx_j$$

$$= 0$$

by Theorem 6.3.3, since the functions φ_k have continuous second partial derivatives.

Case 3. $\omega = f dy_{i_1} \ldots dy_{i_p}$

To begin with,

$$\varphi^*(d\omega) = \varphi^*((df)dy_{i_1} \ldots dy_{i_p})$$

$$= (\varphi^* df)\varphi^* dy_i \ldots \varphi^* dy_{i_p}$$

(by Proposition 6.1)

$$= (d\varphi^* f)\varphi^* dy_{i_1} \ldots \varphi^* dy_{i_p}$$

by Case 1. Thus we need to show that

$$d(\varphi^* f)\varphi^* dy_{i_1} \ldots \varphi^* dy_{i_p} = d((\varphi^* f)(\varphi^* dy_{i_1}) \ldots (\varphi^* dy_{i_p})) \qquad (6.4)$$

Using Proposition 4.2, we see that

$$d((\varphi^* f)(\varphi^* dy_{i_1}) \ldots (\varphi^* dy_{i_p})) = (d(\varphi^* f))(\varphi^* dy_{i_1} \ldots \varphi^* dy_{i_p}) \qquad (6.5)$$

$$+ \sum_{j=1}^{p} (-1)^{j-1} (\varphi^* dy_{i_1}) \ldots \varphi^* dy_{i_{j-1}} d(\varphi^* dy_{i_j})(\varphi^* dy_{i_{j+1}}) \ldots \varphi^* dy_{i_p}$$

However, Case 2 shows that

$$d(\varphi^* dy_{i_j}) = \varphi^* d(dy_{i_j})$$

$$= 0$$

and thus only the first term on the left-hand side of (6.5) remains. Therefore (6.4) does hold, and Case 3 is done.

Case 4. ω an arbitrary p-form.

Since any p-form is the sum of terms of the kind treated in Case 3, this case is a consequence of Proposition 4.1 and Case 3. The Proposition is therefore proved.

EXERCISES

1. Prove equation (6.1) of Proposition 6.1.

2. Prove equation (6.2) of Proposition 6.1.

3. Prove that equation (6.3) of Proposition 6.1 holds for p-forms, $p > 1$, assuming that it holds for $p = 0$ and 1.

4. Give the details of case 4 of the proof of Proposition 6.2.

CHAPTER 16

INTEGRATION OF DIFFERENTIAL FORMS

INTRODUCTION

In this chapter, we give meaning to the various notions of the previous chapter. We show first how to integrate a differential p-form over a so-called "singular p-chain." The result is an integral which includes both line and surface integrals of vector fields as special cases. We then derive a general Stokes' theorem for this integral and show that both Green's Theorem and Stokes' Theorem are special cases of it. As another special case, we prove Gauss' Theorem, which relates certain surface integrals in R^3 to volume integrals, and give an application to hydrodynamics.

1. INTEGRATION OF FORMS

We now define the integral of a p-form over a singular p-cube in R^n. This notion generalizes both line and surface integrals.

To begin with, let ω be a p-form on an open subset $U \subset R^p$ and S a closed subset of U. Then ω can be written $\omega = f dx_1 \ldots dx_p$ and we define the *integral of* ω *over* S by

$$\int_S \omega = \int_S f$$

where the second integral is the usual integral f over S, the one defined in Chapter 11.

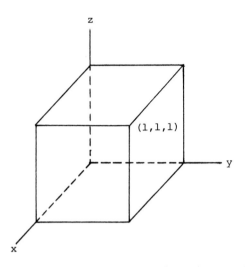

FIGURE 1.1.   The standard 3-cube

In order to define the integral of a p-form in R^n, we need the notion of a singular p-cube. The *standard p-cube* is the subset I^p of R^p defined by

$$I^p = \{(x_1, \ldots, x_p) : 0 \leq x_i \leq 1, \, 1 \leq i \leq p\}$$

if $p > 0$. We set $I^0 = \{0\}$.

For $p > 0$, I_p is simply the rectangle in R^p defined by the origin and the point $(1, 1, \ldots, 1)$. A *singular p-cube* on a subset B of R^n is a differentiable mapping $T : I^p \to B$. For example, we can consider a parametrized curve $\alpha : I^1 \to R^n$ to be a singular 1-cube and a parametrized surface $\alpha : I^2 \to R^n$ to be a singular 2-cube. In general, however, singular cubes can be quite degenerate; the mapping $T : I^p \to R^n$, $T(\vec{v}) = \vec{0}$, is an extreme example.

Let U be an open subset of R^n, ω a p-form on U, and $T : I^p \to U$ a singular p-cube. We define the *integral of ω over T* by

$$\int_T \omega = \int_{I^p} T^* \omega$$

where the integral on the right is the integral of the p-form $T^* \omega$ over the subset $I^p \subset R^p$ defined above. Here are some examples.

1.   Let $p = 1, \omega = f_1 dx_1 + \cdots + f_n dx_n$, and $\alpha : I \to R^n$ a singular 1-cube, $\alpha(t) = (\alpha_1(t), \ldots, \alpha_n(t))$. (We write I for I^1.) Then

$$\int_\alpha \omega = \int_I \alpha^* \omega$$

$$= \int_I (f_1 \circ \alpha)\alpha_1' + \cdots + (f_n \circ \alpha)\alpha_n'$$

$$= \int_I < \vec{F} \circ \alpha, \alpha' >$$

which coincides with our earlier definition of the integral of the vector field $\vec{F} = (f_1, \ldots, f_n)$ along the parametrized curve α.

2.   Let $p = 2$ and $n = 3$. Then ω is a 2-form,

$$\omega = f_1 dx_2 dx_3 - f_2 dx_1 dx_3 + f_3 dx_1 dx_2$$

(note the minus sign; see Section 15.3 for the explanation), and $\alpha : I^2 \to R$, $\alpha(\vec{v}) = (\alpha_1(\vec{v}), \alpha_2(\vec{v}), \alpha_3(\vec{v}))$, is a differentiable singular 2-cube. Suppose we let (x,y) be coordinates in R^2 and (x_1, x_2, x_3) be coordinates in R^3. Then

$$\alpha^* \omega = (f_1 \circ \alpha_1)\alpha^*(dx_2)\alpha^*(dx_3) - (f_2 \circ \alpha_2)\alpha^*(dx_1)\alpha^*(dx_3)$$

$$+ (f_3 \circ \alpha_3)\alpha^*(dx_1)\alpha^*(dx_2)$$

where

$$\alpha^*(dx_1) = \frac{\partial \alpha_1}{\partial x} dx + \frac{\partial \alpha_1}{\partial y} dy$$

$$\alpha^*(dx_2) = \frac{\partial \alpha_2}{\partial x} dx + \frac{\partial \alpha_2}{\partial y} dy$$

$$\alpha^*(dx_3) = \frac{\partial \alpha_3}{\partial x} dx + \frac{\partial \alpha_3}{\partial y} dy$$

It follows that

$$\alpha^* \omega = [(f_1 \circ \alpha_1)(\frac{\partial \alpha_2}{\partial x} \frac{\partial \alpha_3}{\partial y} - \frac{\partial \alpha_2}{\partial y} \frac{\partial \alpha_3}{\partial x}) + (f_2 \circ \alpha_2)(\frac{\partial \alpha_3}{\partial x} \frac{\partial \alpha_1}{\partial y} - \frac{\partial \alpha_3}{\partial y} \frac{\partial \alpha_1}{\partial x})$$

$$+ (f_3 \circ \alpha_3)(\frac{\partial \alpha_1}{\partial x} \frac{\partial \alpha_2}{\partial y} - \frac{\partial \alpha_1}{\partial y} \frac{\partial \alpha_2}{\partial x})] dxdy$$

$$= < \vec{F} \circ \alpha, (D\alpha)\vec{e}_1 \times (D\alpha)\vec{e}_2 > dxdy$$

where $\vec{F}:U \to R^3$ is the vector field given by $\vec{F}(\vec{v}) = (f_1(\vec{v}), f_2(\vec{v}), f_3(\vec{v}))$.
Thus

$$\int_\alpha \omega = \int_{I^2} \alpha^* \omega = \int_{I^2} < F\circ\alpha, (D\alpha)\vec{e}_1 \times (D\alpha)\vec{e}_2 >$$

This coincides with our definition of the surface integral of $\vec{F}$ over
the parametrized surface α.

3.   Let ω be the 3-form on R^4 defined by

$$\omega = x_3^2 x_4 dx_1 dx_3 dx_4 + x_1 x_2 dx_2 dx_3 dx_4$$

and let $T:I^3 \to R^4$ be defined by

$$T(x,y,z) = (z, xy, x, y+z)$$

Then

$$T^* \omega = T^*(x_3^2 x_4) T^*(dx_1) T^*(dx_3) T^*(dx_4)$$
$$+ T^*(x_1 x_2) T^*(dx_2) T^*(dx_3) T^*(dx_4)$$

Now

$$T^*(x_3^2 x_4) = x^2(y + z)$$
$$T^*(x_1 x_2) = xyz$$
$$T^*(dx_1) = dz$$
$$T^*(dx_2) = ydx + xdy$$
$$T^*(dx_3) = dx$$

and

$$T^*(dx_4) = dy + dz$$

Thus

$$T^* \omega = x^2(y + z)dzdx(dy + dz) + xyz(ydx + xdy)dx(dy + dz)$$
$$= (x^2(y + z) - x^2 yz)dxdydz$$

and

$$\int_T \omega = \int_{I^3} x^2(y + z - yz)$$

$$= \int_0^1 \int_0^1 \int_0^1 x^2(y + z - yz)dxdydz$$

$$= \int_0^1 \int_0^1 \frac{1}{3}(y + z - yz)dydz$$

$$= \int_0^1 \frac{1}{3}(\frac{1}{2} + z - \frac{z}{2})dz = \frac{1}{4}$$

We shall have more to say about the integrals of n-forms on singular n-cubes in R^n in Section 3.

EXERCISES

1.   Calculate the integral of the p-form ω over the singular p-cube T, with

(a) $T:I^1 \rightarrow R^3$, $T(s) = (s,s^2,s^3)$ and $\omega = dx + dz$

(b)   T as in (a) and $\omega = xydx - zdy$

(c) $T:I^2 \rightarrow R^3$, $T(s,t) = (s,t,st)$ and $\omega = dxdz$

(d)   T as in (c), $\omega = dydz$

(e)   T as in (c), $\omega = 2dxdz - xdydz$

(f)   T as in (c), $\omega = xy^2dxdy - 2yzdxdz + 4dydz$

(g) $T:I^3 \rightarrow R^4$, $T(s,t,u) = (st^2,tu,s,s + u)$, $\omega = dx_1dx_2dx_4$

(h)   T as in (g), $\omega = x_2dx_1dx_3dx_4 - 3dx_2dx_3dx_4$

(i)   T as in (g), $\omega = (x_2x_3 - x_1^2)dx_1dx_2dx_4 + x_3^2x_4dx_2dx_3dx_4$

2.   Calculate the integral of the 2-form ω over the singular 2-cube $T:I^2 \rightarrow R$ given by $T(s,t) = (st,s^2,t^2,s + t)$ when

(a) $\omega = dx_1dx_3$

(b) $\omega = x_2dx_1dx_2 - 3x_3^2dx_3dx_4$

(c) $\omega = x_1 x_3 dx_1 dx_3 - (x_2^2 + x_3 x_4) dx_2 dx_4$

(d) $\omega = 2x_2 x_3 dx_2 dx_3 + x_1 x_2 x_3 dx_2 dx_4$

3.   Calculate the integral of the p-form ω over the singular p-cube $T: I^3 \to R^4$ where $T(s,t,u) = (stu, s^2 t, t^2 u, u^3)$ and

(a) $\omega = dx_1 dx_3 dx_4$

(b) $\omega = x_2 dx_1 dx_2 dx_4 - 2dx_2 dx_3 dx_4$

(c) $\omega = x_2 x_3^2 dx_1 dx_2 dx_3 + x_1 x_4 dx_2 dx_3 dx_4$

(d) $\omega = x_4 dx_1 dx_2 dx_4 - 2x_2^2 x_3 dx_1 dx_3 dx_4$

2.   THE GENERAL STOKES' THEOREM

In this section, we state the general form of Stokes' Theorem.   This is an extremely important result.   We derive Green's Theorem (Theorem 13.5.1) and Stokes' Theorem (Theorem 14.4.1) from it in Section 3 and give an application to hydrodynamics in Section 4.   The proof is given in Section 5.

We begin with some definitions.

A *singular p-chain* on a subset $B \subset R^n$ is a formal linear combination of singular p-cubes on B;

$$c = a_1 T_1 + a_2 T_2 + \cdots + a_r T_r$$

where $a_1, \ldots, a_r$ are real numbers and $T_1, \ldots, T_r$ singular p-cubes on B. We emphasize that we do not consider c to be a function, just a formal expression.   We shall give a geometric interpretation of certain chains in the next section.   (See also the definition of ∂T below.)

If U is an open subset of R^n, ω a p-form on U, and c a singular p-chain on U, we define the *integral of ω over c by*

$$\int_c \omega = \sum_{j=1}^{r} a_j \int_{T_j} \omega$$

We can add p-chains and multiply them by real numbers in the obvious way; if $c = a_1T_1 + \cdots + a_rT_r$, $c' = b_1T_1' + \cdots + b_mT_m'$, and a is a real number, then

$$c + c' = a_1T_1 + \cdots + a_rT_r + b_1T_1' + \cdots + b_mT_m'$$
$$ac = (aa_1)T_1 + \cdots + (aa_r)T_r$$

Our first result of this section gives some of the elementary properties of the integral of a p-form on a p-chain.

PROPOSITION 2.1. *Let a be a real number, ω and ω' p-forms on $U \subset R^n$, and c,c' singular p-chains on U. Then*

$$\int_c (\omega + \omega') = \int_c \omega + \int_c \omega'$$

$$\int_{c+c'} \omega = \int_c \omega + \int_{c'} \omega$$

$$\int_{ac} \omega = a\int_c \omega = \int_c a\omega$$

The proofs are straightforward; for example, if $c = a_1T_1 + \cdots + a_rT_r$, we have

$$\int_c (\omega + \omega') = \sum_{i=1}^r a_i \int_{T_i} (\omega + \omega')$$

$$= \sum_{i=1}^r a_i \int_{I^p} T_i^*(\omega + \omega')$$

$$= \sum_{i=1}^r a_i \int_{I^p} (T_i^*\omega + T_i^*\omega')$$

$$= \sum_{i=1}^r a_i (\int_{I^p} T_i^*\omega + \int_{I^p} T_i^*\omega')$$

$$= \sum_{i=1}^r a_i \int_{I^p} T_i^*\omega + \sum_{i=1}^r a_i \int_{I^p} T_i^*\omega'$$

$$= \int_c \omega + \int_c \omega'$$

We leave the verification of the remaining two equations to the reader.
For each positive integer p, define mappings

$$\partial_i, \partial_i' : I^{p-1} \to I^p, \quad 1 \le i \le p$$

by

$$\partial_i(t_1, \ldots, t_{p-1}) = (t_1, \ldots, t_{i-1}, 0, t_i, \ldots, t_{p-1})$$
$$\partial_i'(t_1, \ldots, t_{p-1}) = (t_1, \ldots, t_{i-1}, 1, t_i, \ldots, t_{p-1})$$

These functions correspond to the inclusion of I^{p-1} into the various
faces of I^p. For instance, if $p = 2$, ∂_2' takes I^1 onto the edge of I^2
between the vertex $(0,1)$ and the vertex $(1,1)$. If $T : I^p \to R^n$ is a singular p-cube, the *boundary of* T is defined to be the singular (p-1)-
chain

$$\partial T = \sum_{i=1}^{p} (-1)^i (T \circ \partial_i - T \circ \partial_i')$$

where $T \circ \partial_i$ and $T \circ \partial_i'$ are the composite functions. This chain is essentially the restriction of T to the boundary of I^p. If $p = 2$,

$$\partial T = T \circ \partial_2 + T \circ \partial_1' - T \circ \partial_2' - T \circ \partial_1$$

which is the sum, with signs, of the restrictions of T to the four
edges of I^2. In fact, if we interpret the terms $-T \circ \partial_2'$, $-T \circ \partial_1'$ as corresponding to traversing these edges in the negative direction (from 1
to 0), then ∂T corresponds to the closed curve obtained by restricting
T to the boundary of I^2, oriented so that the set I^2 is on the left.
(See Figure 2.1.)

If $c = a_1 T_1 + \cdots + a_r T_r$ is an arbitrary singular p-chain, we define
∂c by

$$\partial c = \sum_{i=1}^{r} a_i (\partial T_i)$$

PROPOSITION 2.2. *If* c,c' *are singular* p-chains *and* a,b *real
numbers, then*

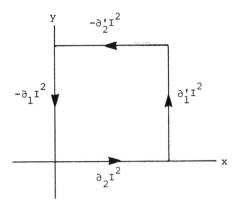

FIGURE 2.1.

$\partial(ac + bc') = a\partial c + b\partial c'$.

Furthermore, $\partial\partial c = 0$ *for any singular* p-*chain* c.

Proof. The proof of the first part of the proposition is trivial and left as an exercise.

The proof of the second part of the proposition is somewhat complicated. However, since we do not need this result in what follows, we also leave its proof as an exercise. (See Exercises 8-11).

We can now state the general form of Stokes' Theorem. The proof is given in Section 5.

THEOREM 2.3. *Let* U *be an open subset of* R^n, ω *a* (p-1) *form on* U, *and* c *a singular* p-*chain on* U. *Then*

$$\int_{\partial c} \omega = \int_c d\omega$$

In the next section, we shall see that this result is indeed a generalization of Green's and Stokes' Theorems.

EXERCISES

1.   Let $T_1, T_2, T_3 : I^2 \to R^3$ be given by

$$T_1(s,t) = (s,t,s+t)$$
$$T_2(s,t) = (st,s^2,t^2)$$
$$T_3(s,t) = (s-t,s+t,st)$$

and let $\omega = ydxdz$ be a 2-form on R^3. Compute the integral of ω over the singular 2-chain c where

   (a) $c = T_1 - T_2$

   (b) $c = 2T_1 + T_2 + T_3$

   (c) $c = T_3 - 4T_1$

   (d) $c = 3T_1 - 2T_2 + 5T_3$

2.   (a)-(d). Let $\omega = 2xydxdy - y^2dydz$ and compute the integral of ω over the singular 2-chains of Exercise 1.

3.   (a)-(d). Let $T_1, T_2, T_3 : I^3 \to R^4$ be given by

$$T_1(s,t,u) = (s,t+u,t,s-u)$$
$$T_2(s,t,u) = (st,su,tu,s+t+u)$$
$$T_3(s,t,u) = (s-u,t^2,tu,st)$$

and let $\omega = dx_1 dx_2 dx_4$. Compute the integral of ω over each of the singular 3-chains c defined by the formulas of (a)-(d) of Exercise 1.

4.   (a)-(d). Let $\omega = x_1 x_3 dx_1 dx_2 dx_3 + x_3^2 dx_2 dx_3 dx_4$ and compute the integral of ω over the singular 3-chains of Exercise 3.

5.   Prove the second and third assertions of Proposition 2.1.

6.   Prove the first assertion of Proposition 2.2.

7. Show that

$$(\partial_i^*)dx_j = (\partial_i^!)^* dx_j = \begin{cases} dx_j, & \text{if } j < i \\ 0, & \text{if } j = i \\ dx_{j-1}, & \text{if } j > i \end{cases}$$

8. Let T be a singular p-cube. Prove that

$$\partial(\partial T) = \sum_{i=1}^{p} (-1)^i \sum_{j=1}^{p-1} (-1)^j (T \circ \partial_i \circ \partial_j - T \circ \partial_i \circ \partial_j^! - T \circ \partial_i^! \circ \partial_j + T \circ \partial_i^! \circ \partial_j^!)$$

9. Prove that, for any i,j, $1 \le i \le j \le p - 1$, we have

$$\partial_i \circ \partial_j = \partial_{j+1} \circ \partial_i$$

$$\partial_i \circ \partial_j^! = \partial_{j+1}^! \circ \partial_i$$

$$\partial_i^! \circ \partial_j = \partial_{j+1} \circ \partial_i^!$$

$$\partial_i^! \circ \partial_j^! = \partial_{j+1}^! \circ \partial_i^!$$

10. Let T be a singular p-chain. Prove

$$\sum_{i=1}^{p} \sum_{j=1}^{p-1} (-1)^{i+j} T \circ \partial_i \circ \partial_j = 0$$

$$\sum_{i=1}^{p} \sum_{j=1}^{p-1} (-1)^{i+j} (T \circ \partial_i \circ \partial_j^! + T \circ \partial_i^! \circ \partial_j) = 0$$

$$\sum_{i=1}^{p} \sum_{j=1}^{p-1} (-1)^{i+j} T \circ \partial_i^! \circ \partial_j^! = 0$$

11. Let c be a singular p-chain. Prove that $\partial \partial c = 0$.

3. GREEN'S THEOREM AND STOKES' THEOREM

We now derive both Green's Theorem (Theorem 13.5.1) and Stokes' Theorem (Theorem 14.4.1). We begin with an interpretation of the integral of an n-form in R^n over a particular kind of singular n-cube.

Let $T: I^n \to R^n$ be a singular n-cube which also happens to be an orientation preserving diffeomorphism onto its image B. Suppose that ω is an n-form on an open set containing B, $\omega = f dx_1 \ldots dx_n$. Then

$$\int_T \omega = \int_{I^n} T^* \omega$$

$$= \int_{I^n} (f \circ T)(\text{Det } DT) dx_1 \ldots dx_n$$

(by Proposition 15.2., see also Exercise 7, Section 15.5)

$$= \int_{I^n} (f \circ T)(\text{Det } DT)$$

(by the definition of the integral of an n-form over a subset of R^n). However

$$\int_{I^n} (f \circ T)(\text{Det } DT) = \int_B f$$

by Theorem 12.6.2. Thus

$$\int_T \omega = \int_B f \tag{3.1}$$

A similar calculation shows that if T reverses orientation, then

$$\int_T \omega = - \int_B f$$

Because of this dependence on orientation, we refer to the integral of ω over T as an *oriented integral*.

We can use this observation to derive Green's Theorem from Theorem 2.3.

Let Γ be a simple closed curve in R^2 enclosing a region B in such a way that Γ is oriented so that B is on the left. Let $T: I^2 \to R^2$ be a singular 2-cube which is an orientation preserving diffeomorphism of I^2 onto B. (Such diffeomorphisms can be shown to exist.) Then ∂T can be thought of as defining a piecewise differentiable parametrization of Γ consistent with its orientation. Now, if

$$\omega = f_1 dx_1 + f_2 dx_2$$

is a 1-form on B, Theorem 2.3 gives us the equation

$$\int_{\partial T} \omega = \int_T d\omega \qquad (3.2)$$

If $\vec{F} = (f_1, f_2) : B \to R^2$, so that $\vec{F}$ corresponds to ω (see Section 15.3), then

$$d\omega = (\frac{\partial f_2}{\partial x_1} - \frac{\partial f_1}{\partial x_2}) dx_1 dx_2$$

and (3.2) becomes

$$\int_\Gamma \vec{F} = \int_B (\frac{\partial f_2}{\partial x_1} - \frac{\partial f_1}{\partial x_2}) \qquad (3.3)$$

which is Green's Theorem.

Suppose now that $\beta : I^2 \to R^3$ is a simple parametrized surface, which we think of as a singular 2-cube. Let

$$\omega = f_1 dx_1 + f_2 dx_2 + f_3 dx_3$$

be a 1-form defined on an open set containing the image of β. Then

$$\int_{\partial\beta} \omega = \int_\beta d\omega$$

by Theorem 2.3. As we saw earlier, $\partial\beta$ can be considered as defining an oriented curve Γ and, by definition, we have

$$\int_{\partial\beta} \omega = \int_\Gamma \vec{F} \qquad (3.4)$$

where $\vec{F}(\vec{v}) = (f_1(\vec{v}), f_2(\vec{v}), f_3(\vec{v}))$. Furthermore,

$$\int_\beta d\omega = \int_{I^2} \beta^* d\omega$$

and

$$\beta^* d\omega = \beta^* [(\frac{\partial f_2}{\partial x_1} - \frac{\partial f_1}{\partial x_2}) dx_1 dx_2 + (\frac{\partial f_3}{\partial x_1} - \frac{\partial f_1}{\partial x_3}) dx_1 dx_3$$
$$+ (\frac{\partial f_3}{\partial x_2} - \frac{\partial f_2}{\partial x_3}) dx_2 dx_3]$$

$$= \sum_{1 \leq i < j \leq 3} (\frac{\partial f_j}{\partial x_i} \circ \beta - \frac{\partial f_i}{\partial x_j} \circ \beta) (\frac{\partial \beta_i}{\partial x} \frac{\partial \beta_j}{\partial y} - \frac{\partial \beta_i}{\partial y} \frac{\partial \beta_j}{\partial x}) dxdy$$

$$= < (\text{curl } \vec{F}) \circ \beta, (D\beta)\vec{e}_1 \times (D\beta)\vec{e}_2 > dxdy$$

(See also Section 15.5, Exercise 8.) Here (x,y) are coordinates in I^2. Thus

$$\int_\beta d\omega = \int_{I^2} < (\text{curl } \vec{F}) \circ \beta, (D\beta)\vec{e}_1 \times (D\beta)\vec{e}_2 > \tag{3.5}$$

This is just the surface integral of the vector field curl $\vec{F}$ over the parametrized surface β.

Combining equations (3.4) and (3.5), we have

$$\int_\Gamma \vec{F} = \int_\beta < (\text{curl } \vec{F}) \circ \beta, (D\beta)\vec{e}_1 \times (D\beta)\vec{e}_2 > \tag{3.6}$$

where Γ is the boundary curve of the parametrized surface β and $\vec{F}:U \rightarrow R^3$ is a differentiable function defined on the open set U containing the image of β. This is the Stokes' Theorem (Theorem 14.5.1) of Chapter 14.

4. THE GAUSS THEOREM AND INCOMPRESSIBLE FLUIDS

The Gauss Theorem is a special case of the general Stokes' Theorem when $n = p = 3$. It relates the surface integral of a vector field $\vec{F}$ over a closed surface with the ordinary integral of the divergence of $\vec{F}$ over the region enclosed by the surface.

Suppose that $T:I^3 \rightarrow R^3$ is a singular 3-cube which is also an orientation preserving diffeomorphism onto its image $D \subset R^3$. Let ω be a 2-form on D,

$$\omega = f_1 dx_2 dx_3 - f_2 dx_1 dx_3 + f_3 dx_1 dx_2$$

(Note the - sign!)   Theorem 2.3 states that

$$\int_{\partial T} \omega = \int_T d\omega \qquad\qquad (4.1)$$

Set $\vec{F} = (f_1, f_2, f_3) : D \to R^3$. Then, using equation (3.1) and the discussion at the end of Section 15.3, we have

$$\int_T d\omega = \int_D \text{div } \vec{F}$$

By definition, the left side of (4.1) is

$$\sum_{i=1}^{3} \int_{T \circ \partial_i} \omega - \sum_{i=1}^{3} \int_{T \circ \partial_i'} \omega$$

The functions $T \circ \partial_i$, $T \circ \partial_i' : I^2 \to R^3$ can be considered as defining oriented surfaces Σ_i, Σ_i', $1 \le i \le 3$, and

$$\int_{T \circ \partial_i} \omega = \int_{\Sigma_i} \vec{F}, \quad \int_{T \circ \partial_i'} \omega = \int_{\Sigma_i'} \vec{F}$$

In fact, these surfaces fit together to make up the boundary surface Σ of the region D. (See Figure 4.1.) Furthermore, the signs are

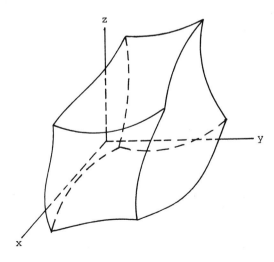

FIGURE 4.1.   The image of a singular 3-cube

arranged so that the orientations of the surfaces Σ_i, Σ_i' define an orientation of Σ. Thus, it makes sense to write the left hand side of equation (3.5) as

$$\int_\Sigma \vec{F}$$

We state the consequence of this discussion as a theorem.

THEOREM 4.1. (The Gauss Theorem) *Let D be a region in* R^3 *with boundary surface* Σ *oriented so that the normal to* Σ *points out of* D. *Let* $\vec{F}:D \to R^3$ *a differentiable function. Then*

$$\int_\Sigma \vec{F} = \int_D \text{div } \vec{F}$$

For an application of this result, we return to the study of hydrodynamics. (See Section 14.5.) We suppose that $U \subset R^3$ is an open set in which a fluid is in motion and $\vec{F}:U \to R^3$ is the corresponding velocity field.

Let Σ be the surface of a closed ball $B \subset U$. Then

$$\int_\Sigma \vec{F}$$

represents the net flow of fluid through the surface Σ. We say that the fluid is *incompressible* if this integral is zero for any ball $B \subset U$. As a consequence of Theorem 4.1, we have

$$\int_B \text{div } \vec{F} = 0$$

for any ball $B \subset U$. It follows that div $\vec{F} = 0$ on U. Thus the fluid is incompressible if and only if div $\vec{F} = 0$.

Suppose further that U is an open ball and that $\vec{F}$ is irrotational. (See Section 14.5.) Then $\vec{F} = \text{grad } \varphi$ for some C^2 function $\varphi:U \to R$. (See Exercise 4, Section 14.5.) If the fluid is also incompressible, we have

$$\text{div } \vec{F} = \text{div grad } \varphi = \frac{\partial^2 \varphi}{\partial x^2} + \frac{\partial^2 \varphi}{\partial y^2} + \frac{\partial^2 \varphi}{\partial z^2} = 0$$

This equation is called the *Laplace equation* and any function satisfying it is said to be *harmonic*.

Thus we see that the velocity field of an irrotational, incompressible fluid in a simply connected subset of R^3 has a harmonic potential function.

The Gauss Theorem can also be applied to electromagnetism. Let
$\vec{E}$ and $\vec{H}$ be vector fields on the open set $U \subset R^3$, $\vec{E}$ the *electric field*
and $\vec{H}$ the *magnetic field*. We assume that these vector fields are
(differentiable) functions of time t. These two vector fields satisfy
the *Maxwell equations*:

$$\text{div } \vec{E} = 4\pi\rho \qquad \text{div } \vec{H} = 0$$
$$\text{curl } \vec{E} = -\frac{1}{c}\frac{\partial \vec{H}}{\partial t} \qquad \text{curl } \vec{H} = \frac{1}{c}\frac{\partial \vec{E}}{\partial t}$$

where c is a universal constant and ρ is the *charge density*. If $\vec{H} = \vec{0}$
(this is the electrostatic case), then $\vec{E}$ is independent of time and

$$\text{curl } \vec{E} = 0$$

Thus, if U is simply connected (see Exercise 4, Section 14.4),

$$\vec{E} = -\text{grad } \varphi$$

for some function $\varphi : U \to R$. We then have (from the first Maxwell equation)

$$\text{div grad } \varphi = -4\pi\rho$$

In particular, φ is harmonic if U is free of electrical charge (that
is, if $\rho = 0$).

Suppose now that D is a region in U with boundary surface Σ. The
total charge contained in D is the integral

$$\int_D \rho = \frac{1}{4\pi} \int_D \text{div } \vec{E}$$

(again using the first Maxwell equation). According to the Gauss Theorem, this last integral is equal to

$$\frac{1}{4\pi} \int_{\Sigma} \vec{E}$$

This quantity is called the *flux across the surface* Σ. Thus we have proved that the *total charge contained in a region* D *is equal to the flux across the boundary of* D.

It is possible to apply Stokes' Theorem to prove an analogue of Theorem 4.1 for regions D which are not the image of a singular 3-cube. To deal with this situation, we break D up into subregions each of which is the image of a singular 3-cube. For example, if D is the *solid torus* (or doughnut) pictured in Figure 4.2, then each of the subregions D_1 and D_2 are images of singular 3-cubes $T_1, T_2: I^3 \to R^3$.

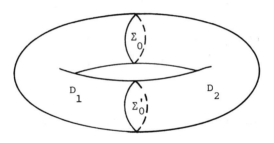

FIGURE 4.2. A solid torus

Let Σ_1 and Σ_2 be the boundary surfaces of D_1 and D_2. Then, if $\vec{F}$ is a vector field on D, we can apply Theorem 4.1 to obtain

$$\int_{\Sigma_1} \vec{F} = \int_{D_1} \text{div } \vec{F}, \quad \int_{\Sigma_2} \vec{F} = \int_{D_2} \text{div } \vec{F}$$

Adding these two equations, we have

$$\int_{\Sigma_1} \vec{F} + \int_{\Sigma_2} \vec{F} = \int_{D_1} \text{div } \vec{F} + \int_{D_2} \text{div } \vec{F} \qquad (4.2)$$

Now the right side of equation (4.2) is just the integral of div $\vec{F}$ over the region D. To determine the left side of equation (4.2), we note the surfaces Σ_1 and Σ_2 have the two pieces Σ_0 and Σ_0' in common.

(See Figure 4.2.) Furthermore, Σ_0 and Σ_0' inherit orientations from both Σ_1 and Σ_2 and the orientation inherited from Σ_1 is the negative of that inherited from Σ_2. It follows that the integrals of $\vec{F}$ over Σ_0 and Σ_0' cancel so that the right side of equation (4.2) can be identified with the integral of $\vec{F}$ over Σ. Thus we have

$$\int_\Sigma \vec{F} = \int_D \text{div } \vec{F}$$

in this case also.

Any region U in R^3 with a smooth boundary can be divided into subregions such that the new boundary surfaces created by the division cancel out (as Σ_0 and Σ_0' did in the example above). Therefore the divergence theorem applies to such regions. Proving that such a division exists is not easy, however. We would need to prove that examples like the Mobius band (see Figure 3.4 of Chapter 14) cannot happen when we deal with open sets in R^3 rather than surfaces. Such matters are studied in topology, and we shall not attempt any further discussion here.

EXERCISES

1.   Let S be the surface of the sphere $x^2 + y^2 + z^2 = 9$. Evaluate

$$\int_S x^2 dydz + y^2 dzdx + z^2 dxdy$$

by using Gauss' Theorem.

2.   Let S be the surface of the sphere $(x - 2)^2 + (y - 3)^2 + (z - 1)^2 = 25$. Evaluate

$$\int_S x^2 dydz + y^2 dzdx + z^2 dxdy$$

3.   Let U be a compact region in R^3, and suppose that Gauss' Theorem can be applied to U and ∂U. Show that the volume of U is given by

$$\text{Vol}(U) = \int_{\partial U} xdydz = \int_{\partial U} ydzdx = \int_{\partial U} zdxdy$$

4.   Let U be as in the previous problem. Show that the center of
mass of U (see Exercise 10 of Section 12..3) is given by

$$\bar{x} = \frac{1}{2\,Vol(U)}\ \int_U x^2 dydz$$

$$\bar{y} = \frac{1}{Vol(U)}\ \int_U xydydz$$

$$\bar{z} = \frac{1}{Vol(U)}\ \int_U xzdydz$$

(Note: the density of U is assumed to be 1 everywhere.)

*5.   Prove *Green's identities*: if f and g:$R^3 \to$ R have two continuous
derivatives in the region U (where U is as above), then

$$\int_U (f\ \Delta g\ +\ <\ \nabla f, \nabla g\ >\ )\ =\ \int_{\partial U} f\ \nabla g$$

and

$$\int_U (f\ \Delta g\ -\ g\ \Delta f)\ =\ \int_{\partial U} (f\ \nabla g\ -\ g\ \nabla f)$$

Here

$$\Delta f\ =\ \frac{\partial^2 f}{\partial x^2}\ +\ \frac{\partial^2 f}{\partial y^2}\ +\ \frac{\partial^2 f}{\partial z^2}$$

is the Laplacian of f and ∇f = grad f. (Hint: let $\vec{F}$ = f∇g in Gauss'
Theorem.)

6.   (a) Let U be any compact region of R^3 which does *not* include
the origin, and let $\vec{F}(\vec{v}) = \dfrac{\vec{v}}{\|\vec{v}\|^3}$.   Show that

$$\int_{\partial U} \vec{F}\ =\ 0$$

(b) Let V be a ball in R^3 centered at the origin, and let F be
as in (a). Show that

$$\int_{\partial V} \vec{F}\ =\ 4\pi$$

*7.   (The heat equation). Suppose that a body occupies a region U of
space, and let T(x,y,z,t) be the temperature at the point (x,y,z) $\in$ U

at time t.   The *specific heat*, c, measures the amount of heat per unit
mass the body can absorb: if V is a subset of U and ρ is the density
of the body, then

$$\int_V c\rho \frac{\partial T}{\partial t}$$

is the amount of heat absorbed by V.   (Both c and ρ may be functions
of x,y, and z.)

Now fix a time t_0, and regard $T(x,y,z,t_0)$ as a function of x,y,
and z only.   The usual law of heat conduction says that the flow of
heat through a surface S is given by

$$\int_S (-k \ \nabla T)$$

where k is a constant.

(a)   Let V be a subset of U such that Gauss' Theorem holds for
V and ∂V.   Show that

$$\int_V c\rho \frac{\partial T}{\partial t} - k \ \mathrm{div}(\nabla T) = 0$$

(Note: heat entering V is flowing in the direction opposite to the
outward normal.)

(b)   Conclude (assuming that everything is continuous) that T
satisfies the *heat equation*:

$$c\rho \frac{\partial T}{\partial t} = k\left(\frac{\partial^2 T}{\partial x^2} + \frac{\partial^2 T}{\partial y^2} + \frac{\partial^2 T}{\partial z^2}\right)$$

5.   PROOF OF THE GENERAL STOKES' THEOREM

We now prove Theorem 2.3.   The proof is divided into three cases, the
first one being the crucial one.

Case 1.   Suppose first of all that n = p, so that

$$\omega = \sum_{i=1}^{p} f_i dx_1 \ldots dx_{i-1} dx_{i+1} \ldots dx_p$$

Suppose further that c is the singular p-chain represented by $I^p \subset R^p$. (More precisely, $c = T : I^p \to R^p$ where T is the *inclusion* mapping, $T(\vec{v}) = \vec{v}$.) Then

$$\int_c d\omega = \int_c \sum_{i=1}^p (df_i) dx_1 \ldots dx_{i-1} dx_{i+1} \ldots dx_p$$

$$= \sum_{i=1}^p (-1)^{i-1} \int_c \frac{\partial f_i}{\partial x_i} dx_1 \ldots dx_p$$

$$= \sum_{i=1}^p (-1)^{i-1} \int_{I^p} \frac{\partial f_i}{\partial x_i} \qquad\qquad (5.1)$$

Using the Fubini Theorem (Theorem 11.1.2), we have

$$\int_{I^p} \frac{\partial f_i}{\partial x_i} = \int_{I^{p-1}} (\int_0^1 \frac{\partial f_i}{\partial x_i} dx_i)$$

$$\qquad\qquad (5.2)$$

$$= \int_{I^{p-1}} (f_i \circ \partial_i^1 - f_i \circ \partial_i)$$

Now

$$(\partial_i)^* dx_j = (\partial_i^1)^* dx_j = \begin{cases} dx_j & \text{if } j < i \\ 0 & \text{if } j = i \\ dx_{j-1} & \text{if } j > i \end{cases}$$

(See Exercise 7, Section 2.) Thus

$$\int_{\partial_i^1} \omega = \int_{I^{p-1}} (\partial_i^1)^* \omega$$

$$= \int_{I^{p-1}} \sum_{j=1}^n (f_j \circ \partial_i^1) \partial_i^{1*} (dx_1 \ldots dx_{j-1} dx_{j+1} \ldots dx_p)$$

$$= \int_{I^{p-1}} f_i \circ \partial_i^1$$

(if dx_i appears, then $\partial_i^{1*}(dx_1 \ldots dx_{j-1} dx_{j+1} \ldots dx_p) = 0$). Similarly,

$$\int_{\partial_i} \omega = \int_{I^{p-1}} f_i \circ \partial_i$$

Thus, from (5.1) and (5.2),

$$\int_c d\omega = \sum_{i=1}^{p} (-1)^{i-1} \int_{\partial_i^! - \partial_i} f_i$$

$$= \int_{\Sigma(-1)^i (\partial_i - \partial_i^!)} f_i = \int_{\partial c} \omega$$

using Proposition 2.1.

The remaining cases are now direct consequences of Case 1 and the definitions, though the computation may seem confusing.

Case 2. Suppose now that $c = T : I^p \to R^n$ is an arbitrary singular p-cube. Then

$$\int_T d\omega = \int_{I^p} T^*(d\omega)$$

$$= \int_{I^p} d(T^*\omega)$$

by Proposition 14.6.2. By Case 1 above, we have

$$\int_{I^p} d(T^*\omega) = \int_{\partial I^p} T^*\omega = \sum_{i=1}^{p} (-1)^i \int_{I^{p-1}} (\partial_i^*(T^*\omega) - \partial_i^{!*}(T^*\omega))$$

$$= \sum_{i=1}^{p} (-1)^i \int_{I^{p-1}} ((T \circ \partial_i)^*\omega - (T \circ \partial_i^!)^*\omega)$$

(from Proposition 15.6.2)

$$= \int_{\partial T} \omega$$

Case 3. If c is any singular p-chain, then c is a linear combination of singular p-cubes. Stokes' Theorem in this case follows from Case 2 (above) and Proposition 2.1.

APPENDIX 1

THE EXISTENCE OF DETERMINANTS

This appendix is devoted to proving that the determinant of a matrix really does exist (Theorem 10.3.3) and that the determinant of a matrix is equal to the determinant of its transpose (Proposition 10.4.6).

We begin by recalling the definition of the determinant given in Section 10.3. Let A be an $n \times n$ matrix. We define Det A inductively on n as follows. If $n = 1$ (so that $A = (a_{11})$), we set Det A $= a_{11}$. Suppose that the determinant is defined for all $n \times n$ matrices with $n \leq k$ and let A be a $(k+1) \times (k+1)$ matrix. Let A_{ij} be the $k \times k$ matrix obtained from A by deleting the ith row and the jth column. Pick any i with $1 \leq i \leq k + 1$ and define

$$\text{Det } A = \sum_{j=1}^{k+1} (-1)^{i+j} \text{Det } A_{ij}$$

Note that A_{ij} is a $k \times k$ matrix and Det A_{ij} is defined by induction.

To prove Theorem 10.3.3, we must show that properties (3.1), (3.2), and (3.3) of Section 10.3 hold for $(n+1) \times (n+1)$ matrices, assuming that they hold for $k \times k$ matrices with $k \leq n$. Explicitly, we must show that, if A is an $(n+1) \times (n+1)$ matrix, then

If all but one of the columns of A are held fixed, then
Det is linear in the remaining column   (1.1)

Det A = 0 if two of the columns of A are equal   (1.2)

If A is the identity matrix, then Det A = 1   (1.3)

The easiest property to check is (1.3). Suppose $A = I_{n+1}$, the (n+1) × (n+1) identity matrix. If $i \neq j$, the elements a_{ii} and a_{jj}, both of which are equal to 1, are deleted in forming A_{ij}. Since I_{n+1} has exactly (n+1) nonzero entries, A_{ij} has (n-1) nonzero entries in this case. Since A_{ij} has n columns, one of the columns consists entirely of zeros. Thus, if $i \neq j$, Det $A_{ij} = 0$ by property (1.1). If $i = j$, then $a_{ii} = 1$ and $A_{ii} = I_n$, the n × n identity matrix (as one sees easily by trying an example). Therefore

$$\text{Det}(I_{n+1}) = \sum_{j=1}^{n+1} (-1)^{i+j} a_{ij} \text{Det}(A_{ij})$$

$$= (-1)^{i+i} a_{ii} \text{Det}(A_{ii})$$

$$= \text{Det}(I_n) = 1$$

as required.

Property (1.1) is not too difficult, either. To simplify notation, we shall prove only that Det is linear in the first column. Let B be the matrix

$$B = \begin{pmatrix} b_{11} & a_{12} & a_{13} & \cdots & a_{1,n+1} \\ b_{21} & a_{22} & & \cdots & a_{2,n+1} \\ \vdots & \vdots & & & \\ b_{n+1,1} & a_{n+1,2} & & \cdots & a_{n+1,n+1} \end{pmatrix}$$

and let

$$C = \begin{pmatrix} a_{11}+b_{11} & a_{12} & a_{13} & \cdots & a_{1,n+1} \\ a_{21}+b_{21} & a_{22} & & & a_{2,n+1} \\ \vdots & \vdots & & \cdots & \\ a_{n+1,1}+b_{n+1,2} & a_{n+1,2} & & \cdots & a_{n+1,n+1} \end{pmatrix}$$

We need to show that

$$\text{Det } C = \text{Det } A + \text{Det } B$$

Let B_{ij} be the matrix obtained by deleting the ith row and j^{th} column from B, and define C_{ij} similarly. Then our definition says that

$$\text{Det } C = (-1)^{i+1}(a_{i1} + b_{i1})\text{Det}(C_{i1}) + \sum_{j=2}^{n+1} (-1)^{i+j}a_{ij}\text{Det}(C_{ij})$$

(1.4)

Since the first column is deleted to obtain A_{i1}, B_{i1}, and C_{i1}, it is clear that they are identical:

$$A_{i1} = B_{i1} = C_{i1}$$

If $j > 1$, the matrices A_{ij}, B_{ij}, and C_{ij} agree except for the first column; the first column of C_{ij} is the sum of the first columns of A_{ij} and B_{ij}. Because (1.1) holds for n × n determinants,

$$\text{Det } C_{ij} = \text{Det } A_{ij} + \text{Det } B_{ij} \text{ if } j > 1$$

Substituting these results in (1.4), we find that

$$\text{Det } C = [(-1)^{i+1}a_{i1}\text{Det } A_{i1} + \sum_{j=2}^{n+1} (-1)^{i+j}a_{ij}\text{Det } A_{ij}]$$

$$+ [(-1)^{i+1}b_{i1}\text{Det } B_{i1} + \sum_{j=2}^{n+1} (-1)^{i+j}a_{ij}\text{Det } B_{ij}]$$

$$= \text{Det } A + \text{Det } B$$

(the last line comes from the formula for Det A and Det B).

To finish the proof that (1.1) holds, we need to show that multiplying any column by a constant multiplies the determinant by the same constant. This proof is just like the one we have just done; we leave the details as an exercise.

We are left with (1.2). Again, we shall do a special case: if the first two columns agree, Det A = 0. Thus we are assuming that

$$A = \begin{pmatrix} a_{11} & a_{11} & a_{13} & \cdots & a_{1,n+1} \\ a_{21} & a_{21} & a_{23} & \cdots & a_{2,n+1} \\ \cdot & \cdot & \cdot & \cdots & \\ a_{n+1,1} & a_{n+1,1} & a_{n+1,3} & \cdots & a_{n+1,n+1} \end{pmatrix}$$

Now it is easy to see that $A_{i1} = A_{i2}$ and

if $j > 2$, the first two columns of A_{ij} are the same   (1.5)

Hence

$$Det\ A = \sum_{j=1}^{n+1} (-1)^{i+j} a_{ij} Det(A_{ij})$$

$$= (-1)^{i+1} a_{i1} Det(A_{i1}) + (-1)^{i+2} a_{i2} Det(A_{i1})$$

$$+ \sum_{j=3}^{n+1} (-1)^{i+j} a_{ij} Det(A_{ij})$$

The first two terms have opposite signs, and therefore sum to 0. The others are all 0 by the inductive hypothesis and statement (1.5) above. Therefore Det A = 0 and the proof of Theorem 10.3.3 is finished.

This result does not depend on the value of i - that is, on the row we used to expand the determinant. Consequently, the following corollary holds.

COROLLARY. Det *is linear in each row; that is, if we fix every row of* A *but one,* Det A *is linear in that remaining row.*

Proof. To see that Det A is linear in the ith row, simply use the formula:

$$Det\ A = \sum_{j=1}^{\ell+1} (-1)^{i+j} a_{ij} Det(A_{ij})$$

The matrices A_{ij} have no entries from the ith row of A, and are there-fore fixed. It is now clear that Det A is a linear function of the numbers a_{ij}, which proves the corollary.

We can use this corollary to derive Proposition 10.4.6. Let A be an n x n matrix and define $\Delta(A)$ by

$$\Delta(A) = Det(A^t)$$

where A^t is the transpose of A. We show that the function Δ satisfies properties (1.1), (1.2), and (1.3) so that

Det A = $\Delta(A)$

     = Det(A^t)

by Proposition 10.3.2.

    Property (1.3) is again the easiest: since $I^t = I$,

$\Delta(I)$ = Det(I^t)

    = Det I

    = 1

Property (1.1) holds by the corollary to Theorem 10.3.3 stated above. To prove property (1.2) we suppose that two columns of A are identical. Then clearly

Rank(A) < n

and, using Theorem 7.4.3 and the remark following Proposition 7.4.2, we have

Rank(A^t) = Rank(A)

      < n

Therefore A^t is not invertible (by Proposition 7.4.4), so that

$\Delta(A)$ = Det(A^t) = 0

by Proposition 9.4.3. Thus property (1.2) holds for Δ and the proof of Proposition 10.4.6 is complete.

EXERCISES

1. Complete the proof that property (1.1) holds for the determinant.

*2. Prove the last part of Proposition 9.4.6 directly from the definition of the determinant (as given in Theorem 9.3.3), without using Proposition 9.4.3.

*3. Give the general proof of property 1.2. (Hint: show that it suffices to give the proof in the case of adjacent columns.)

THE FUNDAMENTAL THEOREM OF ALGEBRA

The Fundamental Theorem of Algebra (Theorem 10.1.2) states that any polynomial equation $a_n z^n + a_{n-1} z^{n-1} + \cdots + a_0 = 0$, where $n > 0$ and $a_n \neq 0$, has a complex root. The idea behind the proof which follows is due to the French mathematician D'Alembert (1717-1783).

There is no harm in assuming that $a_n = 1$, and we do so. Let

$$P(z) = z^n + a_{n-1} z^{n-1} + \cdots + a_0$$

and set

$$S = 1 + |a_0| + \sum_{j=0}^{n-1} |a_j|$$

Then $S \geq 1$, so $1 \leq S \leq S^2 \leq \cdots \leq S^n$, and

$$S^n > S^{n-1}(|a_0| + \cdots + |a_{n-1}|) + |a_0|$$

$$\geq \left(\sum_{j=0}^{n-1} |a_j| S^j \right) + |a_0|$$

So if $|z| = S$, then

$$|P(z)| \geq |z^n| - \sum_{j=0}^{n-1} |a_j z^j|$$

$$= S^n - \sum_{j=0}^{n-1} |a_j| S^j$$

$$> |a_0| = P(0)$$

713

Let F be the subset of C defined by

$$F = \{z \in C : |z| \le S\}$$

Then F is compact (since it is closed and bounded; Theorem 5.6.1) and the function $|P(z)|$ is continuous on F. Therefore $|P(z)|$ has a minimum value on F (by Theorem 5.5.3). The calculation above shows that $|P(0)| < |P(z)|$ if $|z| = S$; therefore the minimum is not reached when $|z| = S$. That is, the minimum value of $|P(z)|$ on F is attained on an interior point of F.

Let z_0 be the point at which the minimum is attained. Since $|z_0| < S$, there is a number $\varepsilon > 0$ such that $B_\varepsilon(z_0) \subset F$; thus $|P(z_0)| \le |P(z)|$ whenever $|z - z_0| < \varepsilon$. We shall use this fact to show that $P(z_0) = 0$. The algebra is a bit simpler when $z_0 = 0$, and we now change coordinates in order to have this simplified situation.

Set $z - z_0 = w$, or $z = w + z_0$. Then $P(z) = P(w + z_0) = Q(w)$, say, where

$$Q(w) = w^n + b_{n-1}w^{n-1} + \cdots + b_0$$

Furthermore, the condition on z_0 now says that if $|w| < \varepsilon$, then $|Q(w)| > |Q(0)|$. We want to prove that $Q(0) = 0$ - that is, that $b_0 = 0$.

Suppose otherwise. Let k be the smallest positive integer with $b_k \neq 0$. (Since b_n, the coefficient of w^n, is 1, k certainly exists.) Then

$$Q(w) = b_0 + b_k w^k + \sum_{j=k+1}^{n} b_j w^j$$

$$= b_0 + b_k w^k + w^{k+1} Q_1(w) \tag{1.1}$$

where

$$Q_1(w) = \sum_{j=k+1}^{n} b_j w^{j-k-1}$$

Our hope is to show that an adroit choice of w will make $|Q(w)| < |Q(0)|$. To simplify our work, we first make a calculation which shows that the last part of the above expression is rather small. We have

$$w^{k+1} Q_1(w) = (b_k w^k) \frac{w}{b_k} Q_1(w)$$

When $w = 0$, $\frac{w}{b_k} Q_1(w) = 0$; by continuity, we can choose $\delta > 0$ such that if $|w| < \delta$, then

$$\left| \frac{w}{b_k} Q_1(w) \right| < \frac{1}{2}$$

This means that if $|w| < \delta$, then

$$|w^{k+1} Q_1(w)| < \frac{1}{2} |b_k w^k| \tag{1.2}$$

To reduce the size of $|Q(w)|$, we pick w so that b_0 and $b_k w^k$ point in opposite directions. Write $\frac{b_0}{b_k}$ in polar form (see Exercise 4 of Section 10.1):

$$\frac{b_0}{b_k} = r_0 e^{i\theta}$$

Then $r_0 = \left| \frac{b_0}{b_k} \right| > 0$. Define by

$$\alpha = \frac{\theta + \pi}{k}$$

Then

$$e^{ik\alpha} = e^{i(\theta+\pi)} = - e^{i\theta}$$

and, if $w = re^{i\alpha}$ (with $r > 0$),

$$b_0 + b_k w^k = b_k \left(\frac{b_0}{b_k} + w^k \right)$$

$$= b_k (r_0 e^{i\theta} + r^k e^{ik\alpha})$$

$$= b_k (r_0 e^{i\theta} - r^k e^{i\theta})$$

$$= b_k (r_0 - r^k) e^{i\theta}$$

Restrict r so that $r^k < r_0$.   Then

$$|b_0 + b_k w^k| = |b_k(r_0 - r^k)e^{i\theta}|$$

$$= |b_k||r_0 - r^k|$$

$$= |b_k|(r_0 - r^k)$$

$$= |b_k|r_0 - |b_k||w^k|$$

$$= |b_0| - |b_k w^k| \qquad (1.3)$$

Now it is easy to complete the proof.   If we pick $w = re^{i\alpha}$ such that $|w| < \varepsilon$, $|w| < \delta$, and $0 < r^k < r_0$, then (1.1), (1.2), and (1.3) imply that

$$|Q(w)| \le |b_0 + b_k w^k| + |w^{k+1}Q_1(w)|$$

$$\le |b_0| - |b_k w^k| + \frac{1}{2}|b_k w^k|$$

$$= |b_0| - \frac{1}{2}|b_k w^k|$$

$$< |b_0| = Q(0)$$

This contradicts our assumption that $|Q(w)| \ge |Q(0)|$ when $|w| < \varepsilon$. Hence $w_0 = 0$, and the theorem is proved.

APPENDIX 3

JORDAN CANONICAL FORM

INTRODUCTION

Let V be a finite dimensional complex vector space and T:V → V a linear transformation. We saw in Section 10.7 that V has a basis of eigenvectors for T whenever T is a normal linear transformation. In this
Appendix, we show that, for arbitrary T, one can choose a basis for
V so that the matrix A = (a_{ij}) of T (relative to this basis) is "almost
diagonal"; that is $a_{ij} = 0$ unless i = j or i = j + 1. The matrix A
is called the Jordan Canonical Form of T.

1. GENERALIZED EIGENVECTORS

Let V be a finite dimensional complex vector space, and let T:V → V
be a linear transformation. A vector $\vec{v} \in V$ is called a *generalized
eigenvector* (with *eigenvalue* λ) if there is a positive integer n such
that

$$(T - \lambda I)^n \vec{v} = \vec{0} \tag{1.1}$$

Clearly $\vec{0}$ is a generalized eigenvector for every λ. We shall adopt
the usual convention: λ is called an eigenvalue of T only if λ is an
eigenvalue for some nonzero generalized eigenvector of V. Notice also
that if $\vec{v}$ is an ordinary eigenvector for T, then it is a generalized
eigenvector as well, since $(T - \lambda I)\vec{v} = \vec{0}$.

One reason for the term "eigenvalue" rather than "generalized eigenvalue " is found in the following result.

PROPOSITION 1.1. *If λ is an eigenvalue of T corresponding to some nonzero generalized eigenvector $\vec{v} \in V$, it is also an eigenvalue for some nonzero eigenvector $\vec{v}_0 \in V$.*

That is, introducing generalized eigenvectors does not involve introducing new eigenvalues.

Proof. We know that $(T - \lambda I)^n \vec{v} = \vec{0}$ for some integer n > 0. Choose n_0 such that

$$(T - \lambda I)^{n_0 + 1} \vec{v} = \vec{0}$$

but

$$\vec{v}_0 = (T - \lambda I)^{n_0} \vec{v} \neq \vec{0}$$

(Thus n_0 is the largest integer with $(T - \lambda I)^{n_0} \vec{v} \neq \vec{0}$.) Notice that $(T - \lambda I)^0 \vec{v} = \vec{v} \neq \vec{0}$, so that $n_0 \geq 0$. Also,

$$\begin{aligned}(T - \lambda I)\vec{v}_0 &= (T - \lambda I)^{n_0 + 1} \vec{v} \\ &= \vec{0}\end{aligned}$$

Thus $T\vec{v}_0 = \lambda \vec{v}_0$, and $\vec{v}_0$ is an eigenvector of T with eigenvalue λ.

COROLLARY. *The eigenvalues of T are the roots of the characteristic equation of T.*

Let λ be an eigenvalue of T, and let V_λ be the set of generalized vectors with eigenvalue λ:

$$V_\lambda = \{\vec{v} \in V : \exists n > 0 \text{ with } (T - \lambda I)^n \vec{v} = \vec{0}\}$$

V_λ is called the *generalized eigenspace* for the eigenvalue λ.

PROPOSITION 1.2. *V_λ is a subspace of V, and $T(V_\lambda) \subset V_\lambda$.*

One often says that V_λ is *stable* under T if $T(V_\lambda) \subseteq V$.

Proof. Neither half of this proposition is too difficult. Notice first that $\vec{0} \in V_\lambda$. If $\vec{v},\vec{w} \in V_\lambda$, then there are positive integers m,p with

$$(T - \lambda I)^m \vec{v} = \vec{0}$$
$$= (T - \lambda I)^p \vec{w}$$

Let n = max(m,p), and let a,b be any complex numbers; then

$$(T - \lambda I)^n (a\vec{v} + b\vec{w}) = a(T - \lambda I)^n \vec{v} + b(T - \lambda I)^n \vec{w}$$
$$= \vec{0}$$

and therefore $a\vec{v} + b\vec{w} \in V_\lambda$. Thus V_λ is a subspace.

Next, pick $\vec{v} \in V_\lambda$. Then there is an integer n with $(T - \lambda I)^n \vec{v} = \vec{0}$. Now $(T - \lambda I)^n$ is the sum of terms of the form $\lambda^{n-k} T^k$, so T commutes with $(T - \lambda I)^n$. (See Exercise 2.) Thus

$$(T - \lambda I)^n T\vec{v} = T(T - \lambda I)^n \vec{v}$$
$$= T(\vec{0}) = \vec{0}$$

Thus $T\vec{v} \in V_\lambda$ by definition and the proposition is proved.

We know from the definition of V_λ that, for each $\vec{v} \in V_\lambda$, there is an n (which may depend upon $\vec{v}$) such that

$$(T - \lambda I)^n \vec{v} = \vec{0}$$

The next result shows that we can choose a single n which works for all $\vec{v} \in V_\lambda$.

PROPOSITION 1.3. *Let λ be an eigenvalue for the linear transformation $T:V \to V$. There is an integer n such that*

$$(T - \lambda I)^n V_\lambda = \vec{0}$$

We define the index of λ, $\nu(\lambda)$, by

$$\nu(\lambda) = \min\{n: (T - \lambda I)^n V_\lambda = \vec{0}\}$$

Proof of Proposition 1.3. Let $\vec{v}_1, \ldots, \vec{v}_k$ be a basis for V_λ. For each integer $j=1,\ldots,k$, let n_j be the integer with

$$(T - I)^{n_j} \vec{v}_j = \vec{0}$$

and let $n = \max\{n_1, \ldots, n_k\}$. Then, if $\vec{v} \in V_\lambda$,

$$\vec{v} = \sum_{j=1}^{k} a_j \vec{v}_j$$

since $\vec{v}_1, \ldots, \vec{v}_k$ is a basis for V, and

$$(T - \lambda I)^n \vec{v} = \sum_{j=1}^{k} a_j (T - \lambda I)^n \vec{v}_j$$
$$= \vec{0}$$

since $n \geq n_j$, $1 \leq j \leq k$.

EXERCISES

1.   Find all generalized eigenvectors (with the corresponding eigenvalue) for

(a) $\begin{pmatrix} 3 & 3 \\ 4 & 2 \end{pmatrix}$

(c) $\begin{pmatrix} 0 & 3 & 1 \\ 0 & 1 & 1 \\ 1 & -3 & 3 \end{pmatrix}$

(b) $\begin{pmatrix} -2 & 4 \\ -1 & 2 \end{pmatrix}$

(d) $\begin{pmatrix} 2 & -1 & 1 \\ 4 & 2 & 2 \\ 2 & 1 & 2 \end{pmatrix}$

2.   (a)   Prove by induction that T commutes with $(T - \lambda I)^n$.

(b)   Prove, more generally, that if λ and μ are complex numbers and p and q integers, then

$$(T - \lambda I)^p (T - \mu I)^q = (T - \mu I)^q (T - \lambda I)^p$$

*3. Let $T:V \rightarrow V$ be a linear transformation and let $\vec{v} \in V$ be a non-zero vector. Suppose that there are positive integers n,m and complex numbers λ, μ such that

$$(T - \lambda I)^m \vec{v} = \vec{0}$$
$$= (T - \mu I)^n \vec{v}$$

Prove that $\lambda = \mu$. Thus $V_\lambda \cap V_\mu = \{0\}$ if $\lambda \neq \mu$. (Hint: use double induction on n and m.)

*4. Prove that nonzero generalized eigenvectors of a linear transformation $T:V \rightarrow V$ corresponding to distinct eigenvalues are linearly independent.

*5. Let V be a finite dimensional vector space and let $T:V \rightarrow V$ be a linear transformation. Prove that V has a basis of generalized eigenvectors for T.

2. THE JORDAN CANONICAL FORM

Let V be a finite dimensional vector space and $T:V \rightarrow V$ a linear transformation. We show, in this section, how to choose a basis for V relative to which the matrix for T is "almost diagonal." This matrix is called the *Jordan Canonical Form* of the linear transformation T.

Suppose, to begin with, that T has exactly one generalized eigenspace. That is, $V = V_\lambda$ where λ is the unique eigenvalue for T. Let $k = \nu(\lambda)$ be the index of λ and set $N = T - \lambda I$. Then N^k is the zero transformation.

For example, if $T:R^3 \rightarrow R^3$ is the linear transformation with matrix

$$\begin{pmatrix} 2 & 0 & 0 \\ 2 & 2 & -1 \\ 0 & 0 & 2 \end{pmatrix} \qquad\qquad (2.1)$$

then $\lambda = 2$ is the only eigenvalue for T. In this case, $N = T - 2I$ has matrix

$$\begin{pmatrix} 0 & 0 & 0 \\ 2 & 0 & -1 \\ 0 & 0 & 0 \end{pmatrix}$$

and N^2 is the zero transformation. Therefore,

$k = \nu(2) = 2$

We now define a sequence of vectors $\vec{v}_1,\ldots,\vec{v}_m$ in V as follows. First of all, let

$$\vec{v}_1,\ldots,\vec{v}_{m_1}$$

be a basis for the subspace $\hbar(N) \cap \mathcal{I}(N^{k-1})$ of V. Extend this set to basis

$$\vec{v}_1,\ldots,\vec{v}_{m_1},\ \vec{v}_{m_1+1},\ldots,\vec{v}_{m_1+m_2}$$

for $\hbar(N) \cap \mathcal{I}(N^{k-2})$. Continuing in this way, we finally obtain a basis

$$\vec{v}_1,\ldots,\vec{v}_{m_1},\vec{v}_{m_1+1},\ldots,\vec{v}_{m_1+m_2},\vec{v}_{m_1+m_2+1},\ldots,\vec{v}_{m_1+\ldots+m_k}$$

for $\mathcal{I}(N)$ such that, for each i, $1 \leq i \leq k$, the vectors

$$\vec{v}_1,\ldots,\vec{v}_{m_1},\vec{v}_{m_1+1},\ldots,\vec{v}_{m_1+m_2},\vec{v}_{m_1+m_2+1},\ldots,\vec{v}_{m_1+\cdots m_i}$$

forms a basis for $\hbar(N) \cap \mathcal{I}(N^{k-i})$.

Now, for each j with $1 \leq j \leq m_1$, choose $\vec{u}_j \in V$ such that

$$N^{k-1}\vec{u}_j = \vec{v}_j$$

This is possible since

$$\vec{v}_j \in \hbar(N) \cap \mathcal{I}(N^{k-1}) \subset \mathcal{I}(N^{k-1})$$

Similarly, for each j with

$$m_1 + \cdots + m_i < j \leq m_1 + \cdots + m_{i+1}$$

$i + 1 < k$, let $\vec{u}_j \in V$ be such that

$$N^{k-i-1}\vec{u}_j = \vec{v}_j$$

For instance, if $T: R^3 \to R^3$ is defined by the matrix (2.1), then

$$N(x,y,z) = (0, 2x - z, 0)$$

so that $\eta(N)$ is the subspace of R^3 with basis $\{\vec{e}_2, \vec{e}_1 + 2\vec{e}_3\}$ and $\mathcal{I}(N)$ is the subspace of R^3 with basis $\{\vec{e}_2\}$. Thus, $m_1 = m_2 = 1$ in this case and we may choose $\vec{v}_1 = \vec{e}_2$, $\vec{v}_2 = \vec{e}_1 + 2\vec{e}_3$, $\vec{u}_1 = -\vec{e}_3$.

Next, we consider the following configuration of vectors:

$$
\begin{array}{ll}
\vec{u}_1, \ N\vec{u}_1, N^2\vec{u}_1, \ldots & , N^{k-1}\vec{u}_1 = \vec{v}_1 \\[2pt]
\quad \cdots & \\[2pt]
\vec{u}_{m_1}, N\vec{u}_{m_1}, \ldots & , N^{k-1}\vec{u}_{m_1} = \vec{v}_{m_1} \quad\quad (2.2)\\[2pt]
\vec{u}_{m_1+1}, N\vec{u}_{m_1+1}, \ldots & , N^{k-2}\vec{u}_{m_1+1} = \vec{v}_{m_1+1} \\[2pt]
\quad \cdots & \\[2pt]
\vec{u}_{m_1+m_2}, N\vec{u}_{m_1+m_2}, \ldots & , N^{k-2}\vec{u}_{m_1+m_2} = \vec{v}_{m_1+m_2} \\[2pt]
& \quad \cdots \\[2pt]
\vec{u}_{m_1+\cdots+m_{k-2}+1}, N\vec{u}_{m_1+\cdots+m_{k-2}+1} & = \vec{v}_{m_1+\cdots+m_{k-2}+1} \\[2pt]
& \quad \cdots \\[2pt]
\vec{u}_{m_1+\cdots+m_{k-1}}, N\vec{u}_{m_1+\cdots+m_{k-1}} & = \vec{v}_{m_1+\cdots+m_{k-1}} \\[2pt]
& \vec{v}_{m_1+\cdots+m_{k-1}+1} \\[2pt]
& \quad \cdots \\[2pt]
& \vec{v}_{m_1+\cdots+m_k}
\end{array}
$$

The basic result regarding this configuration is the following:

   PROPOSITION 2.1.   *The vectors in the configuration (2.2) form*
a *basis for* V.

   We prove this result at the end of this section.

   To determine the matrix of T relative to this basis, we begin
by considering the first row of the configuration (2.2).   Define
$\vec{w}_j \in V$, $1 \leq j \leq k$, by

$$\vec{w}_j = N^{j-1} \vec{u}_1$$

Then

$$\vec{Nw}_j = \begin{cases} \vec{w}_{j+1} & \text{if } j < k \\ \vec{0} & \text{if } j = k \end{cases}$$

Therefore, the matrix of N on the subspace W of V spanned by the first
row of (2.2) relative to the basis $\{\vec{w}_1, \ldots, \vec{w}_k\}$ is simply

$$\begin{pmatrix} 0 & 0 & 0 & \ldots & 0 \\ 1 & 0 & 0 & \ldots & 0 \\ 0 & 1 & 0 & \ldots & 0 \\ & & \ldots & & \\ 0 & \ldots & 0 & 1 & 0 \end{pmatrix}$$

Since $N = T - \lambda I$, $T = N + \lambda I$, and the matrix of T on the subspace W
is

$$\begin{pmatrix} \lambda & 0 & 0 & \ldots & 0 \\ 1 & \lambda & 0 & \ldots & 0 \\ 0 & 1 & \lambda & \ldots & 0 \\ & & \ldots & & \\ 0 & \ldots & 0 & 1 & \lambda \end{pmatrix} \qquad (2.3)$$

   In fact, the same argument shows that the restriction of T to
the subspace of V spanned by any one of the rows of the configuration

(2.2) has a matrix of the form (2.3). It follows that if we order the basis (2.2) in the obvious way, the matrix of T is

$$
\begin{pmatrix}
A_1 & 0 & 0 & \cdots \\
0 & A_2 & 0 & \cdots \\
& & \cdots & \\
0 & \cdots & 0 & A_r
\end{pmatrix}
\tag{2.4}
$$

where r is the number of rows in the configuration (2.2), each A_i is a square matrix of the form (2.3), and each 0 represents a rectangular zero matrix. This matrix is the *Jordan Canonical form* of T (in the case of a single generalized eigenspace).

For example, if $T: R^3 \to R^3$ is the linear transformation with matrix (2.1), the configuration (2.2) for T is

$$
\vec{u}_1 = -\vec{e}_3 \quad N\vec{u}_1 = \vec{v}_1 = \vec{e}_2
$$
$$
\vec{v}_2 = \vec{e}_1 + 2\vec{e}_3
$$

The matrix of N relative to the basis $\{\vec{u}_1, \vec{v}_1, \vec{v}_2\}$ is

$$
\begin{pmatrix}
0 & 0 & 0 \\
1 & 0 & 0 \\
0 & 0 & 0
\end{pmatrix}
$$

and the matrix of T = N + 2I relative to this basis is

$$
\begin{pmatrix}
2 & 0 & 0 \\
1 & 2 & 0 \\
0 & 0 & 2
\end{pmatrix}
$$

To deal with linear transformations having more than one eigenvalue, we recall the notion of the direct sum. We say that a vector space V is the *direct sum* of subspaces $V_1, \ldots, V_n$ if

(a) $V = V_1 + V_2 + \cdots + V_n$

(b) $V_i \cap \sum_{j \neq i} V_j = \{\vec{0}\}$ for all $i = 1, \ldots, n$

Equivalently, V is the direct sum of $V_1, \ldots, V_n$ if each vector $\vec{v} \in V$ can be uniquely expressed in the form

$$\vec{v} = \vec{v}_1 + \cdots + \vec{v}_n$$

with $\vec{v}_j \in V_j$, $1 \leq j \leq n$. (See Exercise 4.)

If V is the direct sum of the subspaces $V_1, \ldots, V_n$, we write

$$V = V_1 \oplus V_2 \oplus \cdots \oplus V_n$$

THEOREM 2.2. *Let V be a finite dimensional vector space and* $T:V \to V$ *a linear transformation with eigenvalues* $\lambda_1, \ldots, \lambda_p$. *Then V is the direct sum of the generalized eigenspaces* $V_{\lambda_1}, \ldots, V_{\lambda_p}$.

The proof of this theorem is given in Section 5.

We know from Proposition 1.2 that $T(V_{\lambda_j}) \subset V_{\lambda_j}$ for $1 \leq j \leq p$. Let

$$T_j : V_{\lambda_j} \to V_{\lambda_j}$$

be the restriction of T to V_{λ_j}. Then T_j has a single generalized eigenspace and we can choose a basis α_j for V_{λ_j} such that, relative to this basis, T_j has matrix C_j of the form (2.4) with $\lambda = \lambda_j$. In fact, the set

$$\alpha = \alpha_1 \cup \alpha_2 \cup \cdots \cup \alpha_p$$

is a basis for V (see Exercise 3) and, relative to this basis, T has matrix

$$\begin{pmatrix} C_1 & 0 & 0 & \cdots & 0 \\ 0 & C_2 & 0 & \cdots & 0 \\ & & \cdots & & \\ 0 & \cdots & & 0 & C_p \end{pmatrix} \tag{2.5}$$

where each C_j is a square matrix of the form (2.4) (with $\lambda = \lambda_j$) and each 0 is a rectangular zero matrix. This is the *Jordan Canonical Form* of the linear transformation T.

We now give the proof of Proposition 2.1. In fact, we prove more.

PROPOSITION 2.3. *For each j, $1 \leq j \leq k$, vectors in the last j columns of configuration (2.2) form a basis for $\hbar(N^j)$.*

Since $\hbar(N^k) = V$, Proposition 2.1 follows from this result.

Proof. We proceed by induction on j. For $j = 1$, the last column of (2.2) consists of the vectors

$$\vec{v}_1, \ldots, \vec{v}_{m_1 + \cdots + m_k}$$

which form a basis for $\hbar(N)$ by construction. Assume the statement of Proposition 2.3 true for $j - 1$; that is, the vectors in the last $(j-1)$-columns of (2.2) form a basis for $\hbar(N^{j-1})$. We need to show that the vectors in the last j-columns form a basis for $\hbar(N^j)$. We first show these vectors to be linearly independent.

Let $m = m_1 + \cdots + m_k$ and suppose that

$$\vec{w} + \sum_{j=1}^{m} a_j \vec{v}_j = \vec{0} \qquad (2.6)$$

where $\vec{w}$ is a linear combination of the vectors in the next to the last $(j-1)$-columns of (2.2). Applying N to equation (2.6), we see that $N\vec{w} = \vec{0}$ (since $N\vec{v}_j = 0$, $1 \leq j \leq m_1 + \cdots + m_k$). However, $N\vec{w}$ is a linear combination of vectors in the last $(j-1)$-columns of (2.2) so, by induction, $N\vec{w}$ must be the trivial linear combination (since $\vec{w}$ does not involve the vectors $\vec{v}_1, \ldots, \vec{v}_{m_1 + \cdots + m_k}$) and equation (2.6) becomes

$$\sum_{j=1}^{m} a_j \vec{v}_j = \vec{0}$$

Since $\vec{v}_1, \ldots, \vec{v}_{m_1 + \cdots + m_k}$ are linearly independent (by construction), we have $a_j = 0$, $1 \leq j \leq m_1 + \cdots + m_k$, so that (2.6) must be the trivial linear combination. Thus the last j columns are linearly independent.

To see that the last j columns of (2.2) span $h(N^j)$, we suppose $\vec{v} \in h(N^j)$ so that $N^j\vec{v} = \vec{0}$. Then $N^{j-1}(N\vec{v}) = \vec{0}$ so that, by induction, $N\vec{v}$ can be written as a linear combination of the vectors in the last $(j-1)$-columns of (2.2). In particular,

$$N\vec{v} = N\vec{w} + \sum_{j=1}^{m_k} a_j \vec{v}_{m_1 + \cdots + m_{k-1} + j}$$

where w is a linear combination of the vectors in the last j columns of (2.2). However, by definition, no linear combination of the vectors

$$\vec{v}_{m_1 + \cdots + m_{k-1} + 1}, \ldots, \vec{v}_{m_1 + \cdots + m_k}$$

can be in $\mathcal{I}(N)$, so $a_1 = \cdots = a_{m_k} = 0$ and

$$N(\vec{v} - \vec{w}) = \vec{0}$$

Therefore $\vec{v} - \vec{w} \in h(N) \cap \mathcal{I}(N^j)$, so that

$$\vec{v} - \vec{w} = \sum_{j=1}^{m_1 + \cdots + m_k} a_j \vec{v}_j$$

and

$$\vec{v} = \vec{w} + \sum_{j=1}^{m_1 + \cdots + m_k} c_j \vec{v}_j$$

with $\vec{w}$ a linear combination of the vectors in the last j columns of (2.2). This completes the induction step and therefore the proof of Proposition 2.3.

EXERCISES

1.   Determine the vectors $\vec{v}_j$, $\vec{u}_j$ and the Jordan Canonical Form for the linear transformations $T:C^n \to C^n$, $n = 2,3$, defined by the matrices of Exercise 1 of the previous section.

2.   Determine the vectors $\vec{v}_j$, $\vec{u}_j$ and the Jordan Canonical Form for the linear transformations $T:C^n \rightarrow C^n$, $n = 4,5$, defined by the following matrices.

(a) $\begin{pmatrix} 7i & 4 & -6 & 12i \\ 3 & -5i & 6i & 6 \\ 2 & -4i & 5i & 4 \\ -4i & -2 & 3 & -7i \end{pmatrix}$

(c) $\begin{pmatrix} -17 & 20 & 8 & 10 & -12 \\ -1 & 3 & -1 & 1 & 0 \\ -7 & 11 & 0 & 5 & -3 \\ -30 & 36 & 12 & 18 & -20 \\ -6 & 12 & -3 & 5 & -1 \end{pmatrix}$

(b) $\begin{pmatrix} 0 & 1 & 0 & 0 & 0 \\ 0 & 0 & 1 & 0 & 0 \\ 0 & 0 & 0 & 1 & 0 \\ 0 & 0 & 0 & 0 & 1 \\ 16 & -16 & -8 & 8 & 1 \end{pmatrix}$

(d) $\begin{pmatrix} 0 & -18 & -10 & 14 & 6 \\ -2 & 6 & 4 & -4 & -8 \\ -3 & -8 & -4 & 10 & -8 \\ -4 & 0 & 1 & 3 & -13 \\ -1 & 5 & 3 & -3 & -5 \end{pmatrix}$

(The characteristic equations for (a), (c), and (d) are $\lambda^4 + 2\lambda^2 + 1 = 0$, $\lambda^5 - 3\lambda^4 + 3\lambda^3 - \lambda^2 = 0$, and $\lambda^5 = 0$ respectively.)

3.   Suppose that V is the direct sum of the subspaces $V_1,\ldots,V_n$ and let α_j be a basis for V_j, $1 \le j \le n$.   Prove

$$\alpha = \alpha_1 \cup \alpha_2 \cup \cdots \cup \alpha_n$$

is a basis for V.

4.   Prove that V is the direct sum of the subspaces $V_1,\ldots,V_n$ if and only if for each vector $\vec{v} \in V$, there are unique vectors $\vec{v}_j \in V_j$, $1 \le j \le n$, such that

$$\vec{v} = \vec{v}_1 + \vec{v}_2 + \cdots + \vec{v}_n$$

5.   Prove that V is the direct sum of the subspaces $V_1,\ldots,V_n$ if and only if there are linear transformations $E_1,\ldots,E_n$ such that

(a) $E_j^2 = E_j$, $1 \le j \le n$

(b)   for each $\vec{v} \in V$, $\sum_{j=1}^{n} E_j \vec{v} = \vec{v}$

(c) $E_i \circ E_j = 0$ for $i \neq j$

(d) $\mathcal{J}(E_i) = V_i$, $1 \leq i \leq n$

E_j is called the *projection of V onto* V_j

6.   A linear transformation $T : V \to V$ is said to be nilpotent if $T^n = 0$ for some integer n.

(a)   Prove that if T is nilpotent and $T^k \neq 0$, then $\mathcal{J}(T^{k+1})$ is a proper subspace of $\mathcal{J}(T^k)$. (Hint: if not, the restriction of T to $\mathcal{J}(T^k)$ is invertible.)

(b)   Prove that V has a basis relative to which the matrix $A = (a_{ij})$ of T is strictly lower triangular; that is, $a_{ij} = 0$ if $j \geq i$. (Prove this assertion using the Jordan Canonical Form and also (using (a)) without using the Jordan Canonical Form.)

7.   (a)   Let V be the vector space of complex linear combinations of the functions e^{2x}, xe^{2x}, e^{3x}, xe^{3x}, and $x^2 e^{3x}$, and let $D : V \to V$ be defined by $Df = f'$. Describe the Jordan Canonical Form for D.

(b)   Let V be the vector space of complex linear combinations of e^x, sin 2x, cos 2x, x sin 2x, and x cos 2x, and let $D : V \to V$ be defined by $Df = f'$. Describe the Jordan Canonical Form for D.

3.   POLYNOMIALS AND LINEAR TRANSFORMATIONS

In this section, we show how to evaluate a polynomial at a linear transformation. We also determine a necessary condition for a polynomial to "vanish" at a linear transformation and apply this condition to prove the Cayley-Hamilton Theorem. Except for this last theorem, the results of this section are needed in Section 5 to prove Theorem 2.4.

Let $P(z)$ be a complex polynomial of degree n;

$$P(z) = a_0 + a_1 z + a_2 z^2 + \cdots + a_n z^n$$
$$= \sum_{j=0}^{n} a_j z^j$$

where $a_0, \ldots, a_n$ are complex numbers and $a_n \neq 0$. The sum and product of complex polynomials are defined in the obvious way and are again complex polynomials.

The *derivative* of a complex polynomial

$$P(z) = \sum_{j=0}^{n} a_j z^j$$

is defined to be the complex polynomial

$$P'(z) = \sum_{j=1}^{n} j a_j z^{j-1}$$

For example, if

$$P(z) = 3i - (1 - 3i)z + 4z^2 + iz^3$$

then

$$P'(z) = -(1 - 3i) + 8z + 3iz^2$$

The higher derivatives of a complex polynomial are defined in the obvious way :

$$P^{(j)}(z) = (P^{(j-1)})'(z)$$

Our first result gives the elementary properties of the derivative.

PROPOSITION 3.1. *Let $P(z)$ and $Q(z)$ be complex polynomials, a and b complex numbers and j an integer ≥ 1. Then*

(a) $(aP + bQ)^{(j)}(z) = aP^{(j)}(z) + bQ^{(j)}(z)$

(b) $(PQ)'(z) = P'(z)Q(z) + P(z)Q'(z)$

This result follows by direct computation and is left as Exercise 1.

The next result shows that a complex polynomial can be expressed in terms of its higher derivatives. (Compare this result with Taylor's Theorem, Theorem 6.4.1.)

PROPOSITION 3.2. *Let* $P(z)$ *be a complex polynomial of degree* n *and let* λ *be any complex number. Then*

$$P(z) = \sum_{j=0}^{n} \frac{P^{(j)}(\lambda)}{j!} (z - \lambda)^j$$

Proof. Using statement (a) of Proposition 3.1, it is enough to prove Proposition 3.2 when $P(z) = z^k$. In this case,

$$P^{(j)}(z) = k(k-1)\cdots(k-j+1)z^{k-j} \tag{3.1}$$

by direct computation. On the other hand, using the binomial theorem and Equation 3.1, we have

$$z^k = ((z - \lambda) + \lambda)^k$$

$$= \sum_{j=0}^{k} \binom{k}{j}\lambda^{k-j}(z - \lambda)^j$$

$$= \sum_{j=0}^{k} \frac{P^{(j)}(\lambda)}{j!} (z - \lambda)^j$$

This proves Proposition 3.2.

Let k be a positive integer and λ a complex number. We say that a complex polynomial $P(z)$ has a *zero of order at least* k at λ if there is a complex polynomial $Q(z)$ such that

$$P(z) = (z - \lambda)^k Q(z)$$

PROPOSITION 3.3. *The polynomial* $P(z)$ *has a zero of order at least* k *at* λ *if and only if* $P^{(j)}(\lambda) = 0$ *for* $0 \le j \le k$.

Proof. If $P^{(j)}(\lambda) = 0$ for $0 \le j \le k$, then

$$P(z) = \sum_{j=k}^{n} \frac{P^{(j)}(\lambda)}{j!} (z - \lambda)^j$$

$$= (z - \lambda)^k \sum_{j=k}^{n} \frac{P^{(j)}(\lambda)}{j!} (z - \lambda)^{j-k}$$

so $P(z)$ has the required form with

$$Q(z) = \sum_{j=k}^{n} \frac{P^{(j)}(\lambda)}{j!} (z - \lambda)^{j-k}$$

On the other hand, suppose

$$P(z) = (z - \lambda)^{k}Q(z)$$

We prove by induction on j that $P^{(j)}(z)$ has a factor $(z - \lambda)^{k-j}$; that is,

$$P^{(j)}(z) = (z - \lambda)^{k-j}Q_{j}(z)$$

for some complex polynomial $Q_{j}(z)$. It certainly follows then that $P^{(j)}(\lambda) = 0$ for $0 \leq j \leq k$.

For $j = 0$, our inductive statement holds with $Q_{0}(z) = Q(z)$. Suppose the assertion true for $j - 1$; that is

$$P^{(j-1)}(z) = (z - \lambda)^{k-j+1}Q_{j-1}(z)$$

Then

$$\begin{aligned}
P^{(j)}(z) &= (P^{(j-1)}(z))' \\
&= ((z - \lambda)^{k-j+1}Q_{j-1}(z))' \\
&= (k-j+1)(z - \lambda)^{k-j}Q_{j-1}(z) \\
&\quad + (z - \lambda)^{k-j+1}Q'_{j-1}(z) \\
&= (z - \lambda)^{k-j}((k-j+1)Q_{j-1}(z) + (z - \lambda)Q'_{j-1}(z))
\end{aligned}$$

and Proposition 3.3 follows.

Suppose V is a complex vector space, $T:V \to V$ a linear transformation, and

$$P(z) = \sum_{j=0}^{n} a_{j}z^{j}$$

a complex polynomial. We define P(T) to be the linear transformation

$$P(T) = \sum_{j=0}^{n} a_j T^j$$

where $T^0 = I$ and $T^j = T \circ T \circ \cdots \circ T$ is the j-fold composite of T with itself.

PROPOSITION 3.4.   *Let T:V → V be a linear transformation, P(z) and Q(z) complex polynomials, and a and b complex numbers. Then*

$$(aP + bQ)(T) = aP(T) + bQ(T)$$
$$(PQ)(T) = P(T) \circ Q(T)$$

The proof of this result is straightforward and left as Exercise 2.

The principal result of this section is the following.

THEOREM 3.5.   *Let V be a finite dimensional vector space, T:V → V a linear transformation, and P(z) a complex polynomial. Then P(T) = 0 if and only if P(z) has a zero of order at least $\nu(\lambda)$ at λ for each eigenvalue λ of T.*

We prove this result in the next section.

COROLLARY (The Cayley-Hamilton Theorem).   *Let V be a finite dimensional vector space, T:V → V a linear transformation, and P(z) = Det(T - zI) the characteristic polynomial of T. Then P(T) = 0.*

Proof. If we use the Jordan Canonical Form of T to compute P(z) = Det(T - zI), we see immediately that P(z) has a zero of order at least λ for each eigenvalue λ of T. The corollary now follows from Theorem 3.5.

Remark. Strictly speaking, the Cayley-Hamilton Theorem will only be proved after we establish the existence of the Jordan Canonical Form. There is no difficulty with giving it here, however, since the Cayley-Hamilton Theorem is not used in the proof of Theorem 2.4.

EXERCISES

1.   Prove Proposition 3.1.

2.   Prove Proposition 3.4.

*3.   (a) Let W be a complex finite-dimensional vector space, and
let $T:W \to W$ be a linear transformation whose only eigenvalue is 1.
Show that T^{-1} can be written as a polynomial in T. (Hint: use the
Jordan Canonical Form to show that $T = I - S$ where $S^m = 0$, some m.
Recall also that $(1 - z^m)/(1 - z) = 1 + z + z^2 + \cdots + z^{m-1}$.)

     (b) Generalize the result in (a) to the case where T has only
one eigenvalue $\lambda \neq 0$.

4.   Let V be a finite dimensional vector space and $T:V \to V$ a linear
transformation. Let M be the subspace of $\mathcal{L}(V,V)$ spanned by $I = T^0$,
$T^1, T^2, \ldots$. Show that $\dim M \leq \dim V$.

5.   Let M be as in the previous exercise and let $p = \dim M$.
     (a) Show that there is a monic polynomial $P(z)$ of degree p such
that $P(T) = 0$. (A polynomial $P(z) = \Sigma_{j=0}^{p} a_j z^j$ of degree p is said to
be *monic* if $a_p = 1$.) Prove that, if $Q(z)$ is a polynomial of degree
less than p, then $Q(T) \neq 0$. ($P(z)$ is called the *minimum polynomial*
for T.)
     (b) Prove that, if $p = \dim V$, then $\pm P(z)$ is the characteristic
polynomial of T.

     Let V be a finite-dimensional vector space, and let $T:V \to V$ be
a linear transformation. For $\vec{v} \in V$, let $V_{\vec{v}}$ be the subspace of V
spanned by $\vec{v}_0 = I\vec{v}, T\vec{v}, T^2\vec{v}, \ldots$. The vector v is called a *cyclic
vector* for T if $V_{\vec{v}} = V$. Note that if $\vec{v}$ is an eigenvector for T, then
$V_{\vec{v}}$ is 1-dimensional; thus cyclic vectors and eigenvectors are in some
sense at opposite extremes in their behavior.
     The next few problems deal with cyclic vectors.

6.   Show that the linear transformations on C^2 defined by the matrices

$$\begin{pmatrix} 1 & 1 \\ 0 & 1 \end{pmatrix} \quad \begin{pmatrix} 1 & 0 \\ 0 & 2 \end{pmatrix}$$

have cyclic vectors but the one given by

$$\begin{pmatrix} 2 & 0 \\ 0 & 2 \end{pmatrix}$$

does not.

7.   Show that if T has a cyclic vector, then the subspace M of Exercise 4 is p-dimensional.

8.   (a) Show that if T has a cyclic vector, then there is a basis for V such that the matrix for T with respect to this basis is of the form

$$\begin{pmatrix} 0 & 0 & 0 & \dots & 0 & a_1 \\ 1 & 0 & 0 & \dots & 0 & a_2 \\ 0 & 1 & 0 & \dots & 0 & a_3 \\ & & \dots & & & \\ 0 & \dots & & 0 & 1 & a_p \end{pmatrix}$$

(b) Show that the characteristic polynomial of T is

$$\lambda^p - a_p \lambda^{p-1} - \dots - a_2 \lambda - a_1$$

*9.   Let $\lambda_1, \dots, \lambda_s$ be the eigenvalues of $T: V \to V$. We know from Proposition 1.2 that each V_{λ_j} is stable under T. Prove that V has a cyclic vector for T if and only if each V_{λ_j} does, $1 \le j \le s$. (Hint: use the projections E_j of V onto V_{λ_j} which exist by Theorem 2.2 and Exercise 5 of Section 1.)

*10.   Suppose $T: V \to V$ has only one eigenvalue. Show that V has a cyclic vector for T if and only if the Jordan Canonical Form (a_{ij}) has the property that $a_{j+1,j} = 1$ for $1 \le j < \dim V$. (Hint: for the "if," half, produce a cyclic vector; for "only if", use Exercise 7.)

4. THE PROOF OF THEOREM 3.5

We begin with a lemma.

LEMMA 4.1. *Let V be a finite dimensional vector space and*
$T:V \rightarrow V$ *a linear transformation with eigenvalues* $\lambda_1,\ldots,\lambda_s$. *Then
there is a complex polynomial* $g(z)$,

$$g(z) = \prod_{j=1}^{s} (z - \lambda_j)^{b_j}$$

with $0 < b_j \leq \nu(\lambda_j)$, $1 \leq j < s$, *and* $g(T) = 0$.

Proof. Since V is finite-dimensional, we know that $\mathcal{L}(V,V)$ is
finite-dimensional (see Proposition 7.5.7 for example) so that, for
n large enough (in particular, $n > \dim \mathcal{L}(V,V)$), the linear transformations

$$I,T,T^2,\ldots,T^n$$

are linearly dependent. Suppose that

$$\sum_{j=0}^{n} a_j T^j = 0$$

is a dependence relation with $a_n = 1$. Define the complex polynomial
$P(z)$ by

$$P(z) = \sum_{j=0}^{n} a_j z^j$$

Then $P(T) = 0$.

Using the Fundamental Theorem of Algebra (Theorem 10.1.2) we
can write

$$P(z) = \prod_{j=1}^{r} (z - \mu_j)^{s_j}$$

Suppose that $\mu_1,\ldots,\mu_p$ are eigenvalues for T and $\mu_{p+1},\ldots,\mu_r$ are not.
Define $P_1(z)$ and $P_2(z)$ by

$$P_1(z) = \prod_{j=1}^{p} (z - \mu_j)^{s_j}$$

$$P_2(z) = \prod_{j=p+1}^{r} (z - \mu_j)^{s_j}$$

Thus

$$P(z) = P_1(z)P_2(z)$$

Now, if μ is not an eigenvalue for T, we know that

$$(T - \mu I)$$

is invertible. Thus $P_2(T)$ is invertible and

$$P(T) = 0$$
$$= P_1(T) \circ P_2(T)$$

implies that $P_1(T) = 0$.

Next, suppose that λ is an eigenvalue of T with eigenvector $\vec{v} \neq \vec{0}$. Then

$$T^j(\vec{v}) = \lambda^j \vec{v}$$

for any $j \geq 0$ and it follows that

$$\vec{0} = P_1(T)(\vec{v}) = P_1(\lambda)\vec{v}$$

Since $\vec{v} \neq 0$, we must have $P_1(\lambda) = 0$, which implies that

$$\lambda \in \{\mu_1, \ldots, \mu_p\}$$

Thus we can write

$$P_1(z) = \prod_{j=1}^{s} (z - \lambda_j)^{t_j}$$

with $t_j > 0$, $1 \leq j \leq s$.

Finally, suppose that some $t_j > \nu(\lambda_j)$, say $t_1 > \nu(\lambda_1)$. Then

$$P_1(T) = (T - \lambda_1 I)^{t_1} \circ (\prod_{j=2}^{s} (T - \lambda_j I)^{t_j})$$
$$= 0$$

so that

$$\mathcal{I}(\prod_{j=2}^{s} (T - \lambda_j I)^{t_j}) \subset V_{\lambda_1}$$

However,

$$(T - \lambda_1 I)^{\nu(\lambda_1)} V_{\lambda_1} = \{0\}$$

so

$$(T - \lambda_1 I)^{\nu(\lambda_1)} \circ (\prod_{j=2}^{s} (T - \lambda_j I)^{t_j}) = 0$$

Therefore, if we define b_j, $0 \leq j \leq s$, by

$$b_j = \min\{\nu(\lambda_j), t_j\}$$

and $g(z)$ by

$$g(z) = \prod_{j=1}^{s} (z - \lambda_j)^{b_j}$$

then $g(T) = 0$ and Lemma 4.1 is proved.

We now return to the proof of Theorem 3.5. Let $\lambda_1, \ldots, \lambda_s$ be the eigenvalues of T ordered so that

$$\nu(\lambda_1) \geq \nu(\lambda_2) \geq \cdots \geq \nu(\lambda_s)$$

Suppose that $P(z)$ is a complex polynomial having a zero of order at least $\nu(\lambda_j)$ at λ_j for $j=1,\ldots,s$. It follows that

$$P(z) = (z - \lambda_1)^{\nu(\lambda_1)} Q_1(z)$$

and

$$P(\lambda_2) = P'(\lambda_2) = \cdots = P^{(\nu(\lambda_2)-1)}_{(\lambda_2)} = 0$$

We claim that there is a $Q_2(z)$ such that

$$Q_1(z) = (z - \lambda_2)^{\nu(\lambda_2)} Q_2(z)$$

According to Proposition 3.2, this is equivalent to the assertion that

$$Q_1^{(j)}(\lambda_2) = 0 \qquad\qquad (4.1)$$

for $1 \le j < \nu(\lambda_2)$. We prove this by induction on j. For $j = 0$, it follows from the fact that $P(\lambda_2) = 0$. Suppose that $Q_1^{(j)}(\lambda_2) = 0$ for $0 \le j < k < \nu(\lambda_2)$. Then

$$0 = P^{(k)}(\lambda_2)$$

$$= \sum_{j=0}^{k} \binom{k}{j} \nu(\lambda_1)(\nu(\lambda_1)-1)\ldots(\nu(\lambda_1)-j+1)(\lambda_2-\lambda_1)^{\nu(\lambda_1)-j} Q^{(k-j)}(\lambda_2)$$

by the Leibniz formula (see Exercise 6(e), Section 1.3)

$$= (\lambda_2 - \lambda_1)^{\nu(\lambda_1)} Q^{(k)}(\lambda_2)$$

by (4.1) for $j < k$. Since $\lambda_1 \ne \lambda_2$, we must have $Q_1^{(k)}(\lambda_2) = 0$, and our claim is proved. Thus

$$P(z) = (z - \lambda_1)^{\nu(\lambda_1)} (z - \lambda_2)^{\nu(\lambda_2)} Q_2(z)$$

Repeating this argument, we obtain

$$P(z) = \Big(\prod_{j=1}^{s} (z - \lambda_j)^{\nu(\lambda_j)} \Big) Q_s(z)$$

$$= g(z) \Big(\prod_{j=1}^{s} (z - \lambda_j)^{\nu(\lambda_j)-b_j} \Big) Q_s(z)$$

where $g(z)$ is as in Lemma 4.1. Now $P(T) = 0$ since $g(T) = 0$ and one half of Theorem 3.5 is proved.

To prove the other half of Theorem 3.5, suppose $P(z)$ is a complex polynomial with $P(T) = 0$. We wish to show that $P(z)$ has a zero of order $\nu(\lambda_i)$ at λ_i for each $i=1,\ldots,s$.

Just as in the proof of Lemma 3.6, we can show that $P(z) = P_1(z)P_2(z)$ with

$$P_1(z) = \prod_{j=1}^{s} (z - \lambda_j)^{t_j}$$

with $t_j > 0$, $1 \le j \le s$, and $P_1(T) = 0$. Suppose $t_j < \nu(\lambda_j)$ for some j, say $j = 1$. Then by the definition of $\nu(\lambda_1)$,

$$(T - \lambda_1 I)^{t_1} V_{\lambda_1} \ne \{0\}$$

In fact, since

$$(T - \lambda_1 I)^{\nu(\lambda_1)} V_{\lambda_1} = \{\vec{0}\}$$

we must have

$$\dim(T - \lambda_1 I)^{t_1 + 1} V_{\lambda_1} < \dim(T - \lambda_1 I)^{t_1} V_{\lambda_1}$$

(see Exercise 6, Section 2). Therefore, there is a $\vec{v} \ne \vec{0}$ in $(T - \lambda_1 I)^{t_1} V_{\lambda_1}$ with

$$(T - \lambda_1 I)\vec{v} = \vec{0}$$

or, equivalently, $T\vec{v} = \lambda_1 \vec{v}$. Write $\vec{v} = (T - \lambda_1 I)^{t_1}\vec{u}$, $\vec{u} \in V_{\lambda_1}$. Then

$$\vec{0} = P_1(T)(\vec{u}) = (\prod_{j=2}^{s} (T - \lambda_j I)^{t_j})(T - \lambda_1 I)^{t_1}\vec{u}$$
$$= (\prod_{j=2}^{s} (T - \lambda_j I)^{t_j})\vec{v}$$

However, $T^i\vec{v} = \lambda_1^i \vec{v}$ for all i, so that

$$(\prod_{j=2}^{s} (T - \lambda_j I)^{t_j})\vec{v} = (\prod_{j=2}^{s} (\lambda_1 - \lambda_j)^{t_j})\vec{v}$$
$$\ne \vec{0}$$

since $\vec{v} \ne \vec{0}$ and $\lambda_j \ne \lambda_1$ for $j > 1$. Thus $t_j \ge \nu(\lambda_j)$, $1 \le j \le s$, and Theorem 3.5 is proved.

5.   THE PROOF OF THEOREM 2.2

Let V be a finite dimensional vector space and $T:V \to V$ a linear transformation with eigenvalues $\lambda_1,\ldots,\lambda_s$. We shall define linear transformations

$$E_j : V \to V$$

for $j=1,\ldots,s$ such that

$$E_j^2 = E_j \tag{5.1}$$

$$E_i \circ E_j = 0 \text{ if } i \neq j \tag{5.2}$$

$$\mathcal{I}(E_i) = V_{\lambda_i} \tag{5.3}$$

$$\sum_{j=1}^{s} E_j = I \tag{5.4}$$

Then

$$V = V_{\lambda_1} + \cdots + V_{\lambda_s}$$

since

$$\vec{v} = \sum_{j=1}^{s} E_j \vec{v}$$

for any $\vec{v} \in V$ by (5.4) and $E_j\vec{v} \in V_{\lambda_j}$ by (5.3). Furthermore, if
$\vec{v} \in V_{\lambda_i}$ and $\vec{v} \in \sum_{j \neq i} V_{\lambda_j}$, then

$$\vec{v} = E_i \vec{u}$$

$$= \sum_{j \neq i} E_j \vec{w}_j$$

for some $\vec{u}, \vec{w}_1, \ldots, \vec{w}_{i-1}, \vec{w}_{i+1}, \ldots, \vec{w}_s \in V$. However

$$E_i \vec{u} = E_i(E_i \vec{u}) \qquad \text{(by (5.1))}$$

$$= \sum_{j \neq i} E_i E_j \vec{w}_j$$

$$= \vec{0}$$

by (5.2). Therefore $\vec{v} = E_i\vec{u} = \vec{0}$ and

$$\vec{V} = \vec{V}_{\lambda_1} \oplus \ldots \oplus \vec{V}_{\lambda_s}$$

Thus Theorem 2.2 will be proved once we define E_j, $1 \le j \le s$, satisfying (5.1) through (5.4). This is done using the following:

LEMMA 5.1. *There exist complex polynomials* $P_1(z),\ldots,P_s(z)$ *such that*

(a) *for each* $j=1,\ldots,s$, $P_j(z) - 1$ *has a zero of order at least* $v.(\lambda_j)$ *at* λ_j

(b) *for each* $i,j=1,\ldots,s$, $i \ne j$, $P_j(z)$ *has a zero of order at least* $v(\lambda_i)$ *at* λ_i

Before proving this lemma, we show how to use it to define linear transformations $E_1,\ldots,E_s:V \to V$ satisfying (5.1)-(5.4).

Let $E_j:V \to V$ be defined by

$$E_j = P_j(T)$$

Then, according to Theorem 3.5,

$$E_1 + \cdots + E_s - I = 0$$

since (using (a) and (b) of Lemma 5.1) the polynomial

$$P_1(z) + \cdots + P_s(z) - 1$$

has a zero of order at least $v(\lambda_j)$ at λ_j for each $i=1,\ldots,s$. Therefore, $E_1,\ldots,E_s$ satisfy (5.4). Similarly, the polynomials

$$P_j^2(z) - P_j(z) = P_j(z)(P_j(z) - 1)$$

$$P_i(z)P_j(z)$$

for $i \ne j$ have zeros of order at least $v(\lambda_k)$ at λ_k for $k = 1,\ldots,s$ so (5.1) and (5.2) hold (again using Theorem 3.5).

To verify (5.3), we note that, for each $j=1,\ldots,s$, the polynomial

$$(z - \lambda_j)^{\nu(\lambda_j)} P_j(z)$$

has a zero of order $\nu(\lambda_i)$ at λ_i for all $i=1,\ldots,s$. Thus

$$(T - \lambda_j I)^{\nu(\lambda_j)} P_j(T) = (T - \lambda_j I)^{\nu(\lambda_j)} E_j$$

$$= 0$$

so that $\mathcal{J}(E_j) \subset V_{\lambda_j}$. On the other hand, if $\vec{v} \in V_{\lambda_j}$, we know from (5.4) that

$$\vec{v} = \sum_{i=1}^{n} E_i \vec{v}$$

Now, for each $i \neq j$, we can find a complex polynomial $Q_i(z)$ such that

$$P_i(z) = Q_i(z)(z - \lambda_j)^{\nu(\lambda_j)}$$

(using (b) of Lemma 5.1). It follows that

$$E_i \vec{v} = P_i(T)\vec{v}$$

$$= Q_i(T)(T - \lambda_j I)^{\nu(\lambda_j)}\vec{v}$$

$$= \vec{0}$$

since $\vec{v} \in V_{\lambda_j}$. Therefore $\vec{v} = E_j \vec{v}$ and $V_{\lambda_j} \subset \mathcal{J}(E_j)$. This proves (5.3).

We now prove Lemma 5.1. We shall define $P_1(z)$; the definitions of $P_2(z),\ldots,P_s(z)$ are analogous and are left to the reader.

Let $g_1(z)$ be the complex polynomial

$$g_1(z) = \prod_{j=2}^{s} (z - \lambda_j)^{\nu(\lambda_j)}$$

and set

$$P_1(z) = g_1(z) \sum_{j=0}^{\nu(\lambda_1)} a_j (z - \lambda_1)^j$$

where a_j, $0 \leq j \leq \nu(\lambda_1)$ are constants to be determined. Note that condition (b) of Lemma 5.1 holds for $P_1(z)$ no matter how the a_j are defined. We proceed by induction on k to show that a_k, $0 \le k \le \nu(\lambda_1)$, can be chosen so that

$$P_1(\lambda_1) = 1 \qquad\qquad\qquad (5.5)$$
$$P_1^{(i)}(\lambda_1) = 0$$

for $1 \le i \le k$.

Let $a_0 = 1/g_1(\lambda_1)$ (which makes sense since $g_1(\lambda_1) \ne 0$) and suppose that $a_0, \ldots, a_{k-1}$ have been defined so that (5.5) holds for $i < k$. It is easy to see that

$$\frac{d^i}{dz^i}(g_1(z) \sum_{j=0}^{\nu(\lambda_1)} a_j(z - \lambda_1)^j)\Big|_{z=\lambda_1} = \frac{d^i}{dz^i}(g_1(z) \sum_{j=0}^{i} a_j(z - \lambda_1)^j)\Big|_{z=\lambda_1}$$

and it follows that we need only define a_k so that

$$\frac{d^k}{dz^k}(g_1(z) \sum_{j=0}^{k} a_j(z - \lambda_1)^j)\Big|_{z=\lambda_1} = 0$$

Now

$$\frac{d^k}{dz^k}(g_1(z) \sum_{j=0}^{k} a_j(z - \lambda_1)^j)\Big|_{z=\lambda_1} = \frac{d^k}{dz^k}(g_1(z) \sum_{j=0}^{k-1} a_j(z - \lambda_1)^j)\Big|_{z=\lambda_1}$$

$$+ a_k \frac{d^k}{dz^k}(g_1(z)(z - \lambda_1)^k)\Big|_{z=\lambda_1}$$

In addition,

$$\frac{d^k}{dz^k}(g_1(z)(z - \lambda_1)^k)$$

is a sum of terms each of which contains a factor $(z - \lambda_1)$ except for the term $j!g_1(z)$ obtained by differentiating $(z - \lambda_1)^k$ at each stage. Then

$$\frac{d^k}{dz^k}(g_1(z)(z - \lambda_1)^k)\Big|_{z=\lambda_1} = j!g_1(\lambda_1) \ne 0$$

and we define a_k by

$$a_k = \frac{-1}{j! g_1(\lambda_1)} \frac{d^k}{dz^k} (g_1(z) \sum_{j=0}^{k-1} a_j (z - \lambda_1)^j) \Big|_{z=\lambda_1}$$

This completes the definition of $P_1(z)$ satisfying (a) and (b) of Lemma 5.1. The polynomials $P_j(z)$, $j > 1$ are defined in exactly the same way and the proof of Lemma 5.1 is complete.

EXERCISES

1.   Show that $TE_j = E_j T$ for $j=1,\ldots,s$ using only (5.1) through (5.4).

2.   Show that $\dim V_j$ = multiplicity of λ_j as a root of the characteristic polynomial of T.   (Hint: use Jordan Canonical form.)

SOLUTIONS TO SELECTED EXERCISES

CHAPTER 1

Section 1.1, p. 6

1. (a) {1,2,3,4,5,6,7,10,20,21}
    (c) {2,4,6,10,20}
    (e) {1,3,5,7}

Section 1.2, p. 12

5. $(\exists \epsilon > 0)(\forall \delta > 0)(\exists x, y \in R) |x - y| < \delta$ and $|f(x) - f(y)| \geq \epsilon$.

Section 1.4, p. 19

4. (a) $(-1, \frac{13}{3})$
    (c) $(-\frac{2}{3}, 2)$

5. (b) $(\frac{11}{7}, \infty) \cup (-\infty, -\frac{17}{7})$

Section 1.5, p. 22

1. (b) and (e) are equivalence relations (assuming in (e) that a ~ b
only if *both* parents are the same).

CHAPTER 2

Section 2.1, p. 28

1. (a) (3,12)   (i) $\vec{0}$
    (c) (0,6)   (k) (10,11)
    (e) (0,0) = $\vec{0}$ (m) (3,2)
    (g) (-6,21)   (o) (2,22)

Section 2.2, p. 34

4. (c) and (g) are vector spaces.

Section 2.3, p. 40

8. (a) $\vec{v}$ = (2,1,0)
    (c) $\vec{v}$ = (3/4, 5/4, 9/4)

Section 2.4, p. 47

1. (b), (c), and (f) are subspaces.

13. (a) No; yes.

Section 2.5, p. 56

2.   (a) Nullspace = $\{c(3,-2):c \in R\}$; image = R^1
     (c) Nullspace = $\{0\}$; image = R^2

9.   (b) $(3x_1 + 3x_2, x_1 - x_2, x_1 + 2x_2)$.

Section 2.6, p. 65

1.   (a) $(1 \quad -2)$

     (c) $\begin{pmatrix} 0 & 1 \\ 1 & -1 \\ 1 & 1 \end{pmatrix}$

     (e) $\begin{pmatrix} 0 & 0 & 1 \\ 1 & 0 & 0 \\ 0 & 1 & 0 \end{pmatrix}$

2.   (a) $\begin{pmatrix} 26 & 39 \\ 17 & 40 \end{pmatrix}$ (g) $\begin{pmatrix} -65 & 50 & 63 \\ -24 & 21 & 29 \end{pmatrix}$

     (c) $\begin{pmatrix} -12 & 4 & 7 \\ -16 & 52 & 41 \end{pmatrix}$ (i) $\begin{pmatrix} 6 & 4 & 2 & 8 \\ -18 & 12 & 15 & -6 \\ 21 & 7 & 42 & -56 \end{pmatrix}$

     (e) $\begin{pmatrix} 4 & 17 \\ 4 & 2 \\ 4 & 11 \end{pmatrix}$

3.   (a) $\begin{pmatrix} 7 & 15 \\ 10 & 22 \end{pmatrix}$ (g) makes no sense

     (c) makes no sense   (i) $\begin{pmatrix} 26 & -38 \\ 18 & -23 \end{pmatrix}$

     (e) $\begin{pmatrix} 9 & 19 \\ 8 & 14 \\ -9 & -15 \end{pmatrix}$

11. (a) $(9,5,-3)$
     (c) $(-3,-4,27)$

12. (a) $(6,0,0,8,1)$
     (c) $(1,-7,18,37,45)$

19. (a) $\begin{pmatrix} -4 & 3 \\ 3 & -2 \end{pmatrix}$

    (c) $\begin{pmatrix} -1 & 3/2 \\ 1 & -1 \end{pmatrix}$

CHAPTER 3

Section 3.1, p. 74

4. (c), (e), and (g) define norms.

Section 3.2, p. 80

3. (b), (c), (f) are open; (a), (d), (g) are closed; (e), (h) are neither.

Section 3.7, p. 115

1. (a) $\dfrac{\partial f}{\partial x_1} = 2x_1 x_2$, $\dfrac{\partial f}{\partial x_2} = x_1^2 + 4x_2$

    (c) $\dfrac{\partial f}{\partial x_1} = \dfrac{1}{x_1^2 + x_2^2} - \dfrac{2x_1(x_1 + x_2)}{(x_1^2 + x_2^2)^2}$, $\dfrac{\partial f}{\partial x_2} = \dfrac{1}{x_1^2 + x_2^2} - \dfrac{2x_2(x_1 + x_2)}{(x_1^2 + x_2^2)^2}$

    (e) $\dfrac{\partial f}{\partial x_1} = e^{x_2 + x_1 x_2}(x_2\sqrt{x_1} - \dfrac{1}{2\sqrt{x_1}})$, $\dfrac{\partial f}{\partial x_2} = \sqrt{x_1}(x_1 + 1)e^{x_2 + x_1 x_2}$,

$$\dfrac{\partial f}{\partial x_3} = 0.$$

2. (a) $\begin{pmatrix} -\sin x_1 \\ \cos x_1 \end{pmatrix}$

    (c) $\begin{pmatrix} x_2 & x_1 \\ 2x_1 + x_2^2 & 2x_1 x_2 \\ 2x_1 x_2 & x_1^2 \end{pmatrix}$

CHAPTER 4

Section 4.1, p. 125

1.   (b), (g), and (h).

3.   (a)   a = 1, b = 6
      (c)   all a,b with ab = 0

Section 4.2, p. 130

There are many correct answers to each exercise; we give only one.

1.   (a) $\{(1,0,2),(1,1,1),(0,0,1)\}$
      (c) $\{(3,0,4,1),(2,0,4,1),(0,0,1,2),(6,1,1,5)\}$

2.   (a)   S ∪ $\{4,5\}$ (e)   S ∪ $\{(0,0,0,1)\}$
      (c)   S ∪ $\{(0,0,1)\}$ (g)   S ∪ $\{1, 2x + x^2\}$
                                              (i)   S ∪ $\{(\begin{smallmatrix}1 & 1\\4 & 3\end{smallmatrix}),(\begin{smallmatrix}0 & 4\\7 & -6\end{smallmatrix})\}$

5.   (a) $\{(-2,1,1,3),(7,1,0,-11)\}$

6.   (a)   Kernel: $\{(3,-1,-2,1),(7,1,0,5)\}$; image: $\{(2,-1,0,1),$
                                                        $(1,7,-5,3)\}$

Section 4.3, p. 138

There are many linear transformations which have the required properties in each exercise; we give the matrix of only one.

1.   (a) $\begin{pmatrix} 1 & 3 \\ 2 & 0 \\ -2 & 5 \end{pmatrix}$ (c) $\begin{pmatrix} 1 & 1 \\ 2 & 3 \\ 1 & 2 \end{pmatrix}$
      (e) $\begin{pmatrix} 1 & -2 \\ 2 & 1 \\ 1 & 4 \end{pmatrix}$

2. (a) $\begin{pmatrix} 2 & 2 & 1 \\ 3 & 3 & 1 \\ 3 & 3 & 1 \end{pmatrix}$

(c) $\begin{pmatrix} 1 & 0 & 1 \\ 0 & 1 & -2 \\ 4/3 & 1/3 & 4/3 \end{pmatrix}$

(e) $\begin{pmatrix} 4 & -1 & -1 \\ -5 & 2 & 4 \\ -4 & 2 & 4 \end{pmatrix}$

3. (a) $\begin{pmatrix} 2 & 0 & 0 \\ 1 & 0 & 0 \\ -4 & 0 & 0 \end{pmatrix}$, $\begin{pmatrix} 2 & 1 & 1 \\ 1 & 1 & 1 \\ -4 & 1 & 1 \end{pmatrix}$

(c) $\begin{pmatrix} 1 & 6 & 1 \\ 1 & 0 & 0 \\ 1 & -1 & 2 \end{pmatrix}$, $\begin{pmatrix} 0 & 7 & 2 \\ 0 & 1 & 1 \\ 1 & -1 & 2 \end{pmatrix}$

Section 4.4, p. 145

12. (a) 2

   (c) 2

Section 4.5, p. 151

1. (d), (e)

Section 4.7, p. 163

1. (a) $\{(\frac{1}{\sqrt{2}}, \frac{1}{\sqrt{2}}), (-\frac{1}{\sqrt{2}}, \frac{1}{\sqrt{2}})\}$

(c) $\{(\frac{1}{\sqrt{6}}, \frac{2}{\sqrt{6}}, \frac{1}{\sqrt{6}}), (-\frac{7}{\sqrt{174}}, -\frac{2}{\sqrt{174}}, \frac{11}{\sqrt{174}}), (\frac{4}{\sqrt{29}}, -\frac{3}{\sqrt{29}}, \frac{2}{\sqrt{29}})\}$

(e) $\{(\frac{2}{\sqrt{6}}, 0, \frac{1}{\sqrt{6}}, \frac{1}{\sqrt{6}}), (\frac{1}{\sqrt{3}}, 0, -\frac{1}{\sqrt{3}}, -\frac{1}{\sqrt{3}}), (0,1,0,0), (0,0,-\frac{1}{\sqrt{2}}, \frac{1}{\sqrt{2}})\}$

3.   Many answers; one is $(\frac{1}{\sqrt{2}},0,0),(0,1,0),(\frac{1}{\sqrt{2}},0,-\sqrt{2})$.

Section 4.8, p. 169

1.   (a) (1,2,-2)   (e) 2
     (c) (2,2,-2)   (g) (-2,1,-1)

CHAPTER 5

Section 5.1, p. 177

1.   (a) diverges   (e) converges to 1
     (c) converges to $\frac{1}{2}$ (g) diverges

Section 5.4, p. 194

1.   (a) none; 2   (e) log 3; none
     (c) $\frac{\pi}{6}$; none   (g) 0; none

Section 5.6, p. 203

1.   A, D, E, G, H, I, and K are compact.

Section 5.8, p. 212

1.   B, C, E, F, J, and K are connected.

CHAPTER 6

Section 6.1, p. 221

1.   (a) $(2xy^2(y^2 + y - 1) \quad x^2y(4y^2 + 3y - z))$

(c) $(4x^3y^2 - 2xy^4 \quad 2x^4y - 4x^2y^3)$

(e) $\begin{pmatrix} yz^3e^{2xy} + 2xy^2z^3e^{2xy} & xz^3e^{2xy} + 2x^2yz^3e^{2xy} & 3xyz^2e^{xy} \\ yz^2e^{xy} + xy^2z^2e^{xy} & xz^2e^{xy} + x^2yz^2e^{xy} & 2xyze^{xy} \end{pmatrix}$

(g) $2xyz(yz \quad zx \quad xy) + ze^{xy}(yz + xy^2z \quad xz + x^2yz \quad 2xy)$

$\qquad\qquad\qquad\qquad\qquad\qquad\qquad\qquad\qquad - 2e^{2xy}(yz^2 \quad xz^2 \quad x)$

(i) $\begin{pmatrix} 3x^2y - y^3 & x^3 - 3xy^2 \\ ye^{x^2-y^2}(1+2x) & xe^{x^2-y^2}(1-2y) \end{pmatrix}$

5. $\dfrac{\partial F}{\partial \rho} = \dfrac{\partial f}{\partial x} \sin \varphi \cos \theta + \dfrac{\partial f}{\partial y} \sin \varphi \sin \theta + \dfrac{\partial f}{\partial z} \cos \varphi;$

$\dfrac{\partial F}{\partial \theta} = -\dfrac{\partial f}{\partial x} \rho \sin \varphi \sin \theta + \dfrac{\partial f}{\partial y} \rho \sin \varphi \cos \theta;$

$\dfrac{\partial F}{\partial \varphi} = \dfrac{\partial f}{\partial x} \rho \cos \varphi \cos \theta + \dfrac{\partial f}{\partial y} \rho \cos \varphi \sin \theta - \dfrac{\partial f}{\partial z} \rho \sin \varphi$

Section 6.3, p. 236

1.   (a) $2y; 2x; 0$

(c) $(4x^3 + 6x)ye^{x^2}; (2x^2 + 1)e^{x^2}; 0$

(e) $-\dfrac{2}{9} y^2(xy + y^2)^{-5/3}; \dfrac{1}{3}(xy + y^2)^{-2/3}(1 - \dfrac{2y(x+2y)}{3(xy+y^2)});$

$\qquad\qquad \dfrac{2}{9}(xy + y^2)^{-8/3}[\dfrac{5}{3}(x + 2y)^2y - 2(x + 3y)(xy + y^2)]$

(g) $\dfrac{2xy}{\sqrt{1-x^4y^2}}(3 + \dfrac{2x^4y^2}{1-x^4y^2}); \dfrac{x^2}{\sqrt{1-x^4y^2}}(3 + \dfrac{2x^4y^2}{1-x^4y^2});$

$\qquad\qquad\qquad\qquad \dfrac{x^6y}{(1-x^4y^2)^{3/2}}(7 + \dfrac{6x^4y^2}{1-x^4y^2})$

2. (a) $\begin{pmatrix} 0 & 0 \\ 0 & 0 \end{pmatrix}$ (e) undefined

   (c) $\begin{pmatrix} 0 & 1 \\ 1 & 0 \end{pmatrix}$ (g) $\begin{pmatrix} 0 & 0 \\ 0 & 0 \end{pmatrix}$

3. (a) $\begin{pmatrix} 0 & 1 & 0 \\ 1 & 0 & 0 \\ 0 & 0 & 0 \end{pmatrix}$

   (c) $\begin{pmatrix} 0 & 0 & 0 \\ 0 & 0 & 0 \\ 0 & 0 & 0 \end{pmatrix}$

   (e) $\begin{pmatrix} 0 & 0 & 1 & 0 \\ 0 & 0 & 0 & 1 \\ 1 & 0 & 0 & 0 \\ 0 & 1 & 0 & 0 \end{pmatrix}$

5. Let $t = g(x,y)$, and evaluate all derivatives of g at (x,y):

$$\frac{\partial^2}{\partial x^2}(f \circ g) = f''(t)\left(\frac{\partial g}{\partial x}\right)^2 + f'(t)\frac{\partial^2 g}{\partial x^2}$$

$$\frac{\partial^2}{\partial y^2}(f \circ g) = f''(t)\left(\frac{\partial g}{\partial y}\right)^2 + f'(t)\frac{\partial^2 g}{\partial y^2}$$

$$\frac{\partial^2}{\partial y \partial x}(f \circ g) = f''(t)\frac{\partial g}{\partial x}\frac{\partial g}{\partial y} + f'(t)\frac{\partial^2 g}{\partial y \partial x} = \frac{\partial^2}{\partial x \partial y}(f \circ g)$$

7. $$\frac{\partial^2 F}{\partial r^2} = \cos^2\theta\frac{\partial^2 f}{\partial x_1^2} + 2\sin\theta\cos\theta\frac{\partial^2 f}{\partial x_1 \partial x_2} + \sin^2\theta\frac{\partial^2 f}{\partial x_2^2}$$

$$\frac{\partial^2 F}{\partial \theta^2} = r^2\sin^2\theta\frac{\partial^2 f}{\partial x_1^2} + r^2\cos^2\theta\frac{\partial^2 f}{\partial x^2} - 2r^2\sin\theta\cos\theta\frac{\partial^2 f}{\partial x_1 \partial x_2} - $$

$$- r\cos\theta\frac{\partial f}{\partial x_1} - r\sin\theta\frac{\partial f}{\partial x_2}$$

$$\frac{\partial^2 F}{\partial \theta \partial r} = \frac{\partial^2 F}{\partial r \partial \theta} = -r\sin\theta\cos\theta\frac{\partial^2 f}{\partial x_1^2} + r\sin\theta\cos\theta\frac{\partial^2 f}{\partial x_2^2}$$

$$+ r(\cos^2\theta - \sin^2\theta)\frac{\partial^2 f}{\partial x_1 \partial x_2}$$

$$- \sin\theta\frac{\partial f}{\partial x_1} - \sin\theta\frac{\partial f}{\partial x_2}$$

(We assume that mixed partials are equal; see Section 6.8.)

10. $\Delta f = \dfrac{\partial^2 F}{\partial \rho^2} + \dfrac{1}{\rho^2 \sin^2 \varphi} \dfrac{\partial^2 F}{\partial \theta^2} + \dfrac{1}{\rho^2} + \dfrac{1}{\rho} \dfrac{\partial F}{\partial \rho} + \dfrac{1}{\rho \sin \varphi} \dfrac{\partial F}{\partial \theta}$

Section 6.4, p. 243

1. (a) $\displaystyle\sum_{j=1}^{m} \dfrac{(-1)^{j-1}}{j} x^j ; \quad \dfrac{(-1)^m}{m+1} \dfrac{x^{m+1}}{(1+c)^{m+1}}$

   (c) If $m = 2n$, $\displaystyle\sum_{j=0}^{n-1} \dfrac{(-1)^j x^{2j+1}}{2j+1}$; if $m = 2n + 1$, $\displaystyle\sum_{j=0}^{n} \dfrac{(-1)^j x^{2j+1}}{2j+1}$.

The remainders have no simple expressions.

   (e) $e^2 [1 + (x-2) + \dfrac{(x-2)^2}{2} + \dfrac{(x-2)^3}{6} + \dfrac{(x-2)^4}{24} + \dfrac{(x-2)^5}{120}];$

$$\dfrac{e^c (x-2)^6}{720}$$

   (g) $1 + rx + \dfrac{r(r-1)}{2} x^2 + \dfrac{r(r-1)(r-2)}{6} x^3 :$

$$\dfrac{r(r-1)(r-2)(r-3)}{24} (1+c)^{r-4} x^4$$

2. (a) .199

   (c) 1.414

   (e) .732

Section 6.5, p. 250

1. (a) $3 + x_1 x_2 + x_2^3$

   (c) $-27 + -3(x_1 - 1) + 28(x_2 + 3) + (x_1 - 1)(x_2 + 3)$
$$- 9(x_2 + 3)^2 + (x_2 + 3)^3$$

   (e) 1

   (g) $2 + [(1 + \dfrac{\pi}{2})(x_1 - 2) + 8(x_2 - \dfrac{\pi}{8})] + [(\dfrac{\pi}{4} + \dfrac{\pi^2}{16})(x_1 - 2)^2$
$$+ (8 + 2\pi)(x_1 - 2)(x_2 - \dfrac{\pi}{8}) + 16(x_2 - \dfrac{\pi}{8})^2]$$

Section 6.6, p. 256

1. (a) $x_1 x_2 x_3$

   (c) $-2 + [-2(x_1 - 1) - 1(x_2 - 2) + 2(x_3 + 1)]$

   $+ [(x_2 - 2)(x_3 + 1) + 2(x_1 - 1)(x_3 + 1) - (x_1 - 1)(x_2 - 2)]$

   (e) $2x_2 + [(x_1 - 1)x_2 + \frac{1}{2}(x_3 - 2)x_2] + (x_1 - 1)(x_3 - 2)x_2$

   (g) x_1

CHAPTER 7

Section 7.1, p. 271

1. $\begin{pmatrix} 0 & 1 & 6 & 0 \\ 0 & 0 & 2 & 18 \\ 0 & 0 & 0 & 3 \end{pmatrix}$

3. $\begin{pmatrix} \dfrac{\sqrt{3}}{2} & -\dfrac{1}{2} \\ \dfrac{1}{2} & \dfrac{\sqrt{3}}{2} \end{pmatrix}$ (counterclockwise rotation)

5. $\begin{pmatrix} \dfrac{3}{5} & \dfrac{4}{5} \\ \dfrac{4}{5} & -\dfrac{3}{5} \end{pmatrix}$

10. $\{(1,-1),(-1,2)\}$

11. One answer is $\alpha = \{(1,0),(0,1)\}$; $\beta = \{(1,1,1),(1,-1,1),(1,1,0)\}$; there are others.

Section 7.2, p. 277

8. $\begin{pmatrix} 1 & -a & ac-b \\ 0 & 1 & -c \\ 0 & 0 & 1 \end{pmatrix}$.

Section 7.3, p. 283

1. $\begin{pmatrix} 36 & -6 \\ -4 & 7 \\ 3 & 9 \end{pmatrix}$

13. Yes; no.

Section 7.4, p. 289

1. (a) 2   (c) 1   (e) 3

Section 7.6, p. 306

There is some choice in P and Q; here are possible answers.

2. (a) $P = I$, $Q = \begin{pmatrix} 1 & 0 \\ -\dfrac{2}{3} & \dfrac{1}{3} \end{pmatrix}$

   (c) $P = \begin{pmatrix} \dfrac{1}{2} & 0 & 0 \\ 1 & -2 & 0 \\ 2 & 0 & 1 \end{pmatrix}$, $Q = \begin{pmatrix} 1 & 0 & -2 & -24 \\ 0 & 1 & 0 & 0 \\ 0 & 0 & 1 & 0 \\ 0 & 0 & 0 & 1 \end{pmatrix}$

(e)

$$P = \begin{pmatrix} 1 & 0 & 0 & 0 & 0 \\ 9 & 1 & 0 & 0 & 0 \\ 0 & 0 & \frac{1}{5} & 0 & 0 \\ 9 & 1 & -3 & 1 & 0 \\ -63 & 7 & \frac{112}{5} & 0 & 1 \end{pmatrix}, \quad Q = \begin{pmatrix} 1 & 0 & -1 \\ 0 & 1 & -16 \\ -1 & 0 & 2 \end{pmatrix}$$

3.

$$\begin{pmatrix} 1 & 2 & -1 \\ -2 & -3 & 2 \\ 7 & 9 & -6 \end{pmatrix}$$

Section 7.7, p. 313

1. (a) $x = -1$, $x_2 = x_3 = 1$
    (c) $x_1 = \frac{13}{2}$, $x_2 = 4$, $x_3 = \frac{9}{2}$, $x_4 = -3$
    (d) $x_1 = 1$, $x_2 = 2x_3$
    (e) no solution

3.

$$\begin{pmatrix} 2 & 0 & -1 \\ -\frac{1}{2} & -\frac{1}{2} & \frac{1}{2} \\ -\frac{5}{2} & \frac{1}{2} & \frac{3}{2} \end{pmatrix}$$

CHAPTER 8

Section 8.1, p. 320

1. (a) Interior $= (a,b)$; boundary $= \{a,b\}$
    (c) Interior $= $; boundary $= \{(x,y) \in R^2 : xy = 1\}$
    (e) Interior $= \phi$; boundary $= \{(x,y) \in R^2 : x > 0 \text{ and } y = \sin(1/x)$
or $x = 0$ and $-1 \le y \le 1\}$
    (g) Interior $= \{(x,y,z) \in R^3 : x + y + z < 0\}$; boundary $=$
$\{(x,y,z) \in R^3 : x + y + z = 0\}$.

4. (a) 1; 2   (g) (0,1)

    (c) $\frac{\pi}{2}$ + nπ, n ∈ Z   (i) all (x,y) with x + y = nπ

    (e) (0,0)   for some n ∈ Z

5. (a) local minimum

    (c) neither

7. $\sqrt{17}$

11. $\sqrt{21}$

Section 8.2, p. 330

1. (a) $x_1^2 + 4x_1x_2 + x_2^2$

    (c) $3x_1^2 + 4x_1x_2 + 7x_2^2$

    (e) $-2x_1^2 + 2x_1x_2 - 3x_2^2$

2. (a) indefinite

    (c) positive definite

    (e) negative definite

3. (a) $x_1^2 + x_2^2 + x_3^2$

    (c) $4x_1^2 + 4x_1x_2 + 5x_2^2 + 6x_2x_3 + 5x_3^2$

    (e) $-3x_1^2 + 2x_1x_2 - 6x_2^2 + 2x_2x_3 + 7x_3^2$

4. (a) positive definite

    (c) positive definite

    (e) indefinite

5. (a) $\begin{pmatrix} 1 & 3 \\ 3 & -3 \end{pmatrix}$ (c) $\begin{pmatrix} 1 & \frac{1}{2} \\ \frac{1}{2} & 1 \end{pmatrix}$ (e) $\begin{pmatrix} 6 & -\frac{5}{2} \\ -\frac{5}{2} & 6 \end{pmatrix}$

6. (a) indefinite

(c) positive definite

(e) positive definite

7. (a) $\begin{pmatrix} 1 & 0 & 0 \\ 0 & -2 & 0 \\ 0 & 0 & 4 \end{pmatrix}$

(c) $\begin{pmatrix} 0 & 1 & 2 \\ 1 & 0 & 3 \\ 2 & 3 & 0 \end{pmatrix}$

(e) $\begin{pmatrix} 1 & -3 & 0 \\ -3 & 1 & \frac{1}{2} \\ 0 & \frac{1}{2} & -1 \end{pmatrix}$

8. (a) indefinite

(c) indefinite

(e) indefinite

Section 8.3, p. 339

1. (a) local minimum at $(2,0)$

(c) local maximum at $(0,0)$

(e) local minimum at $(2,1)$

(g) local maximum at $(-3,7)$, local minimum at $(1,-1)$, and saddles at $(0,-2)$ and $(2,2)$

7. $x = \frac{4}{9}$, $y = \frac{2}{3}$, $z = \frac{4}{3}$

9. $x = 4$, $y = 8$, $z = 12$

11. Bob picks 4, Larry picks -3, and Larry pays Bob $5.

Section 8.4, p. 347

1.   36

3. $(-\frac{1}{3}, -\frac{2}{3}, \frac{2}{3})$

9. $\frac{2}{\sqrt{3}} \times \sqrt{3} \times \frac{5}{\sqrt{3}}$

Section 8.5, p. 353

1.   -375

13.   1

15.   (a)   radius = $10(2\pi)^{-1/3}$ cm., height = $20(2\pi)^{-1/3}$ cm.

17.   length = width = $10\sqrt[3]{3/2}$ cm., height = $10\sqrt[3]{12}$ cm.

19.   Width = $\frac{2r}{\sqrt{3}}$, thickness = $2r\sqrt{2/3}$.

Section 8.6, p. 361

1.   Maximum = $\frac{1}{3}$ at $(-\frac{2}{3}, -\frac{2}{3}, \frac{1}{3})$; minimum = -1, at $(0,0,-1)$

3.   -2

5. $(\frac{4}{3}, \frac{4}{3}, \frac{4}{3})$

7. $\frac{\sqrt{5}}{4}$

CHAPTER 9

Section 9.1, p. 373

2.   (a) $g(y) = (y-6)^{1/2}$, $y \in (10,42)$; $g'(y) = \frac{1}{2}(y-6)^{-1/2}$

     (c)   no inverse

     (e) $g(y) = (y+1)^{1/3}$, $y \in (-344,999)$; $g'(y) = \frac{1}{3}(y+1)^{-2/3}$ for
$y \neq -1$

     (g) $g(y) = \cos^{-1}y$, $y \in (-1,1)$; $g'(y) = -(1-x^2)^{-1/2}$

     (i)   no inverse

     (k)   g exists (on $[-3/8, 3/8]$); no simple formula gives it; if
$f(x) = y$, $g'(y) = \dfrac{1}{3x^2-1}$

     (m) $g(y) = \sinh^{-1}y = \log(y + \sqrt{y^2+1})$, $y \in R$; $g'(y) = (y^2+1)^{-1/2}$

4.   (a) $(-\infty, 7/2)$
     (c) $(\frac{\pi}{2}, \frac{3\pi}{2})$
     (e) $(-\frac{\pi}{2}, \frac{\pi}{2})$
     (g) $(-\infty, 0)$

5.   (a) $-1/5$
     (c) $\sec 3$
     (e) 1
     (g) $\dfrac{-1}{2e}$

Section 9.4, p. 393

1.   (a) $-\dfrac{4}{13}$

     (c) $\dfrac{1}{3}$

     (e) $\dfrac{3}{8}$

3.   Near $(\pm 1, 0)$, y is not a function of x; near $(0, \pm 1)$, x is not a function of y.

5.   (a)   x = y = 0

(b)   all points except $(0,0)$ and $(a\sqrt{2},0)$; $\dfrac{dy}{dx} = \dfrac{x(x^2+y^2-a^2)}{y(x^2+y^2+a^2)}$

Section 9.5, p. 401

1.   (a) $\dfrac{\partial z}{\partial x} = 3$, $\dfrac{\partial z}{\partial y} = -3/2$

(c) $\dfrac{\partial z}{\partial x} = \dfrac{\partial z}{\partial y} = 0$

(e)   no implicit function

2.   (a) $\dfrac{\partial z}{\partial x} = 6$, $\dfrac{\partial z}{\partial y} = -\dfrac{5}{2}$, $\dfrac{\partial w}{\partial x} = -2$, $\dfrac{\partial^2 z}{\partial x \partial y} = -18$

(c)   no implicit function

3.   Any two can be regarded as functions of the third.

CHAPTER 10

Section 10.1, p. 407

1.   (a)   6 - 4i

(c)   10 + 2i

(e)   -1297 - 2329i

(g) $\dfrac{-43}{29} - \dfrac{6i}{29}$

(i) $\dfrac{65}{58} + \dfrac{71i}{29}$

Section 10.3, p. 422

1.   (a)   -17
      (c)   -11
      (e)   14
      (g)   -8

Section 10.4, p. 431

1.   (a)   x = 1, y = 2
      (c)   x = -1, y = 2, z = 1

3. $\dfrac{1}{ad-bc}$ $\begin{pmatrix} d & b \\ -c & a \end{pmatrix}$ (when ad - bc $\neq$ 0)

Section 10.5, p. 440

1.   (a)   (1,-3), 2; (3,1),12
      (c)   (i,1), $\dfrac{3}{5} - \dfrac{4}{5}i$; (-i,1), $\dfrac{3}{5} + \dfrac{4}{5}i$

      (e)   (1,-1),1; (4,1),6
      (g)   (i+i,-2), i;(1-i,-2),-i
      (i)   (1,-1,0),1; (2,2,1),10; (1,0,-2),1 (other answers are
                                                    possible)

Section 10.6, p. 446

1.   (a) $\{(\tfrac{1}{\sqrt{2}}, \tfrac{1}{\sqrt{2}}), (\tfrac{1}{\sqrt{2}}, - \tfrac{1}{\sqrt{2}})\}$
      (c) $\sqrt{\dfrac{2}{5+\sqrt{5}}}(1, \dfrac{-1-\sqrt{5}}{2}), \sqrt{\dfrac{2}{5+\sqrt{5}}}(1, \dfrac{-1+\sqrt{5}}{2})$

(e) $\{(\frac{1}{\sqrt{2}}, \frac{i}{\sqrt{2}}), (\frac{1}{\sqrt{2}}, -\frac{i}{\sqrt{2}})\}$

2.   (a) $\{(\frac{1}{\sqrt{2}}, 0, -\frac{1}{\sqrt{2}}), (\frac{1}{2}, -\frac{\sqrt{2}}{2}, \frac{1}{2}), (\frac{1}{2}, \frac{\sqrt{2}}{2}, \frac{1}{2})\}$

     (c) $\{(\frac{2}{3}, \frac{2}{3}, -\frac{1}{3}), (\frac{2}{3}, -\frac{1}{3}, \frac{2}{3}), (-\frac{1}{3}, \frac{2}{3}, \frac{2}{3})\}$

Section 10.7, p. 450

3. $\{(\frac{1}{\sqrt{3}}, \frac{1}{\sqrt{3}}, \frac{1}{\sqrt{3}}), (-\frac{1}{\sqrt{3}}, \frac{1}{2\sqrt{3}} - \frac{i}{2}, \frac{1}{2\sqrt{3}} + \frac{i}{2}), (-\frac{1}{\sqrt{3}}, \frac{1}{2\sqrt{3}} + \frac{i}{2}, \frac{1}{2\sqrt{3}} - \frac{i}{2})\}$

CHAPTER 11

Section 11.2, p. 481

2.   (a) $\frac{2}{3}(5\sqrt{5} - 2\sqrt{2})$

     (c) $e - 1$

     (e) $\frac{\pi}{8}$

4.   (a) 1

     (c) $\frac{\pi}{4} - \frac{1}{2} \log 2$

     (e) $e^{\frac{5\pi}{6}}(1 + \frac{\sqrt{3}}{4}) + \frac{1}{2}$

Section 11.3, p. 492

5.   (a) $\frac{3}{2}$

     (c) $\frac{2}{3}$

9. $\frac{1}{3}$

Section 11.4, p. 499

5. (a) 2
    (c) 1

9. $\frac{1}{8}$

CHAPTER 12

Section 12.1, p. 528

1. (a) 2   (e) $\frac{(e^2-1)(e^4-1)}{2e^2}$
    (c) 12
                                   (g) 2 log 2-1

2. (a) 288
    (c) $\sqrt{3} - 1$
    (e) $\frac{3\sqrt{3}}{4}$

3. (a) 2   (g) 288
    (c) $\frac{16}{3}$ (i) $\frac{\sqrt{3} - 1}{4}$
    (e) 2 log 2-1   (k) 0

Section 12.2, p. 536

1.   (a) $\dfrac{1006}{105}$ (g) $\dfrac{1}{24}$

    (c) $\dfrac{568}{945}$ (j) $\dfrac{55}{72}$

    (e) $\dfrac{2}{15}$

2.   (a) $\dfrac{1}{24}$

    (c) 0

    (e) $\dfrac{2971}{3780} \sqrt{2}$

3.   (a) 0

    (c) -2

Section 12.3, p. 544

1.   (a) $-\dfrac{57}{4}$

    (c) $\dfrac{1}{4}(2 \sin 4 - \sin 8)$

    (e) $\dfrac{1}{2}$

    (g) $\dfrac{3}{2}\log(1 + \sqrt{2})$

2.   (a) $\dfrac{1}{8}$ (e) 243π

    (c) $\dfrac{\pi}{2}$ (g) $\dfrac{128}{15} a^5$

5. $\dfrac{185}{24}$

7. 144

9. $16 - 8\sqrt{2}$

11. $(0,0,\dfrac{8}{15})$

Section 12.5, p. 553

1. (a) $\dfrac{x \sin x + \cos x - 1}{x^2}$

   (c) $\dfrac{-1}{x^2 + 5x + 6}$

5. $-e^{x^3 + y^3}$

7. $\dfrac{\partial z}{\partial x} = \dfrac{\partial z}{\partial y} = -1$

Section 12.6, p. 560

1. (a) $\dfrac{15\pi}{32}$

   (c) $\dfrac{15\pi}{256}$

2. (c) $\dfrac{31}{60}(4 - \sqrt{2})$

5. (a) $12\pi \log 2$

   (c) $\dfrac{3}{2}\pi \log 2$

6. (a) 0

8. (c) $\dfrac{\pi}{5}(2 - \sqrt{2})(25\sqrt{5} - 1)$

9. (a) 0

   (c) 0

CHAPTER 13

Section 13.2, p. 584

1. (a) $\dfrac{1211\sqrt{2}}{3}$

    (c) $\dfrac{1}{20}(37^{5/2} - 17^{5/2})$

    (e) $\dfrac{-219}{10}$

2. (a) $-\dfrac{117}{2}\sqrt{6}$

    (c) $\dfrac{2e^{10}}{15} + \dfrac{e^8}{12} + \dfrac{2e^7}{7} + \dfrac{e^5}{5} - \dfrac{59}{84}$

    (e) $\dfrac{59}{30}$

3. (a) $4\sqrt{2}$

    (c) $4 + 2\sqrt{2}$

4. (a) $\dfrac{49}{4}$ units

5. (a) $\dfrac{\sqrt{2}}{6}$

    (c) 0

Section 13.3, p. 529

1. (a) -7

    (c) $\dfrac{19}{3}$

    (e) $\dfrac{1044}{5}$

2. (a) $\dfrac{1}{2}$

    (c) $-\pi$

3.   (a) $\frac{11}{15}$

      (c) 2

4.   (a) 0

      (c) $\frac{825}{8}$

5.   (a) $\frac{-549}{20}$

      (c) $\frac{16}{3}$

Section 13.4, p. 602

1.   (a) $x^2 y^3$; -4

      (c) $x^2 + xy + z$; 2

      (e) $\sin xyz$; $\frac{-\sqrt{3}}{2}$

      (g) $\frac{1}{3}(x + 2y)^3$; 0

(One can add an arbitrary constant to the potential functions.)

Section 13.5, p. 609

1.   (a) 6
      (c) -6
      (f) 0

2.   (a) not conservative
      (c) not conservative
      (e) xy

CHAPTER 14

Section 14.1, p. 625

2.   (a) $h(s,t) = (s^{2/3}, t^{2/3}), 8 \leq s,t \leq 27; h^{-1}(u,v) = (u^{3/2}, v^{3/2})$

      (c) $h(s,t) = (s-t, s+t), s^2 + t^2 \leq 1; h^{-1}(u,v) = (\frac{u+v}{2}, \frac{v-u}{2})$

3. $\beta(x,y) = \dfrac{4 + \sqrt{4 - 3x^2 - 3y^2}}{x^2 + y^2 + 4}(x,y,2) - (0,0,2)$

5. $z = 2x + 4y - 5$ (or $\beta(s,t) = (1+s, 2+t, 5+2s+4t)$)

Section 14.2, p. 635

3. $\frac{2}{3}\pi (2\sqrt{2} - 1)$

4.   (c) $16\pi\sqrt{5}$

7. $\dfrac{49\pi}{3}$

9. $4\pi^2 ar$

Section 14.3, p. 644

1.   (a) $\dfrac{293}{768}\sqrt{5} - \dfrac{1}{24} + \dfrac{1}{512}\log(2 + \sqrt{5})$

      (c) $2 + \dfrac{\pi}{2}$

2.   (a) $-\dfrac{1}{3}$

      (c) $1 + \dfrac{\pi}{2}$

Section 14.4, p. 650

3. (a) $(xz - y, -yz, -x)$

    (c) $(1, \sin x - 1, \cos x)$

CHAPTER 15

Section 15.1, p. 657

1. (a) $2x_3^2 dx_1 + x_1 x_2 dx_2 - x_2 dx_3$

    (b) $(2x_3^3 - x_1)dx_1 + x_1 x_2 x_3 dx_2$

    (c) $-x_1 dx_1 dx_2 + dx_1 dx_3 - x_2 x_3 dx_2 dx_3$

    (d) $(-2x_2^2 x_3 - 4x_1 x_3^2)dx_1 dx_2 + 4x_3^2 dx_1 dx_3$

    (e) $2x_2 x_3^3 dx_1 dx_2 dx_3$

    (f) $2(x_1 - 2x_2^2)x_2 x_3^3 dx_1 dx_2 dx_3$

Section 15.3, p. 667

1. (a) $ydx + xdy + 2zdz$

    (b) $dxdy - dydz$

    (c) $(y-x)dxdy + yz^2 dydz + (xz^2 + 2y)dydz$

    (d) $x(7 - \cos xy)dxdydz$

    (e) $(z^2 + 2xy)dxdydz$

Section 15.5, p. 675

1. (a) $4x^2 y^2(x^2 + y^2)$

(b) $4x^2ydx + 4xy^2dy$

(c) $(8x^3y^2 - 2x^2y - 2y^3)dx + (8x^2y^3 - 2x^3 - 2xy^2)dy$

(d) $4(x^2 - y^2)dxdy$

(e) $4(x^2 - y^2)\sin 2xy(x^2 + y^2) \, dxdy$

3. (a) $x_1x_3(x_2 - x_3)(2x_1 + x_2)$

(b) $x_3dx_1 + x_1dx_3$

(c) $2x_1x_3dx_1 + (x_1x_3 - 2x_1 - x_2)dx_2 + (2x_1 + x_2)dx_3$

(d) $2dx_1dx_2 - 2dx_1dx_3 - dx_2dx_3$

(e) $[x_3\sin(2x_1 + x_2) - 2(2x_1 + x_2)^3$

$$- x_3(x_2 - x_3)^2(2x_1 + x_2)]dx_1dx_2$$

$$+ [2x_1(2x_1 + x_2)(x_2 - x_3)^2 + 2(2x_1 + x_2)^3$$

$$- x_3\sin(2x_1 + x_2)]dx_1dx_3$$

$$+ [x_1(2x_1 + x_2)(x_2 - x_3)^2 + (2x_1 + x_2)^3$$

$$- x_1\sin(2x_1 + x_2)]dx_2dx_3$$

(f) $(x_3 - 2x_1)dx_1dx_2dx_3$

(g) $(2x_1^2x_3 + x_1x_2x_3 + 2x_2x_3 - x_2^2 - x_3^2)(x_3 - 2x_1)dx_1dx_2dx_3$

CHAPTER 16

Section 16.1, p. 687

1. (a) 2   (g) $\frac{1}{2}$

(c) $\frac{1}{2}$

                                       (i) $-\frac{13}{63}$

(e) $\frac{5}{4}$

2. (a) $\frac{2}{3}$

    (c) $-\frac{23}{36}$

3. (a) $\frac{2}{5}$

    (c) $\frac{17}{120}$

Section 16.2, p. 692

1. (a) $\frac{5}{18}$ (b) $\frac{33}{18}$ (c) $\frac{-5}{6}$ (d) $\frac{62}{9}$

3. (a) $-\frac{5}{6}$ (b) $\frac{-3}{2}$ (c) $\frac{14}{3}$ (d) $\frac{2}{3}$

Section 16.4, p. 701

1. 0

APPENDIX 3

Section 1, p. 720

1. (a) 6, (1,1); -1, (3,-4)
    (c) 2,(2,1,1); 1, (1,1,1),(3,1,0)

Section 2, p. 728

The bases are as in the answers for Exercise 1, Section 1.

(a) $\begin{pmatrix} 6 & 0 \\ 0 & 1 \end{pmatrix}$

(c) $\begin{pmatrix} 2 & 0 & 0 \\ 0 & 1 & 0 \\ 0 & 1 & 1 \end{pmatrix}$

2. (a) $(0,1,1,0), (-2,0,0,1), (-3,0,0,2), (0,3,2,0);$ $\begin{matrix} i & 0 & 0 & 0 \\ 1 & i & 0 & 0 \\ 0 & 0 & -i & 0 \\ 0 & 0 & 1 & 1 \end{matrix}$

(c) $(1,0,0,2,0); (2,1,3,4,4), (0,1,2,0,3) (2,0,0,3,0), (-4,1,1,-6,3);$

$$\begin{pmatrix} 1 & 0 & 0 & 0 & 0 \\ 0 & 1 & 0 & 0 & 0 \\ 0 & 1 & 1 & 0 & 0 \\ 0 & 0 & 0 & 0 & 0 \\ 0 & 0 & 0 & 1 & 0 \end{pmatrix}$$

7. (a) With respect to the basis $xe^{2x}, e^{2x}, x^2e^{3x}, 2xe^{3x}, 2e^{3x}$, the matrix is

$$\begin{pmatrix} 2 & 0 & 0 & 0 & 0 \\ 1 & 2 & 0 & 0 & 0 \\ 0 & 0 & 3 & 0 & 0 \\ 0 & 0 & 1 & 3 & 0 \\ 0 & 0 & 0 & 1 & 3 \end{pmatrix}$$

(b) With respect to the basis $e^x, xe^{2ix}, e^{2ix}, xe^{-2ix}, e^{-2ix}$, the matrix is

$$\begin{pmatrix} 1 & 0 & 0 & 0 & 0 \\ 0 & 2i & 0 & 0 & 0 \\ 0 & 1 & 2i & 0 & 0 \\ 0 & 0 & 0 & -2i & 0 \\ 0 & 0 & 0 & 1 & -2i \end{pmatrix}$$

INDEX

A

absolute value, 17
adjoint, 293, 412
antiderivative, 480
Archimedean Principle, 195
arc length, 582
associative law, 27

B

ball
  closed, 79
  open, 76
basis, 128
  dual, 139
  orthonormal, 157
  standard, 59
bijective, 6, 56
binomial theorem, 15
Bolzano-Weierstrass Theorem, 200
bound
  of linear transformation, 224
  lower, 190
  upper, 190
boundary
  of set, 82, 316
  of singular p-chain, 690
  of singular p-cube, 690
bounded function, 93
bounded linear transformation,
    93, 224
bounded sequence, 175
bounded set, 175, 186
Brouwer Fixed Point Theorem, 213

C

Cartesian plane, 23
Cartesian product, 5
Cauchy-Riemann equations, 409
Cauchy sequence, 178, 192
Cayley-Hamilton Theorem, 734
chain rule, 216
change of variable formula, 556
characteristic equation, 437
characteristic function, 565
characteristic value, 435
characteristic vector, 435
circular helix, 581
circulation, 648
closed ball, 79
closed curve, 603
closed interval, 19
closed set, 79, 176
closure of a set, 82
column operation, 301
column rank, 285
common refinement, 467, 486
commutative law, 27
compact set, 185
complement, 3
completeness axiom, 192, 195
complex numbers, 404
  conjugate, 405
  imaginary part, 404
  norm, 404
  polar representation, 407
  real part, 404
complex vector space, 409
component, 25
composite, 58, 61